RELATIONS FOR CONVERSIONS OF UNITS[†]

Acceleration:	1 ft/s^2	= 0.3048 m/s^{2*}
	1 in./s^2	= 25.4 mm/s^{2*}
Area:	1 ft^2	= 0.092903 m^2
	1 in^2	= 645.16 mm^{2*}
Density:	1 lbm/ft^3	= 16.018 kg/m^3
Energy:	1 Btu	= 1055.1 J
	1 ft·lb	= 1.3558 J
Force:	1 kip	= 4.4482 kN
	1 lb	= 4.4482 N
Impulse:	1 lb·s	= 4.4482 N·s
Length:	1 ft	= 0.3048 m*
	1 in.	= 25.4 mm*
	1 mi (= 5280 ft)	= 1.6093 km
	1 NM	= 1852 m*
Mass:	1 lbm	= 0.45359237 kg*
	1 slug	= 14.594 kg
	1 ton (= 2000 lbm)	= 907.18 kg
Moment of Force:	1 lb·ft	= 1.3558 N·m
	1 lb·in.	= 0.11298 N·m
Moment of Inertia, area:	1 in^4	= 416231 mm^4
Moment of Inertia, mass:	1 slug·ft^2 (= 1 lb·ft·s^2)	= 1.3558 kg·m^2
Momentum, linear:	1 lb·s	= 4.4482 N·s
Momentum, angular:	1 lb·ft·s	= 1.3558 kg·m^2/s
Power:	1 ft·lb/s	= 1.3558 W
	1 hp (= 550 ft·lb/s)	= 745.70 W
Pressure, stress:	1 lb/ft^2	= 47.880 Pa
	1 psi (= 1 lb/in^2)	= 6.8948 kPa
Velocity:	1 ft/s	= 0.3048 m/s*
	1 mph (= 1 mi/h)	= 0.44704 m/s*
	1 kn (= 1NM/h)	= 0.51444 m/s
Volume:	1 ft^3	= 0.028317 m^3
	1 in^3	= 16.387 cm^3
	1 gal (= 231 in^3)	= 3.7854 L
Work:	1 ft·lb	= 1.3558 J

[†]Refer to Table 1.1 on p. 9.

*Exact value

Engineering Mechanics
DYNAMICS

Engineering Mechanics
DYNAMICS

I. C. JONG

Professor of Mechanical Engineering
University of Arkansas
Fayetteville, Arkansas

B. G. ROGERS

Professor of Civil Engineering
Lamar University
Beaumont, Texas

Saunders College Publishing

A Division of
Holt, Rinehart and Winston, Inc.

Philadelphia   Chicago   Fort Worth   San Francisco   Montreal   Toronto   London   Sydney   Tokyo

Text typeface: Times Roman
Compositor: The Clarinda Company
Acquisitions Editor: Barbara Gingery
Managing Editor: Carol Field
Project Editor: Marc Sherman
Manager of Art and Design: Carol Bleistine
Art and Design Coordinator: Doris Bruey
Text Designer: Arlene Putterman
Cover Designer: Lawrence R. Didona
Text Artwork: Tech-Graphics
Director of EDP: Tim Frelick
Production Manager: Bob Butler

Cover: © COMSTOCK INC./ Brooklyn Bridge

Printed in the United States of America

ENGINEERING MECHANICS: DYNAMICS

ISBN 0–03–026317–4

Library of Congress Catalog Card Number: 90-053084

0 1 2 3   032   9 8 7 6 5 4 3 2 1

THIS BOOK IS PRINTED ON **ACID-FREE, RECYCLED** PAPER

Preface

This book and the preceding volume, *Engineering Mechanics: Statics*, are designed for the fundamental courses in statics and dynamics, which are usually offered in the sophomore or junior year in engineering curricula. In the development of the text, students are assumed to have a background in algebra, geometry, trigonometry, and basic differential and integral calculus. Nevertheless, students with a prior knowledge in college physics normally have a higher degree of readiness for learning engineering mechanics.

Mechanics has long been recognized as mainly a deductive science; however, the learning process is largely inductive. In the text, simple topics and problems precede those that are more complex and advanced. For example, the computation of kinematic quantities using scalar algebra is presented before that using vector algebra, explanations of the rudiments of vector analysis are given before their applications in mechanics, and exercise problems are arranged in order of increasing difficulty. Overall, the text is written to give a clear and up-to-date presentation of the theory and application of engineering mechanics; it is aimed at helping engineering students develop an ability to apply well-established principles to analyze and solve problems in a logical and effective manner.

Organization and Features

As shown in the contents overview, the first ten chapters cover statics and the remaining ten chapters cover dynamics. The text has been especially organized to help students understand and retain the concepts presented. A more definitive approach is used in developing the text. Each chapter is divided into two or three definitive parts with descriptive headings, and each part contains an appropriate number of sections. The *key concepts* to be covered are listed at the beginning of each chapter, and the list is followed by an introductory paragraph. In addition to sets of traditional problems, concept-reinforcing *developmental exercises* are provided, and concluding remarks as well as a set of review problems are given at the end of each chapter.

Appendices A and B include materials for a *review of basic mathematics* from algebra through calculus, Appendix C verifies the vector characteristics of infinitesimal angular displacements, while Appendices D and E provide general *computer programs* for solving linear and nonlinear equations in mechanics. Yet, omission of computer programs will not prejudice the understanding of the text.

The list of *key concepts* at the beginning of each chapter serves as a summary of major topics presented. By identifying the key concepts in a given chapter, this list helps students focus their efforts and increase their efficiency of study. The introductory paragraph, which follows the list, gives an overview of the subject matter and the relevance of the topics to be covered.

The numerous *developmental exercises* interspersed throughout the text are intended for use as learning exercises. Users may either use or skip them as they see fit. These exercises are designed to (a) prompt students to describe and reinforce their understanding of the concepts, laws, and principles they are learning; and (b) provide opportunities for students to practice problem-solving skills by working through exercises that contain cue questions or similarities to preceding examples.

It is believed that *examples* depicting real-life situations are a useful means of providing motivation for students to learn and to develop the skills needed to reduce actual physical situations to models so that basic principles can be applied. Thus, the text includes many examples that illustrate real-life, as well as theoretical, situations. Each example contains a sample problem and a solution to that problem. The solution stresses the interpretation of the situation and the setting up of equations. Whenever intermediate steps in the solution are routine in nature, they are omitted to avoid distraction from the main solution.

The sets of *problems* provided in each chapter are also designed to create real-life, as well as theoretical, situations. Students can practice applying the principles presented in the text to these problems. Further, the text offers an abundance of problems, which provide the instructor with choices of assignments. Approximately two thirds of the problems have answers given in the back of the book. To alert users to a problem without a given answer, an asterisk (∗) is placed after the problem number.

The last section in each chapter is set aside to give *concluding remarks* on the subject matter in that chapter. Generally, these remarks cover the highlights and special features of the chapter. At the end of each chapter, a set of *review problems* is given. These problems generally illustrate the key concepts listed at the beginning of the chapter. Besides providing additional homework problems, they may be used by students as a test of the knowledge and skills they are expected to have acquired upon completing the study of that chapter.

Systems of Units

In recognition of current trends and practices in the United States, both the SI units *(Le Système International d'Unités)* and the U.S. customary units are used. The conversion between the two systems of units is briefly discussed in Chapter 12. Approximately 60 percent of the problems are stated in SI units, and the remainder retain U.S. customary units.

Optional Sections and Order of Coverage

To allow choices of coverage and to serve a wider group of users, the text includes more than a few optional sections. Each of these sections is marked with a star (★) before the section number for distinction from the other sections that form the core of the basic mechanics course. Optional sections may be skipped without affecting the understanding of the core of the course.

At the discretion of the instructor, some of the materials may be covered in a different sequence without losing continuity. One may cover the kinematics in Chapters 11 and 15 prior to covering the kinetics of particles and rigid bodies.

Computer Programs and Diskettes

Although students may use a *calculator* as a tool to solve linear simultaneous equations and find roots of nonlinear equations, they will often benefit further by using a *computer*, which is nowadays utilized in a wide variety of situations. Therefore, some computer programs are included. These programs are written in BASIC and will run on IBM PC and compatibles. Students using a computer should understand the essential concepts involved in solving the problem and be prepared to exercise good judgement in interpreting the output from the computer. The short programs appearing in some sections of the text are simple application programs, which serve as illustrations. The program in Appendix D is a general program for solving a set of n linear simultaneous equations in n unknowns, while the program in Appendix E is a general program for finding roots of a nonlinear or transcendental equation. Several problems that are amenable to computer solutions are identified in the text by footnotes, hints, or suggestions.

All programs listed in the text and the supplementary software, as described in Appendix F, are stored on computer diskettes, which may be obtained from bookstores or the publisher. The authors wish to thank Professor Lynn D. Wills of the University of Arkansas for his contribution in preparing and programming the supplementary software.

Unique Topic Coverage

This text includes a number of presentations, illustrations, and emphases not usually found in similar texts. In particular, note the following:

1. The *computer solution of simultaneous equations* is illustrated in Appendix D. Opportunities to practice this skill are provided in some problems (e.g., Probs. 14.35, 14.36, etc.).
2. The *computer solution of a transcendental equation* is illustrated in Appendix E. Opportunities to practice using a computer to solve transcendental and nonlinear equations are provided in some problems (e.g., 13.57, 13.58, 14.52, 18.51, etc.).

3. The distinction between *compatible virtual displacement* and *constrained virtual displacement* is made in Sec. 13.13 and illustrated in Examples 13.16, 14.14, 17.8, 17.9, 18.10, and 18.11.

4. *Infinitesimal angular displacements* are shown to be vectors in Appendix C. It establishes the background against which angular velocities and angular accelerations can readily be discerned as vectors.

5. The *displacement center* for finding virtual displacements is illustrated in Example 17.8. This concept provides an alternative to the traditional approach using variational calculus. (The displacement center is located in a manner similar to the velocity center.)

6. A *quantitative definition for potential energy* is stated in Sec. 13.8. This definition makes potential energy as easy to understand and compute as work and kinetic energy.

7. The *inertia force* of a body is interpreted in Sec. 12.3 as the reaction of the body to the action of the resultant force on the body. This interpretation helps demystify inertia forces for dynamics students.

8. The *rigor in the definition of work*, when the force does not act continually on the same particle, is emphasized in Secs. 13.2 and 17.1. It brings home to students why the friction force exerted by the ground on a wheel rolling without slipping does no work on the wheel.

9. *Extensions and generalizations of the principle of virtual work* are presented in Secs. 13.13, 14.10, 17.7, and 18.7 as complements to traditional methods. The virtual work approach often obviates the necessity of solving simultaneous equations in kinetics as well as statics.

10. The *corollary to the theorem of acceleration center* of a rigid body is emphasized in Sec. 15.9 and used in Chapters 15 and 16. This concept greatly reduces the algebraic work required in a traditional solution. Moreover, it alerts students to the frequent mistake of taking a velocity center as an acceleration center.

Acknowledgments

The authors thank their colleagues Jim H. Akin, Charles W. Crook, and Charles R. Fagg for their thoughtful suggestions, and department heads, William F. Schmidt (University of Arkansas) and Enno Koehn (Lamar University), for their encouragement. They appreciate the numerous helpful suggestions and comments made on drafts of the text by the reviewers:

Paul Chan, New Jersey Institute of Technology
Sahib Chehl, Southern University
Rick Gill, University of Idaho
Richard Golembiewski, Milwaukee School of Engineering
Reese Goodwin, Brigham Young University
David Hartman, Northern Arizona University
Robert Howland, Notre Dame University
Richard Longman, Columbia University
Scott Martin, Youngstown State University
Luis Clemente Mesquita, University of Nebraska
Saeed Niku, California State Polytechnic University
Semih Oktay, University of Maryland
Rick Reimer, University of California, Berkeley
Eugene Stolarik, University of Minnesota
Robert Swaim, Oklahoma State University
William Van Arsdale, University of Houston
Carl Vilmann, Michigan Technological University

Special thanks are extended to Barbara Gingery (Acquisitions Editor), Marc Sherman (Project Editor), and Saunders College Publishing for their effort and support during the various stages of the project.

I. C. Jong
B. G. Rogers
June 1990

A Note for Students

As students, your goal here is learning mechanics as well as developing a high level of analytical capability. It is going to require some of your *time* and *effort*. However, many people find that it may be more efficiently attained by noting the following hints:

- **Study methodically.** As you begin each new chapter, try first to gain an overview. This may be achieved by doing the following: (a) thumb through the chapter to look it over, (b) go over the key concepts at the beginning of the chapter, (c) read the introduction following the key concepts, and (d) turn to the last section to read the concluding remarks. Then, read rapidly the sections as assigned or desired, and later return for an in-depth study.

- **Draw diagrams.** It is important to realize early on that engineering mechanics is a physical science which heavily involves *geometry*. Pertinent diagrams or sketches are valuable aids in studying the state of rest or motion of physical systems. There is no substitute for the perception and insight gained from the diagrams or sketches drawn.

- **Work problems.** Work the problems assigned or suggested by your instructor. This is not only an effective way to learn the subject but also a way to show to yourself how well you have learned it.

- **Ask questions.** Maintain your inquisitive mind in the study. Most instructors and teaching assistants have a sincere interest in helping students deepen their understanding of the new material.

The authors sincerely hope that the above hints will prove to be helpful, and that you will see the logic, beauty, and relevance of mechanics as you progress in your learning.

List of Symbols

$\mathbf{a}$	Acceleration
a	Constant; radius; distance; axis; semimajor axis
$\bar{\mathbf{a}}$	Acceleration of mass center
$\mathbf{a}_A, \mathbf{a}_B, \ldots$	Accelerations of A, B, ...
$\mathbf{a}_{B/A}$	Acceleration of B in a nonrotating reference frame translating with A; acceleration of B relative to A
$\mathbf{a}_{B/Axyz}$	Acceleration of B in the moving reference frame $Axyz$
$\mathbf{a}_c$	Constant acceleration
$\mathbf{A}, \mathbf{B}, \mathbf{C}, \ldots$	Vectors; forces; reactions from supports and connections
$A, B, C, \ldots$	Points; magnitudes
$A', B', C', \ldots$	Points; magnitudes
A	Area
$\overline{AB}, \overline{AC}, \ldots$	Line segments; distances
$\mathbf{Aimp}_A$	Angular impulse about A
b	Width; distance; axis; semiminor axis
c	Constant; coefficient of viscous damping
C	Centroid; constant; compression; velocity center
d	Distance
$\mathbf{e}_t, \mathbf{e}_n$	Unit vectors in tangential and normal directions
$\mathbf{e}_r, \mathbf{e}_\theta$	Unit vectors in radial and transverse directions
e	Base of natural logarithms; coefficient of restitution
f	Scalar function; cyclic frequency
$\mathbf{F}$	Force; friction force
$\mathbf{F}_k$	Kinetic friction force
$\mathbf{F}_m$	Maximum friction force
$\mathbf{F}_\parallel$	Component of $\mathbf{F}$ parallel to a specific axis
$\mathbf{F}_\perp$	Component of $\mathbf{F}$ perpendicular to a specific axis
g	Gravitational acceleration
G	Constant of gravitation; center of gravity; mass center
h	Height; angular momentum per unit mass
$\mathbf{H}_G, \mathbf{H}_O$	Angular momentum about mass center G, angular momentum about fixed point O
$\mathbf{i}, \mathbf{j}, \mathbf{k}$	Unit vectors along Cartesian coordinate axes
$I, I_x, I_y, \ldots$	Moments of inertia
$I_{xy}, I_{uv}, \ldots$	Products of inertia

$\bar{I}$	Centroidal moment of inertia of an area; central moment of inertia of a body
Imp$_{1\to2}$	Linear impulse exerted during $t_1 \leq t \leq t_2$
J	Polar moment of inertia
k	Spring modulus; constant
k_x, k_y, k_O	Radii of gyration
$\bar{k}$	Centroidal radius of gyration of an area; central radius of gyration of a body
K	Torsional spring modulus; constant
L	Linear momentum
L	Length; span
m	Mass of a body
$\tilde{m}$	Distributed torque; moment per unit length
M	Moment; torque
M	Magnitude of moment or torque; mass of earth
M_{AB}	Moment about axis AB
M$_O$	Moment about point O
M$_O^R$	Resultant moment about point O
n	Normal axis; normal direction
N	Normal component of reaction
O	Origin of coordinates; point
q	Displacement
q$_\parallel$	Component of **q** parallel to a specific force
Q	Time rate of volumetric flow; magnitude; point
$\dot{\mathbf{Q}}$	Time derivative of **Q** in a fixed reference frame (e.g., $OXYZ$)
$(\dot{\mathbf{Q}})_{Axyz}$	Time derivative of **Q** in the moving reference frame $Axyz$
r	Position vector
r$_{B/A}$	Position vector of B relative to A
r	Radius; distance; polar coordinate
R	Resultant force; resultant vector; reaction
R	Radius; radius of earth; magnitude
s	Length of arc; axis
t	Time; thickness; tangential axis; tangential direction
T	Tension; kinetic energy
u	Velocity
u	Axis; polar coordinate; speed
U	Work
$U_{1\to2}$	Work of a force acting on a body from positions 1 to 2
v	Velocity
v	Axis; speed; rectilinear velocity
$\bar{\mathbf{v}}$	Velocity of mass center
v$_A$, **v**$_B$, . . .	Velocities of A, B. . . .
v$_{B/A}$	Velocity of B in a nonrotating reference frame translating with A; velocity of B relative to A
v$_{B/Axyz}$	Velocity of B in the moving reference frame $Axyz$
V	Volume; potential energy
V_a, V_e, V_g	Applied, elastic, and gravitational potential energies
w	Distributed load; force per unit length
W	Weight force

W	Weight
x, y, z	Rectangular coordinates or axes; distances
$\bar{x}, \bar{y}, \bar{z}$	Rectangular coordinates of centroid or center of mass
x', y', z'	Centroidal axes; central axes; auxiliary rectangular coordinate axes
x'', y'', z''	Auxiliary rectangular coordinate axes
X	Amplitude of vibration
Z	Acceleration center
$\boldsymbol{\alpha}$	Angular acceleration
α, β, γ	Angles
β	Angle of contact in belt friction
γ	Specific weight
γ_A, γ_L	Weight per unit area; weight per unit length
δ	Deflection; logarithmic decrement
δ_{st}	Static deflection
$\delta\mathbf{r}$	Virtual displacement vector
$\delta x, \delta y$	Virtual displacements
$\delta\theta$	Angular virtual displacement
$\Delta\theta$	Angular displacement
δU	Virtual work
$\delta U'$	Generalized virtual work
ε	Eccentricity of conic section or of orbit
ζ	Variable analogous to z; axis; damping factor
η	Variable analogous to y; axis; efficiency
θ	Angle; angular coordinate; polar coordinate; Eulerian angle (for nutation)
Θ	Angular amplitude of vibration
λ	Unit vector along a line
μ_k	Coefficient of kinetic friction
μ_s	Coefficient of static friction
ξ	Variable analogous to x; axis
ρ	Mass density; radius of curvature
ρ_A, ρ_L	Mass per unit area, mass per unit length
τ	Variable analogous to t; period of orbit; period of vibration
ϕ	Angle; Eulerian angle (for precession); phase angle (in free vibration)
φ	Phase angle (in forced vibration)
ψ	Angle; Eulerian angle (for spin)
$\boldsymbol{\omega}$	Angular velocity
ω	Angular speed; circular frequency
$\boldsymbol{\Omega}$	Angular velocity of reference frame
*	Asterisk, marking a problem with no answer given
★	Star, marking an optional section
†	Dagger, marking a footnote
⇧	Hollow arrow, representing a moment or torque
⇒	Equipollent to
◀	Answer pointer

Contents Overview

[†]The texts for *Statics* and *Dynamics* are also available in a combined volume, *Engineering Mechanics: Statics and Dynamics*.
[††]Chapter 7 may be covered after Chap. 9 without affecting continuity.

Contents

Chapter 13
Kinetics of Particles: Work and Energy   552

Chapter 14
Kinetics of Particles:
Impulse and Momentum   594

Kinematics of Particles   Chapter 11

KEY CONCEPTS

- Rectilinear motion, curvilinear motion, absolute motion, relative motion, kinematic quantity, velocity, speed, acceleration, deceleration, jerk, and initial conditions.
- Rectilinear motion of a particle whose acceleration is a function of (1) time, (2) position, (3) velocity, or (4) a constant.
- Dependent rectilinear motions of particles.
- Graphical solution of rectilinear-motion problems.
- Descriptions of the curvilinear motion of a particle using (1) rectangular components, (2) tangential and normal components, (3) radial and transverse components, or (4) cylindrical components.

Dynamics is a subject dealing with conditions of bodies under the action of unbalanced force systems. The phase of dynamics which deals with the study of the motion of bodies without considering the cause of motion is called *kinematics*. The phase of dynamics which involves the study of the relation between the motion of a body and the force system causing the motion is called *kinetics*. Nevertheless, kinetics usually contains some kinematics. A kinematic quantity is a measurable quantity in kinematics, such as time, displacement, velocity, acceleration, and jerk. The concepts and methods of computation of various kinematic quantities of particles, which move along straight lines or along curved paths, are presented in this chapter. These concepts and methods form a useful background which will facilitate the study of kinetics in the subsequent chapters.

11.1   Types of Motion of a Particle[†]

A particle is said to be in *motion* if it does not remain in a fixed position during any finite interval of time. The motion of a particle may generally be classified according to its path. A *rectilinear motion* is a motion along a straight-line path. A *curvilinear motion* is a motion along a curved path. When a particle moves in a plane, it is said to have a *plane motion*; otherwise, its motion is a *motion in three dimensions*.

[†]A *particle* is an *idealized body* whose size and shape can be disregarded without introducing appreciable errors in the description and prediction of its state of rest or motion (cf. Sec. 1.6).

A motion may also be classified according to the fixity of the reference frame.[†] An *absolute motion* is a motion in a fixed reference frame, while a *relative motion* is a motion in a moving reference frame. In kinematics, any reference frame may be designated as *fixed*; all others not rigidly attached to it are regarded as *moving*.

Developmental Exercise

D11.1[††] Define the terms: (a) rectilinear motion, (b) curvilinear motion, (c) absolute motion, (d) relative motion.

RECTILINEAR MOTION

11.2   Kinematic Quantities

A *kinematic quantity* is a measurable quantity in kinematics, which can be defined without involving force or mass. Examples of kinematic quantities are position, distance, displacement, total distance traveled, time, speed, velocity, acceleration, deceleration, and jerk. We note that some of the kinematic quantities (e.g., displacement, velocity, and acceleration) do possess magnitudes and directions and are vector quantities (cf. Sec. 1.2).

In rectilinear motion of a particle, the *orientations* of the directions of those vector kinematic quantities must, however, *lie along the given straight-line path*. In such a motion, only the *magnitude* and the *sense* of each vector kinematic quantity are the nontrivial information to be described or predicted. Because of this special feature, the kinematics of particles in rectilinear motion can be studied using essentially *scalar analysis* rather than vector analysis.

Developmental Exercises

D11.2   What is a kinematic quantity? Give three examples.

D11.3   Why can the kinematics of particles in rectilinear motion be studied using essentially scalar analysis?

11.3   Position, Displacement, and Velocity

The *position* of a particle is the location of the particle described by using a coordinate system. To facilitate the study of the rectilinear motion of a particle, we usually lay the x axis along the straight-line path of the particle

---

[†]A *reference frame* is a frame (or body) of reference in which one or more coordinate systems may be embedded. For conciseness, the phrase "in a reference frame" is, from time to time, used to mean "with respect to a reference frame." An *inertial reference frame* is a frame of reference in which Newton's laws of motion are valid or highly accurate (cf. Sec. 3.1).

[††]The developmental exercises (D's) may be *utilized* or *skipped* as explained in the Preface. Whenever a developmental exercise asks for the definition of a term or description of a law or principle, students are encouraged to answer in their *own words*, which reflect their understanding of the concept involved. Answers to selected developmental exercises are given at the end of the text.

and choose a convenient point O on the path as the fixed origin, as shown in Fig. 11.1. At the time t, the particle is situated at the point P whose position is given by the *coordinate* x, where $x = x(t)$. Note that the magnitude of x indicates the *distance* between the origin O and the particle at P. The value of x is positive if P is reached from O by moving along the axis in the positive direction; otherwise, it is negative. Since the coordinate x completely defines the position of the particle in rectilinear motion, it is referred to as the *position* (or the *position coordinate*) of the particle. The position x has the dimension *length* and is usually measured in *meters* (m) in the SI and in *feet* (ft) in the U.S. customary system. The time t is usually measured in *seconds* (s), both in the SI and in the U.S. customary system.

Fig. 11.1

The *displacement* of a particle is the change of position of the particle. Suppose that, over the time interval Δt, the particle in rectilinear motion travels from the point P to the point P' as shown in Fig. 11.2. In terms of the positions x and x' of the points P and P', the *displacement*, denoted by Δx, of the particle from P to P' is given by

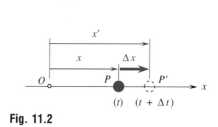

Fig. 11.2

$$\Delta x = x' - x \tag{11.1}$$

The displacement Δx is positive if $\overrightarrow{PP'}$ points in the positive direction of the x axis; otherwise, Δx is negative. The displacement of a particle has both magnitude and direction and is a vector. The *total distance traveled* x_T by a particle is the total length of path traversed by the particle; it is always a positive scalar quantity. Clearly, Δx and x_T have the same dimension and units as those of x.

The *velocity* of a particle is defined as the time rate of change of position (or the time rate of displacement) of the particle. The *average velocity* v_{avg} of a particle having a displacement Δx over a time interval Δt is given by

$$v_{\text{avg}} = \frac{\Delta x}{\Delta t} \tag{11.2}$$

As Δt becomes smaller and approaches zero in the limit, v_{avg} becomes the *instantaneous velocity* v of the particle; i.e.,

$$v = \lim_{\Delta t \to 0} \frac{\Delta x}{\Delta t} \qquad v = \frac{dx}{dt} = \dot{x} \tag{11.3}^\dagger$$

The instantaneous velocity is what we usually call the *velocity*.

A positive value of the velocity v indicates that the value of the position x is increasing algebraically and the particle is moving in the positive direction of the x axis as shown in Fig. 11.3. On the other hand, a negative value of v indicates that x is decreasing algebraically and the particle is moving in the negative direction as shown in Fig. 11.4. The velocity of a particle has both magnitude and direction and is a vector. The magnitude of the velocity is known as the *speed*. If the *total distance traveled* by a particle over the time interval Δt is x_T, the *average speed* $|v|_{\text{avg}}$ of the particle is defined as

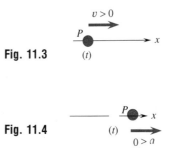

Fig. 11.3

Fig. 11.4

$$|v|_{\text{avg}} = \frac{x_T}{\Delta t} \tag{11.4}$$

---

†A quantity with n dots over it represents the nth derivative of the quantity with respect to the time variable t.

The velocity and the speed have the dimensions *length/time*. Their primary units are m/s in the SI and ft/s in the U.S. customary system.

EXAMPLE 11.1

The position x of a particle in rectilinear motion is related to the time t by the equation $x = t^3 - 6t^2 - 36t + 116$, where x is measured in meters and t in seconds. Over the time interval $0 \le t \le 8$ s, determine Δx, x_T, v_{avg}, $|v|_{avg}$, and the velocity v_7 at $t = 7$ s.

Solution. We compute the displacement Δx of the particle by subtracting its initial position x_0 at $t = 0$ from its final position x_8 at $t = 8$ s as follows:

$$x_0 = x|_{t=0} = 0 - 0 - 0 + 116 = 116$$

$$x_8 = x|_{t=8} = (8)^3 - 6(8)^2 - 36(8) + 116 = -44$$

$$\Delta x = x_8 - x_0 \qquad\qquad \Delta x = -160 \text{ m} \quad \blacktriangleleft$$

The negative sign in the value of Δx indicates that the displacement points in the negative direction during the interval as shown.[†]
To find the total distance traveled x_T, we need to first find out whether the particle has changed its direction of motion over the interval $0 \le t \le 8$ s. The velocity v of the particle is given by

$$v = \dot{x} = 3t^2 - 12t - 36 = 3(t - 6)(t + 2)$$

From the sketch, we note that $v < 0$ (the particle moves in the negative direction) for $0 \le t < 6$ s, $v = 0$ (the particle stops) at $t = 6$ s, and $v > 0$ (the particle moves in the positive direction) for 6 s $< t \le 8$ s. From $t = 0$ to $t = 6$ s, the distance traveled is given by

$$\Delta x_{0 \to 6} = x_6 - x_0 = x|_{t=6} - x|_{t=0} = -100 - 116$$
$$\Delta x_{0 \to 6} = -216 \text{ m} = 216 \text{ m in the negative direction}$$

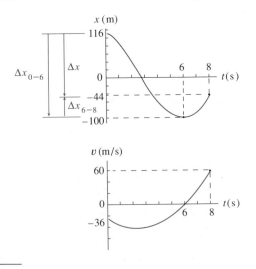

From $t = 6$ s to $t = 8$ s, the distance traveled is given by

$$\Delta x_{6 \to 8} = x_8 - x_6 = x|_{t=8} - x|_{t=6} = -44 - (-100)$$

$$\Delta x_{6 \to 8} = 56 \text{ m in the positive direction}$$

Therefore, the total distance traveled from $t = 0$ to $t = 8$ s is

$$x_T = |\Delta x_{0 \to 6}| + |\Delta x_{6 \to 8}| = 216 + 56 \qquad x_T = 272 \text{ m} \blacktriangleleft$$

From $t = 0$ to $t = 8$ s, we have $\Delta t = 8$ s. To compute v_{avg}, we write

$$v_{\text{avg}} = \frac{\Delta x}{\Delta t} = \frac{-160}{8} \qquad v_{\text{avg}} = -20 \text{ m/s} \blacktriangleleft$$

The negative sign in the value of v_{avg} indicates that the average velocity points in the negative direction.

To find the average speed of the particle from $t = 0$ to $t = 8$ s, we write

$$|v|_{\text{avg}} = \frac{x_T}{\Delta t} = \frac{272}{8} \qquad |v|_{\text{avg}} = 34 \text{ m/s} \blacktriangleleft$$

It should be noted that, in general, $|v_{\text{avg}}| \le |v|_{\text{avg}}$. For the velocity of the particle at $t = 7$ s, we write

$$v_7 = \dot{x}|_{t=7} = 3(7)^2 - 12(7) - 36 \qquad v_7 = 27 \text{ m/s} \blacktriangleleft$$

Developmental Exercises

D11.4   Define for a particle the terms: (a) position, (b) displacement, (c) total distance traveled, (d) velocity, (e) average velocity, (f) speed, (g) average speed.

D11.5   The rectilinear motion of a particle is given by the relation $x = -t^3 + 3t^2 + 24t - 50$, where x is measured in feet and t in seconds. Determine (a) the velocity v_6 of the particle at $t = 6$ s, (b) the time τ at which the velocity will be zero, (c) the position x_τ at $t = \tau$, (d) the total distance traveled x_T from $t = 0$ to $t = 6$ s.

D11.6   A particle moves from the origin O first to the point P and then back to the point Q as shown during a 5-s time interval. Over this time interval, determine Δx, x_T, v_{avg}, and $|v|_{\text{avg}}$.

Fig. D11.6

11.4   Acceleration and Jerk

The *acceleration* of a particle is defined as the time rate of change of velocity of the particle. Suppose that, over the time interval Δt, a particle moves from the point P to the point P' while its velocity changes from v to $v + \Delta v$ as shown in Fig. 11.5. The *average acceleration* a_{avg} of the particle over Δt is given by

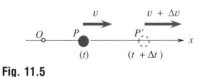

Fig. 11.5

$$a_{\text{avg}} = \frac{\Delta v}{\Delta t} \qquad (11.5)$$

As Δt becomes smaller and approaches zero in the limit, a_{avg} becomes the *instantaneous acceleration* a of the particle; i.e.,

$$a = \lim_{\Delta t \to 0} \frac{\Delta v}{\Delta t} \qquad \boxed{a = \frac{dv}{dt} = \dot{v}} \tag{11.6}$$

The instantaneous acceleration is what we usually call the *acceleration*. Substituting Eq. (11.3) into Eq. (11.6), we write

$$a = \frac{d}{dt}\left(\frac{dx}{dt}\right) \qquad \boxed{a = \frac{d^2x}{dt^2} = \ddot{x}} \tag{11.7}$$

$$a = \frac{dv}{dt} = \frac{dv}{dx}\frac{dx}{dt} = \frac{dv}{dx}v \qquad \boxed{a = v\frac{dv}{dx}} \tag{11.8}^\dagger$$

Note from calculus that when $a = \dot{v} > 0$, the value of v must be increasing algebraically; and it means physically that the particle is either moving faster and faster in the positive direction or moving slower and slower in the negative direction. Furthermore, when $a = \dot{v} < 0$, the value of v must be decreasing algebraically; and it means physically that the particle is either moving slower and slower in the positive direction or moving faster and faster in the negative direction. The term *deceleration* is an alias of acceleration when the speed is decreasing.[††] The acceleration of a particle has both magnitude and direction and is a vector. Moreover, the acceleration has the dimensions *length/time*^{2}, and its primary units are m/s^2 in the SI and ft/s^2 in the U.S. customary system.

The time rate of change of acceleration is referred to as the *jerk*. However, no standard names are given to the second- and higher-order time derivatives of acceleration. The jerk of a vehicle is often used as a measure of discomfort to be experienced by the passengers.

EXAMPLE 11.2

Refer to Example 11.1. Determine (a) the accelerations a_0 and a_7 at $t = 0$ and $t = 7$ s, respectively; (b) a_{avg} over the time interval $0 \le t \le 7$ s.

Solution. The position of the particle in Example 11.1 is

$$x = t^3 - 6t^2 - 36t + 116$$

Thus, we write

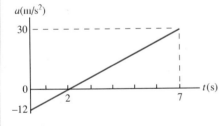

$$v = \dot{x} = 3t^2 - 12t - 36 \qquad a = \dot{v} = 6t - 12$$

$$a_0 = a|_{t=0} = -12 \qquad a_0 = -12 \text{ m/s}^2 \blacktriangleleft$$

$$a_7 = a|_{t=7} = 30 \qquad a_7 = 30 \text{ m/s}^2 \blacktriangleleft$$

To find a_{avg}, we first find Δv and then divide it by Δt as follows:

[†]It can happen that $dv/dx \to \infty$ as $v \to 0$ (e.g., the instant a body is dropped from rest). Thus, it is *false* that $a = 0$ whenever $v = 0$.

[††]Note that *deceleration* of a body is not equal to the negative of the acceleration of the body, just, in a way, as "debar" is not the opposite of "bar," and "inflammable" is not the opposite of "flammable." The deceleration of a body is simply the acceleration of the body whose velocity is opposite to the acceleration at the time considered.

$$v_0 = v|_{t=0} = -36 \qquad v_7 = v|_{t=7} = 27$$

$$\Delta v = v_7 - v_0 = 63 \qquad \Delta t = 7 - 0 = 7$$

$$a_{\text{avg}} = \frac{\Delta v}{\Delta t} = \frac{63}{7} \qquad\qquad a_{\text{avg}} = 9 \text{ m/s}^2 \blacktriangleleft$$

As a check, note that $a_{\text{avg}} = (a_0 + a_7)/2 = 9$ m/s^2.

Developmental Exercise

D11.7   Refer to D11.5. Determine (a) a_0 and a_6 at $t = 0$ and $t = 6$ s, respectively; (b) a_{avg} over the time interval $0 \leq t \leq 6$ s.

EXAMPLE 11.3

The velocity v of a recoiling piston in a cylinder filled with oil, as shown, is given by the relation $v = v_0 - bx$, where x is the position of the piston and v_0 and b are positive constants. Determine the acceleration a of the piston at $x = v_0/4b$.

Solution.   Since v is given as a function of x, we write

$$a = v\frac{dv}{dx} = (v_0 - bx)\frac{d}{dx}(v_0 - bx) = -b(v_0 - bx)$$

At $x = v_0/(4b)$, we obtain

$$a = -\frac{3}{4}bv_0 \blacktriangleleft$$

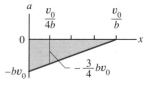

Developmental Exercise

D11.8   Define the *acceleration*, *deceleration*, and *jerk* of a particle.

11.5   Determination of Rectilinear Motion

Problems of the rectilinear motion of a particle may be classified according to the type of acceleration of the particle. In each problem, the conditions that $x = x_0$ and $v = v_0$ at the time $t = 0$ are called the *initial conditions*.

An acceleration may be a function of time, position, velocity, or simply a constant; it may also be a function of several variables.

(1) *Acceleration is a given function of time, $a = f(t)$.*
  For this class of problems, we first use Eq. (11.6) to write

$$dv = a\, dt = f(t)\, dt \qquad \int_{v_0}^{v} dv = \int_{0}^{t} f(t)\, dt$$

$$v = v_0 + \int_{0}^{t} f(t)\, dt \tag{11.9}$$

which expresses v as a function of t. Then, we use Eq. (11.3) to write

$$dx = v\, dt \qquad \int_{x_0}^{x} dx = \int_{0}^{t} v\, dt$$

$$x = x_0 + \int_{0}^{t} v\, dt \tag{11.10}$$

Developmental Exercises

D11.9 The acceleration of a particle moving along a straight line is $a = A \cos \omega t$, where A and ω are constants. The initial conditions are $v = v_0$ and $x = x_0$. Express the velocity v and the position x as functions of the time t.

D11.10 Do D11.9 if $a = A \sin \omega t$ and $v = x_0 = 0$.

(2) *Acceleration is a given function of position, $a = f(x)$.*
  For this class of problems, we first use Eq. (11.8) to write

$$v\, dv = a\, dx = f(x)\, dx \qquad \int_{v_0}^{v} v\, dv = \int_{x_0}^{x} f(x)\, dx$$

$$v^2 = v_0^2 + 2 \int_{x_0}^{x} f(x)\, dx \tag{11.11}$$

which yields v in terms of x. Then, we use Eq. (11.3) to write

$$dt = \frac{dx}{v} \qquad \int_{0}^{t} dt = \int_{x_0}^{x} \frac{dx}{v}$$

$$t = \int_{x_0}^{x} \frac{dx}{v} \tag{11.12}$$

which provides the relation between x and t.

EXAMPLE 11.4

The acceleration of an oscillating weight W, as shown, is $a = -\omega^2 x$, where ω is a constant. If the initial conditions are $x_0 \neq 0$ and $v_0 = 0$, express the position x as a function of the time t.

Solution. The oscillation of the weight W is a rectilinear motion whose acceleration a is given as a function of the position x. Noting that $a = f(x) = -\omega^2 x$, $x_0 \neq 0$, and $v_0 = 0$, we utilize the derived results in Eqs. (11.11) and (11.12) to write

$$v^2 = 0 + 2 \int_{x_0}^{x} (-\omega^2 x)\, dx = \omega^2 (x_0^2 - x^2)$$

$$t = \int_{x_0}^{x} \frac{dx}{v} = \frac{1}{\pm\omega} \int_{x_0}^{x} \frac{dx}{(x_0^2 - x^2)^{1/2}}$$

The last integral can be evaluated by letting $x = x_0 \sin\theta$ and changing the corresponding limits of integration. We write

$$\pm\omega t = \int_{x_0}^{x} \frac{dx}{(x_0^2 - x^2)^{1/2}} = \int_{\pi/2}^{\theta} \frac{x_0 \cos\theta\, d\theta}{x_0 \cos\theta} = \theta - \frac{\pi}{2} = \sin^{-1}\frac{x}{x_0} - \frac{\pi}{2}$$

$$\frac{x}{x_0} = \sin\left(\pm\omega t + \frac{\pi}{2}\right) = \cos(\pm\omega t) = \cos\omega t$$

$$x = x_0 \cos\omega t \quad \blacktriangleleft$$

Equilibrium position

x

W

This result shows that the motion of the weight W is sinusoidal. This type of motion is known as a *simple harmonic motion*.

Developmental Exercises

D11.11 Refer to Example 11.4. Show that the solution $x = x_0 \cos\omega t$ satisfies the initial conditions and yields the acceleration as given.

D11.12 Work Example 11.4 if $x_0 = 0$ and $v_0 \neq 0$.

(3) *Acceleration is a given function of velocity, $a = f(v)$.*
For this class of problems, we first use Eq. (11.6) to write

$$dt = \frac{dv}{a} = \frac{dv}{f(v)} \qquad \int_0^t dt = \int_{v_0}^{v} \frac{dv}{f(v)}$$

$$t = \int_{v_0}^{v} \frac{dv}{f(v)} \qquad (11.13)$$

which expresses t as a function of v. The position x can be related to the velocity v by integrating Eq. (11.8) as follows:

$$dx = \frac{v\, dv}{a} = \frac{v\, dv}{f(v)} \qquad \int_{x_0}^{x} dx = \int_{v_0}^{v} \frac{v\, dv}{f(v)}$$

$$x = x_0 + \int_{v_0}^{v} \frac{v\, dv}{f(v)} \qquad (11.14)$$

EXAMPLE 11.5

The acceleration a of a recoiling piston in a cylinder filled with oil is given by $a = -bv$, where b is a positive constant and v is the velocity of the piston as shown. If $x_0 \neq 0$ and $v_0 \neq 0$, express and plot v and x as functions of the time t.

Solution.   Noting that $a = f(v) = -bv$, $x_0 \neq 0$, and $v_0 \neq 0$, we utilize the derived results in Eqs. (11.13) and (11.14) to write

$$t = \int_{v_0}^{v} \frac{dv}{-bv} = -\frac{1}{b} \ln \frac{v}{v_0}$$

$$\frac{v}{v_0} = e^{-bt} \qquad\qquad v = v_0\, e^{-bt} \quad \blacktriangleleft$$

$$x = x_0 + \int_{v_0}^{v} \frac{v\, dv}{-bv} = x_0 - \frac{1}{b}(v - v_0) = x_0 - \frac{v_0}{b}(e^{-bt} - 1)$$

$$x = x_0 + \frac{v_0}{b}(1 - e^{-bt}) \quad \blacktriangleleft$$

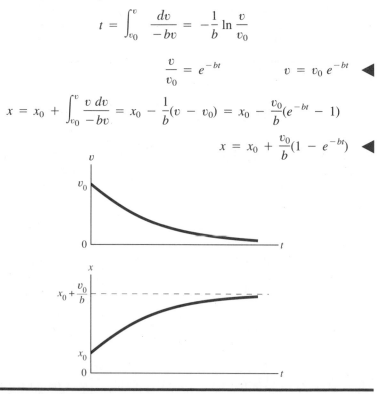

Developmental Exercise

D11.13   The acceleration a of a body immersed in a fluid is given by $a = -bv^2$, where v is the velocity of the body and b is a positive constant. If $x_0 = 0$ and $v_0 \neq 0$, express v and x as functions of t.

(4) *Acceleration is a constant, $a = a_c$.*

For this class of problems, we use Eqs. (11.6) and (11.3) to write

$$dv = a\, dt = a_c\, dt \qquad \int_{v_0}^{v} dv = a_c \int_{0}^{t} dt$$

$$v = v_0 + a_c t \tag{11.15}$$

$$dx = v\, dt \qquad \int_{x_0}^{x} dx = \int_{0}^{t} (v_0 + a_c t)\, dt$$

$$x = x_0 + v_0 t + \tfrac{1}{2} a_c t^2 \tag{11.16}$$

Furthermore, by substituting a_c for $f(x)$ in Eq. (11.11), we obtain

$$v^2 = v_0^2 + 2a_c(x - x_0) \tag{11.17}$$

Equations (11.15) through (11.17) are utilized as formulas when $a = a_c$.

EXAMPLE 11.6

A stone is thrown upward with $v_0 = 40$ ft/s at A off the side of a cliff as shown. Knowing that the stone is subjected to a constant downward acceleration of 32.2 ft/s^2 due to gravity, determine (a) the velocity v and elevation x of the stone as functions of the time t, (b) the maximum height h reached by the stone, (c) the velocity v_D of the stone just before it hits the bottom of the cliff, and (d) the corresponding time t_D.

Solution. We note that the motion of the stone is rectilinear with a constant acceleration $a_c = -32.2$ ft/s^2. By using the point D as the origin of the x axis as given, we have (at $t = 0$) the initial conditions

$$x_0 = 120 \text{ ft} \qquad v_0 = 40 \text{ ft/s}$$

Utilizing Eqs. (11.15) and (11.16), we write

$$v = 40 - 32.2t \qquad x = 120 + 40t - 16.1t^2 \blacktriangleleft$$

where v, x, and t are measured in ft/s, ft, and s, respectively.

As the stone reaches its highest point C, its velocity v becomes zero and its elevation x is equal to h. Utilizing Eq. (11.17), we write

$$0^2 = (40)^2 + 2(-32.2)(h - 120) \qquad h = 144.8 \text{ ft} \blacktriangleleft$$

Just before the stone hits the bottom of the cliff, the elevation x becomes zero and the velocity v is denoted by v_D. Thus, we write

$$v_D^2 = (40)^2 + 2(-32.2)(0 - 120)$$

$$v_D = \pm 96.58 \qquad v_D = -96.6 \text{ ft/s} \blacktriangleleft$$

where we take only the negative value of v_D because the direction of v_D must be downward. To find the corresponding time t_D, we let $v = v_D = -96.58$ ft/s in Eq. (11.15) and write

$$-96.58 = 40 + (-32.2)t_D \qquad t_D = 4.24 \text{ s} \blacktriangleleft$$

REMARK. The value of t_D may also be found by letting $x = x_D = 0$ in Eq. (11.16). This yields $t_D = 4.24$ or -1.757. The negative value of t_D is an extraneous root to be discarded; therefore, the same result of $t_D = 4.24$ s is obtained.

Developmental Exercise

D11.14 Refer to Example 11.6. Determine (a) v_A of the stone as it falls back to the position at A, (b) the corresponding time t_A.

(5) *Acceleration is a given function of time, position,* and *velocity*; $a = f(t, x, v)$.

For this class of problems, the procedure for obtaining the solutions generally depends on the explicit form of the function $f(t, x, v)$ and usually involves the solution of a differential equation as given by $a = f(t, x, v)$. Because of the potential involvement of advanced mathematical skills in obtaining the solutions, a detailed study of this class of problems is beyond the level of this text. However, a sample case in this class is described in Prob. 11.28, which may be solved by those who can solve first-order differential equations.

PROBLEMS†

11.1 and 11.2* The position x of a particle moving along a straight line is given as shown. Determine the time, position, and acceleration when the velocity is zero.

(x in m, t in s)
$$x = t^3 - 9t^2 + 24t - 10$$

Fig. P11.1 and P11.3

(x in mm, t in s)
$$x = t^3 - 12t^2 + 45t + 5$$

Fig. P11.2* and P11.4

11.3 and 11.4 The position x of a particle moving along a straight line is given as shown. Determine the time, position, and velocity when the acceleration is zero.

11.5* The position x of a particle in rectilinear motion is given as shown. Over the time interval $0 \le t \le 8$ s, determine Δx, x_T, v_{avg}, and $|v|_{avg}$.

11.6 The position x of a particle in rectilinear motion is given as shown. Determine (a) the time t at which the velocity is zero, (b) x_T from $t = 0$ to the time at which the acceleration is zero.

11.7 The acceleration of a particle in rectilinear motion is directly proportional to the square of the time t. At $t = 0$, its velocity is -32 m/s. At $t = 2$ s, the particle comes to the origin O and its velocity becomes zero. Express the position x of the particle as a function of t.

(x in in., t in s)
$$x = 4t^3 - 54t^2 + 216t + 17$$

Fig. P11.5* and P11.6

---

†Answers are provided at the end of the text for all problems except those marked with an asterisk.

11.8   A particle is given an acceleration $a = 24t - 72$ along a straight line, where a is measured in in./s^2 and t in seconds. If $v_0 = 0$ and $x_0 = 75$ in., determine (a) the time t_1 when the velocity v is again zero, (b) x_T from $t = 0$ to $t = 8$ s.

11.9*   A particle is subjected to a sinusoidal acceleration $a = A \sin \omega t$ along a straight line, where A and ω are constants. If $v_0 = x_0 = 0$, determine (a) the position x as a function of the time t, (b) the maximum speed of the particle.

11.10*   The position of an oscillating particle is given by $x = X \sin(\omega t + \phi)$, where X, ω, and ϕ are constants and represent the amplitude, the circular frequency, and the phase angle, respectively. Determine the values of X and ϕ in terms of x_0, v_0, and ω.

11.11   The acceleration of an oscillating particle is given by $a = -\omega^2 x$, where ω is a constant. If $x_0 = 3$ m, $v_0 = 0$, and $v = -12$ m/s when $x = 0$, determine (a) the value of ω, (b) the time elapsed during which the particle returns from $x = 3$ m to $x = 0$.

11.12   A particle is given an acceleration $a = 24 - 6x^2$, where a is measured in m/s^2 and x in m. If $x_0 = v_0 = 0$, determine (a) the velocity when $x = 3$ m, (b) the first position where the velocity is again zero, (c) the first position where the velocity is maximum, (d) the first maximum velocity.

11.13*   Solve Prob. 11.12 if $a = 36 - 24x + 3x^2$.

11.14   The acceleration of a particle moving in the vertical direction under the effect of gravity is $a = -gR^2/r^2$, where r is the distance from the center of the earth to the particle, $R = 6.37$ Mm which is the radius of the earth, and $g = 9.81$ m/s^2 which is the gravitational acceleration near the surface of the earth. Based on this formula where air resistance is neglected, determine the height h reached by a bullet fired vertically upward from the surface of the earth with the following muzzle velocities: (a) 500 m/s, (b) 5 km/s, (c) 10 km/s.

11.15*   A particle is released from rest at a height h above the surface of the earth. Using the formula for the acceleration $a = -gR^2/r^2$ as defined in Prob. 11.14, (a) determine the speed of the particle as it strikes the earth, (b) deduce the approximate value of this speed if $h \ll R$ (i.e., $R + h \approx R$), (c) evaluate the time required for the particle to fall to the surface of the earth if $h = R$.

11.16   The *escape velocity* v_{esc} of a particle is the minimum velocity with which a particle projected vertically will not return to the earth; i.e., the velocity of an escaping particle will approach zero when its height approaches infinity as shown. Using the formula $a = -gR^2/r^2$ as defined in Prob. 11.14, determine the value of v_{esc} of a particle projected from the surface of the earth in (a) km/s, (b) mi/s.

Fig. P11.16   **Fig. P11.17**

11.17   The hydroplane shown has a touchdown speed of 90 mi/h upon contact with the water. This speed is to be reduced to 15 mi/h in a distance of 1500 ft. The deceleration of the hydroplane on the water is observed to be proportional to the square of its velocity; i.e., $a = -kv^2$, where k is a constant related to the size and shape of the landing gear vanes. Determine (a) the value of k if a is measured in ft/s^2 and v in ft/s, (b) the time elapsed in slowing down from 90 mi/h to 15 mi/h.

11.18* Due to gravity and air resistance, the downward acceleration of an object is found to be given by $a = 9.81 - 0.0003v^2$, where a is measured in m/s^2 and v in m/s. If the object is released from rest and keeps falling downward as shown, determine its velocity after it has fallen 300 m.

Fig. P11.18*   **Fig. P11.19**   **Fig. P11.20**

11.19 A pellet is fired from an air pistol with a muzzle velocity $v_0 = 350$ ft/s as shown. The muzzle of the pistol is 8 ft above the ground at the time the trigger is pulled. Neglecting air resistance and taking $g = 32.2$ ft/s^2, determine (a) the maximum altitude reached by the pellet, (b) the time elapsed since firing for the pellet to return to the ground.

11.20 An open elevator is moving up at a constant speed $v_E = 15$ ft/s as shown. A ball is released from the elevator and reaches the ground in 3 s. Determine (a) the height at which the ball was released, (b) the speed with which the ball strikes the ground.

11.21 A boy gently pushes a stone off the top of a cliff down to a lake as shown. The time elapsed from the instant the stone is pushed off to the instant the boy hears the splashing sound of the stone is 5 s. Taking the speed of sound in the air to be 332 m/s and $g = 9.81$ m/s^2, determine the height h of the cliff above the water surface.

11.22* The motorcyclist shown is traveling at 72 km/h when a traffic light 300 m ahead of him turns red. The red light is timed to stay for 20 s in his direction. If he wishes to decelerate uniformly so that he reaches the traffic light just as it turns green again, determine (a) the deceleration of the motorcycle, (b) the speed of the motorcycle as it passes the traffic light.

Fig. P11.21

Fig. P11.22*

11.23* The acceleration of a particle moving in a viscous fluid as shown is given by $a = -2v$, where a is measured in ft/s^2 and v in ft/s. If $x_0 = 2$ ft and $v_0 = 10$ ft/s, determine (a) the displacement of the particle before it comes to rest, (b) the time required for its velocity to be reduced to 1 ft/s, (c) the time required for its velocity to become zero.

Fig. P11.23*   **Fig. P11.24**

11.24 The touchdown speed of 180 km/h of an airplane upon contact with the runway as shown is to be reduced to 18 km/h at a constant rate of 4 m/s^2 by operating thrust reversers as well as brakes of the airplane. Determine (a) the time elapsed from touchdown to the time at which the speed is reduced to 18 km/h, (b) the distance on the runway covered by the airplane during that time interval.

11.25* The car shown travels 800 m in 40 s while being accelerated at a constant rate of 0.25 m/s^2. For this interval of travel, determine (a) its initial velocity, (b) its final velocity.

11.26 The car shown is traveling at 90 km/h and is to be stopped over a distance of 125 m through a constant deceleration. Determine (a) the rate of its deceleration, (b) the time required for the car to stop.

Fig. P11.25* and P11.26

11.27 The single-stage rocket shown is launched vertically from the ground, and its thrust is designed to give the rocket a constant upward acceleration of 15 ft/s^2 for 30 s before the fuel is exhausted. Neglecting air resistance, determine the maximum altitude reached by the rocket.

11.28 If the decaying thrust and air resistance are taken into account, the upward acceleration of a rocket during its atmospheric flight is given by $a = ke^{-ct} - bv - g$, where k, c, and b are constants, and the gravitational acceleration g is taken as essentially constant. The rocket is launched from the ground with initial conditions $x_0 = v_0 = 0$. Determine the velocity v and the position x of the rocket as functions of the time t. (*Hint.* Since $a = dv/dt$, we have $dv/dt + bv = ke^{-ct} - g$. Solve this differential equation for v, then integrate to obtain x.)

Fig. P11.27

11.6 Relative Rectilinear Motion

The *relative motion* of any two particles is the *difference of the motions* of those two particles in a given reference frame. The *position of the particle A relative to the particle B*, as shown in Fig. 11.6, is denoted by $x_{A/B}$. We write

$$x_{A/B} = x_A - x_B \quad \text{or} \quad \boxed{x_A = x_B + x_{A/B}} \tag{11.18}$$

The time rate of change of $x_{A/B}$ is called the *velocity of A relative to B* and is denoted by $v_{A/B}$. Differentiating Eq. (11.18), we write

$$\boxed{v_A = v_B + v_{A/B}} \tag{11.19}$$

Fig. 11.6

The time rate of change of $v_{A/B}$ is called the *acceleration of A relative to B* and is denoted by $a_{A/B}$. Differentiating Eq. (11.19), we write[†]

$$a_A = a_B + a_{A/B} \tag{11.20}$$

Note that the same coordinate system and the same time (or clock) have been used in describing the motion of each particle in the system. Thus, when $x_{A/B} > 0$, it means that A is positioned on the positive side of B; and when $v_{A/B} < 0$, it means that A is moving to the negative side of B.

Developmental Exercise

11.15   What does it mean when (a) $x_{A/B} < 0$, (b) $v_{C/D} > 0$, (c) $a_{D/C} < 0$?

EXAMPLE 11.7

$(v_{B/E})_0 = 18$ m/s

15 m

An open elevator is being lowered at a constant rate of 2 m/s. A man in this elevator throws a handball upward with a velocity of 18 m/s relative to the elevator when he is 15 m above the ground. Determine (a) the time t and the position x_E at which the man will be able to catch the ball, (b) the velocity $v_{B/E}$ of the ball relative to the elevator when the ball is caught.

Solution.   Noting that the positive direction is upward, we may write the initial conditions for the ball B and elevator E as follows:

$$(x_B)_0 = 15 \qquad (x_E)_0 = 15 \qquad (v_E)_0 = -2 \qquad (v_{B/E})_0 = 18$$

$$(v_{B/E})_0 = (v_B)_0 - (v_E)_0 \qquad 18 = (v_B)_0 - (-2) \qquad (v_B)_0 = 16$$

The acceleration of the elevator is $a_E = 0$, but the acceleration of the ball is $a_B = -g = -9.81$ m/s². Thus, we write

$$v_E = -2 \qquad\qquad\qquad x_E = 15 - 2t$$

$$v_B = 16 - 9.81t \qquad\qquad x_B = 15 + 16t - \frac{9.81}{2}t^2$$

At the time the ball is caught, we have $x_E = x_B$. Thus, we set

$$15 - 2t = 15 + 16t - \frac{9.81}{2}t^2$$

which yields two roots: $t = 0$ and $t = 3.67$ s. Since $t = 0$ is a trivial solution, the time the ball is caught must be

$$t = 3.67 \text{ s} \blacktriangleleft$$

At this time, we have

$$x_E = 15 - 2(3.67) \qquad\qquad\qquad x_E = 7.66 \text{ m} \blacktriangleleft$$

$$v_B = 16 - 9.81(3.67) = -20$$

$$v_{B/E} = v_B - v_E = -20 - (-2) \qquad v_{B/E} = -18 \text{ m/s} \blacktriangleleft$$

[†]Note in Eqs. (11.18) through (11.20) that the subscript A in the left-hand member is formally equal to the product of the subscripts B and A/B in the right-hand member; i.e., $A = B(A/B)$. This is a useful way of remembering the equations of relative motion.

11.7 Dependent Rectilinear Motions

Through the use of pulleys and cordlike elements (e.g., cords, strings, ropes, wires, cables, belts, or chains), we may have a system of connected particles which execute *dependent rectilinear motions*, as illustrated in Fig. 11.7. We recall that cordlike elements are generally assumed to be perfectly flexible, inextensible, and of negligible mass. Thus, a very important condition to be satisfied here is that the *length of each cordlike element*, as expressed in terms of the position coordinates of the particles, *must be constant*. Such a condition is called the *constraint condition* in the dependent rectilinear motions of particles.

In Fig. 11.7, we note that the length of the cord *CDEFGB* must be constant. Since the lengths of the portions *DE* and *FG* remain constant, it follows that the sum of the lengths of the segments $\overline{CD}$, $\overline{EF}$, and $\overline{GB}$ is constant. Observing that each of the lengths of $\overline{CD}$ and $\overline{EF}$ differs from x_A only by a constant, and that, similarly, the length of $\overline{GB}$ differs from x_B only by a constant, we write

$$2x_A + x_B = \text{constant}$$

Fig. 11.7

Such an equation is a *constraint equation on the positions* of the particles in the system. Because of this constraint equation, only one of the two position coordinates may be chosen arbitrarily. The system in Fig. 11.7 has, therefore, only *one degree of freedom*. By differentiating the above constraint equation on the positions, we obtain the *constraint equations on the velocities and accelerations* of the particles in that system as follows:

$$2\frac{dx_A}{dt} + \frac{dx_B}{dt} = 0 \quad \text{or} \quad 2v_A + v_B = 0$$

$$2\frac{dv_A}{dt} + \frac{dv_B}{dt} = 0 \quad \text{or} \quad 2a_A + a_B = 0$$

where the positive directions of the dependent variables are downward because their directions of increase are downward as shown.

EXAMPLE 11.8

The three cylinders *A, B,* and *C* are connected by cords and pulleys. At the instant shown, the cylinders *A* and *C* have $\mathbf{v}_A = 4$ m/s $\downarrow$, $\mathbf{v}_C = 6$ m/s $\uparrow$, $\mathbf{a}_A = 3$ m/s^2 $\uparrow$, and $\mathbf{a}_C = 8$ m/s^2 $\downarrow$. Determine $\mathbf{v}_B$ and $\mathbf{a}_B$ of the cylinder *B* at this instant.

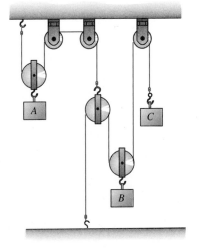

Solution. The position coordinates for the cylinders *A, B, C,* and the pulley *D* are established as shown on the next page. Since the length of the cord *EFGHID* is constant, we write

$$2x_A + x_D = k_1 \tag{1}$$

where k_1 is a constant. Letting the height from the bottom support to the top support be *h*, which is a constant, we see that $\overline{JK} = h - x_D$. Since the length of the cord *JKLMNOPC* is constant, it follows that the sum of the lengths of the segments, $\overline{JK}$, $\overline{LM}$, $\overline{NO}$, and $\overline{PC}$ must be constant. We write

$$(h - x_D) + (x_B - x_D) + x_B + x_C = \bar{k}_2$$

$$2x_B + x_C - 2x_D = k_2 \qquad (2)$$

where $\bar{k}_2$ and k_2 are constants. Multiplying Eq. (1) by 2 and then adding it to Eq. (2), we can eliminate x_D and obtain

$$4x_A + 2x_B + x_C = 2k_1 + k_2$$

Differentiating with respect to time, we write

$$4v_A + 2v_B + v_C = 0 \qquad 4a_A + 2a_B + a_C = 0$$

Since the increasing directions of the dependent variables as shown are downward, their positive directions are all downward. Thus, we have $v_A = 4$ m/s, $v_C = -6$ m/s, $a_A = -3$ m/s^2, and $a_C = 8$ m/s^2. Substituting these values into the above equations, we obtain $v_B = -5$ m/s and $a_B = 2$ m/s^2.

$$\mathbf{v}_B = 5 \text{ m/s} \uparrow \qquad \mathbf{a}_B = 2 \text{ m/s}^2 \downarrow \quad \blacktriangleleft$$

Fig. D11.17

Developmental Exercises

D11.16 How many degrees of freedom does the system in Example 11.8 have?

D11.17 At the instant shown, the cylinders A and C move with $\mathbf{v}_A = 5$ m/s $\downarrow$ and $\mathbf{v}_C = 2$ m/s $\uparrow$. Determine $\mathbf{v}_B$ of the cylinder B at this instant.

PROBLEMS

11.29 through 11.31* At the instant shown, the velocities and accelerations of the cylinders A and C are $\mathbf{v}_A = 2$ ft/s $\uparrow$, $\mathbf{a}_A = 1$ ft/s^2 $\downarrow$, $\mathbf{v}_C = 7$ ft/s $\downarrow$, and $\mathbf{a}_C = 3$ ft/s^2 $\uparrow$, respectively. Determine $\mathbf{v}_B$ and $\mathbf{a}_B$ of the cylinder B at this instant.

Fig. P11.29

Fig. P11.30

Fig. P11.31*

11.32 The block A as shown moves with $\mathbf{v}_A = 1.2$ m/s $\leftarrow$. Determine (a) $\mathbf{v}_B$ of the block B, (b) $\mathbf{v}_C$ and $\mathbf{v}_D$ of the points C and D on the cable, (c) $\mathbf{v}_{A/B}$.

Fig. P11.32 and P11.33*

11.33* The block A as shown starts from rest and moves to the left with a constant acceleration. After 5 s, it is found that $\mathbf{v}_{A/B} = 300$ mm/s $\rightarrow$. Determine $\mathbf{a}_A$ and $\mathbf{a}_B$ at this instant, and Δx_B during the first 4 s.

11.34 The motor at M as shown winds the cable at a constant rate of 700 mm/s. If the block B moves with $\mathbf{v}_B = 100$ mm/s $\leftarrow$, determine $\mathbf{v}_A$.

Fig. P11.34

11.35 The motor at M as shown starts from rest and winds the cable with a constant acceleration of 2 m/s^2 for 4 s. Determine (a) Δx_E of the elevator E over the time interval $0 \le t \le 4$ s, (b) $\mathbf{v}_{E/C}$ of E relative to the counterweight C at $t = 4$ s.

11.36 A crate D is hoisted by a man who holds one end of the rope and walks to the right with a constant speed of 0.5 m/s as shown. The length of the rope ABC is 14 m. Determine $\mathbf{v}_D$ and $\mathbf{a}_D$ of the crate when its bottom passes the corner at E.

Fig. P11.35

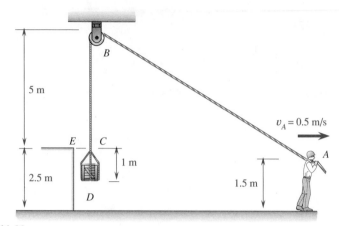

Fig. P11.36

11.37* A truck travels with a constant speed of 54 km/h on a straight road as shown. A car traveling at 90 km/h is 30 m behind the truck when the car's brakes are suddenly applied. If the road condition allows the car to decelerate only at the constant rate of 1.25 m/s², determine (a) whether a rear-end collision could occur, (b) if so, the velocity of the car relative to the truck just before the collision.

Fig. P11.37*

★11.8 Graphical Solution†

When the acceleration or velocity of a particle in rectilinear motion is given by different linear functions over several time intervals, graphical representations of the motion can be very helpful. A *position curve* is a graph obtained by plotting the position x of a particle against the time t. A *velocity curve* is a graph obtained by plotting the velocity v of a particle against the time t. Similarly, an *acceleration curve* is a graph obtained by plotting the acceleration a of a particle against the time t. These three curves are called the *motion curves* of the particle.

By calculus, we deduce the following six *useful relations* for the graphical solution of rectilinear-motion problems:

(1) The *slope of the position curve* at any time is equal to the ordinate of the velocity curve at that time. This is true because we have

$$\frac{dx}{dt} = v \tag{11.21}$$

---

†A five-point star indicates an optional section. Cf. the Preface.

(2) The *slope of the velocity curve* at any time is equal to the ordinate of the acceleration curve at that time. This is true because we have

$$\frac{dv}{dt} = a \tag{11.22}$$

The relations described in (1) and (2) are illustrated in Fig. 11.8.

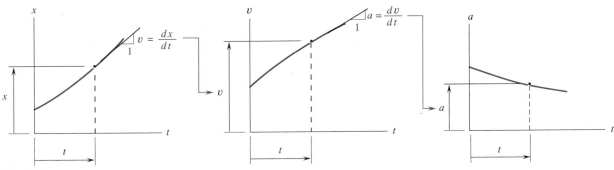

Fig. 11.8

(3) The *change in velocity* from the time t_1 to the t_2 is equal to the algebraic sum of the "areas" bounded by the acceleration curve and the time axis from t_1 to t_2.[†] This is true because, by integrating Eq. (11.22), we have

$$v_2 - v_1 = \int_{t_1}^{t_2} a \, dt \tag{11.23}$$

(4) The *change in position* from the time t_1 to the time t_2 is equal to the algebraic sum of the "areas" bounded by the velocity curve and the time axis from t_1 to t_2.[††] This is true because, by integrating Eq. (11.21), we have

$$x_2 - x_1 = \int_{t_1}^{t_2} v \, dt \tag{11.24}$$

The relations described in (3) and (4) are illustrated in Fig. 11.9.

Fig. 11.9

[†]The dimensions of such "areas" are *length/time*.

[††]The dimension of such "areas" is *length*.

(5) The *concavity of the position curve* is upward if the corresponding acceleration is positive; it is downward if the latter is negative. This is true because, from Eqs. (11.21) and (11.22), we have

$$\frac{d^2x}{dt^2} = a \qquad (11.25)$$

Furthermore, the position curve will have a *point of inflection* at the instant the acceleration changes its sign.

(6) The *total distance traveled* from the time t_1 to the time t_2 is equal to the sum of the absolute values of the "areas" bounded by the velocity curve and the time axis from t_1 to t_2. This is true because the total distance traveled x_T is the total length of the path traversed. In terms of an integral, we write

$$x_T = \int_{t_1}^{t_2} |v| \, dt \qquad (11.26)$$

In terms of the "areas" $A_1, A_2, \ldots, A_n$ bounded by the velocity curve and the time axis, as shown in Fig. 11.10, we write

$$x_T = |A_1| + |A_2| + \cdots + |A_n| \qquad (11.27)$$

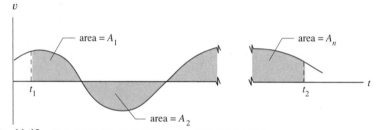

Fig. 11.10   (The dimension of "areas" in this figure is *length*.)

EXAMPLE 11.9

A particle moves along a straight line with the velocity shown. If $x_0 = 20$ ft, (a) draw the acceleration and position curves for $0 \le t \le 14$ s, (b) determine Δx and x_T over the interval $0 \le t \le 12$ s.

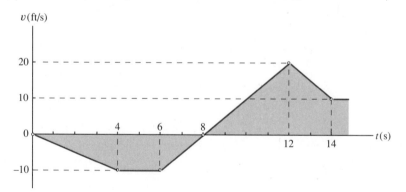

Solution.   Using the first five relations listed in Sec. 11.8, we can readily draw the acceleration and position curves as shown. The given velocity curve is here repeated and lined up in the time scale with these two curves for easy reference. In particular, note in the position curve that (1) the slope

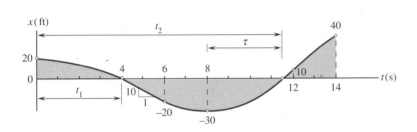

is zero at $t = 0$ and $t = 8$ s, (2) the concavity is downward during $0 < t < 4$ s and 12 s $< t < 14$ s, (3) the concavity is upward during 6 s $< t < 12$ s, (4) a point of inflection exists at $t = 12$ s, and (5) the curve becomes a straight line during 4 s $\leq t \leq 6$ s, in which the acceleration is zero.

Having drawn the position curve, we see that the positions of the particle at $t = 0$ and $t = 12$ s are $x_0 = 20$ ft and $x_{12} = 10$ ft. Thus, the displacement over the time interval $0 \leq t \leq 12$ s is

$$\Delta x = x_{12} - x_0 \qquad\qquad \Delta x = -10 \text{ ft} \blacktriangleleft$$

Alternatively, we can obtain Δx by summing algebraically the "areas" in the v-t diagram from $t = 0$ to $t = 12$ s; i.e.,

$$\Delta x = -\tfrac{1}{2}(4)(10) - 2(10) - \tfrac{1}{2}(2)(10) + \tfrac{1}{2}(4)(20) \text{ ft} = -10 \text{ ft} \quad \text{Q.E.D.}$$

From the position curve drawn, we see that the particle moves from $x = 20$ ft at $t = 0$ to $x = -30$ ft at $t = 8$ s, then back to $x = 10$ ft at $t = 12$ s. Thus, the total distance traveled over that time interval is

$$x_T = |x_8 - x_0| + |x_{12} - x_8| = |-30 - 20| + |10 - (-30)|$$

$$x_T = 90 \text{ ft} \blacktriangleleft$$

Of course, we can alternatively obtain x_T by summing the absolute values of the "areas" in the v-t diagram from $t = 0$ to $t = 12$ s; i.e.,

$$x_T = |-\tfrac{1}{2}(4)(10)| + |-2(10)| + |-\tfrac{1}{2}(2)(10)| + |\tfrac{1}{2}(4)(20)|$$

$$x_T = 90 \text{ ft} \qquad\qquad \text{Q.E.D.}$$

Developmental Exercises

D11.18   Refer to Example 11.9. Over the time interval $0 \le t \le 13$ s, determine Δx and x_T.

D11.19   Refer to Example 11.9. By inspecting the position curve drawn, we see that the particle passes through the origin first at $t = t_1 = 4$ s and later at $t = t_2 = 8 + \tau$ s, as shown. Applying the fourth relation in Sec. 11.8, determine the values of τ and t_2.

EXAMPLE 11.10

A motorcycle is 15 m behind a car when both are traveling at a constant speed of 72 km/h on a straight road as shown. Seeing an obstacle looming ahead, the driver of the car suddenly applies the brakes, causing the car to decelerate at the constant rate of 3 m/s². Two seconds after seeing the car's brake lights activated, the motorcyclist applies the brakes to produce a constant deceleration of a_m and is able to barely avoid a collision. Determine the value of a_m.

Solution.   Let the time be measured from the instant the car's brake lights are activated. Furthermore, let τ be the duration (in seconds) from the instant the brakes of the motorcycle are applied to the instant the motorcycle almost hits the car. Noting that 72 km/h = 20 m/s, we draw the velocity curves for the car and the motorcycle as solid and dashed lines, respectively, as shown.

We know that the motorcycle is 15 m behind the car before the brake lights are activated. The fact that the motorcycle and the car have barely avoided a collision implies that the distance which the motorcycle has traveled more than the car during the braking maneuver must be equal to 15 m.

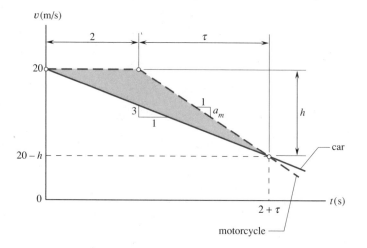

Thus, applying the sixth relation in Sec. 11.8 to the car and motorcycle, we see that the shaded triangular "area" shown in the velocity diagram must be equal to 15 m. From the geometry of this shaded "area" and the slope triangles as indicated, we write

$$\tfrac{1}{2}(2)h = 15 \qquad h = 3(2 + \tau) \qquad h = \tau a_m$$

These equations yield $h = 15$, $\tau = 3$, $a_m = 5$. We write

$$a_m = 5 \text{ m/s}^2 \quad \blacktriangleleft$$

Developmental Exercises

D11.20   Refer to Example 11.10. During the braking maneuver, determine the total distance traveled by (a) the car, (b) the motorcycle.

D11.21   Describe the six useful relations for the graphical solution of rectilinear motion problems.

PROBLEMS

11.38 through 11.40*   The acceleration curve of a particle in rectilinear motion is shown. If $x_0 = -2$ m and $v_0 = -4$ m/s, (a) draw the velocity and position curves of the particle for $0 \le t \le 12$ s, (b) determine the value of t for which the particle passes through the origin.

Fig. P11.38   **Fig. P11.39***   **Fig. P11.40***

11.41 and 11.42*   The velocity curve of a particle in rectilinear motion is shown. If $x_0 = 16$ ft, draw the acceleration and position curves of the particle for $0 \le t \le 15$ s and determine (a) Δx and x_T of the particle over the interval $0 \le t \le 12$ s, (b) the two values t_1 and t_2 of t for which the particle passes through the origin.

Fig. P11.41   **Fig. P11.42***

Fig. P11.43*

Fig. P11.44

11.43* A particle moves in a straight line with the velocity as shown. If $x_0 = -12$ m, draw its position curve for $0 \le t \le 15$ s and determine (a) Δx and x_T of the particle over the interval $0 \le t \le 11$ s, (b) the two values t_1 and t_2 of t for which the particle passes through the origin.

11.44 A motorist is traveling on a straight highway which is being repaired. He reduces his speed from 45 mi/h, when he passes the first flagger, to 16 mi/h 10 s later, when he passes the second flagger. His speed during this interval is recorded as shown. Determine (a) his deceleration at $t = 6$ s, (b) the distance between the first and the second flaggers.

11.45 A motorist traveling at 90 km/h notices that the traffic signal 450 m ahead of him turns red. He knows that it is timed to stay red for 20 s. If he desires to pass the signal at 90 km/h just as it turns green again, determine (a) the smallest common rate of deceleration and acceleration he should have, (b) the minimum speed reached in this maneuver.

11.46 By mistake, cars A and B are traveling toward each other in a single-lane portion of a road with speeds as shown. When $L = 210$ m, both drivers realize the situation and apply their brakes at the same time. If the decelerations of A and B are 1.5 m/s² and 1.2 m/s², respectively, (a) verify that a head-on collision will occur, (b) the velocity of A relative to B just before collision, (c) the distance traveled by each car while decelerating.

Fig. P11.46 and P11.47

11.47 By mistake, cars A and B are traveling toward each other in a single-lane portion of a road with speeds as shown. When $L = 210$ m, both drivers realize the situation and apply their brakes at the same time. If their decelerations are constant and they stop simultaneously just in time to barely avoid a head-on collision, determine (a) the deceleration of each car, (b) the distance traveled by each car while decelerating.

11.48 A car and a truck travel at a constant speed of 72 km/h on a road; the car is 20 m behind the truck as shown. The driver of the car wishes to pass the truck and place his car 20 m in front of the truck, and then resume the speed of 72 km/h. The maximum acceleration and deceleration of the car are 2 m/s² and 2.5 m/s², respectively. Determine the shortest time in which the passing operation can be completed if the car does not exceed a speed of 90 km/h at any time.

Fig. P11.48

11.49 The time rate of change of acceleration is referred to as the *jerk*,[†] which is often regarded as a cause of passenger discomfort. If the jerk of an elevator is limited to ± 2 ft/s^3, determine the shortest time in which an elevator starting from rest can rise (or descend) 13.5 ft and stop.

CURVILINEAR MOTION

11.9 Position Vector, Velocity, and Acceleration

The curvilinear motion of a particle is the motion of the particle along a curved path. The position P occupied by the particle at a given time t is defined by the *position vector* $\mathbf{r}$, which is a vector drawn from the origin O of a fixed coordinate system to the point P as shown in Fig. 11.11. At a later time $t + \Delta t$, the particle occupies the position P' which is defined by the position vector $\mathbf{r}'$. The vector $\Delta \mathbf{r}$, directed from P to P', represents the *change in the position vector* over the time interval Δt; i.e., from Fig. 11.11 and the vector addition (cf. the *triangle rule* in Sec. 2.2), we have

$$\mathbf{r}' = \mathbf{r} + \Delta \mathbf{r} \qquad \boxed{\Delta \mathbf{r} = \mathbf{r}' - \mathbf{r}} \qquad (11.28)$$

This vector $\Delta \mathbf{r}$ is called the *displacement* of the particle over the time interval Δt. Note that $\Delta \mathbf{r}$ accounts for the changes in both the *direction* and the *magnitude* of the position vector over Δt.

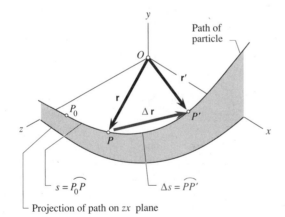

Fig. 11.11

Similar to that in rectilinear motion, the *velocity* of a particle in curvilinear motion in a reference frame is defined as the time rate of change of position (or the time rate of displacement) of the particle in that reference frame. As shown in Fig. 11.12(a), the *average velocity* $\mathbf{v}_{avg}$ of a particle having a displacement $\Delta \mathbf{r}$ over a time interval Δt is given by

$$\mathbf{v}_{avg} = \frac{\Delta \mathbf{r}}{\Delta t} \qquad (11.29)$$

[†]Cf. Sec. 11.4.

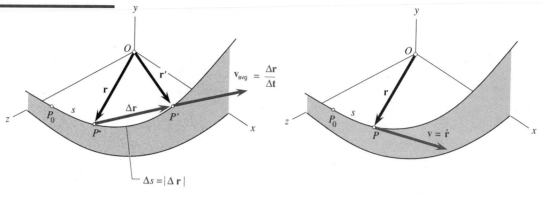

Fig. 11.12   (a)   (b)

Since $\Delta\mathbf{r}$ is a vector and Δt a scalar, the quotient $\Delta\mathbf{r}/\Delta t$ (and hence $\mathbf{v}_{avg}$) is a vector directed from P through P' as shown. As Δt becomes smaller and approaches zero in that limit, $\mathbf{v}_{avg}$ becomes the *instantaneous velocity* $\mathbf{v}$ of the particle as shown in Fig. 11.12(b); i.e.,

$$\mathbf{v} = \lim_{\Delta t \to 0} \frac{\Delta\mathbf{r}}{\Delta t} \qquad \mathbf{v} = \frac{d\mathbf{r}}{dt} = \dot{\mathbf{r}} \qquad (11.30)$$

This instantaneous velocity is what we usually call the *velocity*.

Since P' gets closer to P as Δt becomes smaller, the velocity $\mathbf{v}$ obtained at the limit of $\Delta t \to 0$ must therefore be *tangent to the path* of the particle at P. The magnitude v of the velocity $\mathbf{v}$ is called the *speed* of the particle. Noting that the magnitude of $\Delta\mathbf{r}$ approaches the length Δs of the arc PP' as Δt approaches zero, we write

$$v = \lim_{\Delta t \to 0} \frac{|\Delta\mathbf{r}|}{\Delta t} = \lim_{\Delta t \to 0} \frac{\Delta s}{\Delta t} \qquad v = \frac{ds}{dt} = \dot{s} \qquad (11.31)$$

Similar to that in rectilinear motion, the *acceleration* of a particle in curvilinear motion in a reference frame is defined as the time rate of change of velocity of the particle in that reference frame. Suppose that, over the time interval Δt, a particle moves from the point P to the point P' while its velocity changes from $\mathbf{v}$ to $\mathbf{v}'$ as shown in Fig. 11.13(a). By drawing the

(a)   (b)   (c)

Fig. 11.13

vectors $\mathbf{v}$ and $\mathbf{v}'$ from a common point O' as shown in Fig. 11.13(b), we see that the vector $\Delta\mathbf{v}$, which is drawn from the tip Q of $\mathbf{v}$ to the tip Q' of $\mathbf{v}'$, represents the *change in the velocity* over Δt; i.e.,

$$\Delta\mathbf{v} = \mathbf{v}' - \mathbf{v} \tag{11.32}$$

Note that the vector $\Delta\mathbf{v}$ accounts for the changes in both the *direction* and the *magnitude* of the velocity over Δt. The *average acceleration* $\mathbf{a}_{\text{avg}}$ of the particle over Δt is given by

$$\mathbf{a}_{\text{avg}} = \frac{\Delta\mathbf{v}}{\Delta t} \tag{11.33}$$

which is a vector directed from Q through Q' as shown. As Δt becomes smaller and approaches zero in the limit, $\mathbf{a}_{\text{avg}}$ becomes the *instantaneous acceleration* $\mathbf{a}$ of the particle as shown in Fig. 11.13(c); i.e.,

$$\mathbf{a} = \lim_{\Delta t \to 0} \frac{\Delta\mathbf{v}}{\Delta t} \qquad \boxed{\mathbf{a} = \frac{d\mathbf{v}}{dt} = \dot{\mathbf{v}}} \tag{11.34}$$

This instantaneous acceleration is what we usually call the *acceleration*.

For our study, the *path* of a particle may be viewed as a curve described by the tip of its position vector, which is drawn from a fixed origin O, as illustrated in Fig. 11.14(a). When the velocity vectors of a particle at its various positions are drawn to scale from a fixed origin O', as illustrated in Fig. 11.14(b), the curve described by the tip of the velocity vector is called the *hodograph* of the motion. We readily see from Figs. 11.13 and 11.14 that the acceleration of a particle is a vector tangent to the hodograph of the motion, *not* the path of the particle; the exception is in the case of rectilinear motion.

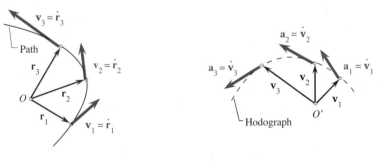

(a)

(b)

Fig. 11.14

The only definable motion of a particle, whose size and shape are to be disregarded, is translation along a path. For two particles A and B in space, as shown in Fig. 11.15, the motion of B in a nonrotating reference frame attached to A (e.g., $Ax'y'z'$)[†] is called the *motion of B relative to A*. The *position vector of B relative to A* is denoted by $\mathbf{r}_{B/A}$. The position vectors $\mathbf{r}_A$ and $\mathbf{r}_B$ of A and B are related to $\mathbf{r}_{B/A}$ in the fixed reference frame $Oxyz$ as follows:[†]

$$\boxed{\mathbf{r}_B = \mathbf{r}_A + \mathbf{r}_{B/A}} \tag{11.35}$$

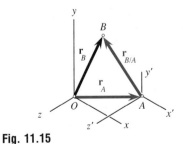

Fig. 11.15

[†]For convenience, a reference frame is often named after the coordinate system which is the only one embedded in it. (Cf. Sec. 11.1.) ·

In terms of the *velocity* and *acceleration of B relative to A,* we write

$$\dot{\mathbf{r}}_B = \dot{\mathbf{r}}_A + \dot{\mathbf{r}}_{B/A} \quad \text{or} \quad \boxed{\mathbf{v}_B = \mathbf{v}_A + \mathbf{v}_{B/A}} \qquad (11.36)$$

$$\dot{\mathbf{v}}_B = \dot{\mathbf{v}}_a + \dot{\mathbf{v}}_{B/A} \quad \text{or} \quad \boxed{\mathbf{a}_B = \mathbf{a}_A + \mathbf{a}_{B/A}} \qquad (11.37)$$

More precisely, the vectors $\mathbf{r}_{B/A}$, $\mathbf{v}_{B/A}$, and $\mathbf{a}_{B/A}$ denote, respectively, the position vector, the velocity, and the acceleration of the particle B in a nonrotating reference frame (e.g., $Ax'y'z'$) attached to the particle A.

Developmental Exercises

D11.22 Define for a particle in curvilinear motion the following terms: (a) position vector, (b) displacement, (c) average velocity, (d) velocity, (e) average acceleration, (f) acceleration, (g) hodograph.

D11.23 Define the vectors (a) $\mathbf{r}_{B/A}$, (b) $\mathbf{v}_{B/A}$, (c) $\mathbf{a}_{B/A}$.

EXAMPLE 11.11

Instruments in an airplane show that the plane is flying north with an airspeed of 250 km/h (relative to the air). Meanwhile, the radar on the ground indicates that the plane is moving at a speed of 220 km/h in a direction 10° west of north. Determine the wind velocity.

Solution. We know from the given data that the velocity of the plane relative to the wind is $\mathbf{v}_{P/W} = 250$ km/h ↑ and that the velocity of the plane relative to the ground is $\mathbf{v}_P = 200$ km/h ⦥ 80°. Denoting the wind velocity by $\mathbf{v}_W$, we have

$$\mathbf{v}_{P/W} + \mathbf{v}_W = \mathbf{v}_P$$

as illustrated. Applying the laws of cosines and sines, we write

$$v_W^2 = (220)^2 + (250)^2 - 2(220)(250) \cos 10°$$

$$v_W = 50.71 \qquad \frac{220}{\sin\theta} = \frac{v_W}{\sin 10°} \qquad \theta = 48.9°$$

$$\mathbf{v}_W = 50.7 \text{ km/h} \ \nearrow \ 41.1° \blacktriangleleft$$

Alternative Solution. Using analytical expressions, we write

$$\mathbf{v}_{P/W} = 250\mathbf{j} \qquad \mathbf{v}_P = 220(-\sin 10° \, \mathbf{i} + \cos 10° \, \mathbf{j})$$

$$\mathbf{v}_W = \mathbf{v}_P - \mathbf{v}_{P/W} = -38.20\mathbf{i} - 33.34\mathbf{j}$$

$$\mathbf{v}_W = 50.7 \text{ km/h} \ \nearrow \ 41.1° \blacktriangleleft$$

Developmental Exercises

D11.24 Name the type of motion of a particle if the hodograph of its motion is (a) a point, (b) a straight line, (c) a circle with radius equal to v.

D11.25 The velocity and acceleration vectors of a car at five separate instants are shown. Which instants indicate that the car is turning a corner?

(a)

(b)

(c)

(d)

(e)

Fig. D11.25

11.10 Derivatives of Vector Functions[†]

If to each value of t (such as time) there corresponds a vector $\mathbf{F}$, we say that $\mathbf{F}$ is a *vector function* of t; i.e., $\mathbf{F} = \mathbf{F}(t)$. We note from Eq. (11.30) that the velocity $\mathbf{v}$ of a particle in curvilinear motion is simply equal to the time derivative of its position $\mathbf{r}$, which is a vector function of the time t. Similarly, we note from Eq. (11.34) that the acceleration $\mathbf{a}$ of a particle is simply equal to the time derivative of its velocity $\mathbf{v}$. Thus, it is quite concise when we use vector calculus, in addition to vector algebra, to derive such equations and formulas. In this section, we shall go over a few rules governing the differentiation of sums and products of vector functions.

For a *scalar function* $f = f(t)$, we recall that its derivative is

$$\frac{df}{dt} = \lim_{\Delta t \to 0} \frac{\Delta f}{\Delta t} = \lim_{\Delta t \to 0} \frac{f(t + \Delta t) - f(t)}{\Delta t} \tag{11.38}$$

For a *vector function* $\mathbf{F} = \mathbf{F}(t)$, we define in a given reference frame that[††]

$$\frac{d\mathbf{F}}{dt} = \lim_{\Delta t \to 0} \frac{\Delta \mathbf{F}}{\Delta t} = \lim_{\Delta t \to 0} \frac{\mathbf{F}(t + \Delta t) - \mathbf{F}(t)}{\Delta t} \tag{11.39}$$

For the *sum of two vector functions* $\mathbf{A}(t)$ *and* $\mathbf{B}(t)$, we similarly define

$$\frac{d(\mathbf{A} + \mathbf{B})}{dt} = \lim_{\Delta t \to 0} \frac{\Delta(\mathbf{A} + \mathbf{B})}{\Delta t}$$

$$= \lim_{\Delta t \to 0} \frac{\mathbf{A}(t + \Delta t) + \mathbf{B}(t + \Delta t) - [\mathbf{A}(t) + \mathbf{B}(t)]}{\Delta t}$$

$$= \lim_{\Delta t \to 0} \frac{\mathbf{A}(t + \Delta t) - \mathbf{A}(t)}{\Delta t} + \lim_{\Delta t \to 0} \frac{\mathbf{B}(t + \Delta t) - \mathbf{B}(t)}{\Delta t}$$

$$= \lim_{\Delta t \to 0} \frac{\Delta \mathbf{A}}{\Delta t} + \lim_{\Delta t \to 0} \frac{\Delta \mathbf{B}}{\Delta t}$$

$$\frac{d(\mathbf{A} + \mathbf{B})}{dt} = \frac{d\mathbf{A}}{dt} + \frac{d\mathbf{B}}{dt} \tag{11.40}$$

[†]Those who are familiar with the derivatives of vector functions may proceed directly to Sec. 11.11.

[††]The time rate of change of a scalar function has the same value in all reference frames. However, the time rate of change of a vector function may have different values in different references frames (cf. Chap. 15).

For the *product of f(t) and* **A**(t), we write

$$\frac{d(f\mathbf{A})}{dt} = \lim_{\Delta t \to 0} \frac{\Delta(f\mathbf{A})}{\Delta t} = \lim_{\Delta t \to 0} \frac{(f + \Delta f)(\mathbf{A} + \Delta\mathbf{A}) - f\mathbf{A}}{\Delta t}$$

$$= \lim_{\Delta t \to 0}\left(\frac{\Delta f}{\Delta t}\mathbf{A}\right) + \lim_{\Delta t \to 0}\left(f\frac{\Delta\mathbf{A}}{\Delta t}\right)$$

$$\frac{d(f\mathbf{A})}{dt} = \frac{df}{dt}\mathbf{A} + f\frac{d\mathbf{A}}{dt} \tag{11.41}$$

The derivatives of the *dot product* and of the *cross product* of two vector functions **A**(t) and **B**(t) may be obtained in a similar manner; i.e.,

$$\frac{d(\mathbf{A} \cdot \mathbf{B})}{dt} = \frac{d\mathbf{A}}{dt} \cdot \mathbf{B} + \mathbf{A} \cdot \frac{d\mathbf{B}}{dt} \tag{11.42}$$

$$\frac{d(\mathbf{A} \times \mathbf{B})}{dt} = \frac{d\mathbf{A}}{dt} \times \mathbf{B} + \mathbf{A} \times \frac{d\mathbf{B}}{dt} \tag{11.43}$$

Note that the order of the factors in Eq. (11.43) must be maintained because the cross product is not commutative.

The unit vectors **i**, **j**, and **k** associated with the x, y, and z axes of a fixed rectangular coordinate system are constant vectors having a magnitude of 1 and fixed directions. Their derivatives are therefore zero. However, note that the derivative of a vector having a constant magnitude but *varying direction* is generally *nonzero*. To find d**A**/dt, we first express the vector **A** in terms of its rectangular components and then apply Eq. (11.41) as follows:

$$\mathbf{A} = A_x\mathbf{i} + A_y\mathbf{j} + A_z\mathbf{k}$$

$$\frac{d\mathbf{A}}{dt} = \frac{dA_x}{dt}\mathbf{i} + \frac{dA_y}{dt}\mathbf{j} + \frac{dA_z}{dt}\mathbf{k} \tag{11.44}$$

It should be pointed out that all equations in this section are valid whether the independent variable t is a time variable or just a parameter.

Developmental Exercises

D11.26 If **A** = $t^2\mathbf{i}$ + 2**k** and **B** = $t^3\mathbf{j}$ − 3t**k**, evaluate at t = 1 (a) $\frac{d}{dt}(\mathbf{A} + \mathbf{B})$, (b) $\frac{d}{dt}(\mathbf{A} \cdot \mathbf{B})$, (c) $\frac{d}{dt}(\mathbf{B} \times \mathbf{A})$.

D11.27 If $\frac{d}{dt}(\mathbf{A} \times \mathbf{B})$ = $-4t^3\mathbf{i}$ − 2t**k** and $\mathbf{B} \times \frac{d\mathbf{A}}{dt}$ = $2t^3\mathbf{i}$ + 2t**k**, what is the value of $\frac{d\mathbf{B}}{dt} \times \mathbf{A}$?

11.11 Rectangular Components

The curvilinear motion of particles may be described by using one of several different coordinate systems embedded in a reference frame. The choice of a particular coordinate system is usually dependent on the geometry of the path, the way the acceleration is generated, and other given data. The rectangular coordinate system is particularly useful when the rectangular components of acceleration are independently generated or determined. The resulting curvilinear motion is, then, obtained by vector additions of the x, y, and z rectangular components of the acceleration, velocity, and position vectors.

Recall that the unit vectors $\mathbf{i}$, $\mathbf{j}$, and $\mathbf{k}$ along the fixed xyz rectangular coordinate axes are constant vectors. For a particle moving along a space curve as illustrated in Fig. 11.16, we may, therefore, represent the position vector $\mathbf{r}$, the velocity $\mathbf{v}$, and the acceleration $\mathbf{a}$ of the particle as follows:

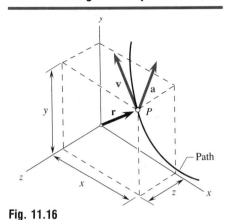

Fig. 11.16

$$\mathbf{r} = x\mathbf{i} + y\mathbf{j} + z\mathbf{k}$$

$$\mathbf{v} = \dot{\mathbf{r}} = \dot{x}\mathbf{i} + \dot{y}\mathbf{j} + \dot{z}\mathbf{k} \qquad (11.45)$$

$$\mathbf{a} = \dot{\mathbf{v}} = \ddot{\mathbf{r}} = \ddot{x}\mathbf{i} + \ddot{y}\mathbf{j} + \ddot{z}\mathbf{k}$$

Thus, the scalar components of $\mathbf{v}$ and $\mathbf{a}$ are

$$v_x = \dot{x} \qquad v_y = \dot{y} \qquad v_z = \dot{z}$$

$$\qquad (11.46)$$

$$a_x = \dot{v}_x = \ddot{x} \qquad a_y = \dot{v}_y = \ddot{y} \qquad a_z = \dot{v}_z = \ddot{z}$$

The magnitudes of $\mathbf{v}$ and $\mathbf{a}$ are

$$v = (v_x^2 + v_y^2 + v_z^2)^{1/2} \qquad a = (a_x^2 + a_y^2 + a_z^2)^{1/2} \qquad (11.47)$$

If the coordinates x, y, and z of a particle are independently generated as functions of the time t, then for any value of t we may combine them and their time derivatives to obtain $\mathbf{r}$, $\mathbf{v}$, and $\mathbf{a}$ as given in Eqs. (11.45). On the other hand, if the acceleration components a_x, a_y, and a_z are given as functions of the time t, we may, by integration and imposition of the given initial conditions, first obtain v_x, v_y, and v_z and then x, y, and z. The curvilinear motion described by using a fixed rectangular coordinate system is, in fact, a combination of three simultaneous rectilinear motions in the x, y, and z directions. Thus, methods developed for studying the rectilinear motion of particles may be applied to each of them separately.

If a particle undergoes curvilinear motion in the xy plane, we have

$$z = \dot{z} = \ddot{z} = 0 \qquad v_z = a_z = 0 \qquad (11.48)$$

and the relations in Eqs. (11.45) through (11.47) may correspondingly be simplified. The curvilinear motion of particles in a plane is further discussed in the next three sections.

EXAMPLE 11.12

The plane curvilinear motion of a particle is defined by the parametric equations $x = 50 - 2t^2$ and $y = 30t - 5t^2$, where x and y are the coordinates measured in meters and t is the parameter measured in seconds. Plot the path traveled by the particle for the interval $0 \le t \le 5$ s and determine its velocity and acceleration at $t = 5$ s.

Solution. By differentiation with respect to t, we have

$$v_x = \dot{x} = -4t \qquad a_x = \ddot{x} = -4$$

$$v_y = \dot{y} = 10(3 - t) \qquad a_y = \ddot{y} = -10$$

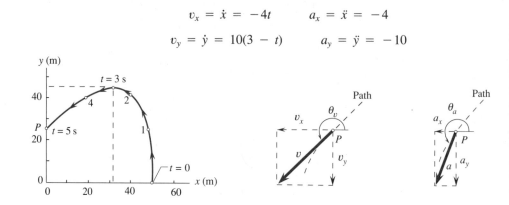

Since $v_y = 0$ at $t = 3$ s, the curved path has a peak at this instant. By calculating corresponding values of x and y for several values of t, we plot the path as shown. The velocity **v** and the acceleration **a** of the particle at P, where $t = 5$ s, are sketched as shown. For $t = 5$ s, we find that

$$v_x = -20 \qquad v_y = -20 \qquad\qquad v = 28.3 \text{ m/s} \blacktriangleleft$$

$$\tan \theta_v = \frac{-20}{-20} \qquad\qquad\qquad \theta_v = 225° \blacktriangleleft$$

$$a_x = -4 \qquad a_y = -10 \qquad\qquad a = 10.77 \text{ m/s}^2 \blacktriangleleft$$

$$\tan \theta_a = \frac{-10}{-4} \qquad\qquad\qquad \theta_a = 248.2° \blacktriangleleft$$

Developmental Exercise

D11.28   Refer to Example 11.12. Determine the velocity and acceleration of the particle at $t = 2$ s.

EXAMPLE 11.13

The motions of the two pegs A and B are constrained in such a way that they remain both in the vertical slot of the member C and in the elliptic path at all times as shown. If the member C moves with a constant velocity of 2 ft/s →, determine $\mathbf{v}_{B/A}$ and $\mathbf{a}_{B/A}$ of B relative to A when $x = 3$ ft.

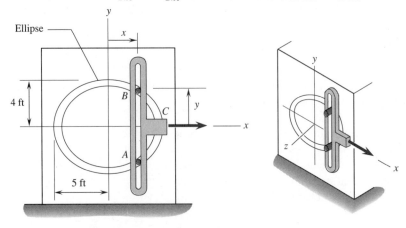

Solution. Since the path of B must be elliptic, we have

$$\frac{x^2}{5^2} + \frac{y^2}{4^2} = 1 \qquad (1)$$

$$\frac{2}{25} x\dot{x} + \frac{1}{8} y\dot{y} = 0 \qquad (2)$$

$$\frac{2}{25} (\dot{x}^2 + x\ddot{x}) + \frac{1}{8} (\dot{y}^2 + y\ddot{y}) = 0 \qquad (3)$$

The constraints require that the x component of the motion of B be the same as that of the member C which has a velocity of 2 ft/s $\rightarrow$ and a zero acceleration. Consequently, we have $\dot{x} = 2$ ft/s, $\ddot{x} = 0$. From Eq. (1), we find that $y = 3.2$ ft when $x = 3$ ft. Substituting these values into Eqs. (2) and (3), we get

$$\dot{y} = -1.2 \text{ ft/s} \qquad \ddot{y} = -1.25 \text{ ft/s}^2$$

By vector addition and symmetry, we write

$$\mathbf{v}_B = 2\mathbf{i} - 1.2\mathbf{j} \qquad \mathbf{v}_A = 2\mathbf{i} + 1.2\mathbf{j}$$

$$\mathbf{v}_{B/A} = \mathbf{v}_B - \mathbf{v}_A = -2.4\mathbf{j} \qquad v_{B/A} = 2.4 \text{ ft/s} \downarrow \quad \blacktriangleleft$$

$$\mathbf{a}_B = -1.25\mathbf{j} \qquad \mathbf{a}_A = 1.25\mathbf{j}$$

$$\mathbf{a}_{B/A} = \mathbf{a}_B - \mathbf{a}_A = -2.5\mathbf{j} \qquad a_{B/A} = 2.5 \text{ ft/s}^2 \downarrow \quad \blacktriangleleft$$

Developmental Exercise

D11.29 Refer to Example 11.13. If $\mathbf{v}_C = 2$ ft/s $\rightarrow$ and $\mathbf{a}_C = 0.5$ ft/s^2 $\rightarrow$ when $x = 3$ ft, determine $\mathbf{a}_{B/A}$ at this instant.

11.12 Free Flight of a Projectile

A *projectile* is a body projected (e.g., thrown, cast, or impelled) through the air or space. The *powered flight* of a projectile is the motion of a body by virtue of the action of a thrust in addition to the inertia of the body and the actions of the gravitational force and air resistance, if any. The *free flight* of a projectile is the motion of a body by virtue of the inertia of the body and the actions of the gravitational force and air resistance, if any. A free flight is a *flight without a thrust* or an *unpowered flight*. The free-body diagrams of space vehicle in powered flight and in free flight are illustrated in Fig. 11.17, where $\mathbf{W}$ is the gravitational force, $\mathbf{F}_T$ is the thrust, and $\mathbf{F}_A$ is the friction force from the air.

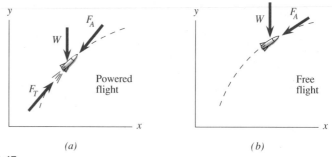

(a)   (b)

Fig. 11.17

The usual *assumptions* for the elementary study of the *free flight of a projectile* are as follows:

1. The projectile travels in a vertical plane near the surface of the earth; and the surface of the earth over which the projectile flies may be taken as flat and horizontal.
2. The friction between the air and the projectile is negligible.
3. The acceleration of the projectile is generated only by gravity; this acceleration points vertically downward and has a magnitude $g = 9.81$ m/s^2 or 32.2 ft/s^2.

Thus, the free flight of a projectile may readily be studied by using an xy rectangular coordinate system fixed to the surface of the earth and noting that the scalar components of its acceleration are

$$a_x = 0 \qquad a_y = -g \qquad (11.49)$$

Fig. 11.18

Let the initial position of the projectile be at (x_0, y_0) and the scalar components of the initial velocity $\mathbf{v}_0$ be $(v_x)_0$ and $(v_y)_0$, as shown in Fig. 11.18. With these conditions and Eqs. (11.49), we can readily show that

$$v_x = \dot{x} = (v_x)_0 \qquad v_y = \dot{y} = (v_y)_0 - gt \qquad (11.50)$$

$$x = x_0 + (v_x)_0 t \qquad y = y_0 + (v_y)_0 t - \tfrac{1}{2} g t^2 \qquad (11.51)$$

$$v_y^2 = (v_y)_0^2 - 2g(y - y_0) \qquad (11.52)$$

We recognize that Eqs. (11.51) form the parametric equations of a parabola; i.e., the trajectory is *parabolic*. Of course, the trajectory would not be precisely parabolic if air resistance were taken into account.

Suppose that a projectile is launched at A with a muzzle velocity $\mathbf{v}_0$ as shown in Fig. 11.19, where the initial conditions are $x_0 = y_0 = 0$, $(v_x)_0 = v_0 \cos\theta$, and $(v_y)_0 = v_0 \sin\theta$. Substituting these conditions into Eqs. (11.51), we obtain

$$x = (v_0 \cos\theta)t \qquad y = (v_0 \sin\theta)t - \tfrac{1}{2} g t^2$$

For the projectile to hit a target at B, we have $y = h$ when $x = L$; i.e.,

$$L = (v_0 \cos\theta)t \qquad h = (v_0 \sin\theta)t - \tfrac{1}{2} g t^2$$

where t is the flying time of the projectile from A to B. Eliminating t between these two equations and simplifying, we get

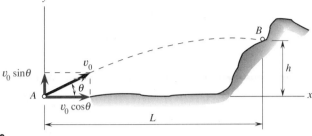

Fig. 11.19

$$\tan^2\theta - \frac{2v_0^2}{gL}\tan\theta + \left(1 + \frac{2v_0^2 h}{gL^2}\right) = 0 \qquad (11.53)$$

We infer from Eq. (11.53) that, for appropriate values of v_0, L, and h, it is possible for the projectile to hit the target at B in Fig. 11.19 in two different ways corresponding to two different values of the firing angle θ. A computer program to find such different values of θ as well as the corresponding flying times is shown in Fig. 11.20.

```
10 REM FIRING ANGLES & FLYING TIMES OF A PROJECTILE   (File Name: PROJECT)
15 REM          Refer to Eq. (11.53) and enter V0 (in meters/second),
20 REM          L (in meters), and h (in meters) as data in Line 100
25 REM  To output to the printer, change PRINT to LPRINT in the program.
30 READ V0,L,H: B=-2*V0^2/(9.81#*L): C=1-B*H/L: D=B^2-4*C: IF D<0 THEN 60
35 E1=(-B+SQR(D))/2:E2=(-B-SQR(D))/2:A1=ATN(E1):A2=ATN(E2):F=18C/3.141593
40 P1=A1*F: P2=A2*F: T1=L/(V0*COS(A1)): T2=L/(V0*COS(A2)): S$="      "
45 PRINT: PRINT "V0 ="; V0; "m/s"; S$; "L ="; L; "m"; S$; "h ="; H; "m"
50 PRINT "Firing angles (in degrees) ="; P1; "and"; P2
55 PRINT "Corresponding flying times (in seconds) ="; T1; "and"; T2: END
60 PRINT: PRINT "The projectile cannot reach the target.": END
100 DATA 300,4000,1000
```

```
V0 = 300 m/s        L = 4000 m        h = 1000 m
Firing angles (in degrees) = 76.15708 and 27.87916
Corresponding flying times (in seconds) = 55.72724 and 15.08405
```

Fig. 11.20 (Cf. the solution in Example 11.14 and the answer to D11.31.)

EXAMPLE 11.14

A projectile is launched at A with a muzzle velocity $\mathbf{v}_0$ as shown. If $v_0 = 300$ m/s, $L = 4$ km, and $h = 1$ km, determine the firing angle θ which will permit the projectile to hit the target at B.

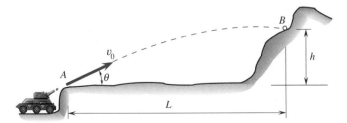

Solution. Substituting $v_0 = 300$ m/s, $L = 4000$ m, $h = 1000$ m, and $g = 9.81$ m/s^2 into Eq. (11.53), we obtain

$$\tan^2\theta - 4.587 \tan\theta + 2.147 = 0$$

which yields $\tan\theta = 4.058$ or 0.529. Thus, we find two solutions: $\theta = 76.16°$ and $\theta = 27.88°$.

$$\theta_1 = 76.2° \qquad \theta_2 = 27.9° \blacktriangleleft$$

This means that the target will be hit if either of these two firing angles is used as shown.

Developmental Exercises

D11.30 What are the usual assumptions for the elementary study of the free flight of a projectile?

D11.31 Refer to Example 11.14. Determine the flying time of the projectile from A to B if the firing angle θ is (a) 76.16°, (b) 27.88°.

D11.32 Refer to Example 11.14. Determine the time t at which the projectile attains the maximum height in its trajectory if the firing angle θ is (a) 76.16°, (b) 27.88°.

D11.33 Using the program in Fig. 11.20, "try" to find θ in Example 11.14 for the same data except that (a) $v_0 = 225$ m/s, (b) $v_0 = 220$ m/s.

PROBLEMS

11.50* The plane curvilinear motion of a particle is defined by $x = 16t - 2t^2$ and $y = 36 - t^2$, where x and y are measured in feet and t in seconds. Plot the *path* traveled by the particle for the interval $0 \leq t \leq 6$ s and determine its velocity and acceleration at $t = 6$ s.

11.51 A particle moves in the xy plane. The x component of its initial velocity is $(v_x)_0 = 18$ ft/s, the x component of its acceleration is $a_x = 6t$, and its y coordinate is $y = 4t^3 - 8t$, where t is measured in seconds, y in feet, and a_x in ft/s^2. Determine the velocity and acceleration of the particle when $t = 2$ s.

11.52 The position vector of a particle is given by $\mathbf{r} = (2 - t^3)\mathbf{i} + (3t - 1)\mathbf{j}$, where t is measured in seconds and r in meters. Determine the velocity and acceleration of the particle at $t = 2$ s.

11.53* The position vector **r** of a particle is given by **r** = $(A \cos \omega t)$**i** + $(B \sin \omega t)$**j**, where t is the time variable, and A, B, and ω are positive constants. It is known that $A > B$. Show that (a) the path of the particle is an ellipse with A and B as the major and minor axes, (b) the acceleration is directed toward the origin, (c) the magnitude of the acceleration is proportional to r, not r^2.

11.54* The motion of a particle is defined by the position vector **r** = $2t$**i** + $(4t^2 + 9t^4)$**j** + $3t^2$**k**, where r is measured in feet and t in seconds. Show that the path of the particle lies on the paraboloid as shown. Determine the magnitudes of the velocity and acceleration when $t = 1$ s.

Fig. P11.54*

×11.55 The motion of a particle is defined by the position vector **r** = 40 $\cos t$**i** + $9t$**j** − 40 $\sin t$**k**, where r is measured in meters and t in seconds, and the path is a circular helix as shown. Determine the magnitudes of the velocity and acceleration of the particle.

×11.56 The motions of two pins A and B are constrained in such a way that they remain in the vertical slot of the member C and in the circular path at all times as shown. If $v_{B/A}$ = 40 mm/s ↓ when x = 100 mm, determine v_C of the member C.

Fig. P11.55

Fig. P11.56

Fig. P11.58*

11.57 Refer to Prob. 11.56. If $v_{B/A}$ = 40 mm/s ↓ and $a_{B/A}$ = 48 mm/s² ↓ when x = 100 mm, determine a_C of the member C.

11.58* The motion of the pin A is constrained in such a way that it remains in the inclined slot of the member BC as well as in the circular path at all times as shown. It is known that the member BC maintains its slope of 3/4 and translates with a constant velocity of 28 mm/s →. At $t = 0$, the y intercept of the inclined slot is 40 mm. (a) Show that the equation of the straight line along the inclined slot at any time t is given by $4y = 3(x − 28t) + 160$. (b) Determine v_A of the pin A when x = 100 mm.

11.59 Refer to Prob. 11.58*. Determine a_A of the pin A when x = 100 mm.

11.60 Water is discharged at A from a pressure tank with a horizontal velocity v_0 as shown. If $L = 4$ m, determine the range of values of v_0 for which the water will enter the drainage opening BC.

11.61 Water is discharged at A from a pressure tank with a velocity $v_0 = 5$ m/s → as shown. Determine the range of values of L for which the water will enter the drainage opening BC.

Fig. P11.60 and P11.61

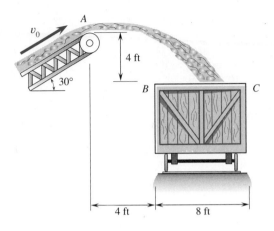

Fig. P11.62

X **11.62**   A freight car is being loaded with coal as shown. Determine the range of values of the constant speed v_0 of the conveyor belt for which the coal will enter the opening BC of the freight car.

11.63   A ball is kicked from A with an initial velocity $\mathbf{v}_0$ which makes an angle θ with the horizontal as shown. Determine (a) the range L, traveled by the ball from A to B, in terms of v_0, θ, and g; (b) the value of θ for which L is maximum; (c) the maximum value of L in terms of v_0 and g; (d) the maximum height, reached in traveling over the maximum range L, in terms of v_0 and g.

Fig. P11.63

11.64*   A ball is kicked from A with an initial velocity $\mathbf{v}_0$ inside a horizontal tunnel of height h as shown. If $v_0^2 > 4gh$, determine in terms of v_0, h, and g, (a) the angle θ for which the range L traveled by the ball from A to B is maximum, (b) the maximum value of L.

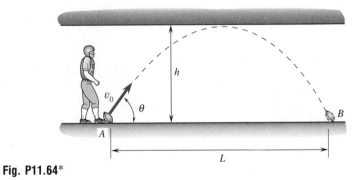

Fig. P11.64*

11.65   Solve Prob. 11.64* if $v_0^2 = 4gh$.

11.66   Solve Prob. 11.64* if $v_0 = 12$ m/s and $h = 5$ m.

11.67   Solve Prob. 11.64* if $v_0 = 15$ m/s and $h = 5$ m.

11.68*   A bale of hay is to be dropped from an airplane A so that it lands at B on a snow-covered ground of a ranch as shown. If the airplane is flying horizontally at an altitude $h = 400$ ft with a speed $v_A = 120$ mi/h, determine the angle θ between the horizontal and the line of sight AB when the bale of hay is dropped.

11.69   Solve Prob. 11.68* if $h = 100$ m and $v_A = 180$ km/h.

11.70   An airplane A is flying horizontally at an altitude of 2 km and with a constant speed of 900 km/h on a path which passes directly over an antiaircraft gun at B. The gun fires a shell with a muzzle velocity of 550 m/s and hits the airplane at C. Knowing that the firing angle of the gun is 60°, determine the angle θ between the horizontal and the line of sight BA when the gun is fired.

Fig. P11.68*

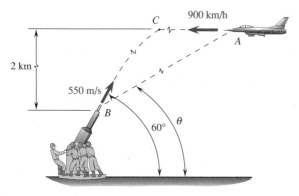

Fig. P11.70

11.71*   Refer to Prob. 11.70. Determine the velocity and acceleration of the shell relative to the airplane at the time of impact at C.

11.72   A rifleman fires a bullet with a muzzle velocity of 1800 ft/s as shown. If the barrel of the rifle is aimed at B and the bullet strikes at C, determine the distance δ between B and C.

Fig. P11.72   **Fig. P11.73**

11.73   A gun fires a shell with a muzzle velocity of 400 m/s as shown. Determine (a) the two values of the firing angle θ with which the shell will hit the target at B, (b) the time of flight of the shell from A to B corresponding to each value of θ.

11.74   An airplane has a cruising airspeed of 130 mi/h. It leaves an airport to fly to another which is 186 mi due north of the first. A wind of 50 mi/h is blowing steadily from the east. Determine the shortest time which the trip will take.

11.75* A 2-km wide river flows at a speed of 6 km/h as shown. A motorboat is used to go from A to B. The boat can travel at a maximum speed of 16 km/h relative to the water. Determine (a) the minimum time which the trip will take, (b) the angle θ at which the boat should be oriented.

Fig. P11.75*

11.76   Solve Prob. 11.75* if the motorboat is used to go from A to C.

11.77   Cars B and C are traveling at constant speeds along straight highways as shown. The car C passes over the point A on the bridge at the time $t = 0$, and the car B passes under the point A at the time $t = 2$ s. Determine (a) the velocity of the car B relative to the car C, (b) the change in position of the car B relative to the car C from $t = 2$ s to $t = 4$ s, (c) the distance between the two cars when $t = 5$ s.

Fig. P11.77

11.13   Tangential and Normal Components

We saw in Sec. 11.9 that the velocity of a particle is a vector tangent to the path of the particle but that the acceleration vector is, in general, not tangent to the path. To facilitate the study of curvilinear motion, the acceleration vector is sometimes resolved into components directed, respectively, along the *tangent* and the *normal* to the path of the particle. We shall denote the unit vector directed along the tangent by $\mathbf{e}_t$ and the unit vector directed along the normal toward the center of curvature of the path by $\mathbf{e}_n$.

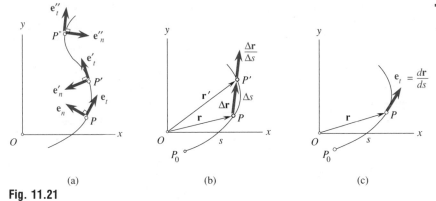

Fig. 11.21

The *tangential unit vector* $\mathbf{e}_t$ and the *normal unit vector* $\mathbf{e}_n$ are considered to be attached to the particle and to move along the path with the particle as seen in Fig. 11.21(a), where the particle moves from P to P' to P''. Unlike the unit vectors $\mathbf{i}$ and $\mathbf{j}$ in a fixed rectangular coordinate system, the tangential and normal unit vectors $\mathbf{e}_t$ and $\mathbf{e}_n$ will generally change their directions as the particle moves from one point to another along the curved path. Thus, $\mathbf{e}_t$ and $\mathbf{e}_n$ are *unit vectors* but *not constant vectors.*[†] From parts (b) and (c) of Fig. 11.21, we see that $\mathbf{e}_t$ may be obtained from the derivative of the position vector $\mathbf{r}$ with respect to the path variable s. We write

$$\Delta\mathbf{r} = \mathbf{r}' - \mathbf{r} \qquad \lim_{\Delta s \to 0} |\Delta\mathbf{r}| = \Delta s$$

$$\mathbf{e}_t = \lim_{\Delta s \to 0} \frac{\Delta\mathbf{r}}{\Delta s} = \frac{d\mathbf{r}}{ds} \qquad \boxed{\frac{d\mathbf{r}}{ds} = \mathbf{e}_t} \qquad (11.54)$$

$$|\mathbf{e}_t| = \lim_{\Delta s \to 0} \frac{|\Delta\mathbf{r}|}{|\Delta s|} = \lim_{\Delta s \to 0} \frac{|\Delta s|}{|\Delta s|} = 1$$

Thus, $\mathbf{e}_t$ as given by Eq. (11.54) is indeed a unit vector and is tangent to the path of the particle at the point P.

Let ρ be the *radius of curvature* of the curved path at the point P as shown in Fig. 11.22(a). In this figure, we note that (1) the tangential unit vectors at P and its neighboring point P' are denoted by $\mathbf{e}_t$ and $\mathbf{e}_t'$, respec-

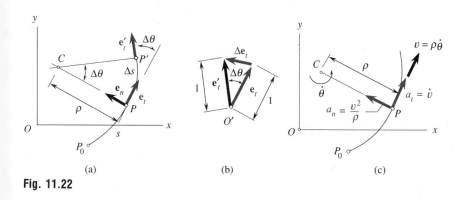

Fig. 11.22

[†]Recall that a vector is changed whenever its *magnitude*, or its *direction*, or *both* are changed.

tively, (2) the normal unit vector $\mathbf{e}_n$ at the point P is directed toward the center of curvature C, and (3) the angle between the radii of curvature at P and P' is denoted by $\Delta\theta$. The angle between $\mathbf{e}_t$ and $\mathbf{e}_t'$ is, of course, also $\Delta\theta$. By drawing $\mathbf{e}_t$ and $\mathbf{e}_t'$ from the point O' as shown in Fig. 11.22(b), we see that the change in the tangential unit vector $\mathbf{e}_t$ as the particle moves from P to P' is a vector $\Delta\mathbf{e}_t$ as shown. From parts (a) and (b) of Fig. 11.22, we write

$$\Delta s = PP' = \rho\,\Delta\theta \qquad \Delta\mathbf{e}_t = \mathbf{e}_t' - \mathbf{e}_t \approx \Delta\theta\,\mathbf{e}_n$$

$$\frac{d\theta}{ds} = \lim_{\Delta s \to 0}\frac{\Delta\theta}{\Delta s} = \lim_{\Delta s \to 0}\frac{\Delta\theta}{\rho\,\Delta\theta} = \frac{1}{\rho}$$

$$\frac{d\mathbf{e}_t}{d\theta} = \lim_{\Delta\theta \to 0}\frac{\Delta\mathbf{e}_t}{\Delta\theta} = \lim_{\Delta\theta \to 0}\frac{\Delta\theta\,\mathbf{e}_n}{\Delta\theta} = \mathbf{e}_n$$

$$\frac{d\theta}{ds} = \frac{1}{\rho} \qquad \frac{d\mathbf{e}_t}{d\theta} = \mathbf{e}_n \tag{11.55}$$

Note that $1/\rho$ represents the curvature κ of the path at the point P.

Using the preceding equations, we derive the velocity $\mathbf{v}$ and the acceleration $\mathbf{a}$ of a particle in curvilinear motion as follows:

$$v = \frac{ds}{dt} = \lim_{\Delta t \to 0}\frac{\Delta s}{\Delta t} = \lim_{\Delta t \to 0}\frac{\rho\,\Delta\theta}{\Delta t} = \rho\frac{d\theta}{dt} = \rho\dot{\theta}$$

$$\mathbf{v} = \frac{d\mathbf{r}}{dt} = \frac{d\mathbf{r}}{ds}\frac{ds}{dt} = \mathbf{e}_t v = v\mathbf{e}_t$$

$$\mathbf{a} = \frac{d\mathbf{v}}{dt} = \frac{d}{dt}(v\mathbf{e}_t) = \frac{dv}{dt}\mathbf{e}_t + v\frac{d\mathbf{e}_t}{dt}$$

$$= \frac{dv}{dt}\mathbf{e}_t + v\frac{d\mathbf{e}_t}{d\theta}\frac{d\theta}{ds}\frac{ds}{dt}$$

$$= \dot{v}\mathbf{e}_t + v(\mathbf{e}_n)\left(\frac{1}{\rho}\right)(v) = \dot{v}\mathbf{e}_t + \frac{v^2}{\rho}\mathbf{e}_n$$

$$= \mathbf{a}_t + \mathbf{a}_n = a_t\mathbf{e}_t + a_n\mathbf{e}_n$$

$$v = \frac{ds}{dt} = \rho\dot{\theta} \qquad a_t = \dot{v} \qquad a_n = \frac{v^2}{\rho} = \rho\dot{\theta}^2 \tag{11.56}$$

$$\mathbf{v} = v\mathbf{e}_t \qquad \mathbf{a} = \dot{v}\mathbf{e}_t + \frac{v^2}{\rho}\mathbf{e}_n \tag{11.57}$$

where a_t and a_n are, respectively, the scalar components of the acceleration in the tangential and normal directions of the path of the moving particle as shown in Fig. 11.22(c). Note in this figure that (1) $\mathbf{v}$ is always *tangent to the path* as the particle moves along the path, (2) the sense of $\mathbf{a}_t$ is the same as that of $\mathbf{v}$ if $\dot{v}$ is positive, (3) $\mathbf{a}_n$ is always directed from P toward the center of curvature C, and (4) $\dot{\theta}$ is geometrically the angular speed of the radius of curvature $\rho = \overline{CP}$. If the path is a circular arc, the radius of curvature ρ becomes a constant equal to the radius of the circular arc, and $\dot{\theta}$ is equal to the angular speed of the radius of curvature joining the center C and the particle at P.

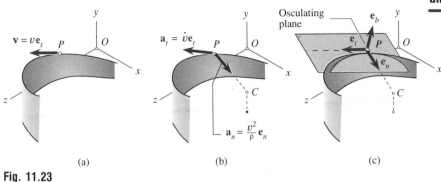

Fig. 11.23

The relations given in Eqs. (11.56) and (11.57) still hold in the case of a particle moving along a space curve as shown in parts (a) and (b) of Fig. 11.23, where $\mathbf{e}_t$ is directed along the tangent to the curve at the point P, and $\mathbf{e}_n$ is perpendicular to $\mathbf{e}_t$ and points toward the center of curvature C of the curve at P. The plane containing $\mathbf{e}_t$ and $\mathbf{e}_n$ is called the *osculating plane* at P as shown in Fig. 11.23(c), where $\mathbf{e}_b$ is called the *binormal unit vector* at P and is given by

$$\mathbf{e}_b = \mathbf{e}_t \times \mathbf{e}_n \qquad (11.58)^\dagger$$

Thus, $\mathbf{e}_t$, $\mathbf{e}_n$, and $\mathbf{e}_b$ form a right-handed trihedral. In distinction from $\mathbf{e}_b$, the vector $\mathbf{e}_n$ is sometimes called the *principal normal unit vector*. It is obvious from Eqs. (11.56) that both the acceleration and the velocity vectors lie entirely in the osculating plane at P, and the acceleration is always directed to the *concave side* of the curved path.

Developmental Exercises

D11.34   Are the unit vectors $\mathbf{e}_t$ and $\mathbf{e}_n$ *constant* vectors?

D11.35   Define the terms: (a) osculating plane, (b) binormal unit vector, (c) principal normal unit vector.

D11.36   The rod AB rotates clockwise inside a fixed pipe as shown. The direction of the velocity of the end B relative to the end A is (a) ↑, (b) ↓, (c) ↙, (d) ↖, (e) →, (f) ←.

Fig. D11.36

EXAMPLE 11.15

A nozzle discharges a stream of water with an initial velocity of 20 m/s as shown. Determine the radius of curvature ρ_A of the stream as it leaves the nozzle at A.

Solution.   Immediately after leaving the nozzle, the stream is subjected to the downward gravitational acceleration $a = g = 9.81$ m/s². Thus, the nor-

---

†In differential geometry, the magnitude of the derivative $d\mathbf{e}_b/ds$ is called the *torsion* τ of the space curve, and $1/\tau$ is called the *radius of torsion* σ.

mal component of acceleration of the stream as it leaves the nozzle at A is

$$a_n = 9.81 \cos 30° = 4.905 \sqrt{3}$$

Since $v = 20$ m/s at A, we write

$$a_n = \frac{v^2}{\rho} \qquad 4.905 \sqrt{3} = \frac{(20)^2}{\rho_A} \qquad \rho_A = 47.1 \text{ m} \quad \blacktriangleleft$$

Developmental Exercises

D11.37   Refer to Example 11.15. Determine the radius of curvature ρ_B of the stream at the point B where the height of the stream is maximum.

D11.38   A particle moves with a *constant speed* of 3 m/s along a circular path of radius 2 m. What is the magnitude of its acceleration?

11.14   Radial and Transverse Components

In some problems of plane motion of particles, it is desirable to use polar coordinates to describe the motion. The position of the particle at P is defined by the polar coordinates r and θ as shown in parts (a) and (b) of Fig. 11.24, where r is a length and θ is an angle in radians. The unit vectors in the *radial and transverse* directions are denoted by $\mathbf{e}_r$ and $\mathbf{e}_\theta$, respectively, as shown.

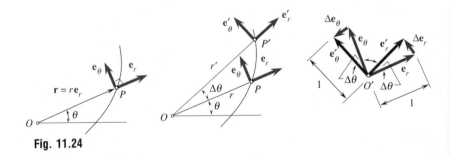

Fig. 11.24

Note that the *radial unit vector* $\mathbf{e}_r$ is directed outward along $\overrightarrow{OP}$, *and the transverse unit vector* $\mathbf{e}_\theta$ is obtained by rotating $\mathbf{e}_r$ through 90° counterclockwise. As the particle moves from the point P to its neighboring point P', the vectors $\mathbf{e}_r$ and $\mathbf{e}_\theta$ change to $\mathbf{e}_r'$ and $\mathbf{e}_\theta'$ by amounts $\Delta\mathbf{e}_r$ and $\Delta\mathbf{e}_\theta$, respectively, as shown in parts (b) and (c) of Fig. 11.24. A derivation similar to the one employed in establishing Eqs. (11.55) will yield the relations

$$\frac{d\mathbf{e}_r}{d\theta} = \mathbf{e}_\theta \qquad \frac{d\mathbf{e}_\theta}{d\theta} = -\mathbf{e}_r \tag{11.59}$$

Expressing the position vector $\mathbf{r}$ of the particle at P as

$$\mathbf{r} = r\mathbf{e}_r \tag{11.60}$$

and using Eqs. (11.59) and (11.60), we derive the velocity and acceleration of the particle as follows:

$$\mathbf{v} = \frac{d\mathbf{r}}{dt} = \frac{dr}{dt}\mathbf{e}_r + r\frac{d\mathbf{e}_r}{dt} = \frac{dr}{dt}\mathbf{e}_r + r\frac{d\mathbf{e}_r}{d\theta}\frac{d\theta}{dt} = \dot{r}\mathbf{e}_r + r\mathbf{e}_\theta\dot{\theta}$$

$$\mathbf{v} = \dot{r}\mathbf{e}_r + r\dot{\theta}\mathbf{e}_\theta \tag{11.61}$$

$$\mathbf{a} = \frac{d\mathbf{v}}{dt} = \ddot{r}\mathbf{e}_r + \dot{r}\frac{d\mathbf{e}_r}{d\theta}\frac{d\theta}{dt} + \dot{r}\dot{\theta}\mathbf{e}_\theta + r\ddot{\theta}\mathbf{e}_\theta + r\dot{\theta}\frac{d\mathbf{e}_\theta}{d\theta}\frac{d\theta}{dt}$$

$$= \ddot{r}\mathbf{e}_r + \dot{r}\mathbf{e}_\theta\dot{\theta} + \dot{r}\dot{\theta}\mathbf{e}_\theta + r\ddot{\theta}\mathbf{e}_\theta + r\dot{\theta}(-\mathbf{e}_r)\dot{\theta}$$

$$\mathbf{a} = (\ddot{r} - r\dot{\theta}^2)\mathbf{e}_r + (r\ddot{\theta} + 2\dot{r}\dot{\theta})\mathbf{e}_\theta \tag{11.62}$$

If we write

$$\mathbf{v} = \mathbf{v}_r + \mathbf{v}_\theta = v_r\mathbf{e}_r + v_\theta\mathbf{e}_\theta \qquad \mathbf{a} = \mathbf{a}_r + \mathbf{a}_\theta = a_r\mathbf{e}_r + a_\theta\mathbf{e}_\theta$$

we see that the r and θ components of the velocity and acceleration are

$$v_r = \dot{r} \qquad v_\theta = r\dot{\theta} \tag{11.63}$$

$$a_r = \ddot{r} - r\dot{\theta}^2 \qquad a_\theta = r\ddot{\theta} + 2\dot{r}\dot{\theta} \tag{11.64}$$

In the case of a particle moving along a circle with its center at the origin O, we have $r = $ constant, $\dot{r} = \ddot{r} = 0$, and the formulas in Eqs. (11.61) and (11.62) degenerate to

$$\mathbf{v} = r\dot{\theta}\mathbf{e}_\theta \qquad \mathbf{a} = -r\dot{\theta}^2\mathbf{e}_r + r\ddot{\theta}\mathbf{e}_\theta \tag{11.65}$$

EXAMPLE 11.16

As the gear A rolls on the fixed gear B, the point P on the gear A describes a cardioid defined by the equations $r = 1 + \cos\pi t$ and $\theta = \pi t$, where r is measured in feet, θ in radians, and t in seconds. When $t = 0$, P is situated at P_0 as shown. Determine the velocity and acceleration of P when $t = \frac{1}{4}$ s.

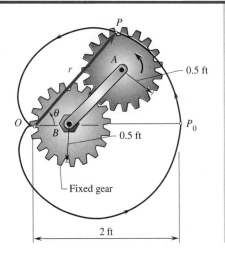

Solution. When $t = \frac{1}{4}$ s, we have

$$r = 1 + \cos\pi t = 1 + 1/\sqrt{2} \qquad \theta = \pi t = \pi/4$$

$$\dot{r} = -\pi\sin\pi t = -\pi/\sqrt{2} \qquad \dot{\theta} = \pi$$

$$\ddot{r} = -\pi^2\cos\pi t = -\pi^2/\sqrt{2} \qquad \ddot{\theta} = 0$$

$$v_r = \dot{r} = -\pi/\sqrt{2} = -2.22 \qquad v_\theta = r\dot{\theta} = (1 + 1/\sqrt{2})\pi = 5.36$$

$$v = (v_r^2 + v_\theta^2)^{1/2} = 5.80 \qquad \beta = \tan^{-1}(v_\theta/|v_r|) = 67.5°$$

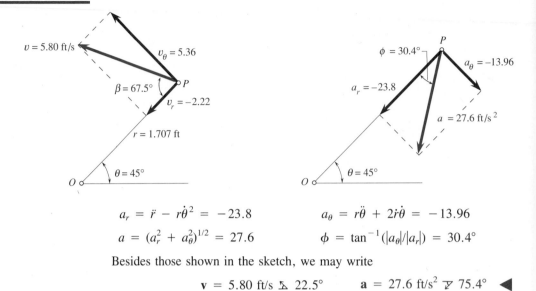

$$a_r = \ddot{r} - r\dot{\theta}^2 = -23.8 \qquad a_\theta = r\ddot{\theta} + 2\dot{r}\dot{\theta} = -13.96$$

$$a = (a_r^2 + a_\theta^2)^{1/2} = 27.6 \qquad \phi = \tan^{-1}(|a_\theta|/|a_r|) = 30.4°$$

Besides those shown in the sketch, we may write

$$\mathbf{v} = 5.80 \text{ ft/s} \;\measuredangle\; 22.5° \qquad \mathbf{a} = 27.6 \text{ ft/s}^2 \;\measuredangle\; 75.4° \blacktriangleleft$$

Developmental Exercises

D11.39 In polar coordinates, is it true that (a) $a_r = \dot{v}_r$, (b) $a_\theta = \dot{v}_\theta$?

D11.40 Refer to Example 11.16. Determine the velocity and acceleration of P when $t = \frac{1}{2}$ s.

11.15 Cylindrical Components

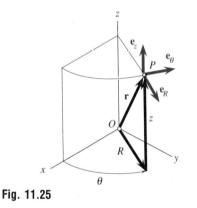

Fig. 11.25

In some cases, the motion of a particle along a space curve is best described by using a cylindrical coordinate system. In such cases, the position of a particle P in space is defined by its cylindrical coordinates R, θ, and z, as shown in Fig. 11.25, where $\mathbf{e}_R$, $\mathbf{e}_\theta$, and $\mathbf{e}_z$ are the *radial*, *transverse*, and *axial unit vectors* attached to P, respectively. Observe that $\mathbf{e}_R$ and $\mathbf{e}_\theta$ are unit vectors parallel to the *xy* plane; they are not constant vectors because their directions may change as P moves from one point to another. However, the axial unit vector $\mathbf{e}_z$, which is identical with the cartesian unit vector $\mathbf{k}$ in the fixed *xyz* coordinate system, is a constant vector because it is constant in direction and magnitude as P moves.

Expressing the position vector $\mathbf{r}$ of the particle P in terms of its radial and axial components, we write

$$\mathbf{r} = R\mathbf{e}_R + z\mathbf{e}_z \tag{11.66}$$

A procedure similar to the one used in deriving Eqs. (11.61) and (11.62) will lead to the formulas for the velocity $\mathbf{v}$ and acceleration $\mathbf{a}$ as

$$\mathbf{v} = \dot{R}\mathbf{e}_R + R\dot{\theta}\mathbf{e}_\theta + \dot{z}\mathbf{e}_z \tag{11.67}$$

$$\mathbf{a} = (\ddot{R} - R\dot{\theta}^2)\mathbf{e}_R + (R\ddot{\theta} + 2\dot{R}\dot{\theta})\mathbf{e}_\theta + \ddot{z}\mathbf{e}_z \tag{11.68}$$

where the coefficients of $\mathbf{e}_R$, $\mathbf{e}_\theta$, and $\mathbf{e}_z$ are respectively the *radial*, *transverse*, and *axial components* of motion of the particle.

EXAMPLE 11.17

The motion of a particle along a spiral path on the surface of a right circular frustrum is defined by the cylindrical coordinates $R = 4 + \frac{1}{4} t^2$, $\theta = 2\pi t$, $z = t^2$, where R and z are measured in meters, θ in radians, and t in seconds. Determine the magnitudes of the velocity and acceleration of the particle when $t = 2$ s.

Solution. The velocity $\mathbf{v}$ and the acceleration $\mathbf{a}$ are given by Eqs. (11.67) and (11.68). When $t = 2$ s, we write

$$R = 4 + \tfrac{1}{4} t^2 = 5 \qquad \theta = 2\pi t = 4\pi \qquad z = t^2 = 4$$
$$\dot{R} = \tfrac{1}{2} t = 1 \qquad \dot{\theta} = 2\pi \qquad \dot{z} = 2t = 4$$
$$\ddot{R} = \tfrac{1}{2} \qquad \ddot{\theta} = 0 \qquad \ddot{z} = 2$$

$$\mathbf{v} = \dot{R}\mathbf{e}_R + R\dot{\theta}\mathbf{e}_\theta + \dot{z}\mathbf{e}_z = \mathbf{e}_R + 10\pi\mathbf{e}_\theta + 4\mathbf{e}_z$$
$$v^2 = (1)^2 + (10\pi)^2 + (4)^2 \qquad v = 31.7 \text{ m/s} \blacktriangleleft$$
$$\mathbf{a} = (\ddot{R} - R\dot{\theta}^2)\mathbf{e}_R + (R\ddot{\theta} + 2\dot{R}\dot{\theta})\mathbf{e}_\theta + \ddot{z}\mathbf{e}_z$$
$$= (\tfrac{1}{2} - 20\pi^2)\mathbf{e}_R + 4\pi\mathbf{e}_\theta + 2\mathbf{e}_z$$
$$a^2 = (\tfrac{1}{2} - 20\pi^2)^2 + (4\pi)^2 + (2)^2 \qquad a = 197.3 \text{ m/s}^2 \blacktriangleleft$$

Developmental Exercises

D11.41 Refer to Example 11.17. At the instant $t = 2$ s, is it true that $\mathbf{e}_R = \mathbf{i}$, $\mathbf{e}_\theta = \mathbf{j}$, and $\mathbf{e}_z = \mathbf{k}$?

D11.42 Refer to Example 11.17 and the answer to D11.41. For the instant $t = 2$ s, determine (a) $\mathbf{v} \times \mathbf{a}$, (b) the binormal unit vector $\mathbf{e}_b$, (c) the angle between $\mathbf{e}_b$ and the z axis, (d) the angle between the osculating plane and the xy plane.

Fig. P11.78*

PROBLEMS

11.78* A ball is thrown upward at the position A from a car traveling at a constant velocity as shown. Determine the radius of curvature of the path of the ball when the ball (a) leaves the position A, (b) passes through its highest point B.

11.79 A car travels at a constant speed of 72 km/h along the curve AB of constant curvature as shown. An accelerometer mounted in the car indicates that the magnitude of acceleration of the car is 2.5 m/s². Determine the change of direction $\Delta\theta$ of the car from A to B.

Fig. P11.79

100 m

Fig. P11.80

250 m   150 m

A   108 km / h   B   C

144 km / h

500 m

Fig. P11.82

Shuttle

v

h = 250 mi

Fig. P11.84

11.80   A car travels with a steady speed of v over a vertical curve whose radius of curvature is 100 m as shown. If the magnitude of acceleration of the car is not to exceed $0.5g$, determine the maximum value of v permitted.

11.81* Upon entering a curve of 200-m radius at a speed of 72 km/h, a motorist applies his brakes to decrease his speed at a constant rate of 2 m/s². Determine the magnitude of the total acceleration of the car when its speed is 54 km/h.

11.82   A racing car travels along the curve ABC of radius 500 m as shown. The speed of the car is increased at a constant rate from 108 km/h at A to 144 km/h at C. Determine the magnitude of the total acceleration of the car when it passes through the point B.

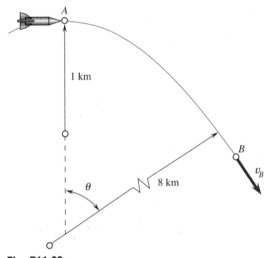

A

1 km

θ

8 km

B

v_B

Fig. P11.83

11.83   A projectile travels along the trajectory AB near the surface of the earth as shown. The radius of curvature at the highest point A of the trajectory is 1 km, and that at the point B is 8 km. Determine the speed v_B and the angle θ between the radii as indicated.

11.84   A spacecraft will coast indefinitely in a circular "parking" orbit around the earth if it travels with $a_n = g(R/r)^2$, where $g = 32.2$ ft/s², R = radius of the earth = 3.96×10^3 mi, and r = distance between the spacecraft and the center of the earth. The altitude h of a space shuttle in one of its flights in such an orbit is 250 mi as shown. Determine (a) the speed v of the shuttle in the orbit, (b) the time τ required for the shuttle to coast in the orbit one revolution.

11.85* The time required for the moon to revolve around the earth once is 27.32 d. Using the information given in Prob. 11.84 and assuming the orbit of the moon around the earth to be a circle,[†] estimate the mean distance between the center of the earth and the center of the moon.

11.86   The position vector of a particle is given by $\mathbf{r} = 12t^2\mathbf{i} + 3t^3\mathbf{j}$, where $\mathbf{r}$ is measured in meters and t in seconds. Determine the radius of curvature of the path for the position at $t = 2$ s.

11.87   A skier travels down a slope as shown, where x and y are measured in feet. If his normal component of acceleration at the position P is equal to $0.5g$, determine his speed at this position.

[†]The orbit of the moon around the earth is close to an ellipse having an eccentricity of 0.055.

Fig. P11.87

11.88 A pin P is confined to move in the quarter circular groove AB and the horizontal slot in the guide CD as shown. The guide CD is being lowered by its lead screw with a constant downward velocity of 10 mm/s. Determine $\mathbf{a}_t$ and $\mathbf{a}_n$ of the pin P when $\theta = 60°$.

Fig. P11.88   **Fig. P11.89**

11.89 A space shuttle is launched vertically by firing its main engine and two booster rockets and is tracked by the radar as shown. When $\theta = \pi/3$ rad, the measurements give $r = 10$ km, $\ddot{r} = 20$ m/s^2, and $\dot{\theta} = 0.02$ rad/s. Determine the velocity and acceleration of the shuttle at this position.

11.90 Determine the velocity of the shuttle in Prob. 11.89 in terms of L, θ, and $\dot{\theta}$.

11.91* Determine the acceleration of the shuttle in Prob. 11.89 in terms of L, θ, $\dot{\theta}$, and $\ddot{\theta}$.

11.92 through 11.94* A particle moves along a curved path defined by the polar coordinates r and θ as shown, where $\theta = (\pi/6)t$, r is measured in meters, θ in radians, and t in seconds. Determine the velocity and acceleration of the particle when it passes through the point P.

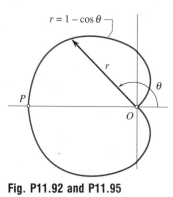

Fig. P11.92 and P11.95

Fig. P11.93 and P11.96

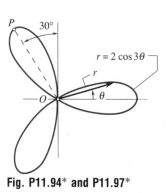

Fig. P11.94* and P11.97*

11.95 through 11.97* A particle moves along a curved path defined by the polar coordinates r and θ as shown, where $\theta = (\pi/12)t^2$, r is measured in feet, θ in radians, and t in seconds. Determine the magnitudes of the velocity and acceleration of the particle when $t = 2$ s.

11.98 The rod BP rotates clockwise at a constant rate of $\pi/4$ rad/s. The pin at P is free to slide along the slot of the arm OA as shown. Determine the values of $\dot{r}$, $\ddot{r}$, $\dot{\theta}$, and $\ddot{\theta}$ at the instant when $\phi = 90°$.

Fig. P11.98   **Fig. P11.99**

11.99 The hydraulic cylinder gives the pin at P a constant velocity $\mathbf{v} = 60$ mm/s $\rightarrow$ as shown. The pin is free to slide along the slot in the arm OA. Determine the values of $\dot{r}$, $\ddot{r}$, $\dot{\theta}$, and $\ddot{\theta}$ at the instant when $\theta = 60°$.

11.100* The motion of a particle P along a circular helix as shown is defined by the cylindrical coordinates $R = 4$, $\theta = \pi t$, $z = t$, where R and z are measured in meters, θ in radians, and t in seconds. Determine the velocity and acceleration of the particle.

11.101 Refer to Prob. 11.100*. Determine the angle between the osculating plane and the xy plane.

Fig. P11.100*   **Fig. P11.102***

11.102* The path of the particle P as shown is defined by the cylindrical coordinates $R = 4 - \frac{1}{2}t$, $\theta = \pi t$, and $z = t$, where R and z are measured in meters, θ in radians, and t in seconds. Determine the magnitudes of the velocity and acceleration of the particle when $t = 4$ s.

11.103 The motion of the end P of the 20-m boom AP of the crane as shown is defined by the cylindrical coordinates $R = 20 \sin(\pi/30) t$, $\theta = (\pi/15) t$, and $z = 1 + 20 \cos(\pi/30)t$, where R and z are measured in meters, θ in radians, and t in seconds. Determine the magnitudes of the velocity and acceleration of the end P when $\theta = \pi/3$ rad.

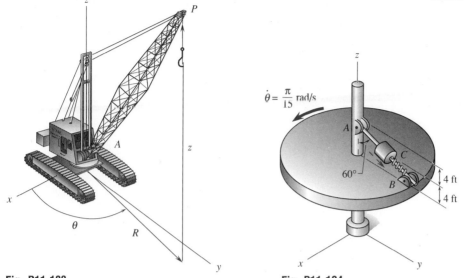

Fig. P11.103

Fig. P11.104

11.104 A collar C oscillates along a rod AB as shown. The turntable rotates with a constant speed $\pi/15$ rad/s about the z axis. At the instant shown, the collar C is moving downward with a velocity of 2 ft/s and an acceleration of 3 ft/s² relative to the rod AB. Determine the magnitudes of the velocity and acceleration of the collar relative to the ground at this instant. (*Hint*. Use cylindrical coordinates.)

11.16 Concluding Remarks

The concepts of position, displacement, total distance traveled, time, velocity, speed, acceleration, deceleration, and jerk are very fundamental. They represent the basic kinematic quantities studied in this chapter. It is pointed out that when the acceleration or velocity of a rectilinear motion is given by different linear functions of time over several intervals, the use of graphical solution (described in Sec. 11.8) may be advantageous. In such a case, the six useful relations for the graphical solution given in that section should be very helpful in achieving effective solutions.

The use of different systems of components to describe the curvilinear motion of particles is presented in the latter half of this chapter. The free flight of a projectile as well as certain constrained motions can conveniently be described by using rectangular components. The use of tangential and normal components is often desirable when velocity, acceleration, and radius of curvature are to be related. The use of the system of radial and transverse components and the system of cylindrical components is largely influenced by the geometry of the path of the particle. We shall see in Chap. 12 that the free flight of a spacecraft is best described using radial and transverse components.

Nowadays, programmable calculators and microcomputers are often utilized in analyzing and solving problems. In addition to finding roots of any nonlinear equations, the *digital root finder* in App. E may be utilized to plot the approximate shapes of some plane curves. For example, we may use the 40-column version of that root finder (named DIGIT40 and stored on the

TUTORIAL diskette as described in App. F) to plot the motion curves in Example 11.1. The *x-t* curve, given by $x = t^3 - 6t^2 - 36t + 116$, may be plotted as shown in Fig. 11.26 by writing Line 1000 of the program as

$$1000 \; Y = X\char`^3 - 6*X\char`^2 - 36*X + 116$$

The *v-t* curve, given by $v = 3t^2 - 12t - 36$, may be plotted as shown in Fig. 11.27 by writing Line 1000 of the program as

$$1000 \; Y = 3*X\char`^2 - 12*X - 36$$

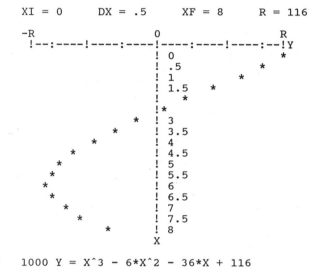

```
XI = 0      DX = .5      XF = 8      R = 116

-R                       0                       R
 !--:----!----:----!----:----!----:----!----:--!Y
                         ! 0                     *
                         ! .5                  *
                         ! 1              *
                         ! 1.5        *
                         !    *
                         !*
                 *       ! 3
               *         ! 3.5
            *            ! 4
          *              ! 4.5
        *                ! 5
       *                 ! 5.5
      *                  ! 6
      *                  ! 6.5
        *                ! 7
          *              ! 7.5
              *          ! 8
                         X
```

Fig. 11.26 $1000 \; Y = X\char`^3 - 6*X\char`^2 - 36*X + 116$

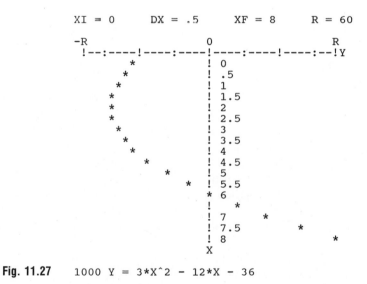

```
XI = 0      DX = .5      XF = 8      R = 60

-R                       0                       R
 !--:----!----:----!----:----!----:----!----:--!Y
           *             ! 0
          *              ! .5
         *               ! 1
       *                 ! 1.5
       *                 ! 2
       *                 ! 2.5
         *               ! 3
          *              ! 3.5
            *            ! 4
              *          ! 4.5
                 *       ! 5
                    *    ! 5.5
                      *  6
                         !     *
                         ! 7        *
                         ! 7.5         *
                         ! 8              *
                         X
```

Fig. 11.27 $1000 \; Y = 3*X\char`^2 - 12*X - 36$

When the program is run, the computer will ask for the values of XI, DX, and XF, which are, respectively, the initial, the incremental, and the final values of the independent variable X of the function Y. Thereafter, the computer takes care of the rest. Note that the Y and X axes in Figs. 11.26 and 11.27† should be understood as the x and t, and the v and t, axes, respectively, in the context of the present application.

REVIEW PROBLEMS

11.105 The acceleration of an oscillator is given by $a = -\omega^2 x$, where ω is a constant. If $v = 6$ m/s when $x = 0$ and $v = 0$ when $x = 2$ m, determine (a) the value of ω, (b) the time elapsed during which the oscillator moves from $x = 2$ m back to $x = 0$.

11.106 The velocities of the cylinders A and C as shown are $\mathbf{v}_A = 1$ m/s ↑ and $\mathbf{v}_C = 6$ m/s ↓, respectively. Determine $\mathbf{v}_B$ of the cylinder B.

Fig. P11.105

Fig. P11.106   **Fig. P11.107**

11.107 A particle moves in a straight line with the velocity as shown. For $x_0 = -21$ ft and over the time interval $0 \leq t \leq 10$ s, determine Δx and x_T, as well as the two values t_1 and t_2 of t for which the particle passes through the origin.

11.108 A basketball is thrown with an initial velocity $\mathbf{v}_0$ and enters the basket at B as shown. Determine the possible values of θ.

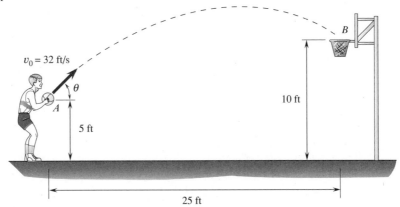

Fig. P11.108

---

†It is true that one may use other plotters to plot these curves. However, most plotters are not intended to find roots to many digits of accuracy as the *digital root finder* in App. E.

11.109 An earth satellite has a velocity $\mathbf{v}_B$ as it passes through the end B of the semiminor axis of its orbit as shown. Knowing that the acceleration of the satellite is $\mathbf{a} = \mathbf{a}_r = -g(R/r)^2\mathbf{e}_r$, where $g = 9.81$ m/s^2, determine the radius of curvature of the orbit at B.

Fig. P11.109   **Fig. P11.110**

11.110 A rocket moves vertically as shown. When $\theta = 30°$, the radar measurements give $r = 8$ km, $\dot{\theta} = 0.025$ rad/s, and $\ddot{\theta} = 0.005$ rad/s^2. Determine the velocity and acceleration of the rocket.

Kinetics of Particles: Force and Acceleration

Chapter 12

KEY CONCEPTS

- Effective force, effective-force diagram, vector-diagram equation, dynamic equilibrium, inertia force, central-force motion, areal speed, eccentricity of orbit, and period of orbit.
- Method of force and acceleration for particles.
- Forces and accelerations in rectilinear motion of particles.
- Forces and accelerations in curvilinear motion of particles using rectangular components, tangential and normal components, radial and transverse components, or cylindrical components.
- Gravitational central-force motion of spacecraft and satellites.

Newton's second law and his law of gravitation are two major fundamental laws in kinetics. They and the fundamental laws used in statics form the physical foundation of kinetics. Early in this chapter, Newton's second law is applied to show that the effective force on a particle is the resultant force on the particle. The effective-force diagram together with the free-body diagram is then employed to formulate the *method of force and acceleration* for particles. Later, Newton's second law and his law of gravitation are jointly applied to study the central-force motion of spacecraft and satellites. We note that present-day space technology involves the applications of space shuttles, geosynchronous satellites, surveying satellites, space stations, etc. Thus, it is easy to see that the applications of central-force motion have significant impacts on our modern living and technology.

NEWTON'S SECOND LAW[†]

12.1 Method of Force and Acceleration

A particle will accelerate if the forces acting on it are not balanced. To be more specific, let $\Sigma\mathbf{F}$ denote the *resultant force* of a system of forces $\mathbf{F}_1$, $\mathbf{F}_2, \ldots, \mathbf{F}_n$ acting on a particle of *mass m*. This particle will move with

---

[†]Newton's work, entitled *Philosophiæ Naturalis Principia Mathematica*, was written in Latin and published in 1687; it was revised in 1713 and 1726. For a modern translation, see Florian Cajori, tran., *Sir Isaac Newton's Mathematical Principles of Natural Philosophy and His System of the World* (Berkeley: University of California Press, 1934).

an *acceleration* **a** if $\Sigma\mathbf{F} \neq \mathbf{0}$. In an inertial reference frame, we may apply *Newton's second law* to the particle to write

$$\Sigma\mathbf{F} = m\mathbf{a} \tag{12.1}$$

which is called the *equation of motion* and is illustrated in Fig. 12.1.

Fig. 12.1   Newton's second law

Note that the *resultant force* of a system of forces acting on a particle is *not* necessarily an actual force applied to the particle. Rather, it is merely an *equivalent force* representing the sum of the forces in the system; a force equal to the resultant force may replace the system of applied forces without changing the net effect on the particle. Thus, it is important to recognize that the *effective force* on a particle is the *resultant force* on the particle. The resultant force $\Sigma\mathbf{F}$ and the vector $m\mathbf{a}$ are, by Eq. (12.1), equivalent to each other. For this reason, the vector $m\mathbf{a}$ is also referred to as the *effective force* on (or of) the particle. Moreover, because of Eq. (12.1), the terms *effective force* and *resultant force* are sometimes interchangeably used. Nevertheless, because of the special *abstract* nature of the quantities $\Sigma\mathbf{F}$ and $m\mathbf{a}$, there are people who refrain from using the word ''force'' in naming them and refer to them simply as the *resultant* and the $m\mathbf{a}$ *vector*, respectively.

The use of the vector $m\mathbf{a}$ as the expression for the *effective force* or the *resultant force* is especially desirable in cases where the acceleration of the particle is known or to be determined. This may be depicted in Fig. 12.2, where the separate diagram showing the applied forces acting on the particle is, as before, called the *free-body diagram*, and the separate diagram showing the effective force $m\mathbf{a}$, or its components, on the particle is referred to as the *effective-force diagram* of the particle.[†] Notice here that the equality of any two diagrams containing vectors, such as those in Figs. 12.1 and 12.2, constitutes a *vector-diagram equation*.

Resolving each force $\mathbf{F}_i$ ($i = 1, 2, \ldots, n$) and the acceleration **a** in Eq. (12.1) into rectangular components, we write

$$\Sigma(F_x\mathbf{i} + F_y\mathbf{j} + F_z\mathbf{k}) = m(a_x\mathbf{i} + a_y\mathbf{j} + a_z\mathbf{k})$$

$$\Sigma F_x = ma_x \qquad \Sigma F_y = ma_y \qquad \Sigma F_z = ma_z \tag{12.2}$$

We refer to Eqs. (12.2) as a set of *scalar equations of motion* of the particle.

[†]Besides the free-body diagram defined in statics, the effective-force diagram mentioned here and the impulse and momentum diagrams to be defined in subsequent chapters are very helpful geometrical aids in the study of plane kinetics. An *effective-force diagram* is also called a *resultant-force diagram*, an *equivalent-force diagram*, *a second free-body diagram*, or a *kinetic diagram*. In some texts, such a diagram is left unnamed.

Free-body   Effective-force
diagram   diagram

Fig. 12.2

The method of solution which directly applies Newton's second law to particles in the form of equations of motion, or a vector-diagram equation consisting of a *free-body diagram* and an *effective-force diagram*, is called the
method of force and acceleration for particles. This method is useful whenever forces and accelerations are to be directly related.

Developmental Exercises

D12.1   Define for a particle the terms: (a) effective force, (b) effective-force
diagram, (c) vector-diagram equation.

D12.2   Describe the *method of force and acceleration* for particles.

12.2   Consistent Systems of Kinetic Units

A *kinetic quantity* is a measurable quantity in kinetics. *Kinetic units* are units
for kinetic quantities. In writing Newton's second law in the form of Eq.
(12.1), we cannot arbitrarily choose the units of length, mass, time, and
force. If the units are arbitrarily chosen, the magnitude of the resultant force
ΣF required to move a particle of mass m with an acceleration $\mathbf{a}$ will *not* be
numerically equal to the product ma; it will only be proportional to this
product. However, the set of units for length, mass, time, and force, which
satisfy Eq. (12.1), is said to form a *consistent system of kinetic units*. Two
consistent systems of kinetic units are in use by American engineers during
the transition period; they are the SI and the U.S. customary system. Since
both systems have been discussed in detail in Chap. 1, we shall describe
them only briefly in this section.

The SI *base units* used in elementary mechanics are those for length,
mass, and time, and are called, respectively, the *meter* (m), *kilogram* (kg),
and *second* (s). The unit *newton* (N) for force is a *derived unit. One newton*
is defined as the net force required to give a body with a mass of 1 kg an
acceleration of 1 m/s^2. By Eq. (12.1), we write

$$1 \text{ N} = 1 \text{ kg} \cdot \text{m/s}^2 \qquad (12.3)$$

In the U.S. customary system of units, the *base units* used in elementary
mechanics are those for length, force, and time, and are called, respectively,
the *foot* (ft), *pound* (lb), and *second* (s). The *second* here is the same as that
in the SI. The foot is defined as

$$1 \text{ ft} = 0.3048 \text{ m} \qquad (12.4)$$

One pound-mass (1 lbm) is legally defined as

$$1 \text{ lbm} = 0.45359237 \text{ kg} \qquad (12.5)$$

One pound of force is equal to the weight of a body with one pound-mass at a location where the measured gravitational acceleration is equal to the standard gravitational acceleration, which is 9.80665 m/s². Thus, we write

$$1 \text{ lb} = 1 \text{ lbm} \cdot \frac{9.80665}{0.3048} \text{ ft/s}^2 \approx 32.2 \text{ lbm} \cdot \text{ft/s}^2 \qquad (12.6)$$

The amount of mass which will accelerate at the rate of 1 ft/s² when it is acted on by a net force of 1 lb is defined as *one slug*. We write

$$1 \text{ lb} = 1 \text{ slug} \cdot \text{ft/s}^2 \qquad (12.7)$$

$$1 \text{ slug} = \frac{9.80665}{0.3048} \text{ lbm} = \frac{9.80665}{0.3048} (0.45359237) \text{ kg} \qquad (12.8)$$

For practical purposes, we write

$$1 \text{ slug} \approx 32.2 \text{ lbm} \approx 14.59 \text{ kg} \qquad (12.9)$$

Note that *slug* is a *derived unit* and the abbreviations for *pound-mass* and *pound* are lbm and lb, respectively. The two consistent systems of kinetic units discussed above are summarized in Table 12.1.

Table 12.1 Consistent Systems of Kinetic Units*

Name of System	Length	Mass	Time	Force
SI	meter (m)	kilogram (kg)	second (s)	newton (N)
U.S. Customary System	foot (ft)	slug	second (s)	pound (lb)

*The SI is termed an *absolute* system since mass is taken as an absolute or base quantity. The U.S. customary system is termed a *gravitational* system since force *as measured from gravity* is taken as a base quantity.

For the relation between weight and mass of a body, we use Eq. (12.1) to write

$$W = mg \qquad (12.10)$$

where W is the weight of a body with a mass m subjected to the gravitational acceleration g. Near the surface of the earth, we take

$$g = 9.81 \text{ m/s}^2 = 32.2 \text{ ft/s}^2 \qquad (12.11)$$

Developmental Exercises

D12.3 Name the units which form a consistent system of kinetic units in (a) the SI, (b) the U.S. customary system.

D12.4 How much does a body on the surface of the earth weigh, in the primary or base unit of the corresponding system of units, if its mass is (a) 10 kg, (b) 10 Mg, (c) 10 mg, (d) 10 g, (e) 10 slugs, (f) 10 lbm?

12.3 Dynamic Equilibrium: Inertia Force

Whether balanced or not, the system of forces acting on a particle is always equivalent to the effective force $m\mathbf{a}$, which is the product of the mass m and the acceleration $\mathbf{a}$ of the particle as illustrated in Fig. 12.2. The negative of the vector $m\mathbf{a}$ is called the *inertia vector* of a particle and is denoted by $(m\mathbf{a})_{\text{rev}}$, where the subscript indicates that it points in the reversed direction of $m\mathbf{a}$; i.e.,

$$(m\mathbf{a})_{\text{rev}} = -m\mathbf{a} \qquad (12.12)$$

If the inertia vector $(m\mathbf{a})_{\text{rev}}$ is added to the free-body diagram, we obtain *a system of balanced vectors* as shown in Fig. 12.3. This is true because, by Eqs. (12.1) and (12.12), the resultant of the vectors in such a system is

$$\Sigma\mathbf{F} + (m\mathbf{a})_{\text{rev}} = \mathbf{0} \qquad (12.13)^{\dagger}$$

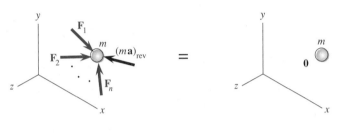

Fig. 12.3   Dynamic equilibrium

A particle perceived as being under the action of the applied forces *and* the inertia vector is said to be in a state of *dynamic equilibrium*, and the equations of equilibrium in statics may similarly be applied. In application, an inertia vector is often called an *inertia force*, as in advanced dynamics and vibration analysis.

Note that the *inertia force* of a body B simply represents (by Newton's third law) the *reaction* of the body B to the action of the resultant force on B.[††] Since such a reaction of B is an action exerted *by* (not *on*) B, the inertia force of B is *not* a force to be included in the free-body diagram of B. If it is included there, the resultant of the system of vectors on B becomes zero, even though the acceleration of B is nonzero. Therefore, a number of people, who stress the fundamental concept of *force* (as defined in Newton's second law) for beginners of dynamics, avoid calling the *inertia vector* an *inertia force*. Furthermore, many people today prefer to think in terms of dynamics (rather than in terms of statics) when they analyze dynamics problems.

[†]Note that d'Alembert's principle (to be covered in Sec. 16.5) was originally written for the motion of *a system of linked masses,* where the sum of interaction forces vanishes. Quotation of Eq. (12.13), which is written for a *single particle,* as d'Alembert's principle has been criticized. For translation of parts of d'Alembert's *Traité de Dynamique,* see R. M. Rosenberg, ''d'Alembert and Others on d'Alembert's Principle,'' *Engineering Education,* April 1968, pp. 959–960.

[††]By Newton's third law, *each action is always matched by a reaction*; i.e., forces always occur in pairs of action and reaction. In this regard, the *inertia force* is clearly a *reactive mechanical force,* rather than a *fictitious* (or *mythical*) force.

Anyhow, one should note that an effective force or a resultant force is simply an *equivalent force*. But, an inertia force is a *reactive force*, which is mainly used in formulating the so-called *dynamic equilibrium* of a body. Although the vectors acting on a body in dynamic equilibrium are balanced, the acceleration of that body is not necessarily zero.

Developmental Exercise

D12.5 Define for a particle the terms: (a) inertia vector, (b) dynamic equilibrium, (c) inertia force.

12.4 Rectilinear Motion of a Particle

The *method of force and acceleration* referred to in Sec. 12.1 may now be used to solve kinetic problems of particles moving along straight paths, where forces and accelerations need to be related. If we choose the x axis to be the straight path of motion of a particle of mass m moving with an acceleration $\mathbf{a}$, as illustrated in Fig. 12.4, we have $a_x = a$, $a_y = a_z = 0$, and Eqs. (12.2) become

$$\Sigma F_x = ma \qquad \Sigma F_y = 0 \qquad \Sigma F_z = 0 \qquad (12.14)$$

Note that the *free-body diagram* of the particle in Fig. 12.4(a) accounts for all the applied forces acting on the particle, while the *effective-force diagram* in Fig. 12.4(b) indicates the effective force (or its components) on the particle. If the sense of any unknown is not known, it may first be assumed and then verified at the end of the solution.[†]

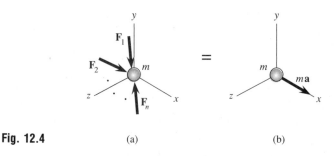

Fig. 12.4   (a)   (b)

In dependent rectilinear motions of particles as discussed in Sec. 11.7, the *constraint equation on the accelerations* of the particles is usually obtained by taking the second derivative of the constraint equation on the positions of the particles with respect to the time. The positive direction of each unknown acceleration in the constraint equation is assumed to point in the positive direction of the corresponding position coordinate of the parti-

---

[†]If the value of an unknown quantity is found to be negative in the solution, the magnitude of this unknown quantity is simply equal to the absolute value of the solved value, but the true sense of this unknown quantity is opposite to that initially assumed. However, this practice may cease to be valid when the unknown quantity is a *friction force* at a location where slipping *impends* or *occurs*.

cle. For consistency, an *unknown effective force* in dependent rectilinear motions *must* also be assumed to point in the positive direction of the corresponding position coordinate if the above-mentioned *constraint equation* is used.[†] Furthermore, the *positive direction* chosen in the summation of components of vectors in a vector-diagram equation must be the *same* for each side of the equal sign. With these precautions, we can avoid most of the errors and difficulties commonly encountered.

EXAMPLE 12.1

A truck is traveling at a constant speed of 72 km/h as shown. Determine the required minimum coefficient of static friction $(\mu_s)_{min}$ between the crate and the flatbed trailer if the truck is to be brought to a stop through a constant deceleration over a distance of 100 m without causing the crate to shift on the flat bed.

72 km/h

Solution. Since the crate is not to shift on the flatbed, the crate and the truck must have the same motion at all times. The initial speed is $v_0 = 72$ km/h $= 20$ m/s. The magnitude a of the constant deceleration may be determined from $v^2 = v_0^2 + 2a_c(x - x_0)$, where $v = 0$, $v_0 = 20$ m/s, $a_c = -a$, and $x - x_0 = 100$ m. Thus, we write

$$0^2 = (20)^2 + 2(-a)(100) \qquad a = 2$$

The acceleration of the crate is, therefore, $\mathbf{a} = 2$ m/s^2 $\leftarrow$.

Assume that the mass of the crate is m. Under the stated condition, slipping of the crate on the flatbed impends. We first equate the free-body diagram to the effective-force diagram as shown. By *equating the sums of the y components of the vectors* (denoted by ΣV_y) in the vector-diagram equation, we write

$$+\uparrow \Sigma V_y\!:^{††} \qquad N - mg = 0 \qquad\qquad (1)$$

Similarly, *equating the sums of the x components of the vectors* (denoted by ΣV_x) in the vector-diagram equation, we write

$$\overset{+}{\leftarrow} \Sigma V_x\!:^{††} \qquad (\mu_s)_{min}N = m(2) \qquad\qquad (2)$$

Equations (1) and (2) yield

$$(\mu_s)_{min} = \frac{2}{g} = \frac{2}{9.81} \qquad (\mu_s)_{min} = 0.204 \;\blacktriangleleft$$

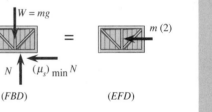

$W = mg$

N $\quad(\mu_s)_{min} N$

m (2)

(FBD)   *(EFD)*

[†] This is illustrated in Example 12.3.

[††] Henceforth, the letter V following the summation sign Σ stands for the vectors being considered.

Developmental Exercise

D12.6   Work Example 12.1 if the constant speed is 54 km/h.

EXAMPLE 12.2

The blocks A and B are released to move from rest as shown. The coefficient of kinetic friction is 0.2 between the cord and the drum C as well as between the block A and the horizontal support. Determine the acceleration of the block B.

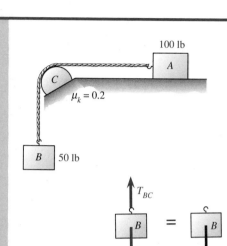

Solution.   Since the cord is considered to be inextensible, the accelerations $\mathbf{a}_A$ and $\mathbf{a}_B$ of the blocks A and B must have the same magnitude. Letting $a_A = a_B = a$, we first draw the following pertinent diagrams:

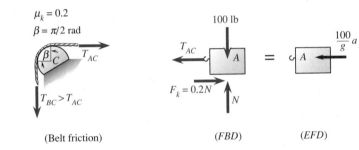

Then, we continue the solution as follows:

Block B.

$$+\downarrow \Sigma V_y: \qquad 50 - T_{BC} = \frac{50}{g}a \qquad (1)$$

Belt friction around the drum C, $T_2 = T_1 e^{\mu_k \beta}$.

$$T_{BC} = T_{AC} e^{0.2(\pi/2)} \qquad (2)$$

Block A.

$$+\uparrow \Sigma V_y: \qquad N - 100 = 0 \qquad (3)$$

$$\overset{+}{\leftarrow} \Sigma V_x: \qquad T_{AC} - 0.2N = \frac{100}{g}a \qquad (4)$$

With $g = 32.2$ (ft/s^2), the above four simultaneous equations yield

$$N = 100 \qquad T_{AC} = 32.1 \qquad T_{BC} = 43.9 \qquad a = 3.90 = a_B$$

$$\mathbf{a}_B = 3.90 \text{ ft/s}^2 \downarrow \quad \blacktriangleleft$$

Developmental Exercise

D12.7   Work Example 12.2 if friction between the block A and its support is negligible.

EXAMPLE 12.3

The blocks A and B are released to move from rest as shown. Neglecting effects of friction, determine the acceleration $\mathbf{a}_A$ of the block A and the tension F in the cord BCD.

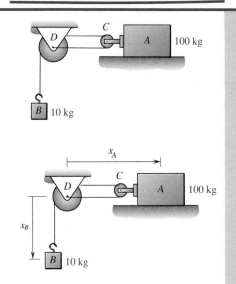

Solution. We first note that the motions of the blocks as shown are subject to the constraint equation on the position coordinates

$$2x_A + x_B = \text{constant}$$

which, through differentiations, yields the *constraint equation on the accelerations*

$$2a_A + a_B = 0 \tag{1}$$

Note in Eq. (1) that the positive sense of a_A is to the right, while that of a_B is downward. For Eq. (1) to hold true, a_A and a_B must have opposite signs. We should let them "run the algebraic course." But we expect a_A to have a negative value in the final solution because the actual acceleration of the block A must be directed to the left. Thus, we continue the solution as follows:

(FBD)   (EFD)   (FBD)   (EFD)

Block A.

$$\xrightarrow{+}\ \Sigma V_x: \qquad -F - F = 100a_A \tag{2}$$

Block B.

$$+\downarrow\ \Sigma V_y: \qquad 98.1 - F = 10a_B \tag{3}$$

Equations (1), (2), and (3) yield

$$a_A = -1.401 \qquad a_B = 2.80 \qquad F = 70.1$$

Since the solved value of a_A is negative (as expected), the true sense of $\mathbf{a}_A$ is to the left. Therefore, we write

$$\mathbf{a}_A = 1.401\ \text{m/s}^2 \leftarrow \qquad F = 70.1\ \text{N} \ \blacktriangleleft$$

REMARK. The positive sense of a_A in Eq. (1) is inherently the same as the positive sense of x_A, which is to the right. Therefore, the sense of the unknown effective force $100a_A$ must be assumed to point to the right in the effective-force diagram for the sake of algebraic consistency with Eq. (1). Furthermore, notice that $F = 70.1$ N and the weight $W_B = 10g = 98.1$ N > 70.1 N. This makes sense because the block B is not in static equilibrium, but accelerating downward.

Developmental Exercise

D12.8 Refer to Example 12.3. Determine the mass of the block B if the mass of the block A remains the same but $\mathbf{a}_A = 2$ m/s^2 ←.

PROBLEMS

12.1* The package A is weighed on a lever scale while the package B is weighed on a spring scale as shown. The scales are fastened to the bar CD which is held horizontally by the force $\mathbf{P}$. When the bar CD is at rest, each scale indicates a reading of 2.5 lb. Determine the reading of each scale when the bar CD is accelerated upward at the rate of 10 ft/s^2.

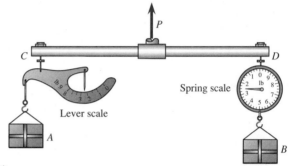

Fig. P12.1*

12.2 Refer to Prob. 12.1*. If the spring scale indicates a reading of 2 lb, determine (a) the acceleration of the bar CD, (b) the reading indicated by the lever scale.

12.3 The device shown is used as an accelerometer. It consists of a 200-mg plunger AB which compresses the spring when the housing of the device is given an acceleration to the right. Assuming that effects of friction are negligible, determine the required spring modulus k which fits the calibration of 1 mm of deflection of the plunger for every 1 m/s^2 of acceleration.

Fig. P12.3

12.4 Determine the acceleration of the cylinder A in each of the three cases illustrated.

Fig. P12.4

Fig. P12.5*

12.5* A 90-kg workman inside a 110-kg cage pulls the hoisting cable with a force of 500 N as shown. Determine the acceleration of the cage.

12.6 A 20-lb block B is suspended from a 5-ft cord attached to a 50-lb cart A which can move freely down the incline. Determine for the instant the system is released from rest (a) the acceleration of the cart, (b) the acceleration of the block, (c) the tension in the cord.

12.7* A car climbs the hill AB at a constant speed as shown. If the driver does not change the throttle setting or shift gears, determine the magnitude of the acceleration of the car immediately after entering (a) the level section BC, (b) the downhill section CD.

Fig. P12.6

Fig. P12.7*

Fig. P12.8*

12.8* If the frame has an acceleration $\mathbf{a}$ and the tensions in the ropes CA and CB are such that $T_{CA} = 4T_{CB}$, determine (a) $\mathbf{a}$, (b) T_{CB}.

12.9 The force $\mathbf{P}$ causes the carriage to have an acceleration $\mathbf{a}$ and an impending loss of contact between the 40-kg pipe and the carriage at A as shown. Determine (a) $\mathbf{a}$, (b) the magnitude of the contact force at B.

12.10 A truck is traveling at a constant speed of 54 km/h down a hill as shown. If $\mu_s = 0.5$ between the crate and the flatbed trailer, determine the shortest distance in which the truck can be brought to a stop without causing the crate to shift on the flatbed.

Fig. P12.9

Fig. P12.10

12.11* and **12.12** If μ_k equals 0.2 between the blocks A and B, and 0.3 between the block B and the fixed support as shown, where the blocks are moving, determine the magnitude of acceleration of each block.

Fig. P12.11*

Fig. P12.12

12.13 Solve Prob. 12.12 if $\theta = 20°$.

12.14 The downward motion of a 200-lb carriage in its smooth vertical guides is controlled by a cable as shown. If $\mu_k = 0.15$ between the cable and the pegs, determine the magnitude of the force **P** which allows (a) a constant velocity downward, (b) a constant acceleration of 3 ft/s² ↓ .

Fig. P12.14

Fig. P12.15

✗12.15 The blocks A and B are sliding down an incline as shown. If μ_k equals 0.2 between the block A and the incline, and 0.3 between the block B and the incline, determine (a) the acceleration of the blocks, (b) the force exerted by the block A on the block B.

12.16* The truck shown suddenly accelerates forward for a moment, and slipping occurs between the crates and the flatbed. If μ_k equals 0.2 between the crate A and the flatbed, and 0.3 between the crate B and the flatbed, determine for the moment in which slipping occurs (a) the minimum acceleration of the truck, (b) the force exerted by A on B.

Fig. P12.16*

Fig. P12.17

12.17 The truck shown is traveling forward at a constant speed when the brakes are applied causing the crates to slide on the flatbed. If the values of μ_k between the flatbed and the crates A, B, and C are 0.3, 0.25, and 0.2, respectively, determine for the period of braking (a) the minimum magnitude of the deceleration of the truck, (b) the force exerted by A on B, (c) the force exerted by C on B.

12.18 and 12.19* Neglecting effects of friction in the system shown, determine (a) the acceleration of each block, (b) the tension in the cable.

Fig. P12.18 and P12.20*

Fig. P12.19* and P12.21

12.20* and 12.21 The blocks shown are moving to the right. If $\mu_k = 0.1$ between the support and the blocks, determine (a) the acceleration of each block, (b) the tension in the cable.

12.22 through 12.24 In the system shown, the cylinder B moves with $\mathbf{a}_B = 4$ ft/s^2 ↓. If the cylinder A weighs 161 lb, determine (a) the weight of the cylinder B, (b) the tension in each cable.

Cable 1

Cable 2

Fig. P12.22 and P12.25

Cable 2

Cable 1

Fig. P12.23* and P12.26

Cable 1

Cable 2

Fig. P12.24 and P12.27*

12.25 through 12.27* In the system shown, the cylinder A moves with $\mathbf{a}_A = 0.6$ m/s^2 ↑. If the mass of the cylinder B is 60 kg, determine (a) the mass of the cylinder A, (b) the tension in each cable.

12.28 and 12.29 The system shown is released from rest. Determine the acceleration of each cylinder and the tension in each cable.[†]

Cable 1

Cable 2

80 kg
B 40 kg
60 kg

Fig. P12.28

Cable 1

Cable 3

A 150 lb

B 400 lb

C 300 lb

D 100 lb

E 250 lb

F 200 lb

Cable 2

Fig. P12.29

12.30* A river boat of mass m is anchored as shown where the water flows with a constant velocity $\mathbf{v}$. The horizontal component of the force exerted on the boat by the anchor chain is $\mathbf{P}$. Suppose that the friction force between the boat and the water is proportional to the velocity of the boat relative to the water. Determine the time required for the boat to attain a speed of $v/3$ if the anchor chain suddenly breaks. (*Hint*. Use the frame fixed to the flowing water as an inertial reference frame.)

Fig. P12.30*

12.31* A steel ball of mass m is released from rest at the surface O of a tank of viscous fluid as shown. During sinking, the ball experiences an upward viscous friction force of magnitude cv, where c is a constant and v is the speed of the ball. Determine the position coordinate x of the ball in terms of m, c, the gravitational acceleration g, and the time t.

Fig. P12.31*

[†]Cf. App. D for computer solution of resulting simultaneous equations.

12.5 Curvilinear Motion of a Particle

We saw in Chap. 11 that the curvilinear motion of a particle may be described by using one of the several sets of components, such as rectangular components, tangential and normal components, radial and transverse components, or cylindrical components. The choice of a particular set of components usually depends on the given conditions in the problem and is one of the basic decisions to be made at the beginning of the solution of a curvilinear-motion problem. Once the choice is made, the *method of force and acceleration* may then be applied to establish the equations necessary to yield the desired solution.

Resolving the forces acting on the particle and the acceleration of the particle into appropriate sets of components, as illustrated in Figs. 12.5 through 12.8, we may write the scalar versions of Eq. (12.1) for the motion of the particle with mass m as follows:

(1) Using *rectangular components* (cf. Fig. 12.5 and Sec. 11.11), we have

$$\Sigma F_x = ma_x \qquad \Sigma F_y = ma_y \qquad \Sigma F_z = ma_z \qquad (12.15)$$

where $a_x = \ddot{x}$, $a_y = \ddot{y}$, and $a_z = \ddot{z}$.

Fig. 12.5

(2) Using *tangential and normal components* (cf. Fig. 12.6 and Sec. 11.13), we have

$$\Sigma F_t = ma_t \qquad \Sigma F_n = ma_n \qquad (12.16)$$

where $a_t = \dot{v}$ and $a_n = v^2/\rho$.

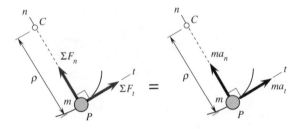

Fig. 12.6

(3) Using *radial and transverse components* (cf. Fig. 12.7 and Sec. 11.14), we have

$$\Sigma F_r = ma_r \qquad \Sigma F_\theta = ma_\theta \qquad (12.17)$$

where $a_r = \ddot{r} - r\dot{\theta}^2$ and $a_\theta = r\ddot{\theta} + 2\dot{r}\dot{\theta}$.

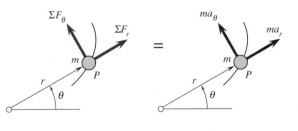

Fig. 12.7

(4) Using *cylindrical components* (cf. Fig. 12.8 and Sec. 11.15), we have

$$\Sigma F_R = ma_R \qquad \Sigma F_\theta = ma_\theta \qquad \Sigma F_z = ma_z \qquad (12.18)$$

where $a_R = \ddot{R} - R\dot{\theta}^2$, $a_\theta = R\ddot{\theta} + 2\dot{R}\dot{\theta}$, and $a_z = \ddot{z}$.

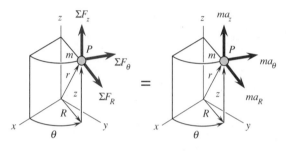

Fig. 12.8

Equations (12.15) through (12.18) are sets of *scalar equations of motion* of a particle in the chosen coordinate systems.

EXAMPLE 12.4

The resultant external force acting on a 2-kg particle in space is $\mathbf{F} = 12t\mathbf{i} - 24t^2\mathbf{j} - 40t^3\mathbf{k}$ N, where t is the time measured in seconds. The particle is at rest at the origin when $t = 0$. Determine the acceleration component $\mathbf{a}_y$, the velocity component $\mathbf{v}_y$, and the coordinate y of the particle when $t = 4$ s.

Solution. The y component of the resultant external force is $F_y = -24t^2$ N, and the mass of the particle is $m = 2$ kg. Referring to the diagrams shown, we write

$$+\uparrow \Sigma V_y: \qquad -24t^2 = 2a_y \qquad a_y = -12t^2$$

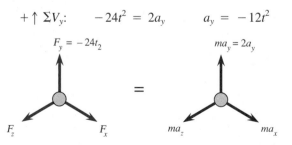

The initial conditions are $v_y = y = 0$ at $t = 0$. Thus, we write

$$\int_0^{v_y} dv_y = \int_0^t a_y \, dt \qquad v_y = \int_0^t (-12t^2) \, dt = -4t^3$$

$$\int_0^y dy = \int_0^t v_y \, dt \qquad y = \int_0^t (-4t^3) \, dt = -t^4$$

At $t = 4$ s, we obtain

$$a_y = -12t^2 = -12(4)^2 \qquad \mathbf{a}_y = -192\mathbf{j} \text{ m/s}^2 \blacktriangleleft$$

$$v_y = -4t^3 = -4(4)^3 \qquad \mathbf{v}_y = -256\mathbf{j} \text{ m/s} \blacktriangleleft$$

$$y = -t^4 = -(4)^4 \qquad y = -256 \text{ m} \blacktriangleleft$$

Developmental Exercises

D12.9 Refer to Example 12.4. Determine for the particle at $t = 4$ s (a) $\mathbf{a}_x$, (b) $\mathbf{v}_x$, (c) x.

D12.10 Refer to Example 12.4. Determine for the particle at $t = 4$ s (a) $\mathbf{a}_z$, (b) $\mathbf{v}_z$, (c) z.

EXAMPLE 12.5

A car traveling at the *rated speed* of a road should have no lateral friction force developed between its tires and the road surface. If a highway curve has a radius of curvature $\rho = 200$ m, a rated speed $v = 90$ km/h, and a banking angle θ as shown, determine the value of θ.

Solution. The rated speed is $v = 90$ km/h $= 25$ m/s. The normal component of acceleration of the car is $a_n = v^2/\rho = (25)^2/200$ m/s². Thus, the

free-body diagram of the car is equated to its effective-force diagram as shown. We write

$$+\uparrow \Sigma V_y: \qquad N\cos\theta - W = 0$$

$$\underset{\leftarrow}{+} \Sigma V_n: \qquad N\sin\theta = \frac{W}{g}\frac{(25)^2}{200}$$

These two equations yield

$$\tan\theta = \frac{(25)^2}{200g} = \frac{(25)^2}{200(9.81)} \qquad \theta = 17.67° \blacktriangleleft$$

Developmental Exercise

EXAMPLE 12.6

A sphere at A is made to revolve in a horizontal circle at a constant speed v by two wires AB and AC, which are fastened to a vertical shaft as shown. Determine the minimum value of the speed v for which both wires are taut.

Solution. Let the center of the horizontal circle be at the point D as shown. The radius r of the circle can be determined as follows:

$$\frac{12}{5}r - \frac{3}{4}r = 3.3$$

Thus, $r = 2$ m. Since the sphere revolves in the horizontal circle with a constant speed, its acceleration components are $a_t = 0$, $a_n = v^2/r = v^2/2$.

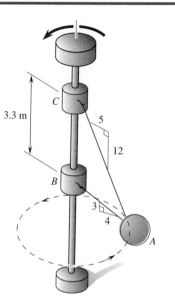

Thus, $a = a_n = v^2/2$. When the speed v takes on a minimum constant value and both wires are still taut, slackening of the wire AB impends and the tension T_{AB} becomes zero. The free-body diagram of the sphere is therefore equated to its effective-force diagram as shown. We write

$$+\uparrow \Sigma V_y: \qquad \frac{12}{13}T_{AC} - W = 0$$

$$\xleftarrow{+} \Sigma V_n: \qquad \frac{5}{13}T_{AC} = \frac{W}{g}\left(\frac{v^2}{2}\right)$$

These two equations yield

$$v^2 = \frac{5g}{6} = \frac{5(9.81)}{6} \qquad v = 2.86 \text{ m/s} \blacktriangleleft$$

Developmental Exercise

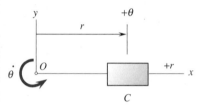

EXAMPLE 12.7

The vertical tube A and the horizontal tube B are connected as shown. The tubes are rotated about the z axis at a constant rate of $\dot{\theta} = 2.5$ rad/s. The radial position r of the 2-kg plug C inside the tube B is controlled by the rope COD which passes around the pulley at O mounted inside the tube A. At the instant shown, $r = 4$ m and the end D is pulled by $\mathbf{P}$ with $\mathbf{v}_D = 3$ m/s $\uparrow$ and $\mathbf{a}_D = 5$ m/s² $\downarrow$. For this instant, determine the tension in the rope and the components of the force $\mathbf{F}$ exerted by the tube B on the plug C if effects of friction are negligible.

Solution. The acceleration of the plug C, which moves in the horizontal plane, can conveniently be determined by using radial and transverse components. At the instant shown, we have from the given data that $r = 4$ m, $\dot{r} = -3$ m/s, $\ddot{r} = 5$ m/s², $\dot{\theta} = 2.5$ rad/s, and $\ddot{\theta} = 0$. Thus,

$$a_r = \ddot{r} - r\dot{\theta}^2 = -20 \text{ m/s}^2 \qquad a_\theta = r\ddot{\theta} + 2\dot{r}\dot{\theta} = -15 \text{ m/s}^2$$

The negative signs in the values of a_r and a_θ indicate that the actual senses of a_r and a_θ are opposite to the positive senses of r and θ, respectively.

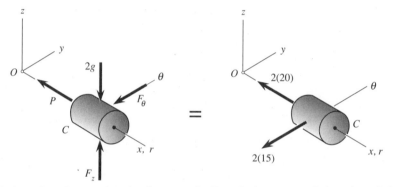

Noting that the tension in the rope is P and the mass of the plug C is $m = 2$ kg, we equate the free-body diagram of the plug C to its effective-force diagram as shown. We write

$$+\searrow \Sigma V_r: \qquad -P = -2(20) \qquad\qquad P = 40 \text{ N} \blacktriangleleft$$

$$+\nearrow \Sigma V_\theta: \qquad -F_\theta = -2(15) \qquad\qquad \mathbf{F}_\theta = -30\mathbf{e}_\theta \text{ N} \blacktriangleleft$$

$$+\uparrow \Sigma V_z: \qquad F_z - 2g = 0 \qquad\qquad \mathbf{F}_z = 19.62\mathbf{k} \text{ N} \blacktriangleleft$$

Developmental Exercise

D12.13 Refer to Example 12.7. (a) Is the z component of the acceleration of the plug C equal to 9.81 m/s² $\downarrow$ or $\mathbf{0}$? (b) What is the z component of the resultant force acting on the plug C?

EXAMPLE 12.8

The path of a 2-kg particle in space is known to be a helix defined by the cylindrical coordinates: $R = 40$, $\theta = (\pi/4)t$, and $z = (9\pi/4)t$, where R

and z are measured in meters, θ is the angle in radians, and t is the time in seconds. Determine the resultant external force $\mathbf{F}$ acting on the particle.

Solution. From the given cylindrical coordinates of the particle, we have

$$R = 40 \qquad \dot{R} = 0 \qquad \ddot{R} = 0$$

$$\theta = \frac{\pi}{4} t \qquad \dot{\theta} = \frac{\pi}{4} \qquad \ddot{\theta} = 0$$

$$z = \frac{9\pi}{4} t \qquad \dot{z} = \frac{9\pi}{4} \qquad \ddot{z} = 0$$

The acceleration $\mathbf{a}$ of the particle is

$$\mathbf{a} = (\ddot{R} - R\dot{\theta}^2)\mathbf{e}_R + (R\ddot{\theta} + 2\dot{R}\dot{\theta})\mathbf{e}_\theta + \ddot{z}\mathbf{e}_z$$

$$= -\frac{5\pi^2}{2}\,\mathbf{e}_R$$

The mass of the particle is $m = 2$ kg. By Newton's second law, we write

$$\mathbf{F} = m\mathbf{a} = 2\left(-\frac{5\pi^2}{2}\,\mathbf{e}_R\right) \qquad \mathbf{F} = -49.3\mathbf{e}_R \text{ N} \quad \blacktriangleleft$$

Developmental Exercise

D12.14   Refer to Example 12.8. Determine (a) the expression of the tangential unit vector $\mathbf{e}_t$ in terms of the transverse unit vector $\mathbf{e}_\theta$ and the axial unit vector $\mathbf{e}_z$, (b) the angle between the velocity vector $\mathbf{v}$ and the xy plane, (c) the radius of curvature ρ of the helical path.

12.6   Systems of Particles

The forces acting on the particle P_i of a system of n particles are illustrated in Fig. 12.9(a), where $\mathbf{F}_i$ is the resultant of the external forces exerted on P_i, and $\mathbf{f}_{i1}, \ldots, \mathbf{f}_{in}$ are the internal forces exerted on P_i by other particles

(a)

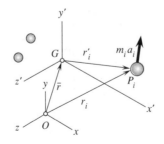

(b)

Fig. 12.9

$P_1, \ldots, P_n$ of the system (where $\mathbf{f}_{ii}$ has no meaning and is set equal to zero). According to Newton's second law, the *resultant* of the external and internal forces acting on the particle P_i is equal to the *effective force* $m_i\mathbf{a}_i$ on that particle, which is illustrated in Fig. 12.9(b). Thus, we write

$$\mathbf{F}_i + \sum_{j=1}^{n} \mathbf{f}_{ij} = m_i\mathbf{a}_i \tag{12.19}$$

Noting that the force vectors in Fig. 12.9(a) are equivalent to the effective force in Fig. 12.9(b), we may take moments about the mass center G of the system of particles and also write

$$\mathbf{r}_i' \times \mathbf{F}_i + \sum_{j=1}^{n} (\mathbf{r}_i' \times \mathbf{f}_{ij}) = \mathbf{r}_i' \times m_i\mathbf{a}_i \tag{12.20}$$

where $\mathbf{r}_i'$ is the position vector of the particle P_i in the nonrotating reference frame $Gx'y'z'$. Since the equivalence relation established in Fig. 12.9 holds true for any particle, it follows that the system of all the external and internal forces acting on the various particles is *equivalent* to the system of effective forces on the particles, as illustrated in Fig. 12.10.

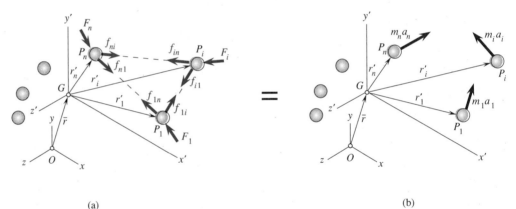

(a)   (b)

Fig. 12.10

By Newton's third law, the internal forces $\mathbf{f}_{ij}$ must occur in collinear pairs which are equal in magnitude and opposite in direction. Thus, adding all the internal forces of the system and summing their moments about the mass center G, respectively, we write

$$\sum_{i=1}^{n}\sum_{j=1}^{n} \mathbf{f}_{ij} = \mathbf{0} \qquad \sum_{i=1}^{n}\sum_{j=1}^{n} (\mathbf{r}_i' \times \mathbf{f}_{ij}) = \mathbf{0} \tag{12.21}$$

We see from Eqs. (12.21) that the sum of the internal forces themselves and the sum of their moments about a given point, such as G, are zero.

Letting $i = 1, 2, \ldots, n$ in Eq. (12.19), we obtain a set of n equations. Adding these n equations member by member, using the first of Eqs. (12.21), and denoting the sum of the external forces by $\Sigma\mathbf{F}$, we get

$$\Sigma\mathbf{F} = \sum_{i=1}^{n} m_i\mathbf{a}_i \tag{12.22}$$

Proceeding similarly with Eq. (12.20) and using the second of Eqs. (12.21), we readily find that the sum of moments about G of the external forces is

$$\Sigma \mathbf{M}_G = \sum_{i=1}^{n} (\mathbf{r}'_i \times m_i \mathbf{a}_i) \qquad (12.23)$$

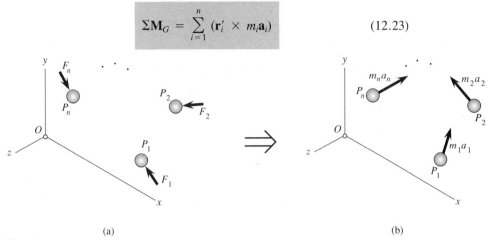

(a)   (b)

Fig. 12.11

Furthermore, it can similarly be shown that the *system of the external forces* $\mathbf{F}_i$ and the *system of effective forces* $m_i \mathbf{a}_i$, as illustrated in Fig. 12.11(a) and (b), have the same resultant force and the same resultant moment about *any* fixed point. In particular, we can readily show that the sum of moments about the fixed origin O of the external forces is

$$\Sigma \mathbf{M}_O = \sum_{i=1}^{n} (\mathbf{r}_i \times m_i \mathbf{a}_i) \qquad (12.24)$$

However, we observe here that the external force $\mathbf{F}_i$ itself acting on a given particle P_i and the effective force $m_i \mathbf{a}_i$ on the same particle P_i are generally *not* equal. Therefore, with internal forces excluded, the two systems of vectors in parts (a) and (b) of Fig. 12.11 do *not* have the same physical effect on the individual particles; these two systems are *equipollent*, not equivalent (cf. Sec. 5.3).

Note in Fig. 12.11 that the sign $\Rightarrow$ denotes *equipollence* between the system of vectors on its left side and the system of vectors on its right side. Although two equipollent systems of forces acting on a rigid body are also *equivalent*, two equipollent systems of forces acting on a set of independent particles are not necessarily equivalent.

Developmental Exercise

D12.15   Is the system of *external forces* acting on a system of particles *equivalent* to the system of *effective forces* on the particles in the system?

12.7   Principle of Motion of the Mass Center

For a system of n particles with masses $m_1, m_2, \ldots, m_n$ and position vectors $\mathbf{r}_1, \mathbf{r}_2, \ldots, \mathbf{r}_n$ as illustrated in Fig. 12.12(a), we may determine

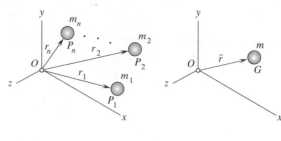

Fig. 12.12 (a) (b)

the total mass m of the system and the position vector $\bar{\mathbf{r}}$ of the center of mass G of the system, as illustrated in Fig. 12.12(b), by applying the *principle of moments* as follows:

$$m = \sum_{i=1}^{n} m_i \qquad m\bar{\mathbf{r}} = \sum_{i=1}^{n} m_i \mathbf{r}_i \tag{12.25}$$

By taking the time derivatives of the second of Eqs. (12.25), we have

$$m\dot{\bar{\mathbf{r}}} = \sum_{i=1}^{n} m_i \dot{\mathbf{r}}_i \qquad m\ddot{\bar{\mathbf{r}}} = \sum_{i=1}^{n} m_i \ddot{\mathbf{r}}_i$$

Denoting the velocities, $\dot{\bar{\mathbf{r}}}$, $\dot{\mathbf{r}}_i$, and accelerations, $\ddot{\bar{\mathbf{r}}}$, $\ddot{\mathbf{r}}_i$, by $\bar{\mathbf{v}}$, $\mathbf{v}_i$, $\bar{\mathbf{a}}$, and $\mathbf{a}$, $(i = 1, 2, \ldots, n)$, respectively, we write

$$m\bar{\mathbf{v}} = \sum_{i=1}^{n} m_i \mathbf{v}_i \qquad m\bar{\mathbf{a}} = \sum_{i=1}^{n} m_i \mathbf{a}_i \tag{12.26}$$

where $\bar{\mathbf{v}}$ and $\bar{\mathbf{a}}$ are the velocity and acceleration of the mass center G, respectively. Thus, from Eq. (12.22) and the second of Eqs. (12.26), we have

$$\Sigma\mathbf{F} = m\bar{\mathbf{a}} \tag{12.27}$$

Equation (12.27) expresses an important property that *the resultant force of all the external forces acting on a system of particles is equal to the total mass of the system times the acceleration of the mass center of the system.* This property, a corollary to Newton's second and third laws, is called the *principle of motion of the mass center* of a system of particles.[†] In other words, the mass center of a system of particles is a point which moves as if the entire mass of the system and all the external forces were concentrated at that point. Since a rigid body may be considered as a system of particles, the principle of motion of the mass center is, of course, applicable to a rigid body.

By Eqs. (12.22) and (12.27), we know that the resultant of the system of external forces in Fig. 12.11(a) and the resultant of the system of effec-

[†]Under the action of internal forces only (i.e., $\mathbf{F}_i = \mathbf{0}$), this principle degenerates into *corollary IV* to Newton's laws of motion (cf. pp. 19–20 of Newton's *Principia* cited in the footnote on p. 507). In some mechanics literature, this principle, or its equivalent, is also called the *generalized Newton's second law* of motion for a mass system, or *Euler's first law*, after Leonhard Euler (1707–1783). For the latter, see C. Truesdell, *Essays in the History of Mechanics* (New York: Springer-Verlag, 1968).

tive forces in Fig. 12.11(b) are both identically equal to the *resultant effective force m**ā***, in magnitude, direction, and line of action, as illustrated in Fig. 12.13. However, the derivation of Eq. (12.27) does not involve the moments of the external forces or the effective forces which are generally nonconcurrent. Thus, the resultant of the external forces, as well as the resultant of the effective forces, *does not*, in general, pass through the mass center G of the system of particles on which they act.

Fig. 12.13

EXAMPLE 12.9

A system consists of three particles P_1, P_2, and P_3. At the instant shown, the particles are acted on by the external forces $F_1 = 10i - 15j$ lb, $F_2 = -13j + 20k$ lb, $F_3 = 18i - 12j + 5k$ lb, and have velocities $v_1 = 5i + 8j$ ft/s, $v_2 = 3i - 6k$ ft/s, $v_3 = 2j - 2k$ ft/s, respectively. For the mass center of the system at the instant shown, determine the analytical expressions of (a) its position vector $\bar{r}$, (b) its velocity $\bar{v}$, (c) its acceleration $\bar{a}$.

Solution. Applying the *principle of moments*, in the form of Eqs. (12.25), we write

$$m = \frac{15}{g} + \frac{13}{g} + \frac{12}{g} = \frac{40}{g}$$

$$\frac{40}{g}\bar{r} = \frac{15}{g}(-5i) + \frac{13}{g}(10i + 8k) + \frac{12}{g}(10i + 6j - 4k)$$

$$\bar{r} = 1.875i + 1.8j + 1.4k \text{ ft} \quad \blacktriangleleft$$

Applying the *principle of moments*, as in the first of Eqs. (12.26), and the *principle of motion of the mass center*, as in Eq. (12.27), we write

$$\frac{40}{g}\bar{v} = \frac{15}{g}(5i + 8j) + \frac{13}{g}(3i - 6k) + \frac{12}{g}(3j - 2k)$$

$$\bar{v} = 2.85i + 3.9j - 2.55k \text{ ft/s} \quad \blacktriangleleft$$

$$(10i - 15j) + (-13j + 20k) + (18i - 12j + 5k) = \frac{40}{g}\bar{a}$$

$$\bar{a} = (0.7i - j + 0.625k)g$$

$$\bar{a} = 22.5i - 32.2j + 20.1k \text{ ft/s}^2 \quad \blacktriangleleft$$

Developmental Exercises

D12.16   Refer to Example 12.9. For the instant shown, determine (a) the moment M_G of the resultant effective force $m\mathbf{a}$ about the mass center G of the system, (b) whether $m\mathbf{a}$ passes through G.

D12.17   Describe the *principle of motion of the mass center*.

PROBLEMS

12.32 and 12.33* The resultant external force acting on the particle P in space is $\mathbf{F} = 240t^2\mathbf{i} - 400t^3\mathbf{j} + 300t^4\mathbf{k}$ N, where t is the time in seconds. The particle is at rest at the position shown when $t = 0$. Determine the position coordinates (x, y, z) of the particle when $t = 3$ s.

Fig. P12.32   **Fig. P12.33***   **Fig. P12.34**

12.34 The resultant external force acting on the particle P in space is $\mathbf{F} = 180t\mathbf{i} + 60t^3\mathbf{j} - 90t^4\mathbf{k}$ lb. When $t = 0$, the particle is situated at the position shown and has an initial velocity $\mathbf{v}_0 = 200\mathbf{j} + 500\mathbf{k}$ ft/s. Determine the position vector $\mathbf{r}$ of the particle when $t = 4$ s.

12.35* A 2-kg particle moves along the curve $x = 4 \sin \pi t$, $y = 4 \cos \pi t$, and $z = 0$, where x and y are measured in meters and t is the time in seconds. Determine the magnitude of the resultant force $\mathbf{R}$ acting on the particle and show that $\mathbf{R}$ is always directed toward the origin O.

Fig. P12.36

X **12.36** A pendulum composed of a thin wire and a bob of mass 5 kg swings in a vertical plane. If the bob has a speed of 6 m/s in the position shown, determine the tension in the wire.

12.37 A small bob of weight 10 lb is revolving in a horizontal circle at a constant speed as shown. Determine (a) the tension in the wire AB, (b) the speed of the bob.

Fig. P12.37   **Fig. P12.38* and P12.39**

12.38* A small bob supported by two wires is revolving in a horizontal circle at a constant speed as shown. Determine the range of values of the speed of the bob for which no wire becomes slack.

X **12.39** A small bob supported by two wires is revolving in a horizontal circle at a constant speed of 9 m/s as shown. Knowing that the mass of the bob is 5 kg, determine the tension in each wire.

Fig. P12.40

12.40 Four cars are all traveling at a constant speed of 72 km/h along a road as shown. If $\mu_k = 0.5$ between all tires and the road and the brakes of each car are suddenly applied and skidding occurs at the respective positions shown, determine the ratios k_A, k_B, k_C, and k_D of the magnitudes of the tangential decelerations of the cars at A, B, C, and D to the magnitude g of the gravitational acceleration, respectively.

Fig. P12.41*   **Fig. P12.42**

12.41* and 12.42 A small sphere is suspended by two wires as shown. Determine (a) the tension in the wire AB before the wire CD is cut, (b) the tension in the wire AB and the acceleration of the sphere immediately after the wire CD is cut.

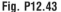12.43 A jet airplane for the training of astronauts flies at high altitude with a constant speed of 600 mi/h, where the gravitational acceleration is 32.1 ft/s². To simulate the zero-contact-force (i.e., apparent weightlessness) condition in the cabin of the airplane for a brief period of time, the pilot maneuvers to fly along a vertical curve as shown. Determine for this brief period (a) the radius of curvature of the path of flight, (b) the rate of change of the orientation of the airplane in degrees per second.

Fig. P12.43

12.44 A small plane flies along a vertical circular path of 150-m radius as shown. As the plane passes through the bottom point B, the pilot experiences an apparent weight equal to three times his actual weight. Determine the speed v_B of the plane as it passes through B. (*Hint.* The pilot's apparent weight is equal to the reaction from his seat.)

Fig. P12.44

12.45* Determine the minimum value of μ_s between the chip P and the vertical wall which will keep the chip from sliding down when the cylindrical vessel is rotating at 1 rev/s.

12.46 A slider S can move freely along a circular ring which is rotating at a constant rate $\dot{\theta}_0$ about the vertical axis OA as shown. If the angle β between the vertical axis and OS is 60°, determine the value of $\dot{\theta}_0$ in revolutions per second.

Fig. P12.45*   **Fig. P12.46**   **Fig. P12.47**

12.47 A conveyor belt moves a series of small objects over an idler of radius 9 in. as shown. If the belt moves at a constant speed of 2 ft/s and no objects slip with respect to the belt, determine the minimum value of μ_s between the objects and the belt.

12.48* A racetrack turn of radius 150 m is banked at an angle as shown. If $\mu_s = 0.8$ between the tires and the pavement, determine the range of values of the speed v at which the race car will not skid sideways when rounding the turn.

12.49 A car rounds a flat unbanked turn of radius 100 m at 72 km/h. Determine the minimum value of μ_s between the tires and the pavement if no sideway skidding occurs.

12.50 A rocket moving in a vertical plane is being propelled by a thrust $\mathbf{T}$ of 10 kips and is experiencing an air resistance $\mathbf{F}$ of 1.9 kips as shown. At this instant, the rocket has a mass m of 150 slugs and a velocity $\mathbf{v}$ of 2 mi/s. If the gravitational acceleration at this altitude is $\mathbf{g} = 28.6$ ft/s² ↓, determine the radius of curvature ρ of the path and the magnitude a_t of the tangential component of acceleration of the rocket at the instant shown. [*Hint.* $m(d\mathbf{v}/dt) = \mathbf{F} + m\mathbf{g} + \mathbf{T}$.]

Fig. P12.48*

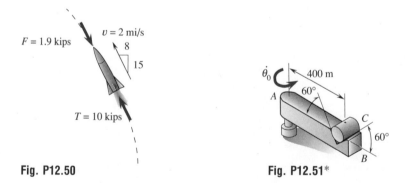

Fig. P12.50   **Fig. P12.51***

12.51* The horizontal arm AB is rotating with a constant angular speed $\dot{\theta}_0$ as shown. If the cylinder C is to remain in the V-shaped groove, determine the maximum allowable value of $\dot{\theta}_0$ in revolutions per second.

12.52 A 5-kg collar C mounted on a horizontal arm AB is connected by a spring to the end A of the arm as shown. The collar is oscillating along the arm while the arm is rotating with a constant angular speed $\dot{\theta}_0 = \frac{1}{2}$ rev/s. At this instant, $r = 1$ m, $\dot{r} = 0.8$ m/s, and $\ddot{r} = -1.2$ m/s^2. If $\mu_k = 0.35$ between the collar and the rod, determine the friction force $\mathbf{F}_f$ and the spring force $\mathbf{F}_s$ acting on the collar at the instant shown.

12.53* Solve Prob. 12.52 if $\dot{\theta} = 1$ rev/s.

12.54 A collar C of mass $m_C = 4$ kg, which is mounted on the frictionless arm AB, is pulled inward by a cord that passes through an eyebolt at the end A as shown. At the instant shown, the arm AB is rotating in a horizontal plane with $\dot{\theta} = 2$ rad/s and $\ddot{\theta} = 3$ rad/s^2, while the end D of the cord is pulled upward with $\dot{y} = 1$ m/s and $\ddot{y} = 2.5$ m/s^2. If the tension in the cord is $T = 34$ N at this instant, determine (a) the radial distance r, (b) the magnitude of the horizontal force $\mathbf{Q}$ acting on C by the arm AB.

12.55* Solve Prob. 12.54 if $m_C = 16.1$ lbm, $\dot{\theta} = 2.5$ rad/s, $\ddot{\theta} = 2$ rad/s^2, $\dot{y} = 3$ ft/s, $\ddot{y} = 5$ ft/s^2, and $T = 15$ lb.

Fig. P12.52

Fig. P12.54   **Fig. P12.56**

12.56 A collar C of weight $W_C = 8.05$ lb, which is mounted on the frictionless arm AB, is pulled inward by a cord that wraps around a fixed drum of radius $b = 3$ in. as the arm AB rotates in a horizontal plane as shown, where $\dot{\theta} = 2$ rad/s and $\ddot{\theta} = 4$ rad/s^2. If $r = 3$ ft at this instant, determine (a) the tension T in the cord, (b) the magnitude of the horizontal force $\mathbf{Q}$ acting on C by the arm AB. (*Hint.* The motion of the collar C relative to the arm AB is given by $\dot{r} = -b\dot{\theta}$ and $\ddot{r} = -b\ddot{\theta}$.)

12.57* Solve Prob. 12.56 if $W_C = 19.62$ N, $b = 80$ mm, $\dot{\theta} = 2.5$ rad/s, $\ddot{\theta} = 5$ rad/s^2, and $r = 1.6$ m.

12.58 A collar C of mass $m_C = 5$ kg is mounted on the spring BC and oscillates along the frictionless rod AB as shown. The rod AB is connected to a horizontal turntable and a vertical shaft OD. The turntable rotates with a constant speed $\dot{\theta}_0 = \pi$ rad/s about the axis of the shaft. At the instant shown, $b = 0.8$ m and the collar C is moving downward with a velocity $\mathbf{v}_{C/AB} = 2$ m/s $\searize$ and an acceleration $\mathbf{a}_{C/AB} = 3$ m/s^2 $\searize$ relative to the rod AB. Determine for this instant (a) the resultant force $\mathbf{F}$ acting on the collar C, (b) the tensile or compressive force $\mathbf{F}_s$ developed in the spring. (*Hint.* Use cylindrical coordinates and note for the instant shown that

$$\mathbf{e}_R = \mathbf{j} \qquad \mathbf{e}_\theta = -\mathbf{i} \qquad \mathbf{e}_z = \mathbf{k}$$

The magnitude of $\mathbf{F}_s$ may be determined by considering $\mathbf{F} \cdot \boldsymbol{\lambda}_{BA}$, where $\boldsymbol{\lambda}_{BA}$ is the unit vector in the direction of $\overrightarrow{BA}$.)

12.59 Solve Prob. 12.58 if $m_C = 48.3$ lbm, $\dot{\theta}_0 = 2\pi$ rad/s, $\mathbf{v}_{C/AB} = 3$ ft/s $\searize$, $\mathbf{a}_{C/AB} = 4$ ft/s^2 $\searize$, and $b = 2$ ft.

12.60* Solve Prob. 12.58 if friction exists between the collar C and the rod AB, where $\mu_k = 0.3$.

Fig. P12.58

Fig. P12.61

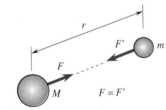

Fig. P12.64*

12.61 A system in space consists of three particles P_1, P_2, and P_3. At the instant shown, the particles are acted on by the external forces $\mathbf{F}_1 = -60\mathbf{j} + 90\mathbf{k}$ N, $\mathbf{F}_2 = -50\mathbf{i} - 100\mathbf{j} - 40\mathbf{k}$ N, $\mathbf{F}_3 = 30\mathbf{i} - 40\mathbf{j}$ N, and have velocities $\mathbf{v}_1 = 4\mathbf{i} - 6\mathbf{j}$ m/s, $\mathbf{v}_2 = 5\mathbf{j} - 3\mathbf{k}$ m/s, $\mathbf{v}_3 = 2\mathbf{i} - 7\mathbf{k}$ m/s, respectively. For the instant shown, determine for the mass center of the system (a) its position vector $\bar{\mathbf{r}}$, (b) its velocity $\bar{\mathbf{v}}$, (c) its acceleration $\bar{\mathbf{a}}$.

12.62* Refer to Prob. 12.61. For the instant shown, determine the analytical expressions of (a) the resultant effective force $m\bar{\mathbf{a}}$ of the system of particles, (b) the moment $\mathbf{M}_G$ of $m\bar{\mathbf{a}}$ about the mass center G of the system.

12.63 Refer to Prob. 12.61. For the instant shown, determine the accelerations $\mathbf{a}_{1/G}$, $\mathbf{a}_{2/G}$, and $\mathbf{a}_{3/G}$ of the particles P_1, P_2, and P_3 relative to the mass center G, respectively.

12.64* A system in space consists of three particles P_1, P_2, and P_3. At the instant shown, the particles are acted on by the external forces $\mathbf{F}_1 = 150\mathbf{i} - 40\mathbf{j}$ lb, $\mathbf{F}_2 = 132\mathbf{i} - 160\mathbf{j} + 240\mathbf{k}$ lb, $\mathbf{F}_3 = 280\mathbf{k}$ lb, respectively. For the mass center of the system at the instant shown, determine the analytical expressions of (a) its position vector $\bar{\mathbf{r}}$, (b) its acceleration $\bar{\mathbf{a}}$.

12.65 Refer to Prob. 12.64*. For the instant shown, determine the accelerations $\mathbf{a}_{1/G}$, $\mathbf{a}_{2/G}$, and $\mathbf{a}_{3/G}$ of the particles P_1, P_2, and P_3 relative to the mass center G, respectively.

CENTRAL-FORCE MOTION

12.8 Gravitational Force

A *gravitational force* is a naturally occurring force of mutual attraction between any two masses in the universe. Such a force is described in *Newton's law of gravitation*, which states that two particles are mutually attracted by a pair of opposite forces that are directed along the line joining the particles and have the same magnitude which is directly proportional to the product of their masses and inversely proportional to the square of the distance between them.[†] For two particles with masses M and m separated by a distance r, as shown in Fig. 12.14, the magnitude of the attractive force may be written as

Fig. 12.14

$$F = G\,\frac{Mm}{r^2} \tag{12.28}$$

where G is the *constant of gravitation*. Experiments show that

$$G = 66.726 \times 10^{-12}\,\frac{\text{m}^3}{\text{kg} \cdot \text{s}^2} \tag{12.29}^{\dagger\dagger}$$

[†]The statement of this law is a synthesis from *Book III, The System of the World* in Newton's *Principia* as cited in the footnote on p. 507. In certain astronomical studies, this law is replaced by the law of gravitation as covered in the *general* theory of relativity (1915) of Albert Einstein (1879–1955). (Cf. Sec. 1.14.)

[††]The accepted value of G as of 1982 is $G = (66.726 \pm 0.005) \times 10^{-12}$ m³/(kg · s²).

Developmental Exercise

D12.18 Using Eqs. (12.10), (12.11), (12.28), and (12.29), *estimate* the mass of the earth whose mean radius is 6.37 megameters.

12.9 Motion Under a Central Force

A *central force* is a force in a central force field, which is a region in space where a body is always acted on by a force directed toward or away from a common fixed point, which is called the *center of force*. A *central-force motion* is the motion of a particle under a central force, such as that shown in Fig. 12.15(a). A *gravitational central-force motion* is a central-force motion where the central force is a gravitational force. The motion of the planets around the sun and the motion of spacecraft and satellites around the earth can be studied as gravitational central-force motions. An *elastic central-force motion* is one where the central force is an elastic spring force. An example of elastic central-force motion is found in the motion of a particle attached to a stretched elastic cord on a frictionless horizontal surface, where the other end of the cord is connected to a fixed point.

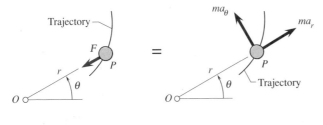

Fig. 12.15   (a)   (b)

For studying a gravitational or elastic central-force motion of a particle P with mass m, we may select the center of force as the origin O and use radial and transverse components to describe the motion as shown in Fig. 12.15. In this case, the transverse component $\mathbf{F}_\theta$ of the central force $\mathbf{F}$ is always zero. Referring to Fig. 12.15, we write

$$+\nwarrow \ \Sigma V_\theta: \qquad 0 = ma_\theta \qquad a_\theta = r\ddot{\theta} + 2\dot{r}\dot{\theta} = \frac{1}{r}\frac{d}{dt}(r^2\dot{\theta}) = 0$$

$$r^2\dot{\theta} = h \tag{12.30}$$

where $m \neq 0$, $r \neq 0$, and h is a constant. Thus, for any given central-force motion, $\dot{\theta}$ must decrease when r is increased, and vice versa. However, h has a different constant value for each different trajectory.[†] Since $v_\theta = r\dot{\theta}$, we may write the equivalent version of Eq. (12.30) as

$$rv_\theta = h \tag{12.31}$$

which is an important equation in the study of central-force motion.

[†]The trajectory of a spacecraft in free flight (i.e., coasting) becomes a different one whenever a noncentral force (e.g., thrust of a rocket) momentarily acts on the ''coasting'' spacecraft.

Fig. 12.16

Either of Eqs. (12.30) or (12.31) lends itself to an interesting geometric interpretation. As shown in Fig. 12.16, the radius vector $\mathbf{r}$ of a particle P moving under a central force $\mathbf{F}$ sweeps an infinitesimal area $dA = \frac{1}{2}r^2\,d\theta = \frac{1}{2}rv_\theta\,dt$ as it rotates about the center of force O through an angle $d\theta$ during the time interval dt. Thus, the time rate at which area is swept by $\mathbf{r}$ is

$$\dot{A} = \frac{dA}{dt} = \frac{1}{2}\,r^2\,\frac{d\theta}{dt} = \frac{1}{2}\,r^2\dot{\theta} = \frac{1}{2}\,rv_\theta$$

By Eq. (12.31), we find that

$$\dot{A} = \frac{1}{2}\,h \tag{12.32}$$

The quantity $\dot{A}$ is called the *areal speed* of the particle, which is a constant as the particle moves along a trajectory in central-force motion. Therefore, *the radius vector of a particle in central-force motion sweeps equal areas in equal times*. This geometric property may be referred to as the *conservation of areal speed* in central-force motion.[†]

Using the cross product of two vectors and Eqs. (12.31) and (12.32), we write

$$h = |\mathbf{r} \times \mathbf{v}| \qquad \dot{A} = \frac{1}{2}|\mathbf{r} \times \mathbf{v}| \tag{12.33}$$

$$\dot{\mathbf{A}} = \frac{1}{2}\,\mathbf{r} \times \mathbf{v} \tag{12.34}$$

where $\dot{\mathbf{A}}$ is called the *areal velocity* of the particle. We find from Eqs. (12.33) and (12.34) that *when a particle is in central-force motion, its areal velocity, as well as its areal speed, is constant*.

The *usual assumptions* made in elementary studies of the central-force motion of a *spacecraft* (which is here used to denote any *space vehicle, space shuttle, satellite,* or *space station*) passing by or revolving around a *planet* (such as the earth) are as follows:

1. The mass m of the spacecraft compared with the mass M of the planet is negligibly small; i.e., $m \ll M$.
2. The altitude of the spacecraft is high enough so that the air drag is negligible.
3. The spacecraft is not too far from the planet so that the dominant external force acting on the spacecraft is exerted by the planet; i.e., other external forces acting on it are negligible.
4. The duration of flight of the spacecraft around the planet is relatively short compared with the period of time it takes the planet to revolve around the sun once.
5. The acceleration of the center of the planet with respect to the sun is very small,[††] and a nonrotating reference frame attached to the center of the planet may serve as an inertial reference frame (cf. Sec. 3.1).

[†] This is *Kepler's second law* in Sec. 12.12. Cf. Eq. (14.23) in Sec. 14.6 regarding the interpretation of h as the *angular momentum per unit mass* of the particle about the center of force O and the *conservation of angular momentum* of a particle in central-force motion.
[††] The average magnitude of the acceleration of the center of the earth relative to the sun is only about 6 mm/s^2.

In the elementary study of the *central-force motion of a spacecraft around the moon*, the moon is treated as if it were a planet and the above usual assumptions are applied accordingly. Of course, the motion of the planets around the sun may similarly be studied.

Developmental Exercises

D12.19   Refer to Eqs. (12.30) and (12.31). (a) Are these two equations true for *any* type of central-force motion? (b) Does the value of h remain the same for *different* trajectories?

D12.20   Describe the *usual assumptions* made in elementary studies of the central-force motion of a spacecraft coasting near a planet.

EXAMPLE 12.10

A spacecraft in free flight around the earth has a velocity $\mathbf{v}_P$ parallel to the surface of the earth as it descends to the lowest point at P as shown. The radius of the earth is $R = 3.96 \times 10^3$ mi. Determine the speed v_Q of the spacecraft as it ascends to the highest point at Q.

Solution.   Since the spacecraft is in free flight, it is under the action of only the gravitational force. Thus, the spacecraft is having a *gravitational central-force motion*. The velocity of the spacecraft is always tangent to its orbit. From the sketch shown, we note that, at the extremities P and Q, the velocity $\mathbf{v}$ of the spacecraft is perpendicular to the radius vector $\mathbf{r}$. Thus, we have

$$(v_\theta)_P = v_P = 5.17 \text{ mi/s} \qquad (v_\theta)_Q = v_Q$$

Referring to the sketch, we write

$$h_P = h_Q: \qquad r_P(v_\theta)_P = r_Q(v_\theta)_Q \qquad r_P v_P = r_Q v_Q$$

$$v_Q = \frac{r_P v_P}{r_Q} = \frac{(3.96 \times 10^3 + 200)(5.17)}{3.96 \times 10^3 + 1800}$$

$$v_Q = 3.73 \text{ mi/s} \blacktriangleleft$$

Developmental Exercises

D12.21   Define the terms: (a) central force, (b) center of force, (c) central-force motion, (d) gravitational central-force motion, (e) elastic central-force motion, (f) areal speed, (g) areal velocity.

D12.22   Refer to Example 12.10. Determine (a) the areal speed $\dot{A}$ of the spacecraft, (b) the semimajor axis a of the elliptic orbit, (c) the distance c between the center C of the elliptic orbit and the center of force O which is a focus of the ellipse, (d) the semiminor axis b of the elliptic orbit, (e) the area A of the plane enclosed by the elliptic orbit, (f) the time τ required for the spacecraft to revolve around the earth once. (*Hint.* $a^2 - b^2 = c^2$, $A = \pi ab$, and $\tau = A/\dot{A}$.)

12.10 Governing Differential Equation

The differential equation defining the family of trajectories of a particle, which is set into central-force motion with various possible initial conditions, is called the *governing differential equation* for the central-force motion of the particle. The actual trajectory of the particle is a branch of the family of trajectories, which corresponds to the solution of the governing differential equation satisfying the actual initial conditions of the particle. We shall here derive such a differential equation valid for any type of central-force motion.

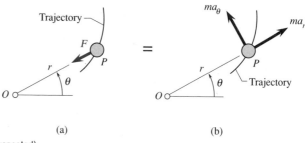

Fig. 12.15 (repeated)

The free-body diagram of a particle P with mass m moving under a central force $\mathbf{F}$ is equated to its effective-force diagram as shown in Fig. 12.15, where $\mathbf{F}$ is directed toward the center of force O. Referring to this figure, we write

$$+ \nearrow \ \Sigma V_r: \qquad -F = ma_r \qquad (12.35)$$

$$+ \nwarrow \ \Sigma V_\theta: \qquad 0 = ma_\theta \qquad (12.36)$$

where $a_r = \ddot{r} - r\dot{\theta}^2$ and $a_\theta = r\ddot{\theta} + 2\dot{r}\dot{\theta}$. We note that Eqs. (12.35) and (12.36) represent a set of *simultaneous* nonlinear ordinary differential equations defining the family of trajectories of a particle in central-force motion. The independent variable in these equations is the time variable t. But, as we saw in Sec. 12.9, Eq. (12.36) leads to Eq. (12.30); i.e.,

$$r^2\dot{\theta} = h \qquad (12.30)$$
(repeated)

where h is a constant. We may write Eq. (12.35) explicitly as

$$m(\ddot{r} - r\dot{\theta}^2) = -F \qquad (12.37)$$

To solve Eq. (12.37), we first introduce a variable u as

$$u = r^{-1} \qquad (r \neq 0) \qquad (12.38)$$

From Eqs. (12.30) and (12.38), we write

$$\dot{\theta} = u^2 h \qquad r = u^{-1} \qquad \frac{dr}{d\theta} = \frac{d}{d\theta}(u^{-1}) = -u^{-2}\frac{du}{d\theta}$$

$$\dot{r} = \frac{dr}{d\theta}\,\dot{\theta} = \left(-u^{-2}\frac{du}{d\theta}\right)(u^2 h) = -h\frac{du}{d\theta}$$

$$\ddot{r} = \frac{d\dot{r}}{d\theta}\,\dot{\theta} = \left(-h\frac{d^2 u}{d\theta^2}\right)(u^2 h) = -h^2 u^2 \frac{d^2 u}{d\theta^2}$$

Substitution of the above expressions into Eq. (12.37) leads to

$$\frac{d^2 u}{d\theta^2} + u = \frac{F}{mh^2 u^2} \tag{12.39}$$

This is the *governing differential equation* for any type of central-force motion, where $u = 1/r$ and the constant $h = r^2\dot{\theta} = rv_\theta$. In deriving Eq. (12.39), the central force **F** was assumed to be an *attractive force* directed toward the center of force O. If **F** is a repulsive force directed away from O, we replace F with $-F$ in Eq. (12.39). If F is a known function of r and hence u, Eq. (12.39) will take the form of a nonhomogeneous second-order ordinary differential equation in u and θ.

Developmental Exercise

D12.23 Is Eq. (12.39) the governing differential equation for either the elastic or the gravitational central-force motion?

12.11 Trajectories of Spacecraft

A spacecraft is in a *powered flight* when its launching rockets or its on-board rockets are producing thrusts to itself.[†] On the other hand, a spacecraft is in a *free flight* when the only external force acting on it is the gravitational attractive force. Thus, a spacecraft is in a *gravitational central-force motion* when it is in a free flight.

Let us consider a spacecraft S with mass m moving under the gravitational central force **F** exerted by the earth of mass M as shown in Fig. 12.17. By Newton's law of gravitation, the magnitude of **F** is given by

$$F = G\frac{Mm}{r^2} = GMmu^2$$

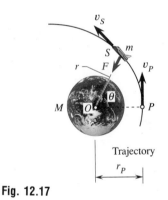

Fig. 12.17

where G is the constant of gravitation and $u = 1/r$. Substituting this expression of F into Eq. (12.39), we obtain

$$\frac{d^2 u}{d\theta^2} + u = \frac{GM}{h^2} \tag{12.40}[††]$$

which is a nonhomogeneous linear second-order ordinary differential equation with constant coefficients.

[†]The launching rockets of a spacecraft may include on-board and booster rockets, and the on-board rockets may include main rockets and auxiliary rockets.

[††]Those who are familiar with the general solution of such a differential equation may skip the ensuing explanation and go directly to Eq. (12.44).

The general solution of Eq. (12.40) is given by the sum of a complementary solution u_c and a particular solution u_p; i.e., $u = u_c + u_p$. The particular solution u_p is any special solution which we can find for Eq. (12.40). Since the derivative of a constant is zero, we readily find that the simplest particular solution is

$$u_p = \frac{GM}{h^2} \tag{12.41}$$

The complementary solution u_c is given by the general solution of the corresponding homogeneous equation; i.e.,

$$\frac{d^2 u_c}{d\theta^2} + u_c = 0 \tag{12.42}$$

Since Eq. (12.42) is a *second-order* homogeneous differential equation, its general solution (which defines a family of curves) should contain *two* arbitrary constants of integration. By writing Eq. (12.42) as $u_c = -d^2 u_c/d\theta^2$, we forthwith infer that any function u_c of θ, which is equal to the negative of the second derivative of the function u_c with respect to θ, is a solution of Eq. (12.42). Noting that the second derivatives of the functions $\cos\theta$ and $\sin\theta$ with respect to θ are equal to $-\cos\theta$ and $-\sin\theta$, respectively, we readily conclude that the general solution of Eq. (12.42) and hence the complementary solution of Eq. (12.40) may be written as

$$u_c = C \cos\theta + D \sin\theta \tag{12.43}$$

where C and D are the two arbitrary constants of integration. Thus, we combine Eqs. (12.41) and (12.43) to obtain the general solution as

$$u = \frac{1}{r} = C \cos\theta + D \sin\theta + \frac{GM}{h^2} \tag{12.44}$$

Using the polar coordinate system established in Fig. 12.17, we note that the radial distance r of the trajectory at P is minimum, and the velocity $\mathbf{v}_P$ of the spacecraft at P is perpendicular to the radius vector $\mathbf{r}_P$. Thus, the *conditions at P* are

$$r = r_P \qquad \dot{r} = 0 \qquad \theta = 0 \qquad \dot{\theta} = \dot{\theta}_P = \frac{v_P}{r_P} \neq 0$$

Differentiating Eq. (12.44) with respect to the time t, we write

$$-r^{-2}\dot{r} = -C\dot{\theta}\sin\theta + D\dot{\theta}\cos\theta \tag{12.45}$$

Imposing the aforementioned *conditions at P* on Eq. (12.45), we get $0 = 0 + D\dot{\theta}_P$. Since $\dot{\theta}_P \neq 0$, we must have $D = 0$. Thus, Eq. (12.44) reduces to

$$\frac{1}{r} = C \cos\theta + \frac{GM}{h^2} \tag{12.46}$$

$$\frac{1}{r} = \frac{GM}{h^2}(1 + \varepsilon \cos\theta) \tag{12.47}^\dagger$$

[†] Note that the $+$ sign in Eq. (12.47) becomes $-$ if the polar coordinate system is chosen in such a way that r has a maximum value when $\theta = 0$.

where

$$\varepsilon = \frac{Ch^2}{GM} \tag{12.48}$$

Recall that Eq. (12.47) represents a *conic section* in polar coordinates, and ε is its *eccentricity*.[†] Therefore, *the trajectory of a spacecraft in free flight is a conic section* (e.g., a circle, ellipse, parabola, or hyperbola). The various shapes of the conic sections are illustrated in Fig. 12.18. Note in Figs. 12.17 and 12.18 that the origin O, which is the center of force, is a *focus* of the conic sections, and the polar axis is an axis of symmetry of the conic sections. Recalling that $-1 \le \cos\theta \le 1$ and referring to Eq. (12.47), we can reason as follows:

1. In the case of $\varepsilon > 1$, r becomes infinite when $\cos\theta = -1/\varepsilon$. Thus, the conic section is a *hyperbola* if $\varepsilon > 1$.
2. In the case of $\varepsilon = 1$, r becomes infinite when $\theta = 180°$. Thus, the conic section is a *parabola* if $\varepsilon = 1$.
3. In the case of $0 < \varepsilon < 1$, r remains finite for any value of θ. Thus, the conic section is an *ellipse* if $0 < \varepsilon < 1$.
4. In the case of $\varepsilon = 0$, r has a constant value of $h^2/(GM)$ for any value of θ. Thus, the conic section is a *circle* if $\varepsilon = 0$.

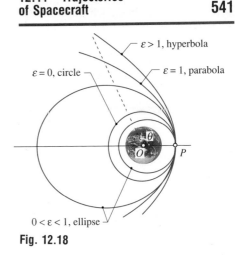

Fig. 12.18

Since the radius vector of an *orbiting* spacecraft must never become infinite, *the trajectory of a spacecraft in orbit must be either an ellipse or a circle.* Such an unopen trajectory is called an *orbit*.

Suppose that a spacecraft S is *orbiting along an elliptic trajectory* around the earth as shown in Fig. 12.19, where the origin O and the point O' are the foci, the point C is the centroid of the ellipse, the point P on the orbit closest to the earth is called the *perigee*, the point Q on the orbit farthest away from the earth is called the *apogee*, the semimajor axis is $\overline{CP} = a$, the semiminor axis is $\overline{CM} = b$, the distance between the centroid C and either focus is c, and $r_P + r_Q = 2a$. Thus, we write

$$a = \frac{1}{2}(r_P + r_Q) \tag{12.49}$$

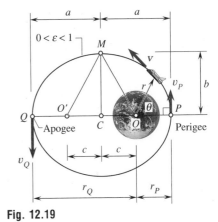

Fig. 12.19

$$c = \overline{CO} = \overline{CP} - \overline{OP} = a - r_P = \frac{1}{2}(r_P + r_Q) - r_P$$

$$c = \frac{1}{2}(r_Q - r_P) \tag{12.50}$$

It can be shown that the sum of the distances from each of the foci to any point of the ellipse is constant. By symmetry, we note that $\overline{OM} = \overline{O'M}$ and $\overline{QO'} = \overline{OP}$. Thus, from Fig. 12.19 and Eqs. (12.49) and (12.50), we write

$$\overline{OM} + \overline{O'M} = \overline{OP} + \overline{O'P}$$

$$2\,\overline{OM} = \overline{QO'} + \overline{O'P} = 2a \qquad \overline{OM} = a$$

$$b^2 = \overline{CM}^2 = \overline{OM}^2 - \overline{CO}^2 = a^2 - c^2 = r_P r_Q$$

$$b = (r_P r_Q)^{1/2} \tag{12.51}$$

[†]Cf. Eq. (12.52) and Secs. B4.2 through B4.6 of App. B.

Since the eccentricity of an ellipse is $\varepsilon = c/a$ and a circle is a special case of an ellipse, the *eccentricity of orbit* is defined to be

$$\varepsilon = (r_Q - r_P)/(r_Q + r_P) \tag{12.52}$$

The elliptic area enclosed by the orbit is $A = \pi ab$, or

$$A = \frac{\pi}{2}(r_P + r_Q)(r_P r_Q)^{1/2} \tag{12.53}$$

The time required for a spacecraft or satellite to travel along a complete orbit is called the *period of orbit* and is denoted by τ. The areal speed of an orbiting spacecraft is $\dot{A} = h/2$. Thus, we write

$$\tau = A/\dot{A}$$

$$\tau = \frac{\pi(r_P + r_Q)(r_P r_Q)^{1/2}}{h} \tag{12.54}$$

When the spacecraft passes through the *perigee P*, we have $r = r_P$ and $\theta = 0$. Imposing these conditions on Eq. (12.47), we write

$$\frac{1}{r_P} = \frac{GM}{h^2}(1 + \varepsilon) \tag{12.55}$$

Moreover, when the spacecraft passes through the *apogee Q*, we have $r = r_Q$ and $\theta = 180°$. Imposing these conditions on Eq. (12.47), we write

$$\frac{1}{r_Q} = \frac{GM}{h^2}(1 - \varepsilon) \tag{12.56}$$

Adding Eqs. (12.55) and (12.56), we get

$$\frac{1}{r_P} + \frac{1}{r_Q} = \frac{2GM}{h^2} \tag{12.57}^\dagger$$

Since $h = rv_\theta$, the constant h in Eq. (12.57) can be written as

$$h = r_P v_P = r_Q v_Q \tag{12.58}$$

The weight W of any body with mass m on the surface of the earth is equal to the magnitude of the attractive force $\mathbf{F}$ exerted on the body by the earth which has a mass M and a radius R. Thus, we apply Newton's second law and Newton's law of gravitation to write

$$W = mg = F = \frac{GM}{R^2}m$$

$$GM = gR^2 \tag{12.59}$$

where $g = 9.81$ m/s^2 or 32.2 ft/s^2 and $R = 6.37$ Mm or 3.96×10^3 mi.[††]

[†]This equation is very useful in subsequent studies.

[††]These values of g and R cannot directly be used to evaluate GM in Eq. (12.59) if M is not the mass of the earth. (Cf. Probs. 12.78 through 12.84.)

The constant h and the eccentricity ε in Eq. (12.47) are the parameters which characterize the trajectory of a spacecraft. These parameters may be determined from the initial conditions of its free flight. In most actual launches, the velocity $\mathbf{v}_P$ of the spacecraft at the end of its powered flight is programmed to be perpendicular to its radius vector $\mathbf{r}_P$ as shown in Fig. 12.20.[†]

Free flight

v_P

r_P

O

P

Burnout or engine shutoff

Launching

Powered flight

Fig. 12.20

If the spacecraft follows a parabolic trajectory, the conditions at P are $r = r_P$, $\theta = 0$, $v_P = v_{\text{par}}$, $h = r_P v_P = r_P v_{\text{par}}$, and $\varepsilon = 1$. Imposing these conditions on Eq. (12.47) and using Eq. (12.59), we write

$$\frac{1}{r_P} = \frac{GM}{r_P^2 v_{\text{par}}^2}(1 + 1) \qquad v_{\text{par}} = (2GM/r_P)^{1/2}$$

$$v_{\text{par}} = (2gR^2/r_P)^{1/2} \tag{12.60}$$

We can easily check that the trajectory becomes hyperbolic if $v_P > v_{\text{par}}$, and elliptic or circular if $v_P < v_{\text{par}}$. Since the value of v_{par} obtained for the parabolic trajectory is the smallest value of v_P for which the spacecraft will not return to its starting point, the velocity with magnitude equal to v_{par} is called the *escape velocity* $\mathbf{v}_{\text{esc}}$ at $r = r_P$; i.e.,

$$v_{\text{esc}} = (2gR^2/r_P)^{1/2} \tag{12.61}$$

If the spacecraft follows a circular trajectory, the conditions at P are $r = r_P$, $\theta = 0$, $v_P = v_{\text{cir}}$, $h = r_P v_P = r_P v_{\text{cir}}$, and $\varepsilon = 0$. Imposing these conditions on Eq. (12.47) and using Eq. (12.59), we write

$$\frac{1}{r_P} = \frac{GM}{r_P^2 v_{\text{cir}}^2} \qquad v_{\text{cir}} = (GM/r_P)^{1/2}$$

$$v_{\text{cir}} = (gR^2/r_P)^{1/2} \tag{12.62}$$

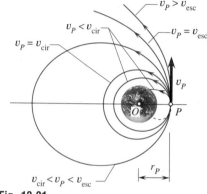

$v_P > v_{\text{esc}}$

$v_P < v_{\text{cir}}$

$v_P = v_{\text{esc}}$

$v_P = v_{\text{cir}}$

v_P

O

P

r_P

$v_{\text{cir}} < v_P < v_{\text{esc}}$

Fig. 12.21

The various trajectories corresponding to different values of the velocity $\mathbf{v}_P$ at the vertex P are illustrated in Fig. 12.21. If $v_P \ll v_{\text{cir}}$, the spacecraft may, of course, reenter the earth's atmosphere. If so, the air drag becomes significant, and the assumption (2) in the latter part of Sec. 12.9 will no longer be valid. The computer program in Fig. 12.22 on the next page is written for an orbiting spacecraft; it may be used to compute the period τ, the eccentricity ε, the maximum speed v_P at the perigee P, and the minimum speed v_Q at the apogee Q of the spacecraft when the altitudes of P and Q in its orbit around the earth are known.

Developmental Exercises

D12.24 Name the conic section if ε is (a) 2, (b) 1, (c) 0.5, (d) 0.

D12.25 Define the terms: (a) orbit, (b) eccentricity of orbit, (c) period of orbit, (d) escape velocity.

D12.26 Derive Eq. (12.61) from Eq. (12.57) by using Eq. (12.59) and letting $r_Q \to \infty$ as well as $h = r_P v_{\text{esc}}$.

D12.27 Derive Eq. (12.62) from Eq. (12.57). (*Hint.* Cf. D12.26.)

[†]Oblique launchings of spacecraft are studied in Chaps. 13 and 14.

```
10 REM         COMPUTING PERIOD, ECCENTRICITY, AND EXTREMUM SPEEDS
15 REM  OF A SPACECRAFT FROM THE ALTITUDES OF ITS PERIGEE AND APOGEE
20 REM   Refer to Eqs. (12.52), (12.54), and (12.59).  Let
25 REM   "ap" and "aq" be the altitudes of the perigee and apogee of
30 REM   the spacecraft orbiting around the earth.  Enter "ap" & "aq"
35 REM   (in miles) as data in Line 100.  To output to the printer,
40 REM   change PRINT to LPRINT in the program.       (File name: SPACFT)
45 READ AP,AQ: R=3960: RP=AP+R: RQ=AQ+R: E=(RQ-RP)/(RQ+RP): G=32.2/5280
50 GM=G*R^2: H=SQR(2*GM*RP*RQ/(RP+RQ)): T=3.141593*(RP+RQ)*SQR(RP*RQ)/H
55 VP=H/RP: VQ=H/RQ: M$="miles": S$="     ": PRINT: PRINT "ap =";AP;M$;
60 PRINT S$;"aq =";AQ;M$: PRINT "Period =";T/3600;"hours";S$;
65 PRINT "Eccentricity =";E: J$="miles/second   or ": K$="miles/hour"
70 PRINT "Vp =";VP;J$;VP*3600;K$: PRINT "Vq =";VQ;J$;VQ*3600;K$
100 DATA 250,500
```

```
ap = 250 miles      aq = 500 miles
Period = 1.610849 hours      Eccentricity = 2.883507E-02
Vp = 4.834349 miles/second  or  17403.66 miles/hour
Vq = 4.563366 miles/second  or  16428.12 miles/hour
```

Fig. 12.22 (Cf. the solution in Example 12.11.)

EXAMPLE 12.11

The altitudes of the perigee P and apogee Q in the orbit of a space shuttle are $a_P = 250$ mi and $a_Q = 500$ mi as shown. For such an orbit, determine (a) the period τ, (b) the eccentricity ε, (c) the maximum speed v_P, (d) the minimum speed v_Q.

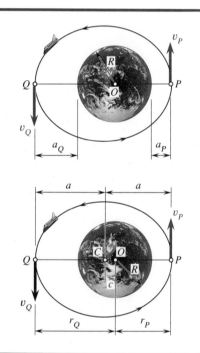

Solution. Since the radius of the earth is $R = 3.96 \times 10^3$ mi, $a_P = 250$ mi, $a_Q = 500$ mi, $g = 32.2/5280$ mi/s^2, and $GM = gR^2$, we write

$$r_P = R + a_P \qquad r_P = 4.21 \times 10^3 \text{ mi}$$

$$r_Q = R + a_Q \qquad r_Q = 4.46 \times 10^3 \text{ mi}$$

$$\frac{1}{r_P} + \frac{1}{r_Q} = \frac{2GM}{h^2} = \frac{2gR^2}{h^2} \qquad h = 20.35 \times 10^3 \text{ mi}^2/\text{s}$$

$$\tau = \frac{\pi(r_P + r_Q)(r_P r_Q)^{1/2}}{h} = 5.799 \times 10^3 \text{ s}$$

$$\tau = 1.611 \text{ hours} \blacktriangleleft$$

$$\varepsilon = (r_Q - r_P)/(r_Q + r_P) \qquad h = r_P v_P = r_Q v_Q$$

$$\varepsilon = 0.0288 \qquad v_P = 4.83 \text{ mi/s} \qquad v_Q = 4.56 \text{ mi/s} \blacktriangleleft$$

Developmental Exercise

D12.28   Refer to Example 12.11. Determine for the orbit (a) the semimajor axis a, (b) the semiminor axis b, (c) the elliptic area A, (d) the areal speed $\dot{A}$, (e) the period τ as computed from $A/\dot{A}$.

EXAMPLE 12.12

A spacecraft is describing a circular orbit of radius 8 Mm around the earth when its on-board rockets are fired momentarily at P to increase its speed from $(v_P)_{cir}$ to $(v_P)_{ell}$ as shown, where $R = 6.37$ Mm. This maneuver sends the spacecraft along an elliptic trajectory to the apogee at Q. Upon reaching Q, its on-board rockets are fired momentarily again to increase its speed from $(v_Q)_{ell}$ to $(v_Q)_{cir}$ to insert itself into a circular orbit of radius 20 Mm as shown. Determine the values of $(v_P)_{cir}$, $(v_P)_{ell}$, $(v_Q)_{ell}$, $(v_Q)_{cir}$, and the time t_{PQ} required for the spacecraft to travel from P to Q.

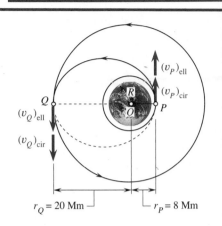

Solution. From the sketch shown, we have $r_P = 8$ Mm $= 8 \times 10^3$ km and $r_Q = 20$ Mm $= 20 \times 10^3$ km. Thus, Eqs. (12.57) and (12.59) are best for the solution. Since the radius of the earth is $R = 6.37$ Mm $= 6.37 \times 10^3$ km and $g = 9.81$ m/s$^2 = 9.81 \times 10^{-3}$ km/s^2, we write

$$\frac{1}{r_P} + \frac{1}{r_P} = \frac{2gR^2}{r_P^2(v_P)_{cir}^2} \qquad (v_P)_{cir} = 7.05 \text{ km/s} \quad \blacktriangleleft$$

At the perigee P and apogee Q of the elliptic trajectory, we write

$$\frac{1}{r_P} + \frac{1}{r_Q} = \frac{2gR^2}{r_P^2(v_P)_{ell}^2} \qquad (v_P)_{ell} = 8.43 \text{ km/s} \quad \blacktriangleleft$$

$$h = r_P(v_P)_{ell} = r_Q(v_Q)_{ell} \qquad (v_Q)_{ell} = 3.37 \text{ km/s} \quad \blacktriangleleft$$

$$\frac{1}{r_Q} + \frac{1}{r_Q} = \frac{2gR^2}{r_Q^2(v_Q)_{cir}^2} \qquad (v_Q)_{cir} = 4.46 \text{ km/s} \quad \blacktriangleleft$$

The time t_{PQ} required to travel from P to Q is equal to one half of the period of orbit along the entire ellipse. Thus, we write

$$t_{PQ} = \frac{\pi(r_P + r_Q)(r_P r_Q)^{1/2}}{2r_P(v_P)_{ell}} = 8.25 \times 10^3 \text{ s}$$

$$t_{PQ} = 2.29 \text{ hours} \quad \blacktriangleleft$$

Developmental Exercise

D12.29   Refer to Example 12.12. Determine the period of orbit of the spacecraft along the circle of radius (a) 8 Mm, (b) 20 Mm.

★12.12   Kepler's Laws of Planetary Motion

The theory developed in preceding sections for central-force motion of spacecraft around the earth is also applicable to the study of the motion of planets around the sun. The properties deducible from Eqs. (12.32) and (12.47), with M representing the mass of the sun, were first discovered by the German astronomer Johann (or Johannes) Kepler (1571–1630) from astronomical observations of the motions of planets. Kepler's discovery served not only as background but also as proving ground for Newton (1642–1727) in

Table 12.2   Some Planetary Data

Planet	Semimajor Axis of Orbit (AU)[†]	Eccentricity of Orbit	Period of Orbit around the Sun	Period of Rotation about own Axis	Mass Relative to the Earth
Mercury	0.3871	0.205629	87.97 d	58.65 d	0.0552
Venus	0.7233	0.006787	224.70 d	−243.0 d[††]	0.8144
Earth	1	0.016721	365.26 d	23 h 56 min 4 s	1[†††]
Mars	1.5237	0.093379	686.98 d	24 h 37 min 23 s	0.1077
Jupiter	5.203	0.0485	11.86 y	9 h 50.5 min (at equator)	317.7
Saturn	9.539	0.0556	29.46 y	0.446 d	95.15
Uranus	19.18	0.0472	84.01 y	−0.717 d[††]	14.4
Neptune	30.06	0.0086	164.8 y	0.67 d	17.2
Pluto	39.44	0.25	247.7 y	6.39 d	0.003

[†]One AU (astronomical unit of distance) is defined to be the length of the semimajor axis of the Earth's orbit around the Sun; i.e.,

$$1 \text{ AU} = 149.5 \text{ Gm} = 149.5 \times 10^6 \text{ km}$$
$$= 92.9 \times 10^6 \text{ mi}$$

[††]The negative sign indicates retrograde motion.

[†††]The mass of the Earth itself is 5.976×10^{24} kg. (The Moon of the Earth has a mass of 7.35×10^{22} kg. Its orbit around the Earth has semimajor axis = 384.398 Mm, eccentricity = 0.055, and period = 27.32 d.)

formulating his law of gravitation as well as laws of motion. For reference, some planetary data are given in Table 12.2, and the mean distances (i.e., the semimajor axes of orbits) of the planets from the sun are illustrated in Fig. 12.23.

Kepler's three *laws of planetary motion* may be stated as follows:

1. *First law*. The orbit of each planet is an ellipse with the sun at one of its foci.
2. *Second law*. The radius vector drawn from the sun to a planet sweeps equal areas in equal times.
3. *Third law*. The squares of the periods of orbits of the planets are proportional to the cubes of the semimajor axes of their orbits.

Mean distances of the planets from the sun (in AU)

Fig. 12.23

Note that Kepler's first law states the particular case of $0 < \varepsilon < 1$ in the result established in Eq. (12.47). His second law is equivalent to the statement that the areal speed of each planet is constant, which is the result established in Eq. (12.32). Kepler's third law may be proved by using Eqs. (12.49), (12.51), (12.54), and (12.57).

Developmental Exercises

D12.30 Do the following two steps to prove Kepler's third law of planetary motion: (a) From Eqs. (12.49), (12.51), and (12.57), show that $h^2 = GMb^2/a$. (b) Using this expression for h^2 and Eq. (12.54), show that $\tau^2 = [4\pi^2/(GM)]a^3$, hence $\tau_i^2/\tau_j^2 = a_i^3/a_j^3$ where the subscripts refer to the ith and the jth planets.

D12.31 Using the data given in Table 12.2, determine the shortest distance r_P and the longest distance r_Q existing between the centers of Mercury and the Sun in miles.

D12.32 Using the data given in Table 12.2, verify Kepler's third law of planetary motion for the planets Mercury and Earth.

PROBLEMS

NOTE. The motion of any spacecraft mentioned in the following problems is measured in a nonrotating reference frame attached to the center of force in the attracting body, such as the earth, and the usual assumptions stated in Sec. 12.9 are implied. Whenever needed, use $R = 6.37$ Mm or 3.96×10^3 mi for the radius of the earth.

12.66 and 12.67*[†] The minimum and maximum altitudes of the orbit of an earth satellite S are indicated as shown. Determine (a) the eccentricity ε of the orbit, (b) the maximum speed $v_{\max}$ of the satellite.

Fig. P12.66 and P12.68* **Fig. P12.67* and P12.69**

12.68* and 12.69 The minimum and maximum altitudes of the orbit of an earth satellite S are indicated as shown. Determine for the satellite (a) its minimum speed, (b) its areal speed, (c) the elliptic area enclosed by its orbit, (d) its period of orbit.

12.70 A *Landsat S* describes a circular orbit around the earth as shown. If its period of orbit is 103 min, determine (a) its altitude a_0 in miles, (b) its speed v_{cir} in miles per second.

[†]Cf. the computer program in Fig. 12.22.

Fig. P12.70 and P12.71

12.71 Determine the altitude a_0 of a geosynchronous satellite in both SI and U.S. customary units.[†]

12.72* Determine the speed of a geosynchronous satellite in both SI and U.S. customary units.[†]

12.73 It is known that the radius of the planet Mars is 2.11×10^3 mi. Making use of the data in Table 12.2, determine the altitude of a synchronous satellite of the planet Mars.

12.74* Making use of the data in Table 12.2, determine the speed of a synchronous satellite of the planet Mars.

12.75* A spacecraft is describing a circular orbit of 12-Mm radius around the earth when its rockets are fired momentarily to increase the speed of the spacecraft by 1 km/s. The spacecraft is thus inserted into an elliptic orbit as shown. For the spacecraft in this elliptic orbit, determine (a) its maximum altitude, (b) its period of orbit.

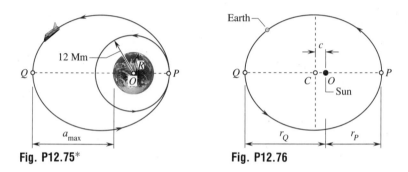

Fig. P12.75*   **Fig. P12.76**

12.76 The earth revolves around the sun in an elliptic orbit as shown. Making use of the data in Table 12.2, determine (a) the distance c between the centroid C of the elliptic orbit and the center O of the sun, (b) the minimum and maximum radial distances r_P and r_Q, (c) the shortest distance existing between the surface of the earth and the surface of the sun if the radius of the sun is 696 Mm.

12.77 Refer to Prob. 12.76. Determine (a) the areal speed $\dot{A}$ of the earth in its orbit around the sun, (b) the ratio of the mass of the sun to the mass of the earth.

12.78 A spacecraft approaches the planet Venus along a parabolic trajectory SQ as shown. As it reaches the vertex Q of the trajectory SQ, its retrorockets are fired momentarily to decrease its speed and insert itself into an elliptic orbit as indicated. The ratio of the mass of Venus to the mass of the earth is known to be 0.8144. Determine the speed of the spacecraft (a) as it approaches Q, (b) after the firing of retrorockets, (c) as it approaches P.

12.79* A spacecraft approaches the planet Mars along a hyperbolic trajectory SQ of eccentricity $\varepsilon = 3$ as shown. As it reaches the vertex Q of the trajectory SQ, its retrorockets are fired momentarily to decrease its speed and insert itself into an elliptic orbit as indicated. The ratio of the mass of Mars to the mass of the earth is known to be 0.1077. Determine for the spacecraft (a) its speed as it approaches Q, (b) its speed after the firing of retrorockets, (c) its period along the elliptic orbit.

Fig. P12.78

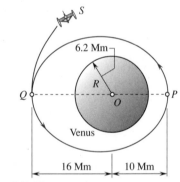

Fig. P12.79*

12.80 In returning from a moon-exploration mission, the lunar excursion module (LEM) is first propelled by its rocket engines from the moon's surface along a curved path to the point P and then travels in free flight along an elliptic path to rendezvous at Q with the command module which is traveling in a circular "parking" orbit around the moon as shown. The velocities of the LEM at the points P and Q are perpendicular to the radius vectors. The mass of the moon is 0.01230 times the mass of the earth. Determine (a) the speed of the LEM as it begins its free flight at P, (b) the speed of the LEM as it reaches Q, (c) the relative speed with which the command module approaches the LEM at Q.

12.81 Refer to Prob. 12.80. Determine (a) the period of orbit of the command module, (b) the time it takes the LEM to travel from P to Q.

12.82 After the rendezvous of the command module and the LEM at the point Q as described in Prob. 12.80, the astronauts in the LEM move into the command module and turn the spacecraft around so that the LEM faces the rear. When the spacecraft reaches the point S as shown, the LEM is separated from the command module and is cast adrift by reducing its speed from $(v_S)_{cir}$ to $(v_S)_{ell}$. Consequently, the LEM crashes on the moon's surface at the point K. Using the mass and radius of the moon given in Prob. 12.80, determine the value of $(v_S)_{ell}$ if K is located at $\theta = 60°$. [*Hint.* The point S is the apogee of the crash trajectory, and the $+$ sign in Eq. (12.47) becomes $-$.]

12.83* Solve Prob. 12.82 if K is located at $\theta = 120°$.

12.84 Refer to Prob. 12.82. If $(v_S)_{ell} = 1.52$ km/s, determine the value of θ at which the LEM will crash on the moon's surface.

12.85* A spacecraft is in free flight along an elliptic orbit of eccentricity ε around the earth as shown. Show that the speeds v_P and v_Q of the spacecraft at the perigee P and the apogee Q may be written as

$$v_P = R[(g/a)(1 + \varepsilon)/(1 - \varepsilon)]^{1/2} = R[gr_Q/(ar_P)]^{1/2}$$
$$v_Q = R[(g/a)(1 - \varepsilon)/(1 + \varepsilon)]^{1/2} = R[gr_P/(ar_Q)]^{1/2}$$

12.86 A spacecraft launched for interplanetary exploration begins its free flight along a conic section of eccentricity ε from the vertex P as shown. For any position S of the spacecraft on its free flight, express in terms of ε, r_P, g, the radius R of the earth, and the angular position coordinate θ for the position S, (a) the areal speed $\dot{A}$ of the spacecraft, (b) the speed of the spacecraft, (c) the angle ψ between $\mathbf{v}$ and the radius vector $\mathbf{r}$. [*Hint.* One may find $\dot{\theta}$ from Eq. (12.30) and $\dot{r}$ by differentiating Eq. (12.47) with respect to the time t.]

12.87 A spacecraft begins its free flight along a parabolic trajectory (i.e., $\varepsilon = 1$) from the vertex P, a distance r_P from the center O of the earth as shown. For any position S of the spacecraft on its trajectory, express the time t elapsed since its flight from P in terms of r_P, g, the radius R of the earth, and the angular position coordinate θ for the position S. [*Hint.* Solve for t using the relation

$$\frac{d\theta}{dt} = \dot{\theta} = \frac{h}{r^2} = \frac{G^2M^2}{h^3}(1 + \cos\theta)^2 = \frac{4G^2M^2}{h^3}\cos^4(\theta/2)$$

and formula (1) in Sec. B5.2 of App. B.]

12.88* Refer to Prob. 12.86. The spacecraft crosses a point on the orbit of geosynchronous satellites when $r = \overline{OS} = 42.1$ Mm. For $\varepsilon = 1$ and $r_P = 6.86$ Mm, determine (a) the speed v_P of the spacecraft at P, (b) the angular position coordinate θ and the speed v of the spacecraft as it crosses the orbit of geosynchronous satellites.

12.89 Refer to Prob. 12.87. For $r = \overline{OS} = 42.1$ Mm = radius of the orbit of geosynchronous satellites and $r_P = 6.86$ Mm, determine the time it takes the spacecraft to fly from P along the parabolic trajectory to S.

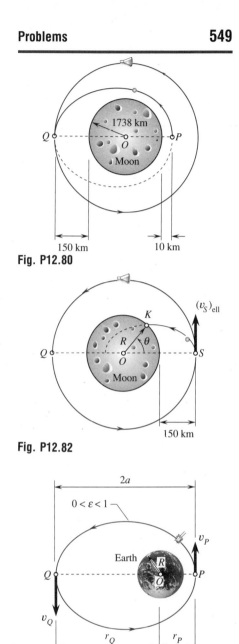

Fig. P12.80

Fig. P12.82

Fig. P12.85*

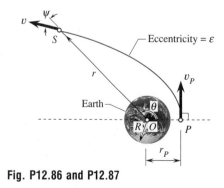

Fig. P12.86 and P12.87

12.13   Concluding Remarks

Newton's second law of motion is a keystone in kinetics. This law is used to formulate the *method of force and acceleration* at the beginning of the chapter. The crux of the *method of force and acceleration* is that the *system of force vectors* in the free-body diagram of a body (which may be a particle or a system of particles) is either equivalent or equipollent to the *system of effective-force vectors* in the effective-force diagram of the body. Those two systems are equivalent if the body is a single particle; otherwise, they are equipollent.

In any case, the said *system of force vectors* and *system of effective-force vectors* must have the *same resultant force* and the *same resultant moment* about any given point. Thus, it is *necessary* that (a) the sum of components of forces in the free-body diagram in any desirable direction is equal to the sum of components of effective forces in the effective-force diagram in the same direction, (b) the sum of moments of the forces in the free-body diagram about any desirable point (or axis) is equal to the sum of moments of the effective forces in the effective-force diagram about the same point (or axis).

For a partial solution of a problem, judiciously repeated applications of one or both of the above two necessary conditions will usually yield equations from which the unknowns may more directly be solved. Note that the resultant effective force vector does not necessarily pass through the mass center of the system of particles. Furthermore, the drawing of the effective-force diagram as well as the free-body diagram is an *enlightening step* in the application of the method of force and acceleration.

The study of gravitational central-force motion of spacecraft is presented in the latter half of the chapter. Such a study is accomplished by joining Newton's law of gravitation with Newton's second law of motion. It should be noted that Eqs. (12.30), (12.31), (12.32), (12.47), (12.52), (12.54), (12.57), and (12.59) are particularly helpful in such a study. However, we should be aware of the *usual assumptions* stated in the latter part of Sec. 12.9 when we use these equations.

REVIEW PROBLEMS

12.90   In the system shown, it is observed that the acceleration of the cylinder *A* is 1 m/s² ↓ . If the mass of the cylinder *B* is 30 kg, determine (a) the mass of the cylinder *A*, (b) the tension in cable 1 as indicated.

Fig. P12.90

12.91 A train consists of an 80-Mg locomotive A, a 45-Mg car B, and a 35-Mg car C as shown. If the train is accelerating to the right at 0.4 m/s^2, determine (a) the magnitude of the tractive friction force between the locomotive A and the tracks, (b) the force in the coupling between A and B, (c) the force in the coupling between B and C.

Fig. P12.91

12.92 A highway curve of radius 100 m is banked at an angle as shown. If $\mu_s = 0.4$ between the tires and the pavement, determine the maximum speed at which the car will not skid sideways when rounding the curve.

Fig. P12.92   **Fig. P12.93**

12.93 A spacecraft in free flight around the earth as shown has a maximum speed of 8 km/s. Determine for the spacecraft (a) its maximum altitude, (b) its minimum speed, (c) the eccentricity of its orbit, (d) its areal speed, (e) its period of orbit.

12.94 A spacecraft approaches the moon along a parabolic trajectory SP as shown. As it reaches the vertex P of the parabola, its retrorockets are fired momentarily to decrease its speed and insert itself into a circular orbit as indicated. The mass of the moon is 0.01230 times the mass of the earth. Determine for the spacecraft (a) its speed as it approaches P, (b) its speed after the firing of retrorockets, (c) its period of orbit.

12.95* Using the data given in Table 12.2, verify Kepler's third law of planetary motion for the planets (a) Venus and Earth, (b) Mars and Earth.

12.96 Using the data given in Table 12.2, determine the maximum speed of the center of the earth with respect to a nonrotating reference frame attached to the center of the sun.

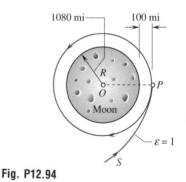

Fig. P12.94

Chapter 13

Kinetics of Particles: Work and Energy

KEY CONCEPTS

- Work of a force, power, efficiency, kinetic energy of a particle, conservative force field, conservative system, potential energy, total mechanical energy, and constrained virtual displacement.
- Principle of work and energy for particles.
- Principle of conservation of energy for particles.
- Principle of virtual work in kinetics for studying forces and accelerations in one-degree-of-freedom systems of particles.

Work and energy are scalar quantities with specific meaning in mechanics. Through the concept of *power*, the subtlety in defining the work of a force is discussed in detail early on in the chapter. Then, the *principle of work and energy* for particles is presented. This principle is useful in solving kinetic problems of particles where displacement and velocity (among others) are to be directly related without initially involving acceleration and time. Later, the concept of potential energy is quantitatively defined. The *principle of conservation of energy* is derived as a special case of the principle of work and energy as it is applied to conservative systems. Furthermore, the *principle of virtual work in kinetics* is shown to be very useful in studying the motion of systems of connected particles with one degree of freedom, where *force* and *acceleration* are to be directly related. It is presented as a complement to the method of force and acceleration studied in Chap. 12.

WORK AND KINETIC ENERGY[†]

13.1 Work of a Force

The *work of a force* **F** *acting on a particle during a differential displacement* d**r** *of the particle* from the point A to a neighboring point A', as shown in

---

[†]The concept of work was presented in Chap. 10. For convenience, we repeat here portions of that material relevant to the current study.

Fig. 13.1, is defined as an infinitesimal scalar quantity dU equal to the dot product of $\mathbf{F}$ and $d\mathbf{r}$; i.e.,

$$dU = \mathbf{F} \cdot d\mathbf{r} \qquad (13.1)^{\dagger}$$

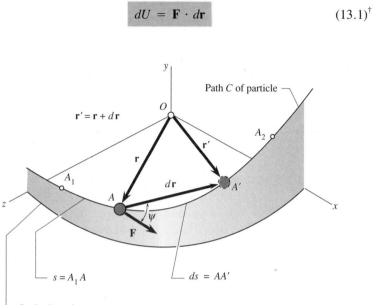

Fig. 13.1

Alternatively, we may write $dU = F \cos\psi\, ds$, where $ds = |d\mathbf{r}|$ and ψ is the angle between $\mathbf{F}$ and $d\mathbf{r}$. Anyway, the *work $U_{1\to2}$ of a force $\mathbf{F}$ acting on a particle during a finite displacement of the particle* from the point A_1 along a path C to the point A_2, as shown in Fig. 13.1, is

$$U_{1\to2} = \int_C \mathbf{F} \cdot d\mathbf{r} \qquad (13.2)$$

where the label C at the integral sign signifies that the work $U_{1\to2}$ done by $\mathbf{F}$ to the particle is generally dependent on the path along which the particle travels from A_1 to A_2. (Cf. D13.4 and D13.5.) If $U_{1\to2}$ is dependent on the coordinates of A_1 and A_2 but not on those of C, then $\mathbf{F}$ is referred to as a *conservative force* and we may write

$$U_{1\to2} = \int_{A_1}^{A_2} \mathbf{F} \cdot d\mathbf{r} \qquad (13.3)$$

The concept of conservative force is further discussed in Sec. 13.8.

If $\mathbf{F}$ is a friction force, which is a nonconservative force, Eq. (13.2) will generally yield a different work for each different path C. By definition, the work of a force is a scalar and has the dimensions of *force* $\times$ *length*. The

†In kinetics, *work* is done *by* a force *to* (or *on*) a body which has a *mass*. It is the *force* (not the *body*) which does work; a body *receives* work from the force acting on it during its displacement from one position to another. If the action of the force on a particle and the displacement of the particle do not occur simultaneously, no work is done by the force to the particle. The *work of a force* acting on a body is the *work done by a force* to a body. Moreover, *no work* can be done to a body which is *stationary* or has *no mass*.

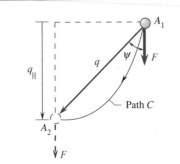

Fig. 13.2

primary units for work are joules (1 joule $= 1$ J $= 1$ N·m) in the SI and foot-pounds (ft·lb) in the U.S. customary system.[†]

If the force **F** is *constant* in magnitude and direction, the work $U_{1\to2}$ of **F** acting on a particle during its motion from the point A_1 along a path C to the point A_2, as shown in Fig. 13.2, is

$$U_{1\to2} = \int_C \mathbf{F} \cdot d\mathbf{r} = \mathbf{F} \cdot \int_C d\mathbf{r} = \mathbf{F} \cdot \mathbf{q}$$

$$U_{1\to2} = Fq\cos\psi \qquad (13.4)$$

where **q** is the displacement of the particle from A_1 to A_2 and ψ is the angle between the positive directions of **F** and **q**. We readily observe that this work $U_{1\to2}$ is positive if ψ is acute and negative if ψ is obtuse. There are three cases which are of special interest: (1) if **F** and **q** have the same direction, then $U_{1\to2} = Fq$; (2) if **F** and **q** have opposite directions, then $U_{1\to2} = -Fq$; (3) if **F** is perpendicular to **q**, then $U_{1\to2} = 0$. Using $\mathbf{q}_\parallel$ to denote the component of the displacement of the particle parallel to the constant force **F**, we may write Eq. (13.4) as

$$U_{1\to2} = Fq_\parallel \qquad (13.5)$$

where

$$q_\parallel = q\cos\psi \qquad (13.6)$$

Note that the value of $q_\parallel$ is negative when $90° < \psi \leq 180°$.

Furthermore, *the work of a force is equal to the sum of the works of its component forces, and conversely.* This property is analogous to *Varignon's theorem* and can be proved by applying the distributive property of dot products of vectors to the definition of the work of a force and the relation between the force and its component forces.[††]

EXAMPLE 13.1

If $\mu_k = 0.2$ between the 100-lb body and the horizontal support as shown, determine the works U_P, U_W, U_N, and U_F done to the body by the 50-lb pushing force **P**, the 100-lb weight force **W** of the body, the normal reaction **N**, and the friction force **F**, respectively, during the 3-ft displacement of the body from A to B.[†††]

[†]The joule (J) is the primary unit of *work* and *energy* in the SI. Note that *moment* is not a form of energy. Although 1 J $= 1$ N·m, the primary units of *moment* of a force are N·m, *not* J, in the SI. In the U.S. customary system, the primary units are usually lb·ft for *moment*, but ft·lb for *work*. [However, some do not strictly follow the conventions for lb·ft and ft·lb, which are in reverse order to those shown in Eqs. (4.18) and (13.2).]

[††]Some refer to this property as the *theorem of work*.

[†††]Cf. the second footnote on p. 553.

Solution. Since the body can move only in the horizontal direction, it is in equilibrium in the vertical direction. We write

$$+\uparrow \Sigma F_y = 0: \quad N - 100 - 50 \sin 30° = 0 \quad N = 125 \text{ lb}$$

$$\text{Friction law:} \quad F_k = \mu_k N \quad\quad\quad F_k = 25 \text{ lb}$$

The angles ψ_P, ψ_W, ψ_N, and ψ_F between the displacement **q** of the body and the forces **P**, **W**, **N**, and $\mathbf{F}_k$, respectively, are

$$\psi_P = 30° \quad \psi_W = 90° \quad \psi_N = 90° \quad \psi_F = 180°$$

Noting that the forces considered are constant and $U_{1\to2} = Fq \cos\psi$, we write

$$U_P = Pq \cos\psi_P = 50(3)(\sqrt{3}/2)$$

$$U_W = Wq \cos\psi_W = 100(3)(0)$$

$$U_N = Nq \cos\psi_N = 125(3)(0)$$

$$U_F = F_k q \cos\psi_F = 25(3)(-1)$$

$$U_P = 129.9 \text{ ft·lb} \quad U_W = U_N = 0 \quad U_F = -75 \text{ ft·lb} \blacktriangleleft$$

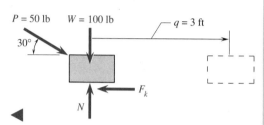

Alternative Solution for U_P. Resolving the 50-lb force **P** into two rectangular components $\mathbf{P}_x = 5 \cos 30°$ lb $\rightarrow$ and $\mathbf{P}_y = 50 \sin 30°$ lb $\downarrow$, we write

$$U_P = P_x q \cos 0° + P_y q \cos 90° = (50 \cos 30°)(3)(1) + 0$$

$$U_P = 129.9 \text{ ft·lb} \blacktriangleleft$$

Developmental Exercises

D13.1 Define the work of a force acting on a particle.

D13.2 A 100-N force **F** acts on a particle P during a 2-m displacement **q** of the particle as shown. Compute the work of **F**.

D13.3 A 260-lb crate moves 10 ft down an incline as shown. Resolving the weight force **W** into two components, one parallel and the other perpendicular to the incline, compute the work of **W**.

(a)   (b)   (c)

Fig. D13.2

Fig. D13.3

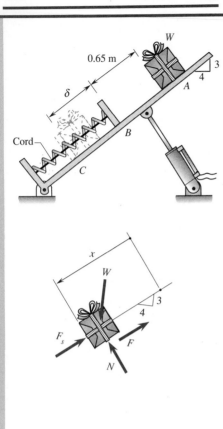

EXAMPLE 13.2

The tension in the cord which holds a spring of modulus $k = 300$ N/m in compression as shown is 60 N. A package slides from the position A down the incline and displaces the right end of the spring from B to C by an amount $\delta = 0.6$ m before it comes to a stop. Determine the work U_s of the spring force $\mathbf{F}_s$ acting on the package during its displacement from A to C.

Solution. We first draw the free-body diagram of the package as shown. Note that the spring force $\mathbf{F}_s$ will act on the package only when the package comes to the region between B and C. The other forces are ever present during the displacement.

Initially, the tensile force in the cord and the compressive force in the spring must have the same magnitude; i.e., $F_{s1} = 60$ N. For a spring modulus $k = 300$ N/m, the magnitude of the initial deflection x_1 of the spring is given by

$$x_1 = F_{s1}/k = 60/300 \qquad x_1 = 0.2 \text{ m}$$

The final deflection x_2 of the spring is given by

$$x_2 = x_1 + 0.6 \qquad x_2 = 0.8 \text{ m}$$

The magnitude of the spring force $\mathbf{F}_s$ in terms of the deflection x of the spring is $F_s = kx = 300x$. The differential displacement $d\mathbf{r}$ of the package and $\mathbf{F}_s$ acting on the package are opposite in sense. Thus, the angle ψ_s between $d\mathbf{r}$ and $\mathbf{F}_s$ is 180°. Noting that $|d\mathbf{r}| = dx$, we write

$$U_s = \int_C \mathbf{F}_s \cdot d\mathbf{r} = -\int_{x_1}^{x_2} F_s \, dx = -\int_{0.2}^{0.8} 300x \, dx$$

$$U_s = -90 \text{ J} \blacktriangleleft$$

REMARK. By the concept of "area" in calculus, the value of U_s may also be obtained from the "area" of the trapezoid in the spring-force diagram as shown. Using the formula for the area of a trapezoid (cf. Sec. B2.3 of App. B), we write

$$U_s = -\frac{1}{2}(60 + 240)(0.6)$$

$$U_s = -90 \text{ J} \blacktriangleleft$$

Developmental Exercises

D13.4 A vertical force $\mathbf{F} = 6x\mathbf{j}$ N acts on a particle during its motion from A_1 to A_2 along the path C defined by $y = x$ as shown. For computing the work $U_{1 \to 2}$ of $\mathbf{F}$ acting on the particle, we note that $\mathbf{r} = x\mathbf{i} + y\mathbf{j}$, $d\mathbf{r} = dx\mathbf{i} + dy\mathbf{j}$. Determine (a) dy in terms of dx along C, (b) $\mathbf{F} \cdot d\mathbf{r}$ in terms of x and dx, (c) the work $U_{1 \to 2}$ using Eq. (13.2) where the integral is evaluated from $x = 0$ at A_1 to $x = 1$ m at A_2.

D13.5 Do D13.4 if the path C is defined by $y = x(2 - x)$ as shown. Is the force $\mathbf{F}$ under consideration a *conservative force*?

Fig. D13.4   **Fig. D13.5**

13.2 Power and the Subtlety in Defining Work

Power is defined as the time rate of doing work. If ΔU is the work of a force **F** during the time interval Δt, the average power during this interval is $P_{\text{avg}} = \Delta U/\Delta t$. As $\Delta t \to 0$, we obtain the power P at any instant as

$$P = \frac{dU}{dt} \tag{13.7}$$

From Eqs. (13.1) and (13.7), we have $P = dU/dt = \mathbf{F} \cdot d\mathbf{r}/dt$. Since $d\mathbf{r}/dt$ is equal to the *velocity* **v** *of the particle* acted on by **F**, we write

$$P = \mathbf{F} \cdot \mathbf{v} \tag{13.8}$$

Such a power is called a *mechanical power*.[†]

As defined above, power is clearly a scalar quantity. In SI, the primary unit of power is the *watt* (W). Note that

$$1 \text{ W} = 1 \text{ J/s} = 1 \text{ N·m/s} \tag{13.9}$$

In the U.S. customary system, power is expressed in ft·lb/s or in *horsepower* (hp). Note that

$$1 \text{ hp} = 550 \text{ ft·lb/s} \tag{13.10}$$

$$1 \text{ hp} = 745.7 \text{ W} = 0.7457 \text{ kW} \tag{13.11}$$

The *efficiency* η of a machine is defined as the ratio of the power output of the machine to the power input to the machine; i.e.,

$$\eta = \frac{P_{\text{output}}}{P_{\text{input}}} \tag{13.12}$$

If the work is done at a constant rate, η may be defined as the ratio of the output work of the machine to the input work to the machine; i.e.,

$$\eta = \frac{U_{\text{output}}}{U_{\text{input}}} \tag{13.13}$$

Efficiency is always less than 1 because every machine operates with some loss of energy due to the negative work done by kinetic friction forces. This negative work is converted into heat energy, which is subsequently dissipated to the surroundings.

Since $dU = P \, dt$, we may use Eq. (13.8) to write the work done $U_{1\to 2}$ during the time interval from t_1 to t_2 as follows:

$$U_{1\to 2} = \int_{t_1}^{t_2} \mathbf{F} \cdot \mathbf{v} \, dt \tag{13.14}$$

where **v** *is the velocity of the particle on which the force* **F** *acts*. Note that **v** here is, in general, *not* the same as the velocity of the point of application of **F** if **F** does *not* act continually on the same particle. The distinction here is a *subtlety* as well as a *necessity* in the definition of work. (Cf. D13.8.)

[†] Note that *electric power = voltage · current*.

This was emphasized and reemphasized by mechanics educators.[†] Since $U_{1 \to 2} = 0$ when $\mathbf{v} = \mathbf{0}$, we see that *no work is done to any body which is stationary during the time the force acts on it.*

For example, let us consider a rigid wheel having a weight W and *rolling without slipping* down a stationary rigid incline, as shown in Fig. 13.3(a). The reaction from the incline to the wheel is denoted by $\mathbf{R}$ as shown in Fig. 13.3(b). We see that the path of each particle on the rim of the wheel is a cycloid with a cusp at the point P where the particle of the wheel comes in contact with the incline, whereas the path of the point of application of $\mathbf{R}$ is a straight line along the incline, as shown in Fig. 13.3(c). The *velocity* $\mathbf{v}$ *of the particle* of the wheel on which $\mathbf{R}$ acts is, at the instant of action by $\mathbf{R}$, equal to *zero*. This is true because $\mathbf{v}$ *reverses its direction* and hence becomes zero at the cusp at P in Fig. 13.3(c).[††] Consequently, according to Eq. (13.14), *the reaction* $\mathbf{R}$ *does no work to the rolling wheel*, despite the fact that the point P of application of the reaction $\mathbf{R}$ moves down the incline with the wheel. On the other hand, the *reaction* $\mathbf{R}$ in Fig. 13.3 *will do negative work to the wheel if the wheel slides down the incline.* In this case, the particle of the wheel at P on which $\mathbf{R}$ acts has, at the instant of action by $\mathbf{R}$, a velocity $\mathbf{v}$ pointing down the incline such that $\mathbf{R} \cdot \mathbf{v} = Rv \cos\psi < 0$, where $90° < \psi < 180°$ as shown in Fig. 13.3(d).

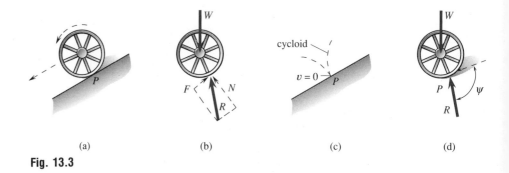

(a)　　　　　(b)　　　　　(c)　　　　　(d)

Fig. 13.3

As another example, let us consider a spinning grinder being slowed down with a stationary rod which is pressed against the grinder at B as shown in Fig. 13.4(a). In such a system, *the kinetic friction force* $\mathbf{F}_B$ *exerted* by the rod at B on the grinder, as shown in Fig. 13.4(b), *does negative work to the grinder* because the particle of the grinder in contact with the rod at B has a velocity $\mathbf{v}$ opposite to $\mathbf{F}_B$ which acts on it. However, *the kinetic friction force* $\mathbf{F}'_B$ *exerted by the grinder on the rod at* B (where $\mathbf{F}'_B = -\mathbf{F}_B$ and $F'_B = F_B$), as shown in Fig. 13.4(c), *does no work to the rod* because the rod has no velocity. Meanwhile, *the normal forces* $\mathbf{N}_B$ *and* $\mathbf{N}'_B$

[†]Osgood, W. F., *Mechanics*, Macmillian, New York, 1937, pp. 261–263; and Langhaar, H. L., "The Role of Rigor in the Teaching of Mechanics," *Mechanics Monograph M-4*, ASEE, 1980, pp. 1–10.

[††]Note that the particle of the wheel at P *does not slide* on the incline and has, at that instant, the same velocity as the incline which is stationary. Furthermore, notice that once the particle of the rolling wheel at P moves upward, the force $\mathbf{R}$ no longer acts on it. Thus, no work is done because the action of the force and the displacement of the particle do not occur simultaneously.

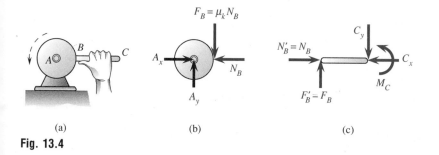

(a)   (b)   (c)

Fig. 13.4

as shown (where $\mathbf{N}_B = -\mathbf{N}'_B$ and $N_B = N'_B$) *do no work to the grinder and the rod*, respectively, because the particles under the action of the normal forces have no velocity components parallel to the normal forces.

Developmental Exercises

D13.6   Define (a) power, (b) efficiency.

D13.7   If a constant force $\mathbf{F}$ does not act continually on the same particle, is the work of $\mathbf{F}$ equal to $\mathbf{F} \cdot \mathbf{q}$ where $\mathbf{q}$ is the displacement of $\mathbf{F}$?

D13.8   An instructor draws a line with a piece of chalk on a chalkboard which is fixed to a wall of the classroom. Determine whether each of the following quantities is *positive*, *zero*, or *negative*: (a) the work done to the chalkboard by the force exerted by the chalk, (b) the work done to the chalk by the reaction from the chalkboard, (c) the work done to the chalk by the force exerted by the instructor.

EXAMPLE 13.3

An electric motor at A is used to hoist a 200-kg crate C. At the instant shown, the crate C has a velocity of 0.4 m/s ↑ and an acceleration of 0.2 m/s² ↑, and the wattmeter at B shows a power input of 850 W. For this instant, determine (a) the velocity of the point D indicated on the cable, (b) the power output of the motor, (c) the efficiency η of the motor.

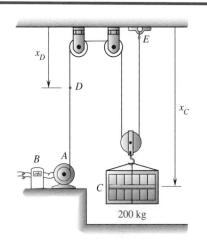

Solution.   At the instant shown, the portion of the cable from D to E must have a constant length. Thus, we write

$$2x_C + x_D = \text{constant} \qquad 2v_C + v_D = 0$$

Since the positive directions are downward, we have

$$v_C = -0.4 \text{ m/s} \qquad v_D = -2v_C = 0.8 \text{ m/s}$$

$$\mathbf{v}_D = 0.8 \text{ m/s} \downarrow \quad \blacktriangleleft$$

Denoting the tension in the cable by F and referring to the free-body and effective-force diagrams drawn, we write

$$+\uparrow \Sigma V_y:^\dagger \qquad F + F - 200g = 200(0.2) \qquad F = 1001 \text{ N}$$

Thus, at the instant shown, the motor exerts a force of $\mathbf{F} = 1001 \text{ N} \downarrow$ on the point D of the cable and moves it with a velocity of $\mathbf{v}_D = 0.8$ m/s $\downarrow$ as shown. We write

$$P_{\text{output}} = \mathbf{F} \cdot \mathbf{v} = (-1001\mathbf{j}) \cdot (-0.8\mathbf{j}) = 800.8$$

$$P_{\text{output}} = 801 \text{ W} \quad \blacktriangleleft$$

$$\eta = \frac{P_{\text{output}}}{P_{\text{input}}} = \frac{800.8}{850} \qquad \eta = 0.942 \quad \blacktriangleleft$$

Developmental Exercise

D13.9 Refer to Example 13.3. Suppose that all the data remain the same except that the acceleration of the crate C is 0.2 m/s^2 $\downarrow$ and the power input to the motor is 800 W. For the instant shown, determine (a) the power output of the motor, (b) the efficiency η of the motor.

13.3 Work of a Gravitational Force

To study the work of a gravitational force acting on a body whose motion involves large changes in altitude in the gravitational force field, let us employ the description using radial and transverse components as presented in Sec. 11.14. We consider a particle of mass m which occupies the position A at a certain instant and is moving under a gravitational force $\mathbf{F}$ exerted at a distance r by a body of mass M at a fixed position O as shown in Fig. 13.5(a), where $m \ll M$. We shall compute the work $U_{1\rightarrow 2}$ of $\mathbf{F}$ acting on the particle during its motion from the position A_1 (where $r = r_1$) to the position A_2 (where $r = r_2$).

Note that the radial and transverse unit vectors are $\mathbf{e}_r$ and $\mathbf{e}_\theta$, respectively. From the first of Eqs. (11.59) and Fig. 13.5, we write

$$d\mathbf{e}_r = d\theta\,\mathbf{e}_\theta \qquad \mathbf{r} = r\mathbf{e}_r \qquad d\mathbf{r} = dr\mathbf{e}_r + r(d\theta\,\mathbf{e}_\theta)$$

$$d\mathbf{r} = dr\,\mathbf{e}_r + r\,d\theta\,\mathbf{e}_\theta \tag{13.15}$$

Applying Newton's law of gravitation, we write

$$\mathbf{F} = -GMmr^{-2}\mathbf{e}_r \tag{13.16}$$

---

†Cf. the second footnote on p. 513.

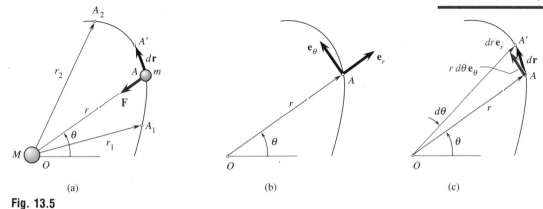

Fig. 13.5

Thus, the work $U_{1 \rightarrow 2}$ of $\mathbf{F}$ acting on the particle during its motion from A_1 to A_2 may be computed as follows:

$$dU = \mathbf{F} \cdot d\mathbf{r} \tag{13.17}$$

$$U_{1 \rightarrow 2} = \int_C \mathbf{F} \cdot d\mathbf{r} = -GMm \int_{r_1}^{r_2} r^{-2}\, dr$$

$$U_{1 \rightarrow 2} = GMm\left(\frac{1}{r_2} - \frac{1}{r_1}\right) \tag{13.18}$$

which shows that the work of a gravitational force acting on a body moving from A_1 to A_2 does not depend on the path connecting A_1 and A_2. Therefore, a gravitational force is a *conservative force*.

Developmental Exercises

D13.10   Refer to Fig. 13.5. (a) Is the position vector $\mathbf{r} = r\mathbf{e}_r$? (b) Is the magnitude $|\mathbf{r}| = r$? (c) Is $|d\mathbf{r}| = dr$? (d) Is $|d\mathbf{r}| = d|\mathbf{r}|$?

D13.11   Determine the work of the gravitational force acting on each 1 kg of mass of a geosynchronous satellite during its flight from the surface of the earth to its orbit, which is 35.8 Mm above the equator.

13.4   **Work of a Spring Force**

The force exerted by an elastic spring on a body attached to the spring is proportional to the deformation in length of the spring.[†] For a body in the position A connected to a fixed point at B by a spring with a modulus k and an elongation x as shown in Fig. 13.6(a), the spring force acting on the body is $\mathbf{F} = kx \leftarrow$; i.e.,

$$F = kx \tag{13.19}{}^{\dagger\dagger}$$

[†]Unless otherwise noted, all springs will be assumed to be linear.

[††]Under dynamic conditions, the error introduced by using the relation $F = kx$ is small if the mass of the spring is small compared with other moving masses in the system.

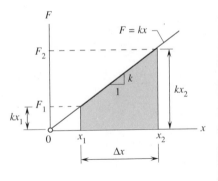

(a)

(b)

Fig. 13.6

The spring modulus k is usually expressed in N/m or kN/m if SI units are used and in lb/ft or lb/in. if U.S. customary units are used. However, care should be taken to express k and x in a consistent system of kinetic units (cf. Table 12.1).

The work $U_{1\rightarrow2}$ of the spring force **F** acting on the attached body during a finite displacement of the body from the position A_1 (where $x = x_1$) to the position A_2 (where $x = x_2$) is obtained as follows:

$$dU = F(-dx) = -kx\,dx \qquad U_{1\rightarrow2} = -k\int_{x_1}^{x_2} x\,dx$$

$$U_{1\rightarrow2} = -\tfrac{1}{2}k(x_2^2 - x_1^2) \tag{13.20}$$

Furthermore, we may obtain $U_{1\rightarrow2}$ by computing the area of the trapezoid shown in the *spring-force diagram* in Fig. 13.6(b), where $F_1 = kx_1$, $F_2 = kx_2$, and $\Delta x = x_2 - x_1$. Since $U_{1\rightarrow2}$ is negative when $|x_2| > |x_1|$, we may use the formula for the area of a trapezoid to write

$$U_{1\rightarrow2} = -\tfrac{1}{2}(F_1 + F_2)\,\Delta x \tag{13.21}$$

It is important to note that the work of the spring force acting on a body is dependent on the changes in length (i.e., x_1 and x_2) of the spring, not on the initial and final orientations of the spring.

EXAMPLE 13.4

A collar P is mounted on the horizontal bar AB and is attached to an elastic cord POD as shown. The elastic cord has a free length $L = 10$ m and a spring modulus $k = 400$ N/m. Using the basic definition of work as given in Eq. (13.2), determine the work $U_{1\rightarrow2}$ of the spring force **F** of the elastic cord acting on the collar during its displacement from A to B.

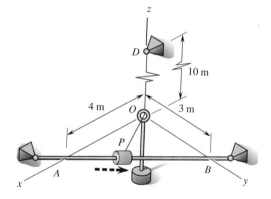

Solution. The path C of the collar P in moving from A to B is given by $y = -\frac{3}{4}x + 3$. The position vector of P is

$$\mathbf{r} = x\mathbf{i} + y\mathbf{j} = x\mathbf{i} + (-\tfrac{3}{4}x + 3)\mathbf{j}$$

Since $k = 400$ N/m, we write

$$\mathbf{F} = -kr\mathbf{e}_r = -k\mathbf{r} = -400x\mathbf{i} + 400(\tfrac{3}{4}x - 3)\mathbf{j}$$

$$d\mathbf{r} = dx\mathbf{i} - \tfrac{3}{4}dx\mathbf{j} = (\mathbf{i} - \tfrac{3}{4}\mathbf{j})\,dx$$

$$\mathbf{F} \cdot d\mathbf{r} = (-625x + 900)\,dx$$

$$U_{1\to2} = \int_C \mathbf{F} \cdot d\mathbf{r} = \int_4^0 (-625x + 900)\,dx = 1400$$

$$U_{1\to2} = 1400 \text{ J} \quad \blacktriangleleft$$

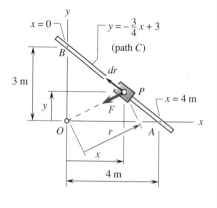

EXAMPLE 13.5

Using a *spring-force diagram*, work Example 13.4.

Solution. Since the free length of the elastic cord is $L = 10$ m and $\overline{OD} = 10$ m, we readily infer that the net elongation of the elastic cord is $r = \overline{OP}$. Therefore, the spring force is $F = kr = 400r$. At the position A,

$r = r_1 = \overline{OA} = 4$ m; and at the position B, $r = r_2 = \overline{OB} = 3$ m. In moving from A to B, r first decreases from $r_1 = 4$ m to $r_{\min} = \frac{3}{5} r_1 = 2.4$ m and then increases to $r_2 = 3$ m. Since $r_2 < r_1$, the total work $U_{1\to2}$ of the spring force acting on the collar during its displacement from A to B must be positive. Thus, by computing the area of the shaded trapezoid in the spring-force diagram shown, we write

$$U_{1\to2} = \tfrac{1}{2}(1200 + 1600)(1) \qquad U_{1\to2} = 1400 \text{ J} \quad \blacktriangleleft$$

Fig. D13.12

Developmental Exercise

D13.12 A spring CD of free length 1 m and modulus 1 kN/m is attached to a sliding collar C of mass 10 kg as shown. Determine the works U_W and U_s of the weight force $\mathbf{W}$ of the collar and the spring force $\mathbf{F}_s$, respectively, acting on the collar during its motion from the position A to the position B.

13.5 Kinetic Energy of a Particle

The *kinetic energy*, denoted by T, of a particle with a mass m and a speed v is defined to be $\frac{1}{2}mv^2$, which is equal to *the total work done to the particle in bringing it from its state of rest to its state of motion with the speed v.* We write

$$T = \tfrac{1}{2}mv^2 \tag{13.22}†$$

For verification, let the resultant force acting on the particle of mass m be $\mathbf{F}$ as shown in Fig. 13.7. Employing the description using tangential and normal components as presented in Sec. 11.13, we write

$$dU = \mathbf{F} \cdot d\mathbf{r} = (F_t\mathbf{e}_t + F_n\mathbf{e}_n) \cdot (ds\,\mathbf{e}_t) = F_t\,ds$$

$$= ma_t\,ds = m\frac{dv}{dt}\,ds = m\frac{dv}{ds}\frac{ds}{dt}\,ds = m\frac{ds}{dt}\,dv$$

$$dU = mv\,dv \tag{13.23}$$

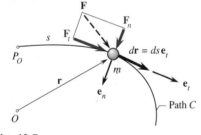

Fig. 13.7

The total work done to the particle in bringing it from a state of rest to a state having a speed v is, therefore,

$$U_{0\to v} = \int_C dU = m\int_0^v v\,dv = \tfrac{1}{2}mv^2 \qquad \text{Q.E.D.}$$

The kinetic energy T as defined in Eq. (13.22) is a *scalar quantity* which is independent of the direction of the velocity and is always *positive* or *zero*. Since a reference frame translating with a constant velocity in an inertial

---

†It may be of interest to note that the relativistic formula for the kinetic energy of a particle with a speed v is $T = m_0c^2[(1 - v^2/c^2)^{-1/2} - 1]$ where c is the speed of light and m_0 is the mass of the particle at rest. By binomial expansion, it can be shown that $T \approx \frac{1}{2}m_0v^2$ if $v \ll c$. This was referred to in Sec. 1.14.

reference frame is also an inertial reference frame, the kinetic energy of a body is clearly a *relative quantity* whose value is dependent on the inertial reference frame chosen for the description. However, *a common inertial reference frame should be used in solving a kinetic problem.*[†]

It can readily be verified that the kinetic energy has the same dimensions and units as the work of a force. This means that the primary units for the kinetic energy are joules (J) in the SI and foot-pounds (ft·lb) in the U.S. customary system.

Developmental Exercises

D13.13   Define the kinetic energy of a particle.

D13.14   Show that the dimensions of kinetic energy are *force × length*.

D13.15   Using Eq. (13.23), determine the total work which must be done to a particle with a mass m and a speed v to bring it to a state of rest.

D13.16   Which of the following are *false*? (a) The kinetic energy points in the same direction as the velocity. (b) Both work and kinetic energy are measured in the same units. (c) Both work and kinetic energy can be positive, zero, or negative. (d) The work is always positive or zero. (e) The kinetic energy is always positive or zero.

13.6   Principle of Work and Energy

A differential work dU done by the resultant force $\mathbf{F}$ to a particle with a mass m and a speed v will result in a differential change dv in the speed of the particle. Such a relation was given by Eq. (13.23). By integrating this differential equation from the position A_1 (where $v = v_1$) to the position A_2 (where $v = v_2$) as shown in Fig. 13.8, we write

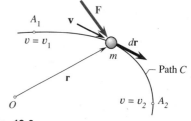

$$U_{1 \to 2} = \int_C dU = m \int_{v_1}^{v_2} v \, dv$$

$$U_{1 \to 2} = \tfrac{1}{2} m v_2^2 - \tfrac{1}{2} m v_1^2 \tag{13.24}$$

Fig. 13.8

where $U_{1 \to 2}$ is the work of $\mathbf{F}$ acting on the particle during its motion from A_1 to A_2. Since the kinetic energies of the particle at A_1 and A_2 are $T_1 = \tfrac{1}{2} m v_1^2$ and $T_2 = \tfrac{1}{2} m v_2^2$, respectively, we may write Eq. (13.24) as

$$\boxed{U_{1 \to 2} = T_2 - T_1} \tag{13.25}$$

Equation (13.25) expresses the *principle of work and energy* for a particle. Since the work of the resultant force is equal to the sum of the works of its component forces (cf. Sec. 13.1), this principle may be stated as follows: *The total work of all forces acting on a particle during an interval of its motion is equal to the change in kinetic energy of the particle during*

[†]Any reference frame which is either fixed or practically translating with a constant velocity relative to the sun is, for most engineering purposes, an inertial reference frame. (Cf. Sec. 3.1.)

that interval. Rearranging the terms in Eq. (13.25), we write

$$T_1 + U_{1\to2} = T_2 \tag{13.26}$$

Thus, the *principle of work and energy* for a particle may alternatively be stated as follows: *The kinetic energy of a particle at A_1 plus the total work of all forces acting on the particle during its motion from A_1 to A_2 is equal to the kinetic energy of the particle at A_2.* Since Newton's second law of motion is a basis upon which this principle is derived, this principle should obviously be applied *only* in an inertial reference frame. For example, the *speed* used to compute the kinetic energy and the *displacement* used to compute the work should be measured in a common inertial reference frame.

We note that the principle of work and energy directly relates *displacement* and *velocity* (among others) without explicitly involving acceleration and time. Furthermore, we refer to the procedure which uses the principle of work and energy to analyze and solve appropriate kinetic problems as the *method of work and energy.*

EXAMPLE 13.6

A 100-lb crate is given an initial speed $v_1 = 4$ ft/s down a chute from the position A_1 as shown. If $\mu_k = 0.3$ between the crate and the chute, determine the speed v_2 of the crate when it reaches the position A_2.

Solution. We note that the crate does not move perpendicularly to the chute; i.e., the crate is in equilibrium in the y direction. Noting in the diagram that $\cos\theta = 24/25$, we write

$$+\nearrow \Sigma F_y = 0: \qquad N - 100\cos\theta = 0 \qquad N = 96 \text{ lb}$$

Friction law: $\qquad F_k = \mu_k N = 0.3N \qquad F_k = 28.8 \text{ lb}$

Kinetic energy at A_1: $\qquad T_1 = \tfrac{1}{2}mv_1^2 = \dfrac{1}{2}\left(\dfrac{100}{32.2}\right)(4)^2$

Kinetic energy at A_2: $\qquad T_2 = \tfrac{1}{2}mv_2^2 = \dfrac{1}{2}\left(\dfrac{100}{32.2}\right)v_2^2$

Total work ($U = Fq\cos\psi$) of **W**, **N**, and **F**:

$$U_{1\to2} = 100(25)\left(\frac{7}{25}\right) + 0 + 28.8(25)(-1)$$

Applying the *principle of work and energy*, $T_1 + U_{1 \rightarrow 2} = T_2$, we write

$$\frac{1}{2}\left(\frac{100}{32.2}\right)(4)^2 + \left[100(25)\left(\frac{7}{25}\right) + 0 + 28.8(25)(-1)\right]$$

$$= \frac{1}{2}\left(\frac{100}{32.2}\right)v_2^2$$

$$v_2 = 1.766 \text{ ft/s} \quad \blacktriangleleft$$

Developmental Exercise

D13.17   Describe the *principle of work and energy* for a particle.

EXAMPLE 13.7

A spacecraft is in free flight around the earth as shown, where $a_1 = 3000$ km and $a_2 = 6000$ km. When the spacecraft passes through the position A_1, its speed is $v_1 = 6.27$ km/s. Determine the speed v_2 of the spacecraft when it passes through the position A_2.

Solution.   The only force acting on the spacecraft in free flight is the gravitational force from the earth whose radius is $R = 6370$ km. We have

$$r_1 = \overline{OA_1} = R + a_1 = 6370 + 3000 \qquad r_1 = 9370 \text{ km}$$

$$r_2 = \overline{OA_2} = R + a_2 = 6370 + 6000 \qquad r_2 = 12370 \text{ km}$$

Let the mass of the spacecraft be m. The work of the gravitational force acting on the spacecraft during its motion from A_1 to A_2 is given by Eq. (13.18), where $GM = gR^2$. Applying the *principle of work and energy*, $T_1 + U_{1 \rightarrow 2} = T_2$, we write

$$\tfrac{1}{2} m v_1^2 + \left[GMm\left(\frac{1}{r_2} - \frac{1}{r_1}\right)\right] = \tfrac{1}{2} m v_2^2$$

$$\tfrac{1}{2} v_1^2 + gR^2\left(\frac{1}{r_2} - \frac{1}{r_1}\right) = \tfrac{1}{2} v_2^2$$

$$\tfrac{1}{2}(6.27)^2 + \frac{9.81}{1000}(6370)^2\left(\frac{1}{12370} - \frac{1}{9370}\right) = \tfrac{1}{2} v_2^2$$

$$v_2 = 4.33 \text{ km/s} \quad \blacktriangleleft$$

Developmental Exercise

D13.18   If the minimum speed of the spacecraft in Example 13.7 is 3.99 km/s, determine the maximum altitude of its orbit.

EXAMPLE 13.8

If the pendulum of weight W is released from rest from a horizontal position OA_1 and swings in a vertical plane to OA_2 as shown, determine (a) the speed v of the bob in terms of g, L, and θ in the range $0 \le \theta \le 90°$, (b) the angle θ at which the tension F in the cord is equal to $2W$, (c) the tension F_2 in the cord when the bob reaches the position A_2.

Solution. As the pendulum swings down, the tensile force **F** exerted by the cord on the bob, as shown in Fig. (a), is always perpendicular to the differential displacement $d\mathbf{r}$ (and hence the path) of the bob. Thus, **F** does no work to the bob. The work is done only by the weight force **W**. Applying the *principle of work and energy*, $T_{A_1} + U_{A_1 \to A} = T_A$, we write

$$0 + [0 + W(L \sin\theta)] = \frac{1}{2}\left(\frac{W}{g}\right)v^2$$

$$v^2 = 2\,gL\,\sin\theta \qquad v = (2gL\,\sin\theta)^{1/2} \blacktriangleleft$$

The normal component of acceleration of the bob at A is

$$a_n = \frac{v^2}{\rho} = \frac{2gL\,\sin\theta}{L} = 2g\,\sin\theta$$

When $\theta = 90°$, we have $a_n = 2g$. These two values of a_n are used in drawing the effective-force diagram as shown in Figs. (b) and (c).

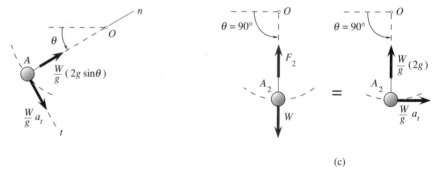

To find the angle θ at which the tension F in the cord is equal to $2W$, we refer to the vector-diagram equation in Fig. (b) and write

$$+\nearrow \Sigma V_n: \qquad 2W - W\sin\theta = \frac{W}{g}(2g\,\sin\theta)$$

$$\sin\theta = \frac{2}{3}$$

$$\theta = 41.8° \blacktriangleleft$$

To find the tension F_2 in the cord when the bob reaches the position A_2, we refer to the vector-diagram equation in Fig. (c) and write

$$+\uparrow \Sigma V_y: \qquad F_2 - W = \frac{W}{g}(2g) \qquad F_2 = 3W \blacktriangleleft$$

Developmental Exercise

D13.19   Refer to Example 13.8. Determine the angle θ at which the tension F in the cord is equal to $2.5W$.

13.7 System of Particles

If a system consists of several particles, we may consider each particle separately and apply the principle of work and energy to each particle. By adding side by side all equations of work and energy, in the form of Eq. (13.26), for all particles in the system, we obtain the *principle of work and energy* for a system of particles as follows:

$$T_1 + U_{1 \to 2} = T_2 \qquad (13.27)$$

where T_1 represents the arithmetic sum of the *initial* kinetic energies of all the particles, T_2 represents the arithmetic sum of the *final* kinetic energies of all the particles, $U_{1 \to 2}$ represents the algebraic sum of the works of all the forces acting on the various particles *during the interval*, and the forces include both *internal* and *external* forces to the system (cf. Sec. 3.3). We know that internal forces occur in pairs of forces equal in magnitude and opposite in direction. If the particles in the system are connected by inextensible cordlike elements or links, such pairs of internal forces in them must be tangent to their centerlines, which do not elongate nor shrink. Consequently, the total work of the internal forces to such a system is zero, and $U_{1 \to 2}$ is equal to the total work of the external forces only.

The kinetic energy T of a system of n particles may be written as

$$T = \frac{1}{2} \sum_{i=1}^{n} m_i v_i^2 \qquad (13.28)$$

where m_i is the mass and v_i is the speed, in the inertial reference frame $Oxyz$, of the particle P_i as shown in Fig. 13.9.[†] Note that

$$\mathbf{v}_i = \bar{\mathbf{v}} + \mathbf{v}_i' \qquad (13.29)$$

where $\bar{\mathbf{v}}$ is the velocity of the mass center G of the system in $Oxyz$, and $\mathbf{v}_i'$ is the velocity of P_i in the reference frame $Gx'y'z'$, which translates with G and keeps the same orientation in $Oxyz$. Since $v_i^2 = \mathbf{v}_i \cdot \mathbf{v}_i$, we may use Eqs. (13.28) and (13.29) to write

$$T = \frac{1}{2} \sum_{i=1}^{n} [m_i(\bar{\mathbf{v}} + \mathbf{v}_i') \cdot (\bar{\mathbf{v}} + \mathbf{v}_i')]$$

$$= \frac{1}{2} \left(\sum_{i=1}^{n} m_i \right) \bar{v}^2 + \bar{\mathbf{v}} \cdot \left(\sum_{i=1}^{n} m_i \mathbf{v}_i' \right) + \frac{1}{2} \sum_{i=1}^{n} m_i v_i'^2$$

We recognize that the first sum is equal to the total mass m of the system. By the *principle of moments*, we see that the second sum is equal to $m\bar{\mathbf{v}}'$, where $\bar{\mathbf{v}}'$ represents the velocity of G in $Gx'y'z'$, which is clearly zero.

Fig. 13.9

[†]Cf. the footnote on p. 479.

Thus, we have

$$T = \tfrac{1}{2} m \bar{v}^2 + \frac{1}{2} \sum_{i=1}^{n} m_i v_i'^2 \qquad (13.30)$$

Equation (13.30) establishes the property that the kinetic energy of a system of particles (including the case of a rigid body) may be perceived to be composed of two parts: (a) *the kinetic energy of the system moving with G in the inertial reference frame Oxyz*, and (b) *the kinetic energy of the system moving relative to G in the translating reference frame Gx'y'z'*.

EXAMPLE 13.9

The blocks A and B are released to move from rest as shown. If $\mu_k = 0.3$ between the block A and the support, determine the velocity $\mathbf{v}_A$ of the block A when the block B strikes the ground.

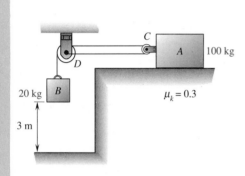

Solution. The free-body diagram of the entire system and the expected motions are first drawn and indicated as shown below. Similar to that in Example 12.3, the constraint condition is

$$2x_A + x_B = \text{constant}$$

from which we have

$$2 \, \Delta x_A + \Delta x_B = 0 \qquad 2v_A + v_B = 0$$

Thus, when the block B has a displacement of $\Delta x_B = 3$ m downward, the displacement of the block A is $\Delta x_A = -\tfrac{1}{2} \Delta x_B = -1.5$ m. The negative sign indicates that the block A displaces to the left as shown. For the block B, we have $v_B = -2v_A$. The block A as shown is in equilibrium in the vertical direction. Hence, we have

$$N_A = W_A = 100g = 981 \text{ N} \qquad F_A = \mu_k N_A = 294.3 \text{ N}$$

$$W_B = 20g = 196.2 \text{ N}$$

During the interval of motion, the forces $\mathbf{W}_A$, $\mathbf{N}_A$, $\mathbf{D}_x$, and $\mathbf{D}_y$ do no work; only the forces $\mathbf{F}_A$ and $\mathbf{W}_B$ do work. We note that the system is initially at

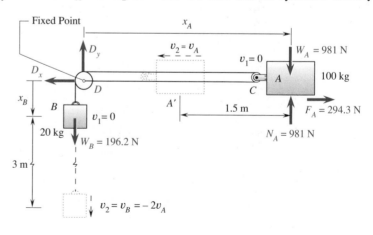

rest. Applying the *principle of work and energy*, $T_1 + U_{1\to2} = T_2$, we write

$$0 + [294.3(-1.5) + 196.2(3)]$$
$$= \tfrac{1}{2}(100)v_A^2 + \tfrac{1}{2}(20)(-2v_A)^2$$
$$v_A = 1.279 \qquad \mathbf{v}_A = 1.279 \text{ m/s} \leftarrow \quad \blacktriangleleft$$

Developmental Exercise

D13.20 Refer to Example 13.9. Determine the displacement of the block *B* when the speed of the block *A* is increased from 0 to 1 m/s.

EXAMPLE 13.10

Two particles *A* and *B* are mounted on a cross, which in turn is mounted on a cart as shown. The cross rotates at a constant angular speed $\dot{\theta}_0 = 4$ rad/s relative to the cart, while the cart moves along the *x* axis at a constant velocity $\mathbf{v}_0 = 3$ ft/s $\searrow$. If each particle weighs 8.05 lb, determine the total kinetic energy *T* of the *particles A* and *B* in the reference frame *Oxyz* fixed to the ground.

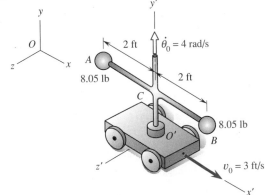

Solution. By symmetry, we know that the mass center of the particles *A* and *B* is located at the point *C* which has a velocity $\mathbf{v} = 3$ ft/s $\searrow$ in *Oxyz*. The speed of each particle relative to the mass center *C* is

$$v' = r\dot{\theta}_0 = 2(4) \qquad v' = 8 \text{ ft/s}$$

The total mass of *A* and *B* is

$$m = m_A + m_B = \frac{8.05 + 8.05}{32.2} \text{ slug} = \frac{16.1}{32.2} \text{ slug}$$

Thus, the total kinetic energies of *A* and *B* in *Oxyz* is [cf. Eq. (13.30)]

$$T = \frac{1}{2}\left(\frac{16.1}{32.2}\right)(3)^2 + \frac{1}{2}\left[\frac{8.05}{32.2}(8)^2 + \frac{8.05}{32.2}(8)^2\right]$$
$$T = 18.25 \text{ ft}\cdot\text{lb} \quad \blacktriangleleft$$

Developmental Exercises

D13.21 Refer to Example 13.10. (a) Determine the total kinetic energy T' of the particles A and B in the reference frame $O'x'y'z'$ which is fixed to the cart. (b) Is $O'x'y'z'$ an inertial reference frame?

D13.22 Refer to Example 13.10. If the cart weighs 50 lb and the weights of the cross and the wheels are negligible, what is the total kinetic energy of the system in $Oxyz$?

PROBLEMS

13.1* Using the *principle of work and energy*, solve Prob. 12.10.

13.2 An astronaut drops a 5-kg rock from a height of 2 m above the surface of the moon, where the gravitational acceleration is 1.618 m/s^2. Later, this same rock is brought back to the earth and dropped from the same height of 2 m above the ground. Determine the kinetic energy T and the speed v of the rock immediately before it strikes the surface of (a) the moon, (b) the earth.

13.3* A 10-kg package is given an initial speed of 1 m/s in the position A down a slope to the position B where its speed becomes 5 m/s as shown. Determine the work U_f of the friction force acting on the package during its motion from A to B.

Fig. P13.3*   **Fig. P13.4**

✗ **13.4** A disk is projected with an initial velocity of 10 m/s up an incline as shown. If $\mu_k = 0.2$ between the disk and the incline, determine (a) the maximum distance L which the disk will travel up the incline, (b) the speed of the disk as it slides back through its original position.

✗ **13.5** A 20-lb collar C is moved from rest from the position A to the position B by a cable CDE which passes around a pulley at D and is pulled downward by a constant force of 50 lb at E as shown. Neglecting effects of friction, determine the speed of the collar as it passes through B.

Fig. P13.5   **Fig. P13.6** B

✗ **13.6** The cord of a pendulum is deflected by a peg at C when it is released from the position A as shown. As it swings to the position B, determine (a) the speed of the bob, (b) the tension in the cord.

13.7* and 13.8 Determine the speed of the block A after it is moved 2 m from rest by the horizontal force **P** as shown.

Fig. P13.7*

Fig. P13.8

Fig. P13.9* and P13.10

13.9* A projectile in the position A_1 has an altitude $a_1 = 80$ km and a velocity $\mathbf{v}_1 = 8$ km/s ↑ as shown. Neglecting air resistance, determine the highest altitude a_2 reached by the projectile.

13.10 At engine burnout, a rocket has an altitude $a_1 = 75$ mi and an upward speed v_1 as shown. If the rocket is to reach a geosynchronous satellite at an altitude $a_2 = 22.2 \times 10^3$ mi and air resistance above the altitude a_1 is negligible, determine the value of v_1.

13.11 and 13.12* A satellite describes an elliptic orbit around the earth as shown. In the position S, it has an altitude a_S and a speed v_S as indicated. Determine the speed v_P of the satellite when it passes through the perigee P.

Fig. P13.11 and P13.13*

Fig. P13.12* and P13.14

13.13* and 13.14 A satellite describes an elliptic orbit around the earth as shown. In the position S, it has an altitude a_S and a speed v_S as indicated. When it passes through the point B, its speed is v_B as indicated. Determine the altitude a_B of the point B.

13.15 Refer to Prob. 13.11. Determine the altitude a_Q of the apogee Q.

13.16 Refer to Prob. 13.12. Determine the altitude a_Q of the apogee Q.

13.17 and 13.18* A spring of free length 2.5 m and modulus 200 N/m is attached to a sliding collar C of mass 20 kg as shown. The collar is released from rest at A to slide downward. Neglecting effects of friction, determine the speed of the collar as it passes through B.

X**13.19** A spring of free length 10 m and modulus 120 N/m is attached to a sliding collar C of mass 20 kg as shown. The collar is released from rest at A. Neglecting effects of friction, determine the speed of the collar as it passes through B.

Fig. P13.17

Fig. P13.18*

Fig. P13.19

Fig. P13.20

Fig. P13.21

13.20 A 60-lb sphere connected to a spring, which has a free length of 4 ft and a modulus of 120 lb/ft, is to be released from rest in the position $x = 0$ as shown. Determine (a) the maximum elongation of the spring, (b) the maximum speed of the sphere. (*Hint*. When the speed is maximum, what is the resultant force acting on the sphere?)

13.21 A 10-kg cylindrical collar is released from rest in the position shown and drops onto the spring which has a free length of 2 m and a modulus of 300 N/m. Determine (a) the maximum speed of the collar, (b) the maximum contraction of the spring.

Fig. P13.22*

Fig. P13.23*

13.22* A spring is precompressed 0.2 m by a light board and restraining cables as shown. The modulus of the spring is 2 kN/m, and $\mu_k = 0.25$ between the 100-kg crate and the incline. If the speed of the crate is 2 m/s when it is 2 m from the spring, determine the maximum additional deflection of the spring.

13.23* The length of the barrel of an air pistol is 250 mm. Upon firing, the magnitude of the net force **F** acting on the pellet varies with the position x of the pellet in the barrel as shown. If the diameter of the bore is 4.5 mm and the mass of the pellet is 500 mg, determine (a) the maximum pressure in the barrel, (b) the muzzle speed of the pellet. (*Hint*. Compute the area by dividing it into a series of trapezoids, or by using the program AREA in the Supplementary Software.)

13.24 Solve Prob. 13.20 if the spring is a hard spring where the magnitude of the spring force **F** developed due to an elongation x is given as shown.

Fig. P13.24

13.25 and 13.26* The blocks are released from rest in the position shown. Determine the speed of the block A when the block B is stopped by impact.

Fig. P13.25

Fig. P13.26*

Fig. P13.27

13.27 The system is released from rest in the position shown. After the 30-lb block B has moved 3 ft downward, the 15-lb collar C is lifted from the block B, which continues to move. Determine the speed of the 20-lb block A just before the block B strikes the bottom of the pit.

13.28 and 13.29 The system is at rest when the 400-N force is applied to the block B as shown. Neglecting effects of friction, determine the speed of the block B when the block A has moved 1.5 m.

Fig. P13.28 and P13.30*

Fig. P13.29 and P13.31

Winch

Fig. P13.32* and P13.33

2470 lb

Fig. P13.35 and P13.36

M

C

100 kg

E

400 kg

Fig. P13.37* and P13.38

13.30* and 13.31 The system is at rest when the 400-N force is applied to the block B as shown. If $\mu_k = 0.25$ between the blocks and the support, determine the speed of the block B when the block A has moved 1.5 m.

13.32* A power winch hoists a 340-kg crate up an incline at a constant speed $v = 0.5$ m/s as shown. If the power input to the winch is 1.5 kW and $\mu_k = 0.4$ between the crate and the incline, determine (a) the power output of the winch, (b) the efficiency of the winch.

13.33 A power winch hoists a 340-kg crate up an incline at a constant speed $v = 1$ m/s as shown. The power output of the winch is known to be 2 kW. Determine μ_k between the crate and the incline.

13.34 Refer to Prob. 13.33. If the power output of the winch is suddenly increased from 2 kW to 3 kW, determine the acceleration which the crate may momentarily experience.

13.35 A 2470-lb car with four-wheel drive climbs an incline as shown. The output power of the car is known to be 100 hp and the wheels do not slip on the incline. Neglecting effects of drag resistance, determine the constant speed v attained by the car.

13.36 A 2470-lb car with four-wheel drive climbs an incline as shown. As it moves forward, it experiences a drag resistance from the wind $F_D = 0.007v^2$ lb, where v is the speed of the car measured in ft/s. The output power of the car is known to be 100 hp, and the wheels do not slip on the incline. Determine the constant speed v attained by the car.[†]

13.37* The elevator E with its freight has a total mass of 400 kg and is hoisted by its 100-kg counterweight C and the motor at M as shown. If the motor has an efficiency of 0.8, determine the power that must be supplied to the motor when the elevator (a) is moving upward at a constant speed of 2 m/s, (b) has an instantaneous velocity of 2 m/s ↑ and an acceleration of 1 m/s² ↑.

13.38 The elevator E with its freight has a total mass of 400 kg and is hoisted by its 100-kg counterweight C and the motor at M as shown. If the output power of the motor is 6 kW at the instant the velocity of the elevator is 2 m/s ↑, determine the acceleration of the elevator.

CONSERVATION OF ENERGY AND SPECIAL TOPIC[††]

13.8 Conservative System and Potential Energy

An *analytic force field* is a region in space where a body experiences a determinable single-valued force or force system at every position in the region. Since the kinetic friction force developed on a body at a given position may have different values in magnitude and direction, a kinetic friction force field is not an analytic force field. A *conservative force field* is an analytic force field in which the work of the force in the field acting on a body traveling in the field is dependent on the initial and final positions of

---

[†]Cf. App. E for computer solution of resulting cubic equation.

[††]The concepts of potential energy and virtual work were presented in Chap. 10. For convenience, we repeat here portions of that material relevant to the current study.

the body, not on the path traveled by the body. The force in a conservative force field is called a *conservative force*.

It can be shown that the necessary and sufficient condition for the force $\mathbf{F} = \mathbf{F}(x, y, z) = F_x\mathbf{i} + F_y\mathbf{j} + F_z\mathbf{k}$ to be conservative is $\nabla \times \mathbf{F} = \mathbf{0}$, where the symbol ∇ is a vector operator called "del." Note that

$$\nabla = \mathbf{i}\,\frac{\partial}{\partial x} + \mathbf{j}\,\frac{\partial}{\partial y} + \mathbf{k}\,\frac{\partial}{\partial z}$$

and the cross product $\nabla \times \mathbf{F}$ is called the *curl* of $\mathbf{F}$. Thus, it can be shown that $\mathbf{F}$ is conservative *if and only if*[†]

$$\frac{\partial F_z}{\partial y} = \frac{\partial F_y}{\partial z} \qquad \frac{\partial F_x}{\partial z} = \frac{\partial F_z}{\partial x} \qquad \frac{\partial F_y}{\partial x} = \frac{\partial F_x}{\partial y} \qquad (13.31)$$

Moreover, one can show that $\mathbf{F} = \nabla\phi$, where $\phi = \phi(x, y, z)$, if $\mathbf{F}$ is conservative. For example, $\mathbf{F} = x\mathbf{i}$ is conservative, but $\mathbf{F} = x\mathbf{j}$ is not. In the rest of this chapter, we shall deal with the following conservative forces: gravitational forces, elastic spring forces, and applied constant loads.

Any force which acts on a body but does no work to the body during a displacement of the body is called a *nonworking force*. For instance, when a body moves on a surface, the normal force exerted by the surface on the body does not work to the body and is a nonworking force. A *conservative system* is a system where the body or assemblage of bodies is subjected to only conservative forces and nonworking forces, if any.

For practical use, the *potential energy of a body* in a conservative force field relative to an appropriate reference datum is defined as *the amount of work which the body will receive from the conservative force if the body travels from its present position to the reference datum.*[††] Thus, the potential energy as well as the kinetic energy has the same dimensions and units as the work of a force. The potential energy of a body situated at the reference datum is, of course, equal to zero.

Developmental Exercises

D13.23 Define (a) analytic force field, (b) conservative force field, (c) conservative force, (d) nonworking force, (e) conservative system, (f) curl of $\mathbf{F}$.

D13.24 Define the potential energy of a body.

13.9 Gravitational Potential Energy

We know that the force exerted on a body in a *gravitational force field* is the weight force of the body. As a body travels from one position to a nearby position in the vicinity of the surface of the earth, the weight force of the body is essentially a downward vertical force of constant magnitude.

[†]Spiegel, M. R., *Vector Analysis and an Introduction to Tensor Analysis*, Schaum's Outline Series, McGraw-Hill, New York, 1959, pp. 89–91.

[††]This is a *quantitative definition*. *Qualitatively*, the potential energy of a body is the energy possessed by the body by virtue of its position relative to the reference datum in the conservative force field. A potential energy is expressible as a function of spatial coordinates only.

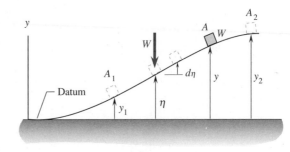

Fig. 13.10

Thus, the *gravitational potential energy* V_g *of a body* with a weight W near the surface of the earth at the position A of altitude y above the reference datum, as shown in Fig. 13.10, is equal to the work which the body will receive from the constant weight force $\mathbf{W}$ of the body if the body travels from its present position at $\eta = y$ to the reference datum at $\eta = 0$. We write

$$V_g = \int_y^0 W\,(-d\eta) \qquad \boxed{V_g = Wy} \qquad (13.32)$$

where V_g is positive only if the datum chosen is situated between the body and the center of the earth. The work $U_{1\to2}$ of the weight force $\mathbf{W} = W\downarrow$ acting on the body traveling from A_1 to A_2 is found to be related to the *change in gravitational potential energy*, ΔV_g, as follows:

$$U_{1\to2} = \int_{y_1}^{y_2} W\,(-d\eta) = -(Wy_2 - Wy_1) = -[(V_g)_2 - (V_g)_1]$$

$$U_{1\to2} = -\Delta V_g \qquad (13.33)$$

In the case of the gravitational central-force motion of a particle (e.g., a spacecraft), we should take into account the variation of the gravitational force with the radial distance r drawn from the center of force O to the particle at the position A as shown in Fig. 13.5. The gravitational potential energy of such a particle is defined relative to *a position at infinity* (i.e., $r = \infty$) *as the reference datum*. Thus, the *potential energy* V_g *of a particle in gravitational central-force motion* at a radial distance r from the center

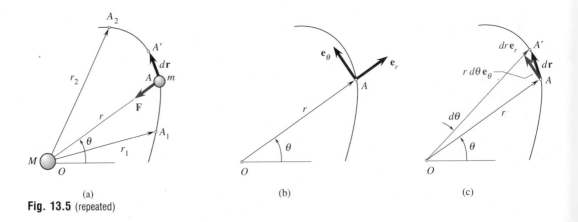

(a)   (b)   (c)

Fig. 13.5 (repeated)

of force O is obtained by computing the work which the particle will receive from the gravitational force $\mathbf{F}$ if the particle travels from its present position to the reference datum at $r = \infty$. Using Eqs. (13.15) through (13.17), we write

$$V_g = \int_r^\infty (-GMmr^{-2}) \, dr \qquad \boxed{V_g = -\frac{GMm}{r}} \qquad (13.34)$$

Note that V_g in Eq. (13.34) is always negative because the work of the gravitational central force $\mathbf{F}$ acting on the particle during its travel from its present position to the reference datum at $r = \infty$ is always negative. The work $U_{1\to 2}$ of such a force $\mathbf{F}$ acting on the particle traveling from A_1 (where $r = r_1$) to A_2 (where $r = r_2$) is found to be related to the change in V_g as follows:

$$\Delta V_g = (V_g)_2 - (V_g)_1 = -\frac{GMm}{r_2} + \frac{GMm}{r_1}$$

$$U_{1\to 2} = \int_{r_1}^{r_2} (-GMmr^{-2}) \, dr = -\left(-\frac{GMm}{r_2} + \frac{GMm}{r_1}\right)$$

$$U_{1\to 2} = -\Delta V_g \qquad (13.35)$$

Note that Eqs. (13.33) and (13.35) are of the same form.

EXAMPLE 13.11

A body of weight W is at an altitude y above the surface of the earth whose radius is R. Choosing the surface of the earth as the reference datum, show that Wy is truly a good approximation for the potential energy V_g of the body if $y \ll R$.

Solution. The weight W of the body is usually measured on the surface of the earth. Let O be the center of the earth, as shown, m be the mass of the body, and M be the mass of the earth. We may apply Newton's law of gravitation and the definition of V_g to write

$$F = \frac{GMm}{r^2} \qquad W = F|_{r=R} = \frac{GMm}{R^2}$$

$$V_g = \int_{R+y}^R F(-dr) = -GMm \int_{R+y}^R r^{-2} \, dr$$

$$= GMm\left(\frac{1}{R} - \frac{1}{R+y}\right) = \frac{GMmy}{R(R+y)}$$

Since $y \ll R$, we have $R + y \approx R$ and

$$V_g \approx \frac{GMm}{R^2} y = Wy \qquad \text{Q.E.D.} \quad \blacktriangleleft$$

Developmental Exercises

D13.25 Define the *gravitational potential energy* of a body.

D13.26 Where is the *reference datum* for the gravitational potential energy of a spacecraft revolving around the earth?

D13.27 Define the *potential energy* of a particle in gravitational central-force motion.

13.10 Elastic Potential Energy

The force exerted by an elastic spring on a body attached to the spring is proportional to the deformation in length of the spring (cf. Sec. 13.4). A body subjected to the forces of an elastic spring is said to be in an *elastic force field*. The *reference datum in an elastic force field* is generally chosen to be the position of the body where the spring attached to the body is neutral. For a body at A connected to B by a spring with a modulus k and an elongation x as shown in Fig. 13.11, the *elastic potential energy V_e of the body* is equal to the work which the body will receive from the spring force $\mathbf{F} = k\xi \leftarrow$ if the body travels from its present position at $\xi = x$ to the reference datum at $\xi = 0$. Thus, we have

$$V_e = \int_x^0 k\xi\,(-d\xi) \qquad \boxed{V_e = \tfrac{1}{2}\,kx^2} \qquad (13.36)^\dagger$$

Fig. 13.11

Similarly, we find that the work $U_{1\to2}$ of the spring force $\mathbf{F}$ acting on the body traveling from A_1 to A_2 is related to the *change in elastic potential energy*, ΔV_e, as follows:

$$U_{1\to2} = \int_{x_1}^{x_2} k\xi\,(-d\xi) = -(\tfrac{1}{2}\,kx_2^2 - \tfrac{1}{2}\,kx_1^2)$$

$$= -[(V_e)_2 - (V_e)_1]$$

$$U_{1\to2} = -\Delta V_e \qquad (13.37)$$

---

† If a body is attached to a *nonlinear spring*, the elastic potential energy of the body may be obtained by integrating the work which the body will receive from the nonlinear spring force if the body travels from its present position to the reference datum. (Cf. Probs. 13.62 and 13.63.)

In the case where a torsional spring of modulus K is used to supply a restoring moment of magnitude $M = K\theta$ to a member AB which carries a body (a lumped mass as a particle) at the end A and undergoes an angular rotation θ from the neutral position of the spring as shown in Fig. 13.12, the *elastic potential energy* V_e *of the member* can similarly be shown to be

$$V_e = \int_\theta^0 K\theta\,(-d\theta) \qquad \boxed{V_e = \tfrac{1}{2} K\theta^2} \qquad (13.38)$$

where the angle θ is measured in radians and the modulus K is measured in moment per radian (e.g., N·m/rad or lb·ft/rad). Of course, Eq. (13.37) is also true in the case of a torsional spring.

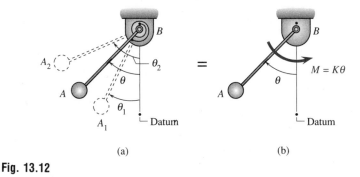

(a)   (b)

Fig. 13.12

Developmental Exercises

D13.28   How is the *reference datum* for the elastic potential energy of a body in an elastic force field generally chosen?

D13.29   Define the *elastic potential energy* of a body.

13.11   Applied Potential Energy

Besides gravitational and elastic force fields, there are other conservative force fields which exist to exert conservative forces on various bodies. A body deflected by an *applied conservative force* $\mathbf{F}$ is said to possess an applied potential energy because the body will receive a definite amount of work from the applied conservative force if it returns to its undeflected state. In this case, we must have $\nabla \times \mathbf{F} = \mathbf{0}$.[†] The *applied potential energy* V_a *of a body* in a deflected state under the action of an applied conservative force is equal to the work which the body will receive from the applied conservative force if the body travels from its present deflected position to its undeflected position of equilibrium, which is the reference datum. However, we shall here focus our attention on applied conservative forces which are constant in magnitude and direction. Note that an *applied constant force* is a constant force which is suddenly or instantaneously applied to a body, and its magnitude as well as direction remains the same thereafter.

Let us now consider a body (a lumped mass as a particle) which is carried by a member AB at its end A and is acted on by an applied constant force $\mathbf{P}$

[†]Cf. Eqs. (13.31).

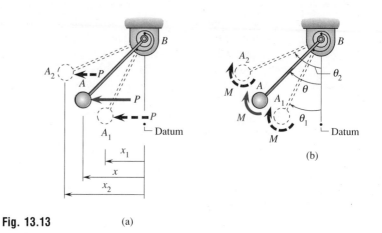

Fig. 13.13   (a)

as shown in Fig. 13.13(a). The *applied potential energy* V_a *of the body* in its deflected position under the action of the applied constant force **P** is equal to the work which the body will receive from **P** if it travels from its present deflected position to its *undeflected neutral equilibrium position*, which is the *reference datum*; i.e.,

$$V_a = \int_x^0 P \, dx \qquad \boxed{V_a = -Px} \qquad (13.39)$$

The work $U_{1 \to 2}$ of the applied constant force **P** acting on the body during its motion from the position where $x = x_1$ to the position where $x = x_2$ is related to the *change in the applied potential energy*, ΔV_a, as follows:

$$U_{1 \to 2} = \int_{x_1}^{x_2} P \, dx = -(-Px_2 + Px_1) = -[(V_a)_2 - (V_a)_1]$$

$$U_{1 \to 2} = -\Delta V_a \qquad (13.40)$$

In the case where an applied constant moment **M** acts on a body which is carried by a member AB at its end A as shown in Fig. 13.13(b), the *applied potential energy* V_a *of the body* can similarly be shown to be

$$\boxed{V_a = -M\theta} \qquad (13.41)$$

Of course, Eq. (13.40) is also true in the case of an applied constant moment **M**.

EXAMPLE 13.12[†]

A constant horizontal force **P** is suddenly applied to a particle of weight W which is attached to a slender rod AB of length L and a spring BC of free length L and modulus k as shown. When the particle is deflected from the position B to the position B' by **P**, determine the gravitational potential

---

[†]To be continued in Example 13.15.

energy V_g, the elastic potential energy V_e, and the applied potential energy V_a of the particle as a function of θ.

Solution. The system is originally in neutral equilibrium in its vertical configuration ABC. When the constant horizontal force **P** is suddenly applied to the particle to deflect it from the point B to the point B', the particle is subjected to an applied constant force field in addition to the gravitational and elastic force fields, and the system is conservative. Thus, we choose the position B as the reference datum for V_g, V_e, and V_a. Referring to the sketch shown, we write

$$V_g = Wy \qquad V_g = WL(1 - \cos\theta) \blacktriangleleft$$

The elongation of the spring is

$$x_s = \overline{CB'} - \overline{CB} = [(2L)^2 + L^2 - 2(2L)L\cos\theta]^{1/2} - L$$

$$= L[(5 - 4\cos\theta)^{1/2} - 1]$$

Thus, we write

$$V_e = \tfrac{1}{2}kx_s^2$$

$$V_e = \tfrac{1}{2}kL^2[(5 - 4\cos\theta)^{1/2} - 1]^2 \blacktriangleleft$$

The force $\mathbf{F}_A$ as shown is directed along the line segment $\overline{B'A}$ because the rod is taken as a massless extension of the particle. The force $\mathbf{F}_A$ does no work and has no contribution to V_a. Thus, we write

$$V_a = -Px \qquad V_a = -PL\sin\theta \blacktriangleleft$$

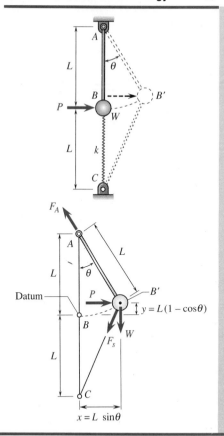

Developmental Exercises

D13.30 Is an applied constant load *gradually* or *suddenly* applied to a body?

D13.31 Define the *applied potential energy* of a body.

13.12 Conservation of Energy

The *potential energy V* of the body in a conservative system is the sum of the gravitational, elastic, and applied potential energies of the body; i.e.,

$$V = V_g + V_e + V_a \qquad (13.42)$$

By definition, we note that the body in a conservative system receives work only from the conservative forces when it undergoes a displacement. From Eqs. (13.32) through (13.42), we readily see that *any work $U_{1\to2}$ of the conservative forces* acting on the body, which moves from the position 1 to the position 2 in the conservative force field, *is equal to the negative of the change in the potential energy of the body*; i.e.,

$$U_{1\to2} = -(V_2 - V_1) \qquad (13.43)$$

This means that, for a body in a conservative system, the principle of work and energy may be expressed in a modified form. Substituting Eq. (13.43) into Eq. (13.26) and rearranging, we obtain

$$T_1 + V_1 = T_2 + V_2 \tag{13.44}$$

The sum of the kinetic energy T and the potential energy V of a body is called the *total mechanical energy* of the body. The modified form of the principle of work and energy for a conservative system, as expressed in Eq. (13.44), is called the *principle of conservation of energy*, which can be stated as follows: *The total mechanical energy of the body in a conservative system remains constant during its motion from one position to another.* For a conservative system consisting of several particles, the energies in Eq. (13.44) are to be understood to represent the sums of the energies of all particles in the system. In particular, V_1 and V_2 should include the potential energies associated with the internal as well as external forces acting on the particles of the system.

EXAMPLE 13.13

Using the *principle of conservation of energy*, work Example 13.7.

Solution. From the given data in Example 13.7, we have $v_1 = 6.27$ km/s, $r_1 = 9370$ km, and $r_2 = 12370$ km. We recognize that the system is conservative. Let the mass of the spacecraft be m. Applying the *principle of conservation of energy* and recalling that $GM = gR^2$, we write

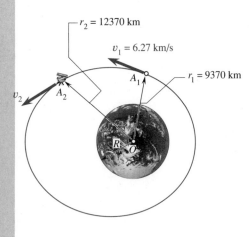

$$T_1 + V_1 = T_2 + V_2: \qquad \tfrac{1}{2}mv_1^2 - \frac{GMm}{r_1} = \tfrac{1}{2}mv_2^2 - \frac{GMm}{r_2}$$

$$\tfrac{1}{2}v_1^2 - \frac{gR^2}{r_1} = \tfrac{1}{2}v_2^2 - \frac{gR^2}{r_2}$$

$$\tfrac{1}{2}(6.27)^2 - \frac{(9.81/1000)(6370)^2}{9370}$$

$$= \tfrac{1}{2}v_2^2 - \frac{(9.81/1000)(6370)^2}{12370}$$

$$v_2 = 4.33 \text{ km/s} \blacktriangleleft$$

Developmental Exercises

D13.32   Define the *total mechanical energy* of a body.

D13.33   Is the *principle of conservation of energy* applicable to a system involving kinetic friction forces?

EXAMPLE 13.14

A 4-kg collar is released from rest in the position A_1 and is slid up a smooth rod by a 50-N horizontal force as shown. The spring attached to the collar has a free length $L = 0.9$ m and a modulus $k = 40$ N/m. Determine the speed v_2 of the collar as it passes through the position A_2.

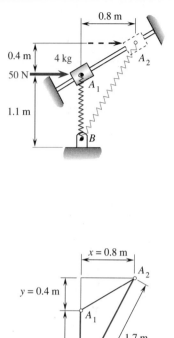

Solution. Since the rod is smooth, the system is conservative. Let us choose the position A_1 as the reference datum for both V_g and V_a. Noting that the elongations of the spring at A_1 and A_2 are $x_1 = (1.1 - 0.9)$ m $= 0.2$ m and $x_2 = (1.7 - 0.9) = 0.8$ m, respectively, we may compute the potential energies V_1 and V_2 of the collar at A_1 and A_2, respectively, as follows:

$$(V_g)_1 = (V_a)_1 = 0$$

$$(V_e)_1 = \tfrac{1}{2}kx_1^2 = \tfrac{1}{2}(40)(0.2)^2 \qquad (V_e)_1 = 0.8 \text{ J}$$

$$(V_g)_2 = Wy = 4g(0.4) \qquad (V_g)_2 = 15.696 \text{ J}$$

$$(V_a)_2 = -Px = -50(0.8) \qquad (V_a)_2 = -40 \text{ J}$$

$$(V_e)_2 = \tfrac{1}{2}kx_2^2 = \tfrac{1}{2}(40)(0.8)^2 \qquad (V_e)_2 = 12.8 \text{ J}$$

$$V_1 = (V_g)_1 + (V_e)_1 + (V_a)_1 = 0.8 \text{ J}$$

$$V_2 = (V_g)_2 + (V_e)_2 + (V_a)_2 = -11.504 \text{ J}$$

The kinetic energies of the collar at A_1 and A_2 are

$$T_1 = 0 \qquad T_2 = \tfrac{1}{2}mv_2^2 = \tfrac{1}{2}(4)v_2^2 = 2v_2^2$$

Applying the *principle of conservation of energy*, we write

$$T_1 + V_1 = T_2 + V_2: \qquad 0 + 0.8 = 2v_2^2 - 11.504$$

$$v_2 = 2.48 \text{ m/s} \quad \blacktriangleleft$$

Developmental Exercise

D13.34   Describe the *principle of conservation of energy*.

EXAMPLE 13.15

Refer to Example 13.12. It is known that the particle comes to a stop when $\theta = 30°$. For $W = 40$ lb, $L = 4$ ft, and $P = 30$ lb, determine the modulus k of the spring if (a) **P** is suddenly applied as stated, (b) **P** is gradually applied (i.e., the magnitude of **P** is gradually increased from 0 to its final value of 30 lb).

Solution. We first consider the case where **P** is suddenly applied. Such a system is clearly conservative. Applying the *principle of conservation of energy* to the particle which moves from B to B' (where the velocity, not

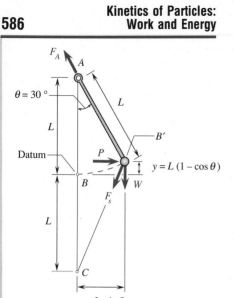

acceleration, is zero), we write

$$T_B + V_B = T_{B'} + V_{B'}$$

Since $T_B = V_B = T_{B'} = 0$ as inferred from the problem, we must have $V_{B'} = 0$. Using the results in Example 13.12, we write

$$
\begin{aligned}
V_{B'} &= V_g + V_e + V_a \\
&= WL(1 - \cos\theta) + \tfrac{1}{2}kL^2[(5 - 4\cos\theta)^{1/2} - 1]^2 - PL\sin\theta \\
&= 40(4)(1 - \cos 30°) + \tfrac{1}{2}k(4)^2[(5 - 4\cos 30°)^{1/2} - 1]^2 \\
&\quad - 30(4)\sin 30° \\
&= 0
\end{aligned}
$$

$$k = 84.17 \qquad\qquad k = 84.2 \text{ lb/ft} \qquad \blacktriangleleft$$

In the case where **P** is gradually applied, **P** is *no longer* an applied constant force; the magnitude of **P** is gradually increased from 0 to its final value of 30 lb at which *static equilibrium* of the particle exists. In the free-body diagram as shown, we have

$$\tan\phi = \frac{L\sin\theta}{L + L(1 - \cos\theta)} = \frac{\sin 30°}{2 - \cos 30°} \qquad \phi = 23.79°$$

From the solution in Example 13.12, we see that the elongation of the spring in the position CB' is

$$
\begin{aligned}
x_s &= \overline{CB'} - \overline{CB} = L[(5 - 4\cos 30°)^{1/2} - 1] \\
&= 4[(5 - 4\cos 30°)^{1/2} - 1]
\end{aligned}
$$

Thus, we have

$$x_s = 0.9573 \text{ ft} \qquad F_s = kx_s = 0.9573k$$

For studying equilibrium of the particle at B', we transmit $\mathbf{F}_s$ to the point C and write

$$
\begin{aligned}
+\circlearrowleft \Sigma M_A = 0: \quad & (L\cos\theta)P - 2L(F_s\sin\phi) \\
& - (L\sin\theta)W = 0
\end{aligned}
$$

$$
\begin{aligned}
(4\cos 30°)(30) &- 2(4)(0.9573k)(\sin 23.79°) \\
&- 4(\sin 30°)(40) = 0
\end{aligned}
$$

$$k = 7.744 \qquad\qquad k = 7.74 \text{ lb/ft} \qquad \blacktriangleleft$$

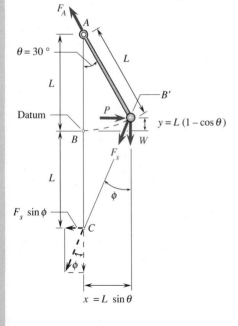

REMARK. The results obtained in this example show a great difference between the values of the modulus k of the spring required to limit the deflection of the structure (i.e., $\theta \le 30°$) when the 30-lb force **P** is allowed to be first suddenly (i.e., dynamically) applied and, the next time, to be gradually (i.e., statically) applied.

Developmental Exercise

D13.35   A 20-lb collar is released from rest in the position A_1 and is slid up a smooth rod by a 30-lb horizontal force as shown. The spring attached to the collar has a free length $L = 2.5$ ft and a modulus $k = 40$ lb/ft. Determine the speed v_2 of the collar as it passes through the position A_2.

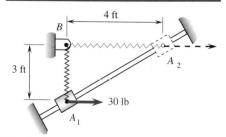

Fig. D13.35

★13.13   Virtual Work in Kinetics: Force and Acceleration[†]

A *compatible virtual displacement* of a body is a set of imaginary *first-order* differential displacements, which conforms to the integrity (i.e., no breakage or rupture) of its free body within the framework of first-order accuracy, where the body may be a particle, a rigid body, or a set of connected particles or rigid bodies. Note that a compatible virtual displacement of a body does not necessarily conform to the constraints at the supports in the space diagram. However, a *constrained virtual displacement* of a body is a compatible virtual displacement of the body which conforms to the constraints at the supports in the space diagram. Note that a constrained virtual displacement of a body may differ from the *actual differential displacement* of the body in *sense*; i.e., the latter is a special case of the former. (Cf. Sec. 10.4.)

We know that when two systems of forces acting on two identical sets of connected particles are equipollent, they must have *equal resultant forces* and *equal resultant moments* about any corresponding given point. Moreover, we can always give a constrained virtual displacement to a one-degree-of-freedom system consistent with its infinitesimal displacement which will actually take place in the next infinitesimal period of time. Thus, it follows logically from the equipollence described in Sec. 12.6 that, *during a constrained virtual displacement of a set of connected particles with one degree of freedom, the total virtual work δU of all the external and internal forces acting on the various particles in the set must be equal to the total virtual work δU_{eff} of all the effective forces on the various particles in the set*; i.e.,

$$\delta U = \delta U_{\text{eff}} \qquad (13.45)$$

If the set of particles under consideration is connected by *inextensible cords* or *links*, the total virtual work of the internal forces is zero, and δU in Eq. (13.45) reduces to the *total virtual work of external forces only*. The property expressed in Eq. (13.45) is here referred to as the *principle of virtual work in kinetics* for particles. Such a principle complements the traditional *method of force and acceleration* in Chap. 12.

Since the system of forces acting on a body in equilibrium is *equivalent* to a zero force and a zero moment (i.e., a zero wrench), the virtual displacement used in the principle of virtual work in statics for studying problems of equilibrium can be *any chosen compatible virtual displacement*. Naturally, any constrained virtual displacement is always a compatible virtual displacement, but the converse is not true.

[†]Cf. Secs. 10.4 and 10.7.

EXAMPLE 13.16

Using the *principle of virtual work in kinetics*, work Example 12.3.

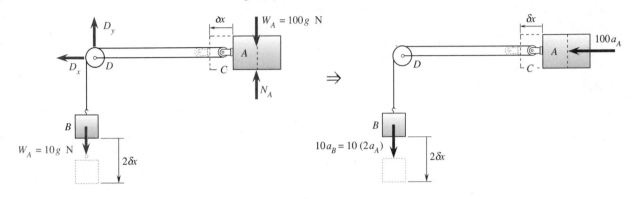

Solution. Let the free-body diagram and the effective-force diagram of the entire system be drawn first, where the pulley at D is assumed to have negligible mass. From the constraints at the supports and the force system in the free-body diagram, we note that the system has only one degree of freedom and the block A will accelerate to the left with a magnitude a_A while the block B will accelerate downward with a magnitude $a_B = 2a_A$. Likewise, the *constrained virtual displacement* consists of $\delta x \leftarrow$ for the block A and $2\,\delta x \downarrow$ for the block B. These are all indicated in the diagrams as shown. Applying the *principle of virtual work in kinetics* and noting that $\mathbf{W}_A$, $\mathbf{N}_A$, $\mathbf{D}_x$, and $\mathbf{D}_y$ do no work during the *constrained virtual displacement*, we write

$$\delta U: \qquad 10g(2\,\delta x) = 100a_A(\delta x) + 10(2a_A)(2\,\delta x) \qquad (1)$$

$$a_A = \frac{g}{7} = \frac{9.81}{7} \qquad \mathbf{a}_A = 1.401 \text{ m/s}^2 \leftarrow \blacktriangleleft$$

To find the tension F in the cord BCD, we give the block B a constrained virtual displacement $\delta x_B \downarrow$ as shown below and apply the *principle of virtual work in kinetics* to the block B to write

$$\delta U: \qquad 10g(\delta x_B) + F(-\delta x_B) = 10(2a_A)(\delta x_B) \qquad (2)$$

$$F = 10g - 20a_A = 70.1 \qquad F = 70.1 \text{ N} \blacktriangleleft$$

REMARK. Equation (1) for finding a_A *cannot* directly be written from the method of force and acceleration, while Eq. (2) for finding F is equivalent to equating the sums of the y components of the vectors in the method of force and acceleration.

Developmental Exercises

D13.36   Define a *constrained virtual displacement* of a system.

D13.37   Describe the *principle of virtual work in kinetics* for particles.

D13.38   Refer to Example 13.9. Assume that $\mu_k = 0.1$ between the block A and the horizontal support. Applying the *principle of virtual work in kinetics*, determine the acceleration of the block A.

PROBLEMS

13.39* through 13.50*   Using the *principle of conservation of energy*, solve the following problems:

13.39* Prob. 13.2.	13.40* Prob. 13.5.
13.41* Prob. 13.9*.	13.42* Prob. 13.10.
13.43* Prob. 13.11.	13.44* Prob. 13.12.
13.45* Prob. 13.13*.	13.46* Prob. 13.14*.
13.47* Prob. 13.17.	13.48* Prob. 13.18*.
13.49* Prob. 13.19.	13.50* Prob. 13.21.

13.51   Neglecting air resistance, determine (a) the magnitude v_{esc} of the escape velocity of a projectile fired from the surface of the earth at the North Pole, as shown, in terms of the gravitational acceleration g on the surface of the earth and the radius R, (b) whether v_{esc} is dependent on θ.

Fig. P13.51

13.52*   A spacecraft of mass m is in free flight around the earth as shown. Show that the speeds v_P and v_Q of the spacecraft at the perigee P and the apogee Q, respectively, are

$$v_P = R\{2gr_Q/[r_P(r_P + r_Q)]\}^{1/2}$$

$$v_Q = R\{2gr_P/[r_Q(r_P + r_Q)]\}^{1/2}$$

and that the total mechanical energy is $E = -gmR^2/(r_P + r_Q)$. (*Hint.* The reference datum for V_g is at $r = \infty$.)

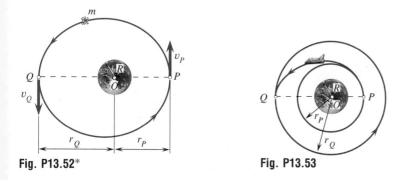

Fig. P13.52*   **Fig. P13.53**

13.53   A spacecraft of mass m is initially in free flight along a circular orbit of radius r_P around the earth. In order to transfer the spacecraft to a larger circular orbit of radius r_Q by placing it on a transitional semielliptic path PQ as shown, amounts of energy ΔE_P and ΔE_Q must momentarily be imparted to the spacecraft at P and Q, respectively. Knowing that the total mechanical energy of the spacecraft traveling along the semielliptic path PQ is $E = -gmR^2/(r_P + r_Q)$, determine the values of (a) ΔE_P, (b) ΔE_Q, (c) the sum $\Delta E'$ of ΔE_P and ΔE_Q. (Cf. Prob. 13.52*.)

Fig. P13.54*

Fig. P13.59 and P13.60

Fig. P13.61*

Fig. P13.62

13.54* The chain is released from rest in the position shown. Neglecting effects of friction, determine the speed of the chain in terms of g and R as its last link leaves the left edge of the semicircular support. (*Hint.* Choose the horizontal line through O as the reference datum for V_g.)

13.55 Solve Prob. 13.54* for (a) $R = 2$ m, (b) $R = 2$ ft.

13.56* A spring of free length 1.2 m is attached to a collar of mass 20 kg. The collar is released from rest in the position A_1 to slide down a smooth vertical rod as shown. If the lowest position reached by the collar is A_2 where $b = 2$ m, determine the modulus k of the spring.

Fig. P13.56* and P13.57

13.57 A spring of free length 1.2 m and modulus 1 kN/m is attached to a collar of mass 20 kg. The collar is released from rest in the position A_1 to slide down a smooth vertical rod as shown. If the lowest position reached by the collar is A_2, determine the distance b. (*Hint.* It leads to a cubic equation in b.)[†]

13.58 Refer to Prob. 13.57. If the collar attains a maximum speed v_{max} as it passes the position A, determine (a) the distance $\overline{A_1 A}$, (b) v_{max}. (*Hint.* It leads to a quartic equation in $\sin \theta$.)[†]

13.59 A 5-kg collar is released from rest in the position A_1 and is slid to the right on a smooth rod by a constant force of 40 N as shown. The spring attached to the collar has a modulus of 50 N/m, and the farthest position reached by the collar is A_2 where $b = 2$ m. Determine the free length L of the spring.

13.60 A 5-kg collar is released from rest in the position A_1 and is slid to the right on a smooth rod by a constant force of 40 N as shown. The spring attached to the collar has a free length of 1.3 m and a modulus of 50 N/m. Determine the speed of the collar as it passes through the position A_2 where $b = 2$ m.

13.61* A bumper composed of three springs is to stop a 40-Mg box car which is traveling at a speed of 1.2 m/s as shown. One of the springs is 80 mm shorter than the other two. The modulus of the shorter spring is 1.5 MN/m, while the modulus of each of the two longer springs is 1 MN/m. Determine the maximum deflection of the bumper.

13.62 A bumper made of a hard spring is to stop a 30-ton gondola car which is traveling at a speed of 4 ft/s as shown. Such a spring develops a restoring force $F = 100x + 12x^3$ kips when it is compressed by an amount of x ft. Determine the maximum deflection of the bumper.

13.63 Solve Prob. 13.62 if the spring is a soft spring which develops a restoring force $F = 100x - 12x^3$ kips when it is compressed by an amount of x ft.

[†]Cf. App. E for computer solution of resulting nonlinear equation.

13.64 A 4-kg slider is connected to *two* identical springs as shown. Each of the two springs has a modulus of 5.1 kN/m, and a force of 240 N is required to hold the slider in equilibrium in the position A_1 as shown. The 240-N force is later suddenly removed to release the slider so that it travels to the right in the smooth horizontal slot. Determine the speed of the slider as it passes through the position A_2.

Fig. P13.64   Fig. P13.65

13.65 A light rod with a fixed 10-kg lumped mass is rotated from rest in the inclined position BA_1C_1 about the hinge at B by a vertical constant force **P** as shown. The vertical spring at D has a modulus of 40 kN/m. Having compressed the spring 50 mm, the rod is brought to a stop in the horizontal position BA_2C_2. Determine (a) the magnitude of **P**, (b) whether P is dependent on b.

13.66* A 10-kg plunger is released from rest in the position shown and is stopped by the nest of two springs. The outer spring has a modulus of 4 kN/m and the inner one a modulus of 3 kN/m. If the inner spring is adjusted to give $b = 100$ mm before the impact, determine the maximum deflection δ of the outer spring.

13.67 The 40-lb sphere as shown is released from rest when $\theta = 0$. Knowing that the spring has a free length of 30 in. and a modulus of 6 lb/in., determine the speed of the sphere when $\theta = 90°$.

Fig. P13.66*

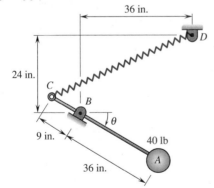

Fig. P13.67

13.68* through 13.77* Using the *principle of virtual work in kinetics*, solve the following problems:

13.14   Concluding Remarks

This chapter covers three principles related to work and energy for studying the motion of particles. They are (1) the *principle of work and energy*, which is derived directly from Newton's second law of motion; (2) the *principle of conservation of energy*, which is a special case of the principle of work and energy as it is applied to conservative systems; and (3) the *principle of virtual work in kinetics*, which is an extension of that in statics. In any case, an *inertial reference frame* should properly be used in describing the various quantities.

Note that a spring is usually idealized as having negligible mass. Having negligible mass, a spring is *not a body* in the usual sense. Rather, a spring is a device which provides an elastic force field for the body attached to it. It is the spring force, rather than the spring, that does the work to the body which is attached to the spring. By the principle of work and energy, *any net amount of work done by forces to a body must be exactly equal to the amount of change of the kinetic energy of the body*. Therefore, *no work can be done to a stationary body or a body with no mass*. If we say that work is done to a spring, we would have to take the mass of the spring into account; otherwise, the requirement that the net amount of work done to the spring be equal to the amount of change of kinetic energy of the spring cannot be satisfied.

Furthermore, note that the potential energy of a body is defined *only* for a body in a conservative force field. The gravitational, elastic, and applied potential energies of a body are *numerically* equal to the amounts of work which the body will receive from the forces in the respective conservative force fields *if* the body travels from its present position to the respective reference datums.

In mechanics, force is exerted by one body on another body; however, it is the *force* (not the *body*) which does *work* to a body having a *mass* and a nonzero component of *displacement* parallel to the force while the force is acting on it. Therefore, phrases such as "the work done by a body," "the work done on a (massless) spring," and "the capacity of a body to do work" are considered as colloquial. Work, kinetic energy, and potential energy are subtle as well as vital concepts in kinetics. A clear understanding of these concepts is essential to the proper applications of principles related to work and energy.

REVIEW PROBLEMS

13.78   A brick is released from rest at A_1 to slide down a chute as shown. The brick leaves the chute at A_2 with a horizontal velocity and strikes the ground at A_3. Determine the coefficient of kinetic friction μ_k between the brick and the chute.

13.79   A 120-lb cylinder C is released from rest when the spring with a modulus of 40 lb/ft is unstressed as shown. Determine (a) the maximum deflection of the cylinder, (b) the maximum speed attained by the cylinder.

13.80*   A satellite is in free flight around the earth as shown. In the position S, it has an altitude and a speed as indicated. Determine the altitude a_Q of its apogee.

Fig. P13.78

Fig. P13.79

Fig. P13.80*

13.81 The chain is released from rest in the position as shown. Neglecting effects of friction, determine the speed of the chain as its last link leaves the right edge of the quarter-circular support.

Fig. P13.81

Fig. P13.82*

13.82* A 10-kg block is released from rest in the position A_1 to be moved up a smooth incline by the constant force **P** of 500 N as shown. The attached spring has a modulus of 400 N/m and is stretched 120 mm when the block is in the position A_1. Determine the speed of the block as it passes through the position A_2.

13.83* Using the *principle of virtual work in kinetics*, solve Prob. 12.90.

13.84 A step-type escalator as shown is designed to transport 120 persons per minute at a constant speed $v = 1.5$ ft/s. The overall efficiency of the escalator is estimated to be 70%, and a 200% *overload* is to be allowed. Assuming that the average weight per person is 160 lb, determine the required capacity of the motor of the escalator.

(a) Side view

Fig. P13.84

(b) Steps

Chapter 14

Kinetics of Particles: Impulse and Momentum

KEY CONCEPTS

- Impulse, momentum, linear impulse, linear momentum, angular impulse, angular momentum, impulsive force, direct central impact, oblique central impact, and coefficient of restitution.
- Principle of impulse and momentum for particles.
- Conservation of angular momentum in central-force motion.
- Direct and oblique central impacts between particles.
- Principle of generalized virtual work for studying impulses and momenta in one-degree-of-freedom systems of particles.
- Kinetics of variable systems of particles.

Impulse and momentum are vector quantities with specific meaning in mechanics. They are first defined and then used in formulating the *principle of impulse and momentum*, which is an effective tool for solving kinetic problems where *time* and *velocity* (among others) are to be directly related without initially involving acceleration and displacement. Furthermore, such a principle can be applied to provide useful equations in analyzing problems of central-force motion of particles and problems of impact. Later, the *principle of generalized virtual work* for particles is propounded as a complement to the principle of impulse and momentum for systems of connected particles with one degree of freedom. The last part of this chapter covers the motions of variable systems (e.g., turbines, jet engines, and rockets). The study of variable systems is a step into the study of advanced engineering systems.

IMPULSE AND MOMENTUM

14.1 Linear Impulse and Linear Momentum

The *linear impulse of a force* **F** acting on a particle during the time interval $t_1 \leq t \leq t_2$ is denoted by $\mathbf{Imp}_{1 \to 2}$ and is defined as the integral of **F** times the differential time dt evaluated over that time interval; i.e.,

$$\mathbf{Imp}_{1\to2} = \int_{t_1}^{t_2} \mathbf{F}\, dt \qquad (14.1)$$

If the force $\mathbf{F}$ acting on the particle remains *constant* during the interval $\Delta t = t_2 - t_1$, we readily obtain

$$\mathbf{Imp}_{1\to2} = \mathbf{F}\,\Delta t \qquad (14.2)$$

The *linear impulse* defined above is often simply referred to as the *impulse*. Physically, the *linear impulse of a force* on a particle during a time interval is a measure of the *cumulative effectiveness or tendency of the force to change the state of motion* of the particle.[†]

Clearly, the dimensions of the linear impulse are *force* × *time*, and its primary units are newton-seconds (N·s) in the SI and pound-seconds (lb·s) in the U.S. customary system. If $\mathbf{F}$ does not have a constant direction, the direction of the linear impulse $\mathbf{Imp}_{1\to2}$ is dependent on the final value of the integral in Eq. (14.1). If $\mathbf{F}$ is constant, the direction of $\mathbf{Imp}_{1\to2}$ is identical with the direction of $\mathbf{F}$.

The *linear momentum* of a particle is denoted by $\mathbf{L}$ and is defined as the product of its mass m and its velocity $\mathbf{v}$; i.e.,

$$\mathbf{L} = m\mathbf{v} \qquad (14.3)$$

The linear momentum defined above is a vector quantity possessed by the particle and is tangent to the path of the particle as shown in Fig. 14.1. Physically, the *linear momentum* of a particle indicates the *strength as well as the direction of motion* of the particle as observed from the inertial reference frame. Note that the dimensions of linear momentum are *mass* × *velocity*. Using Newton's second law, we write

$$\text{Mass} \times \text{Velocity} = \frac{\text{Force} \times (\text{Time})^2}{\text{Length}} \times \frac{\text{Length}}{\text{Time}} = \text{Force} \times \text{Time}$$

This shows that the linear momentum and linear impulse have equivalent dimensions, and they may have the same units. The *linear momentum* defined above is often simply referred to as the *momentum*.

By Newton's second law, the resultant force $\Sigma\mathbf{F}$ acting on a particle of mass m is given by $\Sigma\mathbf{F} = m\mathbf{a}$, where $\mathbf{a}$ is the acceleration of the particle. Noting that $\mathbf{a} = d\mathbf{v}/dt$ and recognizing m as being constant, we may express Newton's second law in terms of the linear momentum of the particle as

$$\Sigma\mathbf{F} = \frac{d}{dt}(m\mathbf{v}) \qquad (14.4)$$

Substituting Eq. (14.3) into Eq. (14.4) and using an overdot to denote the time derivative, we may write *Newton's second law* as

$$\Sigma\mathbf{F} = \dot{\mathbf{L}} \qquad (14.5)$$

Fig. 14.1

[†]Note that *impulse* is exerted *by* a force *on* a body. It is the *force* (not the *body*) which exerts the impulse; a body *receives* impulse from the force acting on it during a time interval whether or not the body has any velocity or displacement during that time interval. Besides, *impulse of a force* and *impulse exerted by a force* are interchangeably used.

Thus, the resultant force acting on a particle is equal to the time rate of change of the linear momentum of the particle. Newton's second law expressed in the form of Eq. (14.4) or (14.5) is *valid* in *newtonian mechanics* for a particle with a constant mass m, but is *required* in *relativistic mechanics* for a particle whose mass m is a function of its speed v.[†] However, Eq. (14.4) or (14.5) should not, in general, be used to solve problems involving the motion of a variable system gaining or losing particles (and hence masses), such as rockets.[††]

We can readily see from Eq. (14.4) that $m\mathbf{v}$ is a constant vector when $\Sigma\mathbf{F} = \mathbf{0}$. In other words, the linear momentum of a particle remains constant in magnitude and direction if the resultant force acting on the particle is zero. This is an alternative statement of *Newton's first law* of motion.

Developmental Exercises

D14.1　Define the terms: (a) linear impulse, (b) linear momentum.

D14.2　The force $\mathbf{F} = 2t\mathbf{i} - 2\mathbf{j} - 3t^2\mathbf{k}$ N acts on a particle during the time interval $0 \leq t \leq 3$ s, where t is the time measured in seconds. Determine the linear impulse of $\mathbf{F}$ during that interval.

D14.3　Determine the linear momentum of a particle if it has (a) a weight of 20 lb and a velocity of $20\mathbf{i} - 25\mathbf{j} + 30\mathbf{k}$ ft/s near the surface of the earth, (b) a mass of 30 mg and a velocity of $1500\,\mathbf{j} - 800\mathbf{k}$ m/s.

14.2　Principle of Impulse and Momentum

The motion of a particle of mass m and velocity $\mathbf{v}$ under the action of the resultant force $\Sigma\mathbf{F}$ of a system of forces $\mathbf{F}_1, \mathbf{F}_2, \ldots, \mathbf{F}_n$ is defined by Eq. (14.4). Multiplying both sides of this equation by the differential time dt and integrating from a time t_1 to a time t_2, we write

$$\Sigma\mathbf{F}\,dt = d(m\mathbf{v}) \qquad \int_{t_1}^{t_2} \Sigma\mathbf{F}\,dt = m\mathbf{v}_2 - m\mathbf{v}_1$$

where $\mathbf{v}_1$ and $\mathbf{v}_2$ are the velocities of the particle at the times t_1 and t_2, respectively. Thus, we have

$$m\mathbf{v}_1 + \int_{t_1}^{t_2} \Sigma\mathbf{F}\,dt = m\mathbf{v}_2 \qquad (14.6)$$

Using the symbols defined in Eqs. (14.1) and (14.3), we may write Eq. (14.6) as

$$\mathbf{L}_1 + \Sigma\mathbf{Imp}_{1\to2} = \mathbf{L}_2 \qquad (14.7)$$

[†]In relativistic mechanics, the mass m of a particle moving with a speed v is given by $m = m_0/(1 - v^2/c^2)^{1/2}$, where c is the speed of light and m_0 is the mass of the particle at rest in an inertial reference frame. The mass m_0 is called the *rest mass*, while the mass m is called the *relativistic mass* of the particle, which is a function of v.

[††]Cf. Eqs. (14.43) and (14.44).

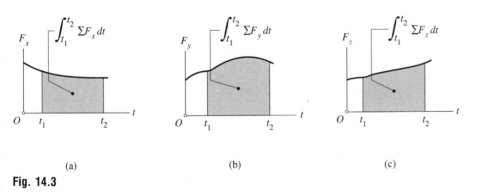

Fig. 14.2

The property expressed in Eq. (14.6) or (14.7) and illustrated in Fig. 14.2 is called the *principle of impulse and momentum* for a particle, which may be stated as follows: *The initial momentum of a particle plus the impulses received by the particle during a time interval is equivalent to the final momentum of the particle.* In other words, the net impulse received by a particle during a time interval is equivalent to the change in momentum of the particle during that time interval.

Unlike work and kinetic energy, impulse and momentum are vector quantities. In obtaining an analytic solution, we may replace Eq. (14.6) with its equivalent component equations

$$(mv_x)_1 + \int_{t_1}^{t_2} \Sigma F_x \, dt = (mv_x)_2$$

$$(mv_y)_1 + \int_{t_1}^{t_2} \Sigma F_y \, dt = (mv_y)_2$$

$$(mv_z)_1 + \int_{t_1}^{t_2} \Sigma F_z \, dt = (mv_z)_2$$

Geometrically, the integrals in the above equations are equal to the shaded areas indicated in the force-time diagrams in Fig. 14.3.

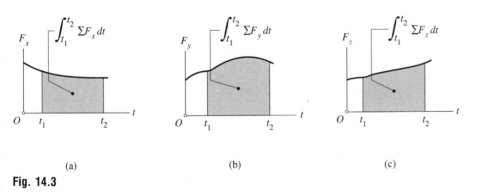

(a)   (b)   (c)

Fig. 14.3

We note that the principle of impulse and momentum directly relates *time* and *velocity* (among others) without initially involving acceleration and displacement. Furthermore, we refer to the procedure which uses the principle of impulse and momentum to analyze and solve a proper type of problem as the *method of impulse and momentum*. In using such a method, it is recommended that appropriate *momentum and impulse diagrams*, such as those shown in Fig. 14.2, *be drawn in formulating the solution*.

EXAMPLE 14.1

36 km/h

8
15

A man on a light-weight racing bicycle is coasting at a speed of 36 km/h down an incline as shown when he applies the brakes. He reduces his speed to 18 km/h in 10 s, and the combined mass of him and the bicycle is known to be 85 kg. Neglecting the rotational effects of the wheels, determine the average magnitude F of the total braking force exerted by the incline on the tires.

Solution. We note that 36 km/h = 10 m/s and 18 km/h = 5 m/s. Since the rotational effects of the wheels are to be neglected, we may treat the man and the bicycle as a *particle*. Applying the *principle of impulse and momentum* to the man and the bicycle during the 10-s interval, we have

| Initial momentum diagram | Impulse diagram | Final momentum diagram |

$$+ \swarrow \Sigma V_x:^\dagger \qquad 85(10) + [\tfrac{8}{17}(85g)(10) - F(10)] = 85(5)$$

$$F = 435 \text{ N} \blacktriangleleft$$

Developmental Exercise

D14.4 Describe the *principle of impulse and momentum* for a particle.

EXAMPLE 14.2

A force of magnitude $F = 38t$ lb is applied to a 100-lb crate as shown, where t is measured in seconds. The crate is at rest when $t = 0$, and the direction of **F** is always parallel to the incline. If $\mu_s = 0.5$ and $\mu_k = 0.4$ between the crate and the incline, determine that the time t_2 at which the speed of the crate is 16.1 ft/s.

100 lb

$F = 38t$ lb

24
7

$\mu_s = 0.5, \mu_k = 0.4$

---

†Cf. the second footnote on p. 513.

Solution. Assuming that slipping of the crate impends at $t = t_1$ and observing that the free-body diagram as shown is in equilibrium, we write

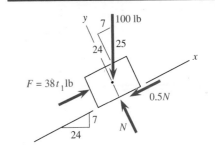

$$+\nwarrow \Sigma F_y = 0: \quad N - \frac{24}{25}(100) = 0 \quad N = 96 \text{ lb}$$

$$+\nearrow \Sigma F_x = 0: \quad 38t_1 - \frac{7}{25}(100) - 0.5N = 0 \quad t_1 = 2 \text{ s}^\dagger$$

Since the crate has no motion perpendicular to the incline, we have $N = 96$ lb, and $F_k = \mu_k N = 0.4(96)$ lb $= 38.4$ lb. Applying the *principle of impulse and momentum* to the crate during $2 \text{ s} < t \le t_2$, we have

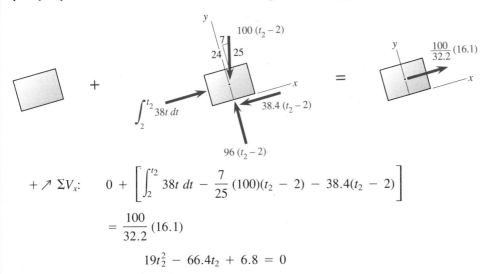

$$+\nearrow \Sigma V_x: \quad 0 + \left[\int_2^{t_2} 38t \, dt - \frac{7}{25}(100)(t_2 - 2) - 38.4(t_2 - 2) \right]$$

$$= \frac{100}{32.2}(16.1)$$

$$19t_2^2 - 66.4t_2 + 6.8 = 0$$

which yields two roots: $t_2 = 3.39$ s and $t_2 = 0.1056$ s. Since $t_2 > 2$ s, we discard the second root and write

$$t_2 = 3.39 \text{ s} \quad \blacktriangleleft$$

Developmental Exercise

D14.5   A package slides down an incline with a speed of 1.5 m/s at the time $t = 0$ as shown. If $\mu_k = 0.35$ between the package and the incline, determine the time t_2 at which the package comes to a stop.

Fig. D14.5

14.3   Angular Impulse and Angular Momentum

To study the motion of a particle about a point, it is useful to know the moments of the forces acting on the particle about that point. The *angular impulse* of a force **F** acting on a particle P about a fixed point A, as shown in Fig. 14.4(a), during the time interval $t_1 \le t \le t_2$ is denoted by $(\textbf{Aimp}_A)_{1 \to 2}$ and is defined as the integral of the moment of **F** about A times

---

$\dagger$The crate starts to move at $t = t_1 = 2$ s. During the interval $0 \le t \le 2$ s, the resultant force acting on the crate is zero.

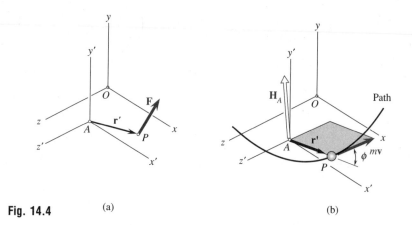

Fig. 14.4   (a)   (b)

the differential time dt evaluated over that interval. We write

$$(\textbf{Aimp}_A)_{1\to2} = \int_{t_1}^{t_2} \textbf{r}' \times \textbf{F} \, dt \tag{14.8}$$

where $\textbf{r}'$ is the position vector of any convenient point, say P, on the line of action of $\textbf{F}$ in the reference frame $Ax'y'z'$ as shown. Denoting the moment of $\textbf{F}$ about A by $\textbf{M}_A$, we may write Eq. (14.8) as

$$(\textbf{Aimp}_A)_{1\to2} = \int_{t_1}^{t_2} \textbf{M}_A \, dt \tag{14.9}$$

If the force $\textbf{F}$, and hence its moment $\textbf{M}_A$, remains *constant* during the interval $\Delta t = t_2 - t_1$, we write

$$\textbf{M}_A = \textbf{r}' \times \textbf{F} = d_s F \, \boldsymbol{\lambda}$$

$$(\textbf{Aimp}_A)_{1\to2} = \textbf{M}_A \, \Delta t \tag{14.10}$$

where d_s is the shortest distance between A and the line of action of $\textbf{F}$, and $\boldsymbol{\lambda}$ is a unit vector pointing in the direction perpendicular to the plane containing A and $\textbf{F}$. An angular impulse is a vector quantity; its dimensions are *force* × *length* × *time*, and its primary units are newton-meter-seconds (N·m·s) in the SI and pound-foot-seconds (lb·ft·s) in the U.S. customary system.

The *angular momentum* of a particle about any point A is denoted by $\textbf{H}_A$ and is defined as the *moment of the linear momentum* $m\textbf{v}$ of the particle about the point A, as shown in Fig. 14.4(b);[†] i.e.,

$$\textbf{H}_A = \textbf{r}' \times m\textbf{v} \tag{14.11}$$

where m and $\textbf{v}$ are the mass and velocity of the particle P, and $\textbf{r}'$ is the position vector of P in the reference frame $Ax'y'z'$.

It is obvious that the dimensions of angular momentum are *length* × *mass* × *velocity*. Using Newton's second law, we write

---

[†]Note that a *hollow arrow* is used, for easy distinction, to represent an *angular momentum*, an *angular impulse*, and a *moment* in this text.

Length $\times$ Mass $\times$ Velocity

$$= \text{Length} \times \frac{\text{Force} \times (\text{Time})^2}{\text{Length}} \times \frac{\text{Length}}{\text{Time}}$$

$$= \text{Force} \times \text{Length} \times \text{Time}$$

Thus, the angular momentum and angular impulse have equivalent dimensions, and they may have the same units. Because of its definition, the *angular momentum* is also called the *moment of momentum*.

EXAMPLE 14.3

A system consists of three particles A, B, and C. At the instant considered, they are located as shown and have velocities $\mathbf{v}_A = 5\mathbf{i} - 8\mathbf{k}$ m/s, $\mathbf{v}_B = 4\mathbf{i} + v_y\mathbf{j} + v_z\mathbf{k}$ m/s, and $\mathbf{v}_C = 6\mathbf{i} + 3\mathbf{j} - 2\mathbf{k}$ m/s, respectively. If the masses of these particles are $m_A = 3$ kg, $m_B = 1$ kg, and $m_C = 2$ kg, respectively, and the resultant angular momentum $\mathbf{H}_O$ of the system about the origin O is parallel to the z axis, determine the value of $\mathbf{H}_O$.

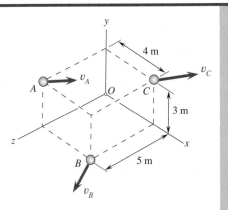

Solution. Since the resultant angular momentum $\mathbf{H}_O$ of the system is equal to the vector sum of the angular momenta of the individual particles in the system, we write

$$\mathbf{H}_O = \mathbf{r}_A \times m_A\mathbf{v}_A + \mathbf{r}_B \times m_B\mathbf{v}_B + \mathbf{r}_C \times m_C\mathbf{v}_C$$

$$= (3\mathbf{j} + 5\mathbf{k}) \times (3)(5\mathbf{i} - 8\mathbf{k})$$

$$\quad + (4\mathbf{i} + 5\mathbf{k}) \times (1)(4\mathbf{i} + v_y\mathbf{j} + v_z\mathbf{k})$$

$$\quad + (4\mathbf{i} + 3\mathbf{j}) \times (2)(6\mathbf{i} + 3\mathbf{j} - 2\mathbf{k})$$

$$= -(5v_y + 84)\mathbf{i} + (111 - 4v_z)\mathbf{j} + (4v_y - 57)\mathbf{k}$$

Since $\mathbf{H}_O$ is parallel to the z axis, we set

$$-(5v_y + 84) = 0 \qquad 111 - 4v_z = 0$$

These two equations yield $v_y = -16.8$ and $v_z = 27.75$. Thus, we have

$$\mathbf{H}_O = (4v_y - 57)\mathbf{k} = [4(-16.8) - 57]\mathbf{k}$$

$$\mathbf{H}_O = -124.2\mathbf{k} \text{ kg·m}^2/\text{s} \quad \blacktriangleleft$$

Developmental Exercises

D14.6 Define the angular impulse about a point as exerted by (a) a force, (b) a moment.

D14.7 Define the angular momentum of a particle about a point.

D14.8 Refer to Example 14.3. Determine the value of $\mathbf{H}_O$ if it is known to be parallel to the x axis at the instant considered.

14.4   System of Particles

The principle of impulse and momentum for a particle, as expressed in Eq. (14.6) and illustrated in Fig. 14.2, may be applied to each individual particle in a system of n particles as illustrated in Fig. 14.5. Note in this figure that $\mathbf{F}_i$ is the resultant force acting on the particle P_i at the time t, and $(m_i\mathbf{v}_i)_1$ and $(m_i\mathbf{v}_i)_2$ are the momenta of P_i at the times t_1 and t_2, respectively.

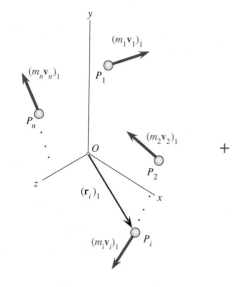

Initial momenta

Fig. 14.5   (a)

Impulses

(b)

Final momenta

(c)

By superposition of the principle of impulse and momentum as applied to the individual particles in a system, we see that

| The *momentum vectors* of a system of particles at the time t_1 | $+$ | The *impulse vectors* exerted on the system from t_1 to t_2 |

$$\Rightarrow \boxed{\text{The } \textit{momentum vectors} \text{ of the system at the time } t_2} \qquad (14.12)^{\dagger}$$

Since internal forces in a system of particles occur in collinear pairs of forces equal in magnitude and opposite in sense, their sum and their resultant moment about any point are zero at all times. This means that only the impulses of the external forces need to be considered in applying Eq. (14.12). Thus, Eq. (14.12) may be simplified and written as

Syst Momenta$_1$ + Syst Ext Imp$_{1\rightarrow 2}$ $\Rightarrow$ Syst Momenta$_2$ (14.13)

---

†Recall that the double arrow sign $\Rightarrow$ indicates *equipollence*.

If no external force acts on the system, Eq. (14.13) reduces to

$$\textbf{Syst Momenta}_1 \quad \Rightarrow \quad \textbf{Syst Momenta}_2 \qquad (14.14)$$

which expresses the *conservation of momentum*.

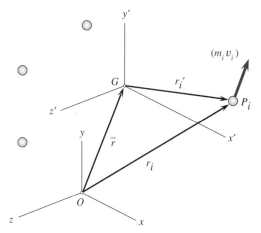

Fig. 14.6

Referring to Fig. 14.6, we define the linear momentum **L** and the angular momentum $\textbf{H}_G$ about the mass center G of the system of particles in the fixed reference frame $Oxyz$ as

$$\textbf{L} = \sum_{i=1}^{n} m_i \textbf{v}_i \qquad \textbf{H}_G = \sum_{i=1}^{n} (\textbf{r}_i' \times m_i \textbf{v}_i) \qquad (14.15)$$

We note that the position vectors $\textbf{r}_i$ and $\bar{\textbf{r}}$ and the velocities $\textbf{v}_i$ and $\bar{\textbf{v}}$ of the particle P_i and the mass center G are related to each other by

$$\textbf{r}_i = \bar{\textbf{r}} + \textbf{r}_i' \qquad \textbf{v}_i = \bar{\textbf{v}} + \textbf{v}_i' \qquad (14.16)$$

Applying Eqs. (12.22) and (14.15), we find that the sum of the external forces acting on the system of particles is

$$\Sigma \textbf{F} = \sum_{i=1}^{n} m_i \textbf{a}_i = \sum_{i=1}^{n} \frac{d}{dt}(m_i \textbf{v}_i) = \frac{d}{dt} \sum_{i=1}^{n} m_i \textbf{v}_i = \frac{d\textbf{L}}{dt}$$

$$\boxed{\Sigma \textbf{F} = \dot{\textbf{L}}} \qquad (14.17)$$

Applying Eqs. (14.15), (14.16), and (12.23), as well as the *principle of moments*, we see that the time derivative of the angular momentum $\textbf{H}_G$ about the mass center G of a system of particles of total mass m taken in $Oxyz$ is

$$\dot{\textbf{H}}_G = \frac{d}{dt} \sum_{i=1}^{n} (\textbf{r}_i' \times m_i \textbf{v}_i) = \sum_{i=1}^{n} (\textbf{v}_i' \times m_i \textbf{v}_i + \textbf{r}_i' \times m_i \textbf{a}_i)$$

$$= \sum_{i=1}^{n} [(\textbf{v}_i - \bar{\textbf{v}}) \times m_i \textbf{v}_i] + \sum_{i=1}^{n} (\textbf{r}_i' \times m_i \textbf{a}_i)$$

$$= \sum_{i=1}^{n} (\textbf{v}_i \times m_i \textbf{v}_i) - \bar{\textbf{v}} \times \sum_{i=1}^{n} m_i \textbf{v}_i + \sum_{i=1}^{n} (\textbf{r}_i' \times m_i \textbf{a}_i)$$

$$= \textbf{0} - \bar{\textbf{v}} \times m\bar{\textbf{v}} + \Sigma \textbf{M}_G = \textbf{0} - \textbf{0} + \Sigma \textbf{M}_G$$

Therefore, the sum of moments about G of the external forces is

$$\Sigma \mathbf{M}_G = \dot{\mathbf{H}}_G \tag{14.18}$$

Defining the angular momentum $\mathbf{H}_O$ about the fixed origin O for the system in Fig. 14.6 as

$$\mathbf{H}_O = \sum_{i=1}^{n} (\mathbf{r}_i \times m_i \mathbf{v}_i) \tag{14.19}$$

and applying Eq. (12.24), we find that the time derivative of $\mathbf{H}_O$ in $Oxyz$ is

$$
\dot{\mathbf{H}}_O = \frac{d}{dt} \sum_{i=1}^{n} (\mathbf{r}_i \times m_i \mathbf{v}_i) = \sum_{i=1}^{n} (\mathbf{v}_i \times m_i \mathbf{v}_i + \mathbf{r}_i \times m_i \mathbf{a}_i)
$$

$$
= \sum_{i=1}^{n} (\mathbf{0} + \mathbf{r}_i \times m_i \mathbf{a}_i) = \sum_{i=1}^{n} (\mathbf{r}_i \times m_i \mathbf{a}_i) = \Sigma \mathbf{M}_O
$$

Thus, the sum of moments about a fixed point O of the external forces is

$$\Sigma \mathbf{M}_O = \dot{\mathbf{H}}_O \tag{14.20}$$

Equations (14.17), (14.18), and (14.20) are fundamental equations of motion expressed in terms of the time derivatives of momenta of a system.

EXAMPLE 14.4

An 80-Mg locomotive A coasting at 0.65 m/s as shown strikes, and is coupled with, a 50-Mg box car B which is at rest with its brakes released. If the coupling is completed in 0.5 s, determine (a) the speed v_2 of the locomotive and the box car after the coupling, (b) the magnitude F of the average force exerted on the box car during the coupling.

Solution. We note that 80 Mg = 80×10^3 kg and 50 Mg = 50×10^3 kg. For simplicity, let us denote the weight forces of the locomotive and the box car by $\mathbf{W}_A$ and $\mathbf{W}_B$, and the resultant reactions from the tracks to the locomotive and the box car by $\mathbf{R}_A$ and $\mathbf{R}_B$, respectively. Applying the *principle of impulse and momentum* to the locomotive and the box car as a system during the 0.5-s interval of coupling, we have

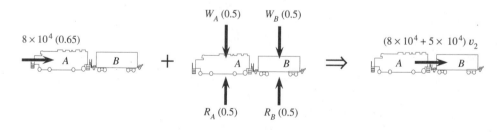

Notice that the coupling forces between A and B are *internal forces* and have no net contribution to the impulse received by A and B as a system. We write

$$\overset{+}{\rightarrow} \Sigma V_x: \qquad (80 \times 10^3)(0.65) + 0 = (80 \times 10^3 + 50 \times 10^3)v_2$$

$$v_2 = 0.4 \text{ m/s} \quad \blacktriangleleft$$

By applying the same principle to the box car alone during the 0.5-s interval of coupling, we have

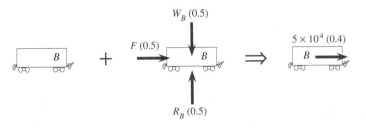

Notice that the coupling force $\mathbf{F}$ is an *external force* to the box car alone and contributes impulse to the box car. We write

$$\overset{+}{\rightarrow} \Sigma V_x: \qquad 0 + F(0.5) = (50 \times 10^3)(0.4)$$

$$F = 40 \times 10^3 \text{ N} \qquad F = 40 \text{ kN} \quad \blacktriangleleft$$

Developmental Exercise

D14.9 Refer to Example 14.4. Using the result that $v_2 = 0.4$ m/s, determine the value of F by considering the locomotive alone.

EXAMPLE 14.5

The blocks A and B as shown are released to move from rest at the time $t = 0$. If $\mu_k = 0.3$ between the block A and the support, determine the speed v_A of the block A when $t = 4$ s.

Solution. From the constraints of the system, we readily see that the velocities of the blocks A and B are $\mathbf{v}_A = v_A \leftarrow$ and $\mathbf{v}_B = 2v_A \downarrow$, respectively. Since the block A has no motion perpendicular to the support, the normal reaction to the block A is $\mathbf{N}_A = 100g$ N $\uparrow$, and the kinetic friction force acting on the block A is $\mathbf{F}_k = 0.3(100g)$ N $\rightarrow$. Denoting the tension in the cord BCD by F and applying the *principle of impulse and momentum*

to the blocks A and B, separately, during the interval $0 \leq t \leq 4$ s, we have

Block A

$$\overset{+}{\leftarrow} \Sigma V_x: \quad 0 + [F(4) + F(4) - 0.3(100g)(4)] = 100v_A \quad (1)$$

Block B

$$+\downarrow \Sigma V_y: \quad 0 + [20g(4) - F(4)] = 20(2v_A) \quad (2)$$

Solving Eqs. (1) and (2), we obtain

$$F = 174.4 \text{ N} \qquad v_A = 2.18 \text{ m/s} \blacktriangleleft$$

Developmental Exercise

D14.10   Work Example 14.5 if the velocity of the block B is 1 m/s $\downarrow$ at $t = 0$.

EXAMPLE 14.6

The cylinders A, B, and C as shown are released from rest at the time $t = 0$. Determine the tension in each cable and the velocity of each cylinder at the time $t = 3$ s.

Solution.   As derived in the solution in Example 11.8, we have the following constraint equation on the velocities:

$$4v_A + 2v_B + v_C = 0 \quad (1)$$

Notice that the positive directions of the unknown velocities are downward and all pulleys are assumed to have negligible mass.

Let the tensions in cables 1 and 2 be F_1 and F_2, respectively. Since the pulley D is taken to have negligible mass, it has a negligible effective force as shown. We write

$$+\uparrow \Sigma V_y: \qquad F_1 - F_2 - F_2 = 0 \qquad (2)$$

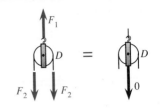

Applying the *principle of impulse and momentum* to each individual cylinder during the interval $0 \le t \le 3$ s, we have

Cylinder A

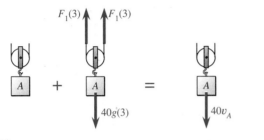

$$+\downarrow \Sigma V_y: \qquad 0 + [40g(3) - F_1(3) - F_1(3)] = 40v_A \qquad (3)$$

Cylinder B

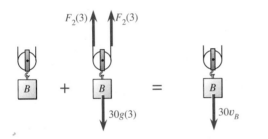

$$+\downarrow \Sigma V_y: \qquad 0 + [30g(3) - F_2(3) - F_2(3)] = 30v_B \qquad (4)$$

Cylinder C

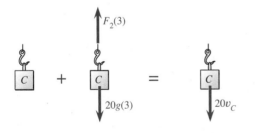

$$+\downarrow \Sigma V_y: \qquad 0 + [20g(3) - F_2(3)] = 20v_C \qquad (5)$$

Solution of the preceding five equations in five unknowns yields

$$F_1 = 235.44 \qquad F_2 = 117.72 \qquad v_A = -5.886$$

$$v_B = 5.886 \qquad v_C = 11.772$$

$$F_1 = 235 \text{ N} \qquad F_2 = 117.7 \text{ N} \qquad \mathbf{v}_A = 5.89 \text{ m/s} \uparrow \quad \blacktriangleleft$$

$$\mathbf{v}_B = 5.89 \text{ m/s} \downarrow \qquad \mathbf{v}_C = 11.77 \text{ m/s} \downarrow \quad \blacktriangleleft$$

Developmental Exercise

D14.11   Work Example 14.6 if the initial velocities of the cylinders A and B are $v_A = 1$ m/s $\downarrow$ and $v_B = 2$ m/s $\downarrow$, respectively, when the cylinders are released at the time $t = 0$.

EXAMPLE 14.7[†]

A projectile of mass $m = 100$ kg is traveling in space. At the time $t = 0$, it is observed to pass through the origin O of an inertial reference frame with a constant velocity $\mathbf{v}_0 = 756\mathbf{i}$ m/s. Following an explosion at $t = t_1$, where $0 < t_1 < 1.3$ s, the projectile separates into three parts, A, B, and C, of masses $m_A = 40$ kg, $m_B = 24$ kg, and $m_C = 36$ kg, respectively. At $t = 1.3$ s, it is observed that the position vectors of B and C are $\mathbf{r}_B = 1218\mathbf{i} - 420\mathbf{j} - 840\mathbf{k}$ m and $\mathbf{r}_C = 698\mathbf{i} - 320\mathbf{j} + 560\mathbf{k}$ m, and that the y component of the velocity $\mathbf{v}_A$ of A is $(\mathbf{v}_A)_y = 675\mathbf{j}$ m/s. Neglecting the effect of gravity, determine the position vector $\mathbf{r}_A$ of A and the velocities $\mathbf{v}_A$, $\mathbf{v}_B$, and $\mathbf{v}_C$ of A, B, and C, respectively, at $t = 1.3$ s.

Solution.   We observe that there is no external force acting on the system.

By the *principle of motion of the mass center* of a system of particles as expressed in Eq. (12.27), we know that the mass center G of the system must move with the constant velocity $\mathbf{v}_0 = 756\mathbf{i}$ m/s along the x axis. At $t = 1.3$ s, the position vector of G must, therefore, be $\bar{\mathbf{r}} = 982.8\mathbf{i}$ m. Applying the *principle of moments*, we write

$$m\bar{\mathbf{r}} = m_A\mathbf{r}_A + m_B\mathbf{r}_B + m_C\mathbf{r}_C$$

$$100(982.8\mathbf{i}) = 40\mathbf{r}_A + 24(1218\mathbf{i} - 420\mathbf{j} - 840\mathbf{k})$$
$$+ 36(698\mathbf{i} - 320\mathbf{j} + 560\mathbf{k})$$

$$\mathbf{r}_A = 1098\mathbf{i} + 540\mathbf{j} \text{ m} \blacktriangleleft$$

Since there is no external force, *both the linear momentum and the angular momentum of the system are conserved*; i.e., the initial momentum $m\mathbf{v}_0$ is equipollent to the system of final momenta as shown. Equating the

---

[†]The complete solution of this example is somewhat lengthy. *Skip* this example and Probs. 14.39 and 14.40* if circumstances warrant. However, the technical intricacy here is pedagogically valuable. Note that the solution involves the direct applications of (a) the principle of motion of the mass center of a system of particles, (b) the principle of moments, (c) the conservation of linear momentum, (d) the conservation of angular momentum, and (e) Newton's first law.

sums of the linear momentum vectors, we write

$$\mathbf{L}_1 = \mathbf{L}_2: \qquad m\mathbf{v}_0 = m_A\mathbf{v}_A + m_B\mathbf{v}_B + m_C\mathbf{v}_C$$

$$100(756\mathbf{i}) = 40[(v_A)_x\mathbf{i} + 675\mathbf{j} + (v_A)_z\mathbf{k}]$$

$$+ 24[(v_B)_x\mathbf{i} + (v_B)_y\mathbf{j} + (v_B)_z\mathbf{k}]$$

$$+ 36[(v_C)_x\mathbf{i} + (v_C)_y\mathbf{j} + (v_C)_z\mathbf{k}]$$

which yields, after simplification, three scalar equations:

$$10(v_A)_x + 6(v_B)_x + 9(v_C)_x = 18900 \qquad (1)$$

$$2(v_B)_y + 3(v_C)_y = -2250 \qquad (2)$$

$$10(v_A)_z + 6(v_B)_z + 9(v_C)_z = 0 \qquad (3)$$

Equating the sums of the angular momentum vectors about O, we write

$$(\mathbf{H}_O)_1 = (\mathbf{H}_O)_2:$$

$$\bar{\mathbf{r}} \times m\mathbf{v}_0 = \mathbf{r}_A \times m_A\mathbf{v}_A + \mathbf{r}_B \times m_B\mathbf{v}_B + \mathbf{r}_C \times m_C\mathbf{v}_C$$

$$\mathbf{0} = 40 \begin{vmatrix} \mathbf{i} & \mathbf{j} & \mathbf{k} \\ 1098 & 540 & 0 \\ (v_A)_x & 675 & (v_A)_z \end{vmatrix} + 24 \begin{vmatrix} \mathbf{i} & \mathbf{j} & \mathbf{k} \\ 1218 & -420 & -840 \\ (v_B)_x & (v_B)_y & (v_B)_z \end{vmatrix}$$

$$+ 36 \begin{vmatrix} \mathbf{i} & \mathbf{j} & \mathbf{k} \\ 698 & -320 & 560 \\ (v_C)_x & (v_C)_y & (v_C)_z \end{vmatrix}$$

which, after expansion and simplification, yields

$$15(v_A)_z + 14(v_B)_y - 7(v_B)_z - 14(v_C)_y - 8(v_C)_z = 0 \qquad (4)$$

$$610(v_A)_z + 280(v_B)_x + 406(v_B)_z - 280(v_C)_x$$
$$+ 349(v_C)_z = 0 \qquad (5)$$

$$300(v_A)_x - 140(v_B)_x - 406(v_B)_y - 160(v_C)_x$$
$$- 349(v_C)_y = 411750 \qquad (6)$$

The above scalar equations (1) through (6) contain eight unknowns. For a complete solution, we need at least two additional equations.

Since no force acts on the parts A, B, and C, immediately after the explosion at $t = t_1$, they must, by *Newton's first law*, move away with *constant* velocities from the point of explosion on the x axis. At $t = 1.3$ s, the y and z components of the position vectors of A, B, and C must, therefore, be equal to the y and z components of the velocities of A, B, and C multiplied by the elapsed time $(1.3 - t_1)$ s since the explosion, respectively. From the given data and the solution for $\mathbf{r}_A$, we have $y_A = 540$ m, $z_A = 0$, $y_B = -420$ m, $z_B = -840$ m, $y_C = -320$ m, and $z_C = 560$ m. Since $(v_A)_y = 675$ m/s is given and the elapsed time since the explosion is common to all parts, we write

$$1.3 - t_1 = \frac{540}{675} = \frac{0}{(v_A)_z} = \frac{-420}{(v_B)_y} = \frac{-840}{(v_B)_z}$$

$$= \frac{-320}{(v_C)_y} = \frac{560}{(v_C)_z} \qquad (7)$$

Solving Eqs. (7), we obtain

$$(v_A)_z = 0 \qquad (v_B)_y = -525 \qquad (v_B)_z = -1050$$

$$(v_C)_y = -400 \qquad (v_C)_z = 700$$

which satisfy Eqs. (2), (3), and (4) and render Eqs. (5) and (6) to

$$(v_B)_x - (v_C)_x = 650 \tag{5'}$$

$$15(v_A)_x - 7(v_B)_x - 8(v_C)_x = 2950 \tag{6'}$$

Solving Eqs. (1), (5'), and (6') simultaneously, we obtain

$$(v_A)_x = 900 \qquad (v_B)_x = 1050 \qquad (v_C)_x = 400$$

Thus, the velocities of parts A, B, and C at $t = 1.3$ s are

$$\mathbf{v}_A = 900\mathbf{i} + 675\mathbf{j} \text{ m/s} \qquad \mathbf{v}_B = 1050\mathbf{i} - 525\mathbf{j} - 1050\mathbf{k} \text{ m/s} \blacktriangleleft$$

$$\mathbf{v}_C = 400\mathbf{i} - 400\mathbf{j} + 700\mathbf{k} \text{ m/s} \blacktriangleleft$$

Developmental Exercise

D14.12   Refer to Example 14.7. Based on the solution, determine the time t_1 and the abscissa x_1 at which the explosion occurred.

14.5   Impulsive Motion

An *impulsive force* is a relatively large force which acts suddenly on a body for a relatively short time interval and causes a significant change in the momentum of the body. Thus, forces due to collision are impulsive forces. The motion of a body caused by impulsive forces is called an *impulsive motion*. *Nonimpulsive forces* include weight forces and those which do not cause a significant change in the momentum of the body during a relatively short time interval.

For instance, when a tennis ball of mass m coming with a velocity of $\mathbf{v}_1$ is hit, the contact between the ball and the racket lasts for only a very short time interval Δt. However, the average value of the contact force $\mathbf{F}$ exerted by the racket on the ball is relatively large, and the impulse $\mathbf{F} \, \Delta t$ is large enough to cause a change in momentum from $m\mathbf{v}_1$ to $m\mathbf{v}_2$ as illustrated in Fig. 14.7, where $W = mg =$ weight of the ball. Here, the contact force $\mathbf{F}$ is an impulsive force, while the weight force $\mathbf{W}$ is a *nonimpulsive force*. Note that *a nonimpulsive force in juxtaposition with an impulsive force is negligible*. Thus, we set $W \, \Delta t \approx 0$ in Fig. 14.7.

Fig. 14.7

If $\Sigma\mathbf{F}$ is the vector sum of the average impulsive forces acting on a particle of mass m during the time interval Δt, Eq. (14.6) becomes

$$m\mathbf{v}_1 + \Sigma\mathbf{F}\,\Delta t = m\mathbf{v}_2 \qquad (14.21)$$

In general, an *impulsive force* is a force having a sudden surge in magnitude during a relatively short time interval Δt, whereas a *nonimpulsive force* is small compared with an impulsive force and remains essentially constant during the time interval Δt.

EXAMPLE 14.8

A 2.59-g rifle bullet B is fired with a velocity of 300 m/s into a 500-g sphere A which is suspended by an inextensible wire and is at rest as shown. Knowing that the bullet becomes embedded in the sphere in 1 ms, determine (a) the velocity $\mathbf{v}_2$ of the sphere immediately after the bullet becomes embedded, (b) the average impulsive tension developed in the wire.

Solution. Since the wire is taken to be inextensible, the velocity $\mathbf{v}_2$ of the sphere must be perpendicular to the wire and point to the left. We note that the mass of the bullet is $m_B = 2.59$ g $= 2.59 \times 10^{-3}$ kg, the mass of the sphere is $m_A = 500$ g $= 0.5$ kg, and the time interval is $\Delta t = 1$ ms $= 10^{-3}$ s. Letting the average impulsive tension in the wire be F and applying the *principle of linear impulse and momentum* to the bullet and the sphere as a system during the time interval $0 \le t \le 10^{-3}$ s, we have

$$\overset{+}{\leftarrow}\ \Sigma V_x: \qquad 2.59 \times 10^{-3}\,(300)\sin 30° + 0$$

$$= (0.5 + 2.59 \times 10^{-3})v_2$$

$$v_2 = 0.773 \qquad \mathbf{v}_2 = 0.773 \text{ m/s} \leftarrow \quad \blacktriangleleft$$

$$+\downarrow \Sigma V_y: \qquad 2.59 \times 10^{-3}\,(300)\cos 30° = F(10^{-3}) = 0$$

$$F = 672.9 \qquad\qquad F = 673 \text{ N} \quad \blacktriangleleft$$

Developmental Exercises

D14.13 Define the terms: (a) impulsive force, (b) impulsive motion.

D14.14 A 5-oz baseball is traveling with a velocity as shown in (a) just before it is hit by a bat. After having been hit, it travels with a velocity as shown in (b). If the impact lasts for 20 ms, determine the average impulsive force exerted on the baseball.

Fig. D14.14

72 km/h

$\mu_k = 0.6$

Fig. P14.2*

170 kg

$F = 50t$ N

$\mu_s = 0.7$
$\mu_k = 0.6$

8 / 15

Fig. P14.4

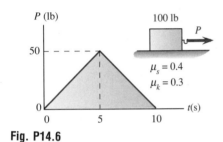

P (lb)

100 lb

P

50

$\mu_s = 0.4$
$\mu_k = 0.3$

0

0   5   10 t(s)

Fig. P14.6

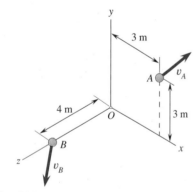

y

3 m

A v_A

4 m

O

3 m

B

z v_B

x

Fig. P14.9

y

A

10 m

5 m

10 m

B O

C

z

5 m

7.5 m

x

Fig. P14.11

PROBLEMS

14.1   A 2400-lb car increases its speed uniformly on a straight road from 0 to 45 mi/h in 8 s. Neglecting the rotational effects of the wheels, determine the resultant friction force acting on the tires of the car.

14.2*   A truck is traveling at 72 km/h as shown when the brakes are fully applied, causing all wheels to skid. Determine the time required for the truck to stop if $\mu_k = 0.6$ between the tires and the road surface.

14.3   Refer to Problem. 14.2*. Assume that the brakes are fully applied for only 2 s and then released. Determine the speed v_2 of the truck when the brakes are released.

14.4   A force of magnitude $F = 50t$ N is applied to a 170-kg crate as shown, where t is measured in seconds. The crate is at rest when $t = 0$, and the direction of **F** is always parallel to the incline. It is known that $\mu_s = 0.7$ and $\mu_k = 0.6$ between the crate and the incline. Determine the time t_2 at which the speed of the crate is 4 m/s.

14.5*   Refer to Prob. 14.4. Determine the speed v_2 of the crate when $t = 12$ s.

14.6   The magnitude of a force **P** varies as shown. The force **P** is applied to a 100-lb block at the time $t = 0$. It is known that the block is at rest when $t = 0$, and that $\mu_s = 0.4$ and $\mu_k = 0.3$ between the block and the support. Determine (a) the time t_1 at which the block starts to move, (b) the time t_2 at which the block attains its maximum speed, (c) the maximum speed attained by the block, (d) the time t_3 at which the block stops again.

14.7   The linear momentum of a 2-kg particle is given by $\mathbf{L} = (28 - 3t^2)\mathbf{i} + (16 + 6t)\mathbf{j} - (8 + 4t)\mathbf{k}$ N·s, where t is measured in seconds. When $t = 2$ s, determine (a) the resultant force **F** acting on the particle, (b) the magnitude of the acceleration of the particle, (c) the speed of the particle.

14.8*   The resultant force acting on a 16.1-lb particle is given by $\mathbf{F} = (8 + 2t)\mathbf{i} + (4 + t^2)\mathbf{j} + (10 - 2t)\mathbf{k}$ lb, where t is measured in seconds. If the initial velocity of the particle is $\mathbf{v}_1 = 24\mathbf{i} - 6\mathbf{j} + 48\mathbf{k}$ ft/s at $t = 0$, determine (a) the time t_2 at which the velocity of the particle is parallel to the xy plane, (b) the velocity $\mathbf{v}_2$ of the particle at $t = t_2$.

14.9   A system consists of a 2-kg particle A and a 3-kg particle B. At the instant shown, the velocities of A and B are $\mathbf{v}_A = 2\mathbf{i} + v_y\mathbf{j} + v_z\mathbf{k}$ m/s and $\mathbf{v}_B = 3\mathbf{i} - 4\mathbf{j} + 5\mathbf{k}$ m/s. For this instant, determine (a) the numerical values of v_y and v_z for which the resultant angular momentum $\mathbf{H}_O$ of the system about O is parallel to the y axis, (b) the corresponding value of $\mathbf{H}_O$.

14.10*   Solve Prob. 14.9 if $\mathbf{H}_O$ is parallel to the x axis.

14.11   A system consists of three particles A, B, and C, which have masses $m_A = 5$ kg, $m_B = 3$ kg, $m_C = 2$ kg, and velocities $\mathbf{v}_A = 3\mathbf{i} + 5\mathbf{j} - 10\mathbf{k}$ m/s, $\mathbf{v}_B = 4\mathbf{i} - 7\mathbf{j} + 8\mathbf{k}$ m/s, $\mathbf{v}_C = 6\mathbf{i} + 2\mathbf{j} - 7\mathbf{k}$ m/s, respectively, at the instant shown. For this instant, determine the resultant angular momentum $\mathbf{H}_G$ of the system about its mass center G.

14.12   A 10-kg object is traveling with a velocity of 40 m/s ↓ when it explodes into a 6-kg fragment A and a 4-kg fragment B as shown. Knowing that immediately after the explosion the fragments travel with velocities as indicated, determine the angles θ_A and θ_B.

14.13   A boy tosses a 2-kg sandbag with a velocity of 5 m/s ↘ onto an 18-kg wagon, which is at rest, as shown. Knowing that the wagon can move freely, determine the speed of the wagon after the sandbag has slid to a relative stop on the wagon.

Fig. P14.12

Fig. P14.13

Fig. P14.14*

14.14* A 48-kg boy steps with a horizontal velocity of 1 m/s onto his 2-kg skateboard which is at rest as shown. Determine the speed of the skateboard when the boy rides steadfastly on it.

14.15 A 5-lb skateboard and a 95-lb boy, who stands on it, are initially at rest. At the instant shown, the boy jumps off the skateboard with a velocity $\mathbf{v}_{B/S}$ of 10 ft/s relative to the skateboard. If the jump is completed in 0.5 s, determine (a) the speed v_S of the skateboard just after the jump, (b) the magnitude of the average horizontal force exerted on the skateboard by the boy, (c) the magnitude of the average total force exerted on the ground by the wheels of the skateboard during the jump.

14.16 Two railroad cars are traveling as shown. Knowing that coupling occurs and is completed in 0.4 s, determine (a) the speed of the coupled cars, (b) the magnitude of the average force exerted by the car A on the car B during the coupling, (c) the percentage of the total loss of kinetic energy of the cars during the coupling.

Fig. P14.15

Fig. P14.16

Fig. P14.17

14.17 The pucks B and C are at rest side by side on a horizontal plane when the puck A strikes the puck B with a velocity of 5 m/s → as shown. After the impact, the pucks are observed to move with velocities as indicated, where $\mathbf{v}_A = .1.5\mathbf{i} - 2\mathbf{j}$ m/s. If the pucks have equal masses, determine the magnitudes of $\mathbf{v}_B$ and $\mathbf{v}_C$.

14.18* A 30-g bullet is fired with a speed v_0 into a 10-kg block which is at rest as shown. If $\theta = 30°$ when the block with the embedded bullet swings to its highest position, determine (a) the speed v_0, (b) the percentage of loss of the kinetic energy during the impact.

14.19 A 1-oz bullet is fired with a speed v_0 into a 15-lb block which is at rest as shown. The block with the embedded bullet is observed to move 2 ft to the right before coming to a stop. If $\mu_k = 0.3$ between the block and the support, determine the speed v_0 of the bullet.

Fig. P14.18*

Fig. P14.19

14.20 A 10-g bullet is fired with a speed $v_1 = 260$ m/s at a 2-kg block as shown. The bullet pierces through the block and causes the block to slide a distance of 0.2 m. If $\mu_k = 0.3$ between the block and the support, determine the speed v_2 with which the bullet leaves the block.

Fig. P14.20

14.21* A 2-kg particle, which is moving on a smooth horizontal surface with a constant velocity of 8 m/s ←, is struck by an impulsive force **F** that acts to the right on the particle for 10 ms. If **F** varies with the time t as shown, determine the resulting velocity of the particle.[†]

14.22* The block A of mass m_A and the block B of mass m_B are held to compress a spring as shown. The spring is not attached to the blocks, and effects of friction between the support and the blocks are negligible. When the blocks are suddenly released, show that $v_A/v_B = T_A/T_B = m_B/m_A$ where v_A and v_B are the speeds after separation, and T_A and T_B are the kinetic energies of the blocks A and B, respectively.

Fig. P14.21*

Fig. P14.22*   **Fig. P14.23**

14.23 A 1-oz bullet strikes a hard surface with a velocity of 2400 ft/s ∡ 30° and incurs a 6.6-in. scratch on the surface before ricocheting with a velocity of 2000 ft/s ⊾ 20° as shown. Assuming an average speed of 2200 ft/s during contact, determine the average impulsive force exerted by the hard surface on the bullet.

14.24 A 60-g tennis ball is traveling with a velocity of 10 m/s ∡ 30° just before it is hit by a racket. After having been hit, it travels with a velocity of 20 m/s ⊾ 30° as shown. If the contact between the ball and the racket lasts for 50 ms, determine the average impulsive force exerted by the racket on the ball.

Fig. P14.24

14.25 and 14.26* The system shown is released from rest at the time $t = 0$. Neglecting friction, determine the speed of the block A when $t = 3$ s.

Fig. P14.25 and P14.27*   **Fig. P14.26* and P14.28**

14.27* A system of blocks is shown. At the time $t = 0$, the velocity of the block B is 2 m/s →. If $\mu_k = 0.1$ between the support and the blocks, determine the speed of the block A when $t = 3$ s.

14.28 A system of blocks is shown. At the time $t = 0$, the velocity of the block B is 5 ft/s →. If $\mu_k = 0.1$ between the support and the blocks, determine the speed of the block B when $t = 3$ s.

[†]Cf. the *hint* in Prob. 13.23*.

14.29 through 14.31 A system of cylinders is shown. The initial velocity of the cylinder B is 2 ft/s ↓ when the system is released at the time $t = 0$. If the cylinder A has a weight of 64.4 lb and a velocity of 3 ft/s ↓ when $t = 4$ s, determine the weight of the cylinder B.

Fig. P14.29 and P14.32

Fig. P14.30* and P14.33

Fig. P14.31 and P14.34*

14.32 through 14.34* A system of cylinders is shown. The initial velocity of the cylinder A is 3 m/s ↑ when the system is released at the time $t = 0$. If the cylinder B has a mass of 40 kg and a velocity of 2 m/s ↑ when $t = 4$ s, determine the mass of the cylinder A.

14.35 and 14.36 The system shown is released from rest at the time $t = 0$. Determine the tension in each cable and the velocity of each cylinder when $t = 2$ s.[†]

Fig. P14.35

Fig. P14.36

14.37 A 100-lb object is falling with a velocity of 100 ft/s ↓ when it explodes into three fragments, A, B, and C, of weights $W_A = 38$ lb, $W_B = 27$ lb, and $W_C = 35$ lb, respectively. Immediately after the explosion, the velocities of the fragments are directed as shown. Determine the speed of each fragment immediately after the explosion.

14.38* A 50-kg shell is traveling with a velocity $\mathbf{v} = 60\mathbf{i} + 40\mathbf{j} + 120\mathbf{k}$ m/s when it explodes into three fragments, A, B, and C, of masses $m_A = 25$ kg, $m_B = 15$ kg, and $m_C = 10$ kg, respectively. Immediately after the explosion, the

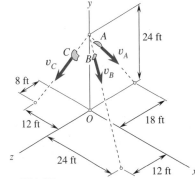

Fig. P14.37

[†]Cf. App. D for computer solution of resulting simultaneous equations.

Fig. P14.39

Fig. P14.40*

Fig. 14.8

Fig. 14.9

velocities of A and B are observed to be $\mathbf{v}_A = -30\mathbf{i} + 240\mathbf{j} + 120\mathbf{k}$ m/s and $\mathbf{v}_B = 180\mathbf{i} - 90\mathbf{j} - 60\mathbf{k}$ m/s. Determine the velocity $\mathbf{v}_C$ of C immediately after the explosion.

14.39   A 55-kg object traveling in space is observed at the time $t = 0$ to pass through the origin O of an inertial reference frame Oxy with a constant velocity $\mathbf{v}_0$ along the x axis. Following an explosion at $t = t_1$, where $0 < t_1 < 4.5$ s, the object separates into three fragments, A, B, and C, of masses $m_A = 17$ kg, $m_B = 8$ kg, and $m_C = 30$ kg, respectively, as shown. At $t = 4.5$ s, it is observed that the position vectors of A and B are $\mathbf{r}_A = 135\mathbf{i} - 102\mathbf{j}$ m and $\mathbf{r}_B = 225\mathbf{i} + 48\mathbf{j}$ m, and the velocities of A, B, and C are $\mathbf{v}_A = 30\mathbf{i} + (v_A)_y\mathbf{j}$ m/s, $\mathbf{v}_B = 60\mathbf{i} + (v_B)_y\mathbf{j}$ m/s, and $\mathbf{v}_C = 22\mathbf{i} + (v_C)_y\mathbf{j}$ m/s. Neglecting the effect of gravity, determine (a) the velocity $\mathbf{v}_0$, (b) the position vector $\mathbf{r}_C$ of C at $t = 4$ s, (c) the time t_1, (d) the numerical values of $(v_A)_y$, $(v_B)_y$, and $(v_C)_y$.

14.40*   A 100-kg object traveling in space is observed at the time $t = 0$ to pass through the origin O of an inertial reference frame $Oxyz$ with a constant velocity $\mathbf{v}_0 = 624\mathbf{i}$ m/s. Following an explosion at $t = t_1$, where $0 < t_1 < 1.75$ s, the object separates into three fragments, A, B, and C, of masses $m_A = 39$ kg, $m_B = 36$ kg, and $m_C = 25$ kg, respectively, as shown. At $t = 1.75$ s, it is observed that the position vectors of A and B are $\mathbf{r}_A = 568\mathbf{i} + 300\mathbf{j} + 75\mathbf{k}$ m and $\mathbf{r}_B = 1118\mathbf{i} - 325\mathbf{j} + 650\mathbf{k}$ m, and that the z component of the velocity $\mathbf{v}_C$ of C is $(v_C)_z = -1053\mathbf{k}$ m/s. Neglecting the effect of gravity, determine the velocities $\mathbf{v}_A$ and $\mathbf{v}_B$ of A and B, respectively, after the explosion.

14.6   Central-Force Motion

We saw in Sec. 12.8 that, when a particle P of mass m moves under a central force $\mathbf{F}$, the line of action of $\mathbf{F}$ always passes through the center of force O as shown in Fig. 14.8. Consequently, the moment, and hence the angular impulse, of $\mathbf{F}$ about O is always zero during the motion. From this fact and Eq. (14.18), we see that, *when a particle moves under a central force, the angular momentum of the particle about the center of force must remain constant during the motion.* The magnitude of this angular momentum about the center of force O is

$$H_O = |\mathbf{r} \times m\mathbf{v}| = mrv \sin\phi = \text{constant} \qquad (14.22)$$

where ϕ is the angle between the radius vector $\mathbf{r}$ and the velocity $\mathbf{v}$ of the particle P. Since $v_\theta = v \sin\phi = r\dot\theta$ and m is a constant, we deduce from Eqs. (14.22) that

$$rv \sin\phi = rv_\theta = r^2\dot\theta = h \qquad (14.23)$$

where h is the same constant appearing in Eqs. (12.30) through (12.33). We saw in Eq. (12.32) that *the constant h is equal to twice the areal speed of the particle P.* On the other hand, we see in Eqs. (14.23) that *the constant h represents the angular momentum per unit mass of a particle moving under a central force about the center of force.* This interpretation provides us with an additional insight into the physical significance of the constant h.

If a particle of mass m moves under a *conservative central force* (e.g., a gravitational central force or an elastic central force) from the position A_1 to the position A_2 as illustrated in Fig. 14.9, we note that (a) the *angular momentum* of the particle about the center of force O is *conserved*, and (b) the *total mechanical energy* of the particle is *conserved*. In this case, we

write

$$h_1 = h_2: \qquad r_1 v_1 \sin\phi_1 = r_2 v_2 \sin\phi_2 \qquad (14.24)$$

$$\text{Energy:} \qquad T_1 + V_1 = T_2 + V_2 \qquad (13.44)$$
$$\text{(repeated)}$$

In gravitational central-force motion, Eq. (13.44) may be written as

$$\frac{1}{2} m v_1^2 - \frac{GMm}{r_1} = \frac{1}{2} m v_2^2 - \frac{GMm}{r_2}$$

$$\frac{1}{2} v_1^2 - \frac{GM}{r_1} = \frac{1}{2} v_2^2 - \frac{GM}{r_2} \qquad (14.25)$$

where M is the mass of the attracting body at the center of force O, $m \ll M$, and G is the constant of gravitation. In the case of an elastic central force, Eq. (13.44) may take the form

$$\tfrac{1}{2} m v_1^2 + \tfrac{1}{2} k x_1^2 = \tfrac{1}{2} m v_2^2 + \tfrac{1}{2} k x_2^2 \qquad (14.26)$$

where k is the spring modulus, and x_1 and x_2 are the changes in length of the spring from its natural length at A_1 and A_2, respectively.

EXAMPLE 14.9

A spacecraft starts its free flight from the position A_1 where $v_1 = 5$ mi/s, $a_1 = 1000$ mi, and $\phi_1 = 75°$ as shown. Determine the maximum and minimum altitudes of the orbit of the spacecraft.

Solution. We recognize that the spacecraft moves under a gravitational central force, and the system is conservative. The center of force is located at O as shown. Let an extremum altitude be in the position A_2, where $\mathbf{v}_2 \perp \mathbf{r}_2$. Recalling that the radius of the earth is $R = 3960$ mi, we write

$$r_1 = R + a_1 = 3960 + 1000 \qquad r_1 = 4960 \text{ mi}$$

$$h_1 = h_2: \qquad r_1 v_1 \sin\phi_1 = r_2 v_2 \sin\phi_2$$

$$4960(5) \sin 75° = r_2 v_2 \sin 90° \qquad (1)$$

$$\text{Energy:} \qquad \frac{1}{2} v_1^2 - \frac{GM}{r_1} = \frac{1}{2} v_2^2 - \frac{GM}{r_2}$$

$$\frac{1}{2} v_1^2 - \frac{gR^2}{r_1} = \frac{1}{2} v_2^2 - \frac{gR^2}{r_2}$$

$$\frac{1}{2} (5)^2 - \frac{32.2(3960)^2}{5280(4960)} = \frac{1}{2} v_2^2 - \frac{32.2(3960)^2}{5280(r_2)} \qquad (2)$$

Elimination of v_2 between Eqs. (1) and (2) leads to

$$6.781 r_2^2 - 9.563 \times 10^4 r_2 + 2.869 \times 10^8 = 0$$

which yields the two roots $r' = 9.77 \times 10^3$ mi and $r'' = 4.33 \times 10^3$ mi. These are the maximum and minimum values of r. Thus, we write

$$a_{\max} = r' - R \qquad a_{\max} = 5.81 \times 10^3 \text{ mi} \blacktriangleleft$$

$$a_{\min} = r'' - R \qquad a_{\min} = 0.37 \times 10^3 \text{ mi} \blacktriangleleft$$

D14.15 Refer to Example 14.9. Determine the angular momentum per unit mass of the spacecraft about the center of the earth.

D14.16 Refer to Example 14.9. Determine the maximum and minimum speeds of the spacecraft in its orbit.

EXAMPLE 14.10

A 1-kg sphere attached to an elastic cord of modulus $k = 100$ N/m moves along a closed path on the horizontal plate as shown. The cord is undeformed when the sphere is situated over the small hole at O. In the position shown, $r_1 = 1$ m, $v_1 = 10$ m/s, and $\phi_1 = 60°$. Knowing that friction is negligible, determine the maximum and minimum distances r_{max} and r_{min} of the sphere from O.

Solution. We recognize that the sphere moves under an elastic central force, and the system is conservative. The center of force is located at O. Let an extremum distance of the sphere from O occur in the position A_2, where $\mathbf{v}_2 \perp \mathbf{r}_2$. We write

$$h_1 = h_2: \quad r_1 v_1 \sin\phi_1 = r_2 v_2 \sin\phi_2$$

$$1(10) \sin 60° = r_2 v_2 \sin 90° \quad (1)$$

$$\text{Energy:} \quad \tfrac{1}{2}mv_1^2 + \tfrac{1}{2}kx_1^2 = \tfrac{1}{2}mv_2^2 + \tfrac{1}{2}kx_2^2$$

$$\tfrac{1}{2}(1)(100)^2 + \tfrac{1}{2}(100)(1)^2 = \tfrac{1}{2}(1)v_2^2 + \tfrac{1}{2}(100)r_2^2 \quad (2)$$

Elimination of v_2 between Eqs. (1) and (2) leads to

$$50r_2^4 - 100r_2^2 + 37.5 = 0$$

which yields $r_2 = \pm 1.225$ m and ± 0.707 m. Thus, we write

$$r_{max} = 1.225 \text{ m} \qquad r_{min} = 0.707 \text{ m} \blacktriangleleft$$

REMARK. It can be shown that the path described by the sphere in an ellipse. However, the center of force is located at the geometric center O, rather than one of the foci, of the ellipse. (Cf. Example 14.9.)

Developmental Exercises

D14.17 Refer to Example 14.10. Determine the angular momentum of the sphere about the point O.

D14.18 Refer to Example 14.10. Determine the maximum and minimum speeds of the sphere moving along its path.

PROBLEMS

14.41* A spacecraft describes an elliptic orbit around the earth as shown. In the position S, its velocity $\mathbf{v}_S$ and the vertical make an angle $\phi_S = 70°$, its altitude is $a_S = 5000$ mi, and its speed is $v_S = 3.2$ mi/s. Determine the maximum and the minimum altitudes of its orbit.

Fig. P14.41* and P14.42

14.42 A spacecraft describes an elliptic orbit around the earth as shown. As it passes through the perigee P, its altitude is $a_P = 3$ Mm and its speed is $v_P = 7.3$ km/s. Determine (a) the angle ϕ_S between its velocity $\mathbf{v}_S$ and the vertical as it passes through the position S where $a_S = 7.5$ Mm, (b) its speed v_S.

14.43 A spacecraft starts its free flight around the earth in the position S at an altitude $a_S = 1500$ mi and with a velocity $\mathbf{v}_S$ of magnitude 4 mi/s as shown. Determine (a) the minimum altitude a_{min} of its orbit if the angle ϕ_S between $\mathbf{v}_S$ and the vertical is 85°, (b) the range of values of ϕ_S if a_{min} is to be no less than 300 mi.

14.44 A spacecraft starts its free flight around the earth in the position S at an altitude $a_S = 1500$ mi and with a velocity v_S of magnitude 4 mi/s as shown. If the minimum altitude a_{min} of its orbit is 150 mi, determine (a) the angle ϕ_S between $\mathbf{v}_S$ and the vertical, (b) the eccentricity of orbit, (c) the period of orbit.

14.45* A *Landsat L* is describing a circular orbit of altitude 900 km around the earth as shown. At an appropriate time, a space shuttle is launched to rendezvous with the *Landsat*. If the shuttle starts its free flight in the position S at an altitude of 200 km and with a velocity $\mathbf{v}_S$ of magnitude 5 km/s, determine the angle ϕ_S with which the trajectory of the shuttle will be tangent at A to the orbit of the *Landsat*.

14.46 A *Landsat L* is describing a circular orbit of altitude 900 km around the earth as shown. At an appropriate time, a space shuttle is launched to rendezvous with the *Landsat*. If the shuttle starts its free flight in the position S at an altitude of 200 km and with a velocity $\mathbf{v}_S$ which makes an angle $\phi_S = 65°$ with the vertical, determine the magnitude of $\mathbf{v}_S$ with which the trajectory of the shuttle will be tangent at A to the orbit of the *Landsat*.

14.47 An elastic cord having a free length $L = 1.2$ m and a modulus $k = 100$ N/m is attached to a fixed point at O and a 0.4-kg disk which is given an initial velocity $\mathbf{v}_1 = 6\mathbf{j}$ m/s at P_1 to move on a horizontal smooth surface in the xy plane as shown. Determine (a) the speed v_2 of the disk when the cord becomes slack at P_2, (b) the shortest distance d_s that the disk will come to O.

14.48* Refer to Prob. 14.47. Determine the magnitude of $\mathbf{v}_1$ as shown if the shortest distance that the disk will come to O is $d_s = 1$ m.

14.49 Refer to Prob. 14.47. Determine the minimum magnitude of $\mathbf{v}_1$ as shown if the elastic cord is to remain taut at all times.

Fig. P14.43 and P14.44

Fig. P14.45* and P14.46

Fig. P14.47

14.50 An elastic cord having a free length $L = 25$ in. and a modulus $k = 1$ lb/in. is attached to a fixed point at O and a 1-lb disk which is given an initial velocity $\mathbf{v}_1 = v_1\mathbf{j}$ at P_1 to move on a horizontal smooth surface in the xy plane as shown. If the maximum distance between O and the trajectory of the disk is $\overline{OP}_2 = 30$ in. as indicated, determine (a) the value of v_1, (b) the radius of curvature of the trajectory at P_2.

Fig. P14.50   **Fig. P14.51***

14.51* A spring having a free length $L = 0.5$ m and a modulus $k = 400$ N/m is mounted on a horizontal cantilever rod as shown. One end of the spring is attached to the fixed point O, and its other end is attached to a 2-kg collar C. The rod can rotate freely about the vertical axis through O. The system is set in motion with $r = 0.3$ m, $v_r = 0$, and $v_\theta = 6$ m/s. Neglecting effects of friction and the mass of the rod, determine the velocity components of the collar when $r = 0.8$ m.

14.52 Refer to Prob. 14.51*. Determine (a) the maximum distance r_{max} between the point O and the collar during the ensuing motion, (b) the speed of the collar when $r = r_{max}$.[†]

IMPACT AND SPECIAL TOPIC

14.7 Impact

An *impact* is a forceful collision or contact between two bodies over a very short period of time. Several types of impacts are illustrated in Figs. 14.10 through 14.13.

The *line of impact* is the straight line normal to the contacting surfaces during the impact. The lines of impact in Figs. 14.10 through 14.13 are indicated by dashed lines. A *direct impact* is an impact where the velocities of the points of contact of the two colliding bodies are both directed along the line of impact. An *oblique impact* is an impact where the velocity of at least one of the points of contact is not directed along the line of impact; it is an impact where one body strikes another body with a glancing blow. A *central impact* is an impact where the mass centers of the two colliding bodies are both located on the line of impact. An *eccentric impact* is an impact where at least one of the mass centers of the two colliding bodies is not located on the line of impact. Direct and oblique *central impacts* involve motions of *particles* and are considered in the next two sections, whereas

---

[†]Cf. App. E for computer solution of resulting quartic equation.

(a)   (b)

Direct central impact

Fig. 14.10

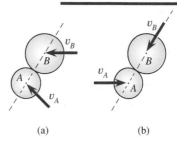

(a)   (b)

Oblique central impact

Fig. 14.11

(a)   (b)

Direct eccentric impact

Fig. 14.12

(a)   (b)

Oblique eccentric impact

Fig. 14.13

direct and oblique *eccentric impacts* involve motions of *rigid bodies* and are considered in Chap. 18.

When the two colliding particles are considered together as a system, we note that there is no external impulsive force acting on the system during the impact. Therefore, *the total linear momentum of any two particles must remain the same immediately before and after the impact*; and we may write Eq. (14.20) for two colliding particles A and B as

$$m_A\mathbf{v}_A + m_B\mathbf{v}_B = m_A\mathbf{v}'_A + m_B\mathbf{v}'_B \tag{14.27}$$

where m_A and m_B are their masses, $\mathbf{v}_A$ and $\mathbf{v}_B$ are their velocities just before the impact, and $\mathbf{v}'_A$ and $\mathbf{v}'_B$ are their velocities just after the impact. If the two particles A and B have a *direct central impact*, they must move along the same straight line; and we have

$$m_A v_A + m_B v_B = m_A v'_A + m_B v'_B \tag{14.28}$$

Developmental Exercise

D14.19   Define the terms: (a) impact, (b) line of impact, (c) direct impact, (d) oblique impact, (e) central impact, (f) eccentric impact.

14.8   Direct Central Impact

In general, the process of *direct central impact* of two particles involves a period of *deformation* and a period of *restitution*, as illustrated in Figs. 14.14 and 14.15. The impulses exerted by the force of impact on either particle during these two periods are generally unequal.

(a) Motion before
impact

(b) Motion at maximum
deformation

(a) Motion after
impact

Fig. 14.14

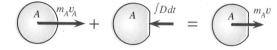

(a) Momenta and impulse during
period of deformation

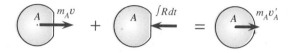

(b) Momenta and impulse during
period of restitution

Fig. 14.15

Letting the impulsive force acting on A by B during the period of deformation be $\mathbf{D}$ and the common velocity at the end of deformation be $\mathbf{v}$, as shown in Fig. 14.15(a), we write

$$\xrightarrow{+}\ \Sigma V_x: \qquad m_A v_A - \int D\, dt = m_A v \qquad (14.29)$$

where the integral extends over the period of deformation. Likewise, letting the impulsive force acting on A by B during the period of restitution be $\mathbf{R}$, as shown in Fig. 14.15(b), we write

$$\xrightarrow{+}\ \Sigma V_x: \qquad m_A v - \int R\, dt = m_A v'_A \qquad (14.30)$$

where the integral extends over the period of restitution. Generally, *the impulse $\int R\, dt$ during restitution* is smaller than *the impulse $\int D\, dt$ during deformation*. The ratio of these two impulses is called the *coefficient of restitution* and is denoted by e.[†] We write

$$e = \frac{\int R\, dt}{\int D\, dt} \qquad (14.31)$$

Solving for the integrals in Eqs. (14.29) and (14.30) and substituting them into Eq. (14.31), we write

$$v - v'_A = e(v_A - v) \qquad (14.32)$$

(a) Momenta and impulse during
period of deformation

(b) Momenta and impulse during
period of restitution

Fig. 14.16

The momenta and impulses during the periods of deformation and restitution of the particle B are depicted in Fig. 14.16, from which we similarly obtain the relation

$$v'_B - v = e(v - v_B) \qquad (14.33)$$

[†]The coefficient of restitution e is usually determined from experiments. It is employed to facilitate the elementary study of the phenomenon of impact. The value of e is highly dependent on the two materials involved; moreover, it is also dependent on the relative velocity and the geometry of the two colliding bodies. Experiments show that e may approach 1 if the relative velocity just before the impact approaches zero.

Adding Eqs. (14.32) and (14.33) side by side, we get

$$v'_B - v'_A = e(v_A - v_B) \qquad (14.34)$$

A problem of direct central impact is usually solved by applying Eqs. (14.28) and (14.34) simultaneously, where the positive sense chosen for each of the quantities v_A, v_B, v'_A, and v'_B must, of course, be the same.

   A *perfectly plastic impact* is an impact where $e = 0$; i.e., the two particles cling together and have a common velocity after the impact. On the other hand, a *perfectly elastic impact* is an impact where $e = 1$; i.e., the magnitude of the relative velocity just before the impact is equal to that just after the impact. In a perfectly elastic impact, it can be shown that the total kinetic energy of the two particles, as well as their total momentum, is conserved. In impacts where $e < 1$, only their total momentum is conserved; their total kinetic energy will suffers a loss.

EXAMPLE 14.11

A 20-lb block A is released from rest in the position shown and hits the 40-lb sphere B which is at rest. If $e = 0.8$ between A and B, determine the velocities of A and B just after the impact.

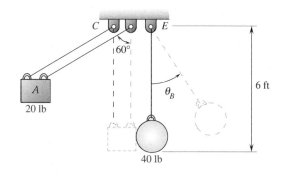

Solution. Letting v_A be the speed of the block A just before it hits the sphere B and applying the *principle of work and energy* to the block A before the impact, we write

$$T_1 + U_{1 \to 2} = T_2: \qquad 0 + 20[6(1 - \cos 60°)] = \frac{1}{2}\left(\frac{20}{32.2}\right)v_A^2$$

$$v_A = 13.90 \qquad \mathbf{v}_A = 13.90 \text{ ft/s} \rightarrow$$

The impact which ensues is a *direct central impact*. The speed of the sphere B just before the impact is $v_B = 0$. Let us choose the positive sense of each velocity to be to the right and assume the velocities of A and B after the impact to be $\mathbf{v}'_A = v'_A \rightarrow$ and $\mathbf{v}'_B = v'_B \rightarrow$. Applying the *principle of impulse and momentum* to A and B as a system during the impact, we have

Notice that the net impulse exerted on A and B as a system is zero. We write

$$\overset{+}{\rightarrow} \Sigma V_x: \qquad \frac{20}{32.2}(13.90) + 0 = \frac{20}{32.2}v'_A + \frac{40}{32.2}v'_B \qquad (1)$$

Since it is given that $e = 0.8$, we apply Eq. (14.34) to write

$$v'_B - v'_A = 0.8(13.90 - 0) \qquad (2)$$

Solving Eqs. (1) and (2), we obtain $v'_A = -2.78$ and $v'_B = 8.34$. Recalling that we have chosen the positive sense to be to the right, we write

$$\mathbf{v}'_A = 2.78 \text{ ft/s} \leftarrow \qquad \mathbf{v}'_B = 8.34 \text{ ft/s} \rightarrow \qquad \blacktriangleleft$$

Developmental Exercises

D14.20 Define the terms: (a) coefficient of restitution, (b) perfectly plastic impact, (c) perfectly elastic impact.

D14.21 Refer to Example 14.11. Determine the angle θ_B when B reaches its highest position.

D14.22 Refer to Example 14.11. If the velocity of the block A just after the impact is zero, what is the value of e?

14.9 Oblique Central Impact

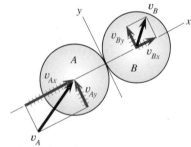

Fig. 14.17

When two particles A and B have an oblique central impact, their velocities $\mathbf{v}_A$ and $\mathbf{v}_B$ just before the impact are not both directed along the line of impact. Let us choose the x axis to be along the line of impact of the particles A and B as illustrated in Fig. 14.17. Then, the y axis, which lies in the plane of motion, is along the common tangent to the surfaces in contact.

The elementary study of the oblique impact between any two bodies is based on the *usual assumption* that the surfaces of contact are perfectly smooth so that *impulses due to friction between surfaces of contact are negligible during the impact*. The only impulsive forces acting on them are the forces of impact between them, which obey Newton's third law.

For analysis, let the particles A and B in Fig. 14.17 have, respectively, masses m_A and m_B, velocities $\mathbf{v}_A$ and $\mathbf{v}_B$ before the impact, and velocities $\mathbf{v}'_A$ and $\mathbf{v}'_B$ after the impact, where

$$\mathbf{v}_A = v_{Ax}\mathbf{i} + v_{Ay}\mathbf{j} \qquad \mathbf{v}_B = v_{Bx}\mathbf{i} + v_{By}\mathbf{j}$$
$$\mathbf{v}'_A = v'_{Ax}\mathbf{i} + v'_{Ay}\mathbf{j} \qquad \mathbf{v}'_B = v'_{Bx}\mathbf{i} + v'_{By}\mathbf{J}$$

During the *oblique central impact* of A and B, we recognize that:
(1) The impulse exerted on either particle in the y direction is zero; hence the momentum of either particle in the y direction is conserved. As the mass of either particle is constant, we have

$$v_{Ay} = v'_{Ay} \qquad v_{By} = v'_{By} \qquad (14.35)$$

(2) The total impulse exerted in the x direction on both particles as a system is, by Newton's third law, equal to zero; hence the total momentum in the x direction is conserved during the impact. We write

$$m_A v_{Ax} + m_B v_{Bx} = m_A v'_{Ax} + m_B v'_{Bx} \qquad (14.36)$$

(3) The components of velocities in the x direction, immediately before and after the impact, are related by the equation

$$v'_{Bx} - v'_{Ax} = e(v_{Ax} - v_{Bx}) \qquad (14.37)^\dagger$$

The unknown quantities in a problem of oblique central impact are usually determined by solving Eqs. (14.35) through (14.37) simultaneously.

Developmental Exercise

D14.23   What is the usual assumption on which the elementary study of the oblique impact between any two bodies is based?

EXAMPLE 14.12

The velocities of the disks A and B just before their impact are shown. If $e = 0.8$, determine the velocities $\mathbf{v}'_A$ and $\mathbf{v}'_B$ of the disks A and B just after the impact.

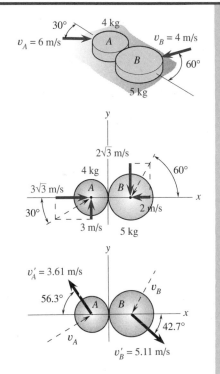

Solution. The impact between A and B is an oblique central impact. Choosing the x axis to be along the line of impact, we may indicate the components of velocities of A and B just before impact as shown; i.e., $v_{Ax} = 3\sqrt{3}$ m/s, $v_{Ay} = 3$ m/s, $v_{Bx} = -2$ m/s, and $v_{By} = -2\sqrt{3}$ m/s. The masses of A and B are $m_A = 4$ kg and $m_B = 5$ kg. Applying Eqs. (14.35) through (14.37), we write

$$3 = v'_{Ay} \qquad -2\sqrt{3} = v'_{By}$$

$$4(3\sqrt{3}) + 5(-2) = 4v'_{Ax} + 5v'_{Bx}$$

$$v'_{Bx} - v'_{Ax} = 0.8[3\sqrt{3} - (-2)]$$

From these four equations, we obtain

$$v'_{Ax} = -2 \qquad v'_{Ay} = 3 \qquad v'_{Bx} = 3.757 \qquad v'_{By} = -3.464$$

Adding vectorially the velocity components of each disk, we obtain

$$\mathbf{v}'_A = 3.61 \text{ m/s} \; \measuredangle \; 56.3° \quad \blacktriangleleft$$

$$\mathbf{v}'_B = 5.11 \text{ m/s} \; \measuredangle \; 42.7° \quad \blacktriangleleft$$

Developmental Exercise

D14.24   Refer to Example 14.12. Determine the percentage of the loss in kinetic energy of the disks due to the impact.

---

†Cf. Eq. (14.34).

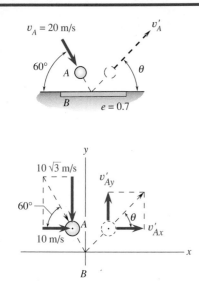

$v_A = 20$ m/s

v'_A

$60°$

A

θ

B

$e = 0.7$

$10\sqrt{3}$ m/s

y

v'_{Ay}

$60°$

A

θ

v'_{Ax}

10 m/s

B

x

EXAMPLE 14.13

A steel ball A strikes an embedded steel plate B with a velocity $\mathbf{v}_A$ of 20 m/s and rebounds from it with a velocity $\mathbf{v}'_A$ as shown. If $e = 0.7$ between the ball and the plate, determine the velocity $\mathbf{v}'_A$.

Solution. We first resolve the velocities before and after the impact into components as shown. Since the surfaces of contact are to be assumed as being smooth, the ball experiences no impulse in the x direction. Its momentum and hence velocity in the x direction are conserved. We write

$$v'_{Ax} = v_{Ax} = 20 \cos 60° \qquad \mathbf{v}'_{Ax} = 10 \text{ m/s} \rightarrow$$

As the steel plate is embedded in the support, the mass of the plate (and the earth) is practically infinite. The conservation of momentum in the y direction will lead to the sensible conclusion that the velocity of the plate is always essentially zero. Therefore, we have

$$v'_{By} = v_{By} = 0 \qquad v_{Ay} = -20 \sin 60° = -10\sqrt{3}$$

The components of velocities in the y direction are, by Eq. (14.34), related to e by $v'_{By} - v'_{Ay} = e(v_{Ay} - v_{By})$. Thus, we write

$$0 - v'_{Ay} = 0.7(-10\sqrt{3} - 0) \qquad \mathbf{v}'_{Ay} = 7\sqrt{3} \text{ m/s} \uparrow$$

$$\mathbf{v}'_A = \mathbf{v}'_{Ax} + \mathbf{v}'_{Ay} = 10\mathbf{i} + 7\sqrt{3}\mathbf{j} \text{ m/s}$$

$$\mathbf{v}'_A = 15.72 \text{ m/s} \measuredangle 50.5° \blacktriangleleft$$

Developmental Exercise

D14.25 Refer to Example 14.13. If the ball rebounds from the plate with a velocity $\mathbf{v}'_A = 10\sqrt{2}$ m/s $\measuredangle 45°$, determine the coefficient of restitution e between the ball and the plate.

★14.10 Generalized Virtual Work: Impulse and Momentum

In Sec. 13.13, we made the first application of the concept of *virtual work* to the study of certain kinetic problems. As an extension, we here define the *generalized virtual work* $\delta U'$ *of a non-force vector* $\mathbf{L}$ (e.g., a linear impulse or linear momentum vector) acting on or associated with a particle during a virtual displacement $\delta\mathbf{r}$ of the particle from the position A to the position A', as shown in Fig. 14.18, as the dot product of $\mathbf{L}$ and $\delta\mathbf{r}$; i.e.,

$$\delta U' = \mathbf{L} \cdot \delta\mathbf{r} = L \cos\psi \, \delta s \qquad (14.38)$$

where $\delta s = |\delta\mathbf{r}|$ and ψ is the angle between $\mathbf{L}$ and $\delta\mathbf{r}$.

By equivalence and equipollence in the *principle of impulse and momentum* in Secs. 14.2 and 14.4 and the nature of *constrained virtual displacement* for a one-degree-of-freedom system in Sec. 13.13, it follows logically

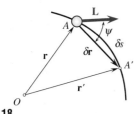

A

$\mathbf{L}$

ψ

$\delta\mathbf{r}$

δs

A'

$\mathbf{r}$

$\mathbf{r}'$

O

Fig. 14.18

that, *during a constrained virtual displacement of a one-degree-of-freedom system of particles, the generalized virtual work of all the momentum vectors in the initial momentum diagram of the system plus the generalized virtual work of all the impulse vectors in the impulse diagram of the system must be equal to the generalized virtual work of all the momentum vectors in the final momentum diagram of the system.* This property is referred to as the *principle of generalized virtual work* for particles, which complements the traditional *principle of impulse and momentum* for particles.

EXAMPLE 14.14

Using the *principle of generalized virtual work*, do Example 14.5.

Solution.   Let the vector-diagram equation for the momenta and impulses during the interval $0 \leq t \leq 4$ s for the entire system be drawn first.

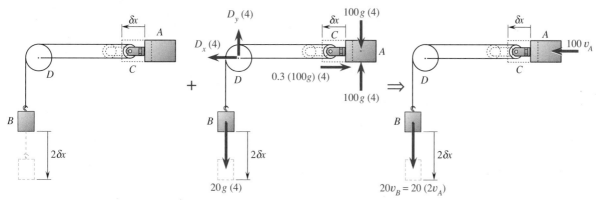

From the constraints of the system, we readily see that $\mathbf{v}_A = v_A \leftarrow$, $\mathbf{v}_B = 2v_A \downarrow$, and the constrained virtual displacements of A and B are $\delta x \leftarrow$ and $2\,\delta x \downarrow$, respectively. Applying the *principle of generalized virtual work*, we write

$$\delta U': \quad 0 + [0.3(100g)(4)(-\delta x) + 20g(4)(2\,\delta x)]$$
$$= 100v_A(\delta x) + 20(2v_A)(2\,\delta x)$$
$$v_A = \tfrac{2}{9}g \qquad\qquad v_A = 2.18 \text{ m/s} \blacktriangleleft$$

REMARK. The contrasting feature between the solutions in Examples 14.5 and 14.14 is that the application of the *principle of generalized virtual work* enables us to *uncouple the unknown* v_A without mathematically involving the tension F in the cord BCD.

Developmental Exercises

D14.26   Describe the *principle of generalized virtual work* for particles.

D14.27   The velocity of the cylinder A shown is $\mathbf{v}_A = 2$ m/s $\downarrow$ when the system is released at the time $t = 0$. Using the *principle of generalized virtual work*, determine the value of $\mathbf{v}_A$ when $t = 2$ s.

Fig. D14.27

PROBLEMS

✗14.53 A ball is thrown vertically downward with an initial speed of v_0 at a height of 2 m above a floor and rebounds from the floor to the same height of 2 m before falling downward again. If $e = 0.7$ between the ball and the floor, determine the value of v_0.

14.54* and **14.55** If $e = 0.8$ between the sliders A and B as shown, determine (a) the velocity of each slider after the impact, (b) the percentage of loss in kinetic energy during the impact.

7 m/s   4 m/s   10 ft/s   5 ft/s

A   B   A   B

2 kg   3 kg   4 lb   6 lb

Fig. P14.54*   **Fig. P14.55**

✗14.56 A sphere A is released from rest in the position shown and strikes the block B which is at rest. If $e = 0.75$ between A and B and $\mu_k = 0.5$ between B and the support, determine (a) the velocity of A just after the impact, (b) the displacement of B after the impact.

2 kg

A

$90°$

C

1.5 m

4 kg

B

1 m

A

$60°$

θ_B

B

Fig. P14.56   **Fig. P14.57***

14.57* The sphere A is released from rest in the position shown and strikes an identical sphere B which is at rest. If $\theta_B = 40°$ when B reaches its highest position after the impact, determine (a) the value of e between A and B, (b) the velocity of A just after the impact.

14.58 The sphere A is released from rest in the position shown and strikes an identical sphere B which is at rest. If the wire suspending B becomes horizontal when B reaches its highest position after the impact, determine (a) the value of e between A and B, (b) the velocity of A just after the impact.

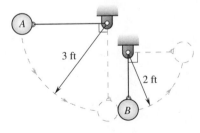

A

3 ft

2 ft

B

Fig. P14.58

Fig. P14.59

Fig. P14.60

14.59 and 14.60 A sphere is traveling with a velocity $\mathbf{v}_A$ when it strikes a sphere B which is at rest as shown. If $e = 0.65$ between A and B, determine the velocity of each sphere just after the impact.[†]

14.61 The block B, which can move freely in the horizontal direction, is at rest when it is struck by a ball A traveling with a velocity of 6 m/s ↓ as shown. If $e = 0.6$ between A and B, determine the velocities of A and B just after the impact.

Fig. P14.61

Fig. P14.62*

14.62* If $e = 0.7$ between the disks A and B as shown, determine the velocities of A and B just after the impact.

14.63 A steel ball A is dropped from rest in the position shown. The ball strikes the embedded steel plate B, rebounds from B, and barely clears the wall C as shown. If $e = 0.8$ between A and B, determine the distance x.

Fig. P14.63

Fig. P14.64*

14.64* The disk A is dropped from rest in the position shown. It falls onto the disk B which is resting on a spring of modulus $k = 120$ lb/ft. If $e = 0$ between A and B, determine the maximum deflection of the disk B.

14.65 The ram A of a pile driver is dropped from rest in the position shown. It falls onto the pile B and drives it 100 mm into the ground. Determine the magnitude of the average resistance of the ground to the penetration by the pile if $e = 0$ between A and B.

Fig. P14.65

[†]Cf. Example 14.8.

14.66* through 14.75* Using the *principle of generalized virtual work*, solve the following problems:

14.66* Prob. 14.25.	14.67* Prob. 14.26*.
14.68* Prob. 14.27*.	14.69* Prob. 14.28.
14.70* Prob. 14.29.	14.71* Prob. 14.30*.
14.72* Prob. 14.31.	14.73* Prob. 14.32.
14.74* Prob. 14.33.	14.75* Prob. 14.34*.

VARIABLE SYSTEMS[†]

*14.11   Variable Systems of Particles

We observe that each of the systems of particles heretofore studied contains the same particles and has the same mass during its motion. However, the operations of many engineering systems (e.g., vanes, pipes, nozzles, turbines, conveyors, fans, propellers, jet engines, chains, and rockets) do involve motions of *systems which are continuously acquiring or ejecting particles, or doing both at the same time*. Such systems are called *variable systems of particles*. A variable system of particles whose total mass remains constant is called a *system with steady mass flow*. On the other hand, a variable system of particles whose total mass is increasing or decreasing is called a *system with variable mass*.

Each variable system of particles constantly involves the *flow of mass*. The flow of mass, in turn, involves the change of momentum of the particles as well as the exertion of impulse on the particles in the flow. Accordingly, the *principle of impulse and momentum*, among others, should be very useful in the analysis of the motion of variable systems of particles. However, there is a catch: the principles in kinetics heretofore established are directly applicable to only *constant systems* which neither gain nor lose particles. Therefore, we employ a strategy which seeks to analyze the motion of an *auxiliary constant system* isolated from the variable system during a short time interval Δt of its motion. The specific procedure for the analysis is explained and illustrated in Secs. 14.12 and 14.13 for two major types of variable systems.

Developmental Exercises

D14.28   What is a variable system of particles? Give three examples.

D14.29   How does a system with steady mass flow differ from a system with variable mass?

[†]The topics related to *variable systems* in Secs. 14.11 through 14.13 may be studied at any later time, or omitted if circumstances warrant, without affecting continuity to the rest of this text.

*14.12 Systems with Steady Mass Flow

A *steady mass flow* is a condition of flow of mass where the time rate at which mass enters a given volume is equal to the time rate at which mass leaves the same volume. In other words, when there exists a steady mass flow through a volume, *the total mass in the volume remains unchanged, although the particles occupying the volume keep changing.* The volume in question may be the space in any fixed or moving rigid container, the space between blades, or a certain space associated with a vane. Let us consider *a steady mass flow* through the rigid container BC as shown in Fig. 14.19. The continuity of flow requires that the *time rates of mass flow dm/dt* through cross sections B and C be equal; i.e.,

$$\frac{dm}{dt} = \rho_B A_B v_B = \rho_C A_C v_C \tag{14.39}$$

Fig. 14.19

where ρ_B and ρ_C are the mass densities of the stream of flow at B and C, A_B and A_C are the (right) cross-sectional areas at B and C, and v_B and v_C are the speeds of the stream at B and C, respectively.

In order to examine the resultant force $\Sigma\mathbf{F}$ exerted on the particles in contact with the container, we consider the particles situated in the container and denote them by S as shown in Fig. 14.19. We see that S is *a variable system of particles* because it continuously gains and loses particles (at the same rate). Thus, the principles in kinetics established so far are *not* directly applicable to S. We need to define and isolate *an auxiliary constant system of particles* for analysis.

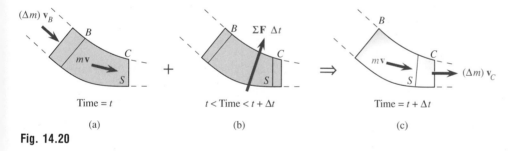

Fig. 14.20

As illustrated in Fig. 14.20(a), we consider an auxiliary system of particles, which is *a system consisting of two parts*: (1) the particles which are situated in the container BC at the time t and have a total mass m, and (2) the particles which will enter the container BC during the next short time interval Δt and have a total mass Δm. Since the flow through the container is a *steady mass flow*, the total amount of mass of the particles situated in the container at the time $t + \Delta t$ must still be equal to m, and the total amount of mass of the particles which have left the container from the time t to the time $t + \Delta t$ must be equal to Δm. At the time $t + \Delta t$, the auxiliary system has a configuration as illustrated in Fig. 14.20(c). Clearly, the same particles are involved in the auxiliary system in its motion during the interval Δt. The principle of impulse and momentum may, therefore, be applied to the foregoing *auxiliary constant system of particles* during the interval Δt, as shown in Fig. 14.20, where $\mathbf{v}_B$ and $\mathbf{v}_C$ are the velocities of the stream of flow at B and C, respectively, the resultant momentum of the particles situated in the container is represented by $m\mathbf{v}$ and is the same in (a) as in (c), and the resultant impulse $\Sigma\mathbf{F}\,\Delta t$ in (b) does not necessarily pass through the center of mass of the system.

Note that, unless otherwise stated, fluid friction in the system is assumed to be negligible, and the impulse of weight forces are assumed to be also negligible. As indicated in Fig. 14.20, we write

$$(\Delta m)\mathbf{v}_B + m\mathbf{v} + \Sigma\mathbf{F}\,\Delta t = m\mathbf{v} + (\Delta m)\mathbf{v}_C$$

$$(\Delta m)\mathbf{v}_B + \Sigma\mathbf{F}\,\Delta t = (\Delta m)\mathbf{v}_C \tag{14.40}$$

Referring to Fig. 14.20 and Eq. (14.40), we conclude that the system of vectors consisting of the momentum $(\Delta m)\mathbf{v}_B$ of the particles entering S during the time interval Δt and the impulses of the forces exerted on S during Δt is *equipollent* to the momentum $(\Delta m)\mathbf{v}_C$ of the particles leaving S during the same time interval Δt.

Dividing both sides of Eq. (14.40) by Δt and letting $\Delta t \to 0$, we find that the resultant force acting on S is

$$\Sigma\mathbf{F} = \frac{dm}{dt}(\mathbf{v}_C - \mathbf{v}_B) \tag{14.41}$$

where dm/dt is given by Eqs. (14.39) for systems with steady mass flow. Note that *a consistent system of kinetic units* must be used in Eqs. (14.40) and (14.41), and $\mathbf{v}_B$ and $\mathbf{v}_C$ must be measured in *an inertial reference frame*, which may be fixed or translating with a constant velocity. Furthermore, we may express

$$\frac{dm}{dt} = \rho Q = \rho A v \tag{14.42}$$

where ρ is the mass density of the stream of flow, Q is the time rate of volumetric flow of the stream (in volume per unit time), A is the cross-sectional area of the stream, and v is the speed of the stream.

We can see that the principle of impulse and momentum is very useful in determining the force developed when a stream of water or other fluid is deflected by a blade or a vane. *The mass of fluid striking the blade during the time interval Δt* depends on the cross-sectional area and velocity of the stream, the mass density of the fluid, the velocity of the blade, and whether a single blade or a series of blades mounted on the periphery of a wheel is involved.

For example, if the stream strikes a fixed blade as shown in Fig. 14.21(a), the mass of fluid striking the blade during the interval Δt is

Fig. 14.21   (a)   (b)   (c)

$\Delta m = \rho A v_S \, \Delta t$, where v_S is the speed of the stream. If *a single blade* is moving such that the stream strikes the blade with a speed u relative to the blade as shown in Fig. 14.21(b), then $\Delta m = \rho A u \, \Delta t$. When streams of fluid strike *a series of blades* mounted on the periphery of a wheel as shown in Fig. 14.21(c), all the fluid in each stream strikes the blades during the interval Δt, and $\Delta m = \rho A v_S \, \Delta t$ from each stream.

The vector diagram equation for the momenta and impulses during the interval Δt for the *auxiliary constant system of particles* is a very helpful tool for gaining geometric and physical perception into the problem situation of variable systems. We shall use it in the examples.

EXAMPLE 14.15

A nozzle discharges a stream of water of cross-sectional area $A = 100 \text{ mm}^2$ with a speed $v = 50$ m/s, and the stream is deflected by a fixed vane as shown. The mass density of water is $\rho = 1 \text{ Mg/m}^3$. Determine the resultant force $\mathbf{F}$ exerted on the stream by the fixed vane.

Solution. The system in the problem is a system with steady mass flow. The mass density of water is $\rho = 1 \text{ Mg/m}^3 = 10^3 \text{ kg/m}^3$, the cross-sectional area of the stream is $A = 100 \text{ mm}^2 = 10^{-4} \text{ m}^2$, and the speed of the stream is $v = 50$ m/s. Thus, the time rate of mass flow in the stream is

$$\frac{dm}{dt} = \rho A v = 10^3 (10^{-4})(50) \qquad \frac{dm}{dt} = 5 \text{ kg/s}$$

Let the *auxiliary constant system of particles* be a system consisting of (1) the particles of the stream which is in contact with the vane at the time t and has a total mass m, and (2) the particles of the stream which will be in contact with the vane during the next short time interval Δt and has a total mass Δm. Applying the *principle of impulse and momentum* to such a system during the interval Δt, we have

Noting that $v_B = v_C = v = 50$ m/s and canceling the momentum $m\mathbf{v}$ which appears on both sides, we write

$$\overset{+}{\rightarrow} \Sigma V_x: \qquad -(\Delta m)v_B + F_x \, \Delta t = -(\Delta m)v_C \cos 60°$$

$$F_x = \frac{dm}{dt} v(1 - \cos 60°) = 5(50)(1 - \cos 60°) = 125$$

$$+\uparrow \Sigma V_y: \qquad F_y \, \Delta t = (\Delta m)v_C \sin 60°$$

$$F_y = \frac{dm}{dt} v \sin 60° = 5(50) \sin 60° = 216.5$$

$$\mathbf{F} = F_x \mathbf{i} + F_y \mathbf{j} \qquad \mathbf{F} = 250 \text{ N} \measuredangle 60° \quad \blacktriangleleft$$

Developmental Exercises

D14.30   In a system with steady mass flow, how do you define and isolate an *auxiliary constant system of particles* for analysis?

D14.31   Work Example 14.15 if the vane moves with a constant velocity of 10 m/s ←. (*Hint*. Use the vane as the inertial reference frame.)

EXAMPLE 14.16[†]

600 mi/h

C   B

A jet airplane is cruising at a speed of 600 mi/h in level flight. Each of its jet engines scoops in air at the rate of 180 lb/s and ejects it with a speed of 2000 ft/s relative to the airplane. Neglecting effect of fuel consumption, determine for each jet engine (a) the magnitude of the propulsive force generated, (b) the total horsepower generated relative to the ground.

Solution.   With effect of fuel consumption neglected, the system may be treated as one with steady mass flow. Furthermore, we may use the *jet engine* as an inertial reference frame because it moves at a constant velocity. Thus, the velocities of the air at B and C are

$$\mathbf{v}_B = -600\mathbf{i} \text{ mi/h} = -880\mathbf{i} \text{ ft/s} \qquad \mathbf{v}_C = -2000\mathbf{i} \text{ ft/s}$$

The time rate of mass flow through the jet engine is

$$\frac{dm}{dt} = \frac{180}{32.2} \text{ slugs/s}$$

Let the *auxiliary constant system of particles* be a system consisting of (1) the air particles in the jet engine at the time t which have a total mass m, and (2) the air particles which will enter the jet engine during the next short time interval Δt and have a total mass Δm. Applying the *principle of impulse and momentum* to such a system during the interval Δt, we have

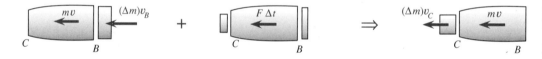

Notice that the force $\mathbf{F}$ acting on the air is equal in magnitude but opposite in direction to the propulsive force acting on the engine. We write

$$\xleftarrow{+} \Sigma V_x: \qquad mv + (\Delta m)v_B + F \, \Delta t = (\Delta m)v_C + m\overset{\text{)}}{v}$$

$$F = \frac{dm}{dt}(v_C - v_B) = \frac{180}{32.2}(2000 - 880)$$

$$F = 6260.9 \text{ lb} \qquad\qquad F = 6.26 \text{ kips} \blacktriangleleft$$

Relative to the ground, the total horsepower P generated consists of two parts: (1) the horsepower P_1 to propel the airplane, and (2) the horsepower P_2 to eject the air particles out of the engine. The velocities of the airplane and the ejected air are $\mathbf{v}_A = 880\mathbf{i}$ ft/s and $\mathbf{v}_E = (880 - 2000)\mathbf{i}$ ft/s $= -1120\mathbf{i}$ ft/s. Recalling that 1 hp $= 550$ ft·lb/s, we write

---

[†]Effect of fuel consumption is considered in Example 14.20.

$$P_1 = \frac{Fv_A}{550} = \frac{6260.9(880)}{550} = 10017$$

$$P_2 = \frac{\frac{1}{2}\frac{dm}{dt}v_E^2}{550} = \frac{\frac{1}{2}\left(\frac{180}{32.2}\right)(1120)^2}{550} = 6375$$

$$P = P_1 + P_2 = 16392 \qquad P = 16.39 \times 10^3 \text{ hp} \blacktriangleleft$$

Developmental Exercise

D14.32   Refer to Example 14.16. What is the total horsepower generated by each jet engine relative to the jet airplane?

EXAMPLE 14.17

A nozzle discharges water at the rate $Q = 2$ gal/s with a speed $v = 60$ ft/s, and the stream is deflected by the vane BC as shown. The combined weight force of the vane and the water it supports is $\mathbf{W} = 100$ lb $\downarrow$ acting through the point G. It is known that the specific weight of water is $\gamma = 62.4$ lb/ft^3 and 1 ft^3 = 7.48 gal. Determine the reaction from the support at C.

Solution.   The mass density ρ and the time rate of volumetric flow Q of the stream of water are

$$\rho = \frac{\gamma}{g} = \frac{62.4}{32.2} \text{ slugs/ft}^3 \qquad Q = \frac{2}{7.48} \text{ ft}^3/\text{s}$$

Thus, we write

$$\frac{dm}{dt} = \rho Q = 0.5182 \text{ slug/s} \qquad v_B = v_C = v = 60 \text{ ft/s}$$

Let the *auxiliary constant system of particles* be a system consisting of (1) the stream of water in contact with the vane at the time t which has a total mass m, (2) the particles of the stream which will be in contact with the vane during the next short time interval Δt and have a total mass Δm, and (3) the vane BC, where the combined weight of the vane and the water it supports is $W = 100$ lb. Applying the *principle of impulse and momentum* to such a system during the interval Δt, we have

Canceling the momentum $m\mathbf{v}$ which appears on both sides, we write

$$+\circlearrowleft \Sigma M_B: \quad -0.5\left[\frac{4}{5}(\Delta m)v_B\right] - 1.5(W\,\Delta t) + 4(C_y\,\Delta t)$$

$$= 2[(\Delta m)v_C]$$

$$-0.5\left[\frac{4}{5}\left(\frac{dm}{dt}\right)v\right] - 1.5W + 4C_y = 2\left(\frac{dm}{dt}\right)v$$

$$C_y = 56.15 \qquad\qquad \mathbf{C} = 56.2 \text{ lb} \uparrow \quad \blacktriangleleft$$

Developmental Exercise

D14.33   Refer to Example 14.17. Determine the components of reaction from the support at B.

★14.13   Systems with Variable Mass: Motion of Rockets

A *system with variable mass* is a variable system of particles where the system is gaining mass by continuously absorbing particles or is losing mass by continuously ejecting particles. In other words, the particles, as well as the total mass, in a system with variable mass do not remain constant. In order for the principles in kinetics established earlier to be applicable, we again need to define and isolate an *auxiliary constant system of particles* for analysis.

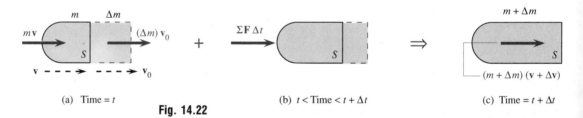

(a) Time = t

(b) $t <$ Time $< t + \Delta t$

(c) Time = $t + \Delta t$

Fig. 14.22

As illustrated in Fig. 14.22(a), we consider an auxiliary system of particles, which is *a system consisting of two parts*: (1) the particles which are contained in the system S at the time t and have a total mass m moving in an inertial reference frame with a velocity $\mathbf{v}$, and (2) the particles which will be absorbed by S during the next short time interval Δt and have a total mass Δm moving in the inertial reference frame with a velocity $\mathbf{v}_0$. During the interval Δt, the resultant $\Sigma \mathbf{F}$ of the external forces acting on the auxiliary constant system (as a whole) will have imparted to the system an impulse of $\Sigma \mathbf{F}\,\Delta t$ as shown in Fig. 14.22(b).[†] At the time $t + \Delta t$, the auxiliary constant system has a velocity $\mathbf{v} + \Delta \mathbf{v}$ and a momentum

---

[†]Note that the forces of interaction between S and the particles being absorbed are internal forces to the *auxiliary constant system*. Clearly, $\Sigma \mathbf{F}$ does *not* include the *internal force* exerted on S by the particle being absorbed.

$(m + \Delta m)(\mathbf{v} + \Delta \mathbf{v})$ as shown in Fig. 14.22(c). Since the same particles are involved, the *principle of impulse and momentum* may, therefore, be applied to the *auxiliary constant system of particles* during the interval Δt, as depicted in Fig. 14.22. We write

$$\overset{+}{\rightarrow} \Sigma V_x: \qquad m\mathbf{v} + (\Delta m)\mathbf{v}_0 + \Sigma \mathbf{F}\,\Delta t = (m + \Delta m)(\mathbf{v} + \Delta \mathbf{v})$$

$$\Sigma \mathbf{F}\,\Delta t = m\,\Delta \mathbf{v} + \Delta m(\mathbf{v} - \mathbf{v}_0) + (\Delta m)\,\Delta \mathbf{v}$$

Neglecting the second-order term, dividing through by Δt, and letting $\Delta t \rightarrow 0$, we obtain the equation of motion

$$\Sigma \mathbf{F} = m\frac{d\mathbf{v}}{dt} + \frac{dm}{dt}\mathbf{u} \qquad (14.43)^{\dagger}$$

where

$$\mathbf{u} = \mathbf{v} - \mathbf{v}_0 \qquad (14.44)^{\dagger}$$

Here, dm/dt represents the *time rate of increase of mass* in S by absorbing particles, and $\mathbf{u}$ is *the velocity of S relative to the particles which are being absorbed by S*. Note that the positive senses of $\Sigma \mathbf{F}$, $\mathbf{v}$, and $\mathbf{u}$ are all the same (e.g., to the right) when the system is in rectilinear motion. By writing Eq. (14.43) as

$$\Sigma \mathbf{F} - \frac{dm}{dt}\mathbf{u} = m\frac{d\mathbf{v}}{dt} \qquad (14.45)$$

we see that the action exerted on S by the particles being absorbed is equivalent to *a thrust of $-(dm/dt)\mathbf{u}$ to slow down the motion of S*.

(a) Time $= t$ (b) $t <$ Time $< t + \Delta t$ (c) Time $= t + \Delta t$

Fig. 14.23

On the other hand, the system S of total mass m, which moves in an inertial reference frame with a velocity $\mathbf{v}$ and is acted on by the resultant external force $\Sigma \mathbf{F}$, may be ejecting mass at the rate of dm_e/dt, where the velocity of the ejected particles in the inertial reference frame is $\mathbf{v}_0$. In this case, we may likewise apply the *principle of impulse and momentum* to the *auxiliary constant system of particles* as illustrated in Fig. 14.23, from which we similarly obtain the relation

$$\Sigma \mathbf{F} = m\frac{d\mathbf{v}}{dt} - \frac{dm_e}{dt}\mathbf{u} \qquad (14.46)$$

where $\mathbf{u}$ is the same as that defined in Eq. (14.44). Note in Eq. (14.46) that dm_e/dt represents the *time rate of mass being ejected* from S, and $\mathbf{u}$ is the velocity of S relative to the particles which have just been ejected from S.

---

†From Eqs. (14.43) and (14.44), it is clear that Newton's second law in the form of Eq. (14.4) is applicable to a system with variable mass *only when $\mathbf{u} = \mathbf{v}$ or $\mathbf{v}_0 = \mathbf{0}$* in the inertial reference frame.

The positive senses of $\Sigma\mathbf{F}$, $\mathbf{v}$, and $\mathbf{u}$ are still all the same (e.g., to the right) when the system is in rectilinear motion. By writing Eq. (14.46) as

$$\Sigma\mathbf{F} + \frac{dm_e}{dt}\mathbf{u} = m\frac{d\mathbf{v}}{dt} \tag{14.47}$$

we readily see that the action exerted on S by the particles being ejected is equivalent to a *thrust of* $(dm_e/dt)\mathbf{u}$ *to speed up the motion of S*. A *rocket* furnishes an example of such a system S. The burning of jet fuel and ejecting it into the exhaust of a *jet airplane* furnishes another example of a system S ejecting mass, although the fuel consumption is usually a small percentage of the air consumption in jet engines.

A comparison of Eqs. (14.43) with (14.46) readily reveals that

$$\frac{dm_e}{dt} = -\frac{dm}{dt} \tag{14.48}$$

By considering the *time rate of mass being ejected* as the negative of the time rate of change of mass, we readily see that Eq. (14.46) is a special case of Eq. (14.43).

EXAMPLE 14.18

A chain of mass ρ_L per unit length and total length L lies in a pile on the floor. Starting from $y = 0$, one end of the chain is lifted vertically with a constant velocity $\mathbf{v}_0$ by a variable force $\mathbf{P}$. Determine the magnitude of $\mathbf{P}$ as a function of y.

Solution. For an open-link chain as shown, each link on the floor acquires its velocity abruptly through the impact with its upper link which lifts it off the floor. Thus, the lifting of such a chain involves an energy loss. We may choose the *auxiliary constant system of particles* to be a system consisting of (1) the part of the chain already off the floor which has a total mass $\rho_L y$ and total weight $\rho_L yg$, and (2) the part of the chain which will be set in motion during the next short time interval Δt and has a total mass $\rho_L v_0 \Delta t$. Applying the *principle of impulse and momentum* to such a system during the interval Δt, we have

$$+\uparrow \Sigma V_y: \quad (\rho_L y)v_0 + P\,\Delta t - (\rho_L yg)\,\Delta t$$
$$= (\rho_L y)v_0 + (\rho_L v_0\,\Delta t)v_0$$
$$P = \rho_L(gy + v_0^2) \blacktriangleleft$$

REMARK. If the open-link chain here were replaced with an ideal rope of the same length L and the same mass ρ_L per unit length as shown, the elements of the rope would acquire their upward velocity in a gradual and continuous manner. No impact would occur and there would be no energy loss. Applying the *principle of conservation of energy* in Sec. 13.12 to this entire ideal rope and choosing the floor to be the reference datum for the potential energies V_g and V_a, we write

$$T + V = T + V_g + V_a = \text{constant}$$

$$\frac{d}{dy}(T + V_g + V_a) = 0$$

where

$$T = \text{kinetic energy} = \tfrac{1}{2}\,\rho_L y v_0^2$$

$$V_g = \text{gravitational potential energy} = \rho_L y g\left(\frac{y}{2}\right)$$

$$V_a = \text{applied potential energy} = \int_y^0 P\,dy$$

Thus, we have[†]

$$\frac{d}{dy}\left(\tfrac{1}{2}\,\rho_L y v_0^2 + \tfrac{1}{2}\,\rho_L g y^2 + \int_y^0 P\,dy\right) = 0$$

$$\tfrac{1}{2}\,\rho_L v_0^2 + \rho_L g y - P = 0$$

$$P = \rho_L(gy + \tfrac{1}{2}\,v_0^2) \blacktriangleleft$$

which is smaller than the result obtained for the open-link chain by an amount of $(\rho_L v_0^2)/2$.

Developmental Exercise

D14.34 Refer to Example 14.18. Determine (a) the work done by **P** to the chain from $y = 0$ to $y = L$, (b) the kinetic energy of the chain when $y = L$, (c) the potential energy of the chain when $y = L$ with the floor being used as a reference datum, (d) the total loss of energy due to successions of impacts between links during the lifting of the chain, (e) the magnitude of the reaction from the floor when $0 < y < L$.

EXAMPLE 14.19

An experimental rocket has a gross mass $m_0 = 1$ Mg, including a total fuel $m_f = 900$ kg. The rocket is fired vertically up from the north pole at the time $t = 0$. The fuel is consumed at a constant rate $q = 15$ kg/s and ejected

---

[†]To differentiate the integral, we apply Leibnitz's rule and write

$$\frac{d}{dy}\int_y^0 P\,dy = \frac{d}{dy}\int_y^0 P(y)\,dy = \frac{d}{dy}\int_y^0 P(t)\,dt$$

$$= 0 + 0 - P(y)\frac{dy}{dy} = -P(y) = -P$$

(Cf. Sec. B5.9 of App. B for Leibnitz's rule.)

with a speed of 2 km/s relative to the rocket. Neglecting the air resistance and the variation of gravity with altitude, determine the velocity of the rocket when (a) half of the fuel is consumed, (b) the total fuel is consumed.

Solution. From the given data, we have $m_0 = 1\ \text{Mg} = 1000\ \text{kg}$, $dm_e/dt = q = 15\ \text{kg/s}$, and $\mathbf{u} = 2\ \text{km/s} \uparrow = 2000\ \text{m/s} \uparrow$. At the time t, the total mass of the rocket is $m = m_0 - qt$, the resultant external force is $\Sigma \mathbf{F} = mg \downarrow = (m_0 - qt)g \downarrow$, the velocity of the rocket is $\mathbf{v} = v \uparrow$, and the velocity of the ejected fuel is $\mathbf{v}_0 = \mathbf{v} - \mathbf{u} = (v - u) \uparrow$. The mass of the ejected fuel during the next short time interval Δt is $\Delta m_e = q\ \Delta t$. Applying the *principle of impulse and momentum* during the interval Δt, we have

$mv = (m_0 - qt)v$

$W\ \Delta t = (m_0 - qt)g\ \Delta t$

$(m - \Delta m_e)(v + \Delta v)$
$= (m_0 - qt - q\ \Delta t)(v + \Delta v)$

$\Delta m_e v_0 = q(\Delta t)(v - u)$

$+\uparrow \Sigma V_y:\qquad (m_0 - qt)v - (m_0 - qt)g\ \Delta t$

$$= (m_0 - qt - q\ \Delta t)(v + \Delta v) + q(\Delta t)(v - u)$$

Dividing through by Δt and letting $\Delta t \to 0$, we write

$$-(m_0 - qt)g = (m_0 - qt)\frac{dv}{dt} - qu \qquad (1)$$

Separating variables and integrating, we write

$$\int_0^t \left(\frac{qu}{m_0 - qt} - g\right) dt = \int_0^v dv$$

$$v = u \ln\left(\frac{m_0}{m_0 - qt}\right) - gt$$

When half of the fuel is consumed, we have $t = t_1$ and $v = v_1$. Thus, $t_1 = (m_f/2)/q = (900/2)/15\ \text{s} = 30\ \text{s}$ and

$$v_1 = 2000 \ln\left[\frac{1000}{1000 - 15(30)}\right] - 9.81(30)$$

$$\mathbf{v}_1 = 901\ \text{m/s} \uparrow \quad \blacktriangleleft$$

When the total fuel is consumed, $t = t_2$ and $v = v_2$. We have $t_2 = m_f/q = 900/15\ \text{s} = 60\ \text{s}$ and

$$v_2 = 2000 \ln\left[\frac{1000}{1000 - 15(60)}\right] - 9.81(60)$$

$$\mathbf{v}_2 = 4.02\ \text{km/s} \uparrow \quad \blacktriangleleft$$

Developmental Exercise

D14.35 Refer to Example 14.19. Show that Eq. (1) in the solution can be obtained by applying Eq. (14.43) to the rocket as described.

EXAMPLE 14.20

Refer to Example 14.16. Assume that each jet engine consumes fuel at the rate of 1.2 lb/s and ejects it into the exhaust with the same speed of 2000 ft/s relative to the airplane. Taking this fuel consumption into account, determine the magnitude F' of the propulsive force developed by each jet engine.

Solution. At any instant, the operations of each jet engine may be considered to consist of two parts: (1) operation with steady mass flow, and (2) operation with variable mass. From the solution in Example 14.16, we know that the operation of each jet engine with steady mass flow of air yields a propulsive force of magnitude $F = 6260.9$ lb. The operation of each jet engine with variable mass involves the loss of mass by consuming fuel at the rate of 1.2 lbm/s and ejecting it into the exhaust with the speed of 2000 ft/s relative to the airplane. By Eq. (14.46) or (14.47), we know that this operation will generate an additional propulsive force of magnitude

$$\Delta F = \frac{dm_e}{dt}\, u = \frac{1.2}{32.2}\,(\text{slug/s})(2000\ \text{ft/s}) = 74.5\ \text{lb}$$

Thus, with fuel consumption taken into account, we write

$$F' = F + \Delta F = 6335.4\ \text{lb} \qquad F' = 6.34\ \text{kips} \blacktriangleleft$$

Note that $\Delta F/F \approx 0.012$ or about 1.2% in the present case.

Developmental Exercise

D14.36 What is the thrust on a rocket which consumes fuel at the rate of 15 kg/s and is ejecting it at a speed of 2 km/s relative to itself?

PROBLEMS

NOTE: When any of the following problems involves a stream of water, use $\rho = 1$ Mg/m^3 for its mass density in SI units and $\gamma = 62.4$ lb/ft^3 for its specific weight (or weight density) in U.S. customary units. Unless otherwise stated, the motion of any rocket in a problem is measured in a nonrotating reference frame attached to the center of the earth, and the variation of gravity with altitude as well as air resistance during the flight of the rocket is assumed to be negligible.

14.76 through 14.78 A stream of water of cross-sectional area $A = 1.96$ in^2 and velocity $\mathbf{v} = 100$ ft/s $\rightarrow$ is split into two equal streams by a vane which is held at a fixed position by a force $\mathbf{R}$ as shown. Determine the magnitude of $\mathbf{R}$.

Fig. P14.76 and P14.79

Fig. P14.77* and P14.80   **Fig. P14.78 and P14.81***

14.79 through 14.81* A stream of water of cross-sectional area $A = 400$ mm^2 and velocity $\mathbf{v} = 50$ m/s $\rightarrow$ is split into two equal streams by a vane which is held by a force $\mathbf{R}$ and moves with a constant velocity of 10 m/s $\rightarrow$. Determine the magnitude of $\mathbf{R}$.

14.82* The nozzle shown discharges a stream of water at the rate of $Q = 4$ L/s with a velocity $\mathbf{v} = 30$ m/s $\rightarrow$. The stream is split into two branches by a smooth vane which is held at a fixed position by a force $\mathbf{R}$ as shown. Determine (a) the magnitude of $\mathbf{R}$, (b) the rates of flow Q_1 and Q_2 of the two branch streams. (*Hint.* The force $\mathbf{R}$ is perpendicular to the vane.)

Fig. P14.82*   **Fig. P14.83**

14.83 A nozzle discharges a stream of water at the rate of $Q = 0.9$ m^3/min with a velocity $\mathbf{v} = 18$ m/s $\leftarrow$, and the stream is deflected by the vane BC as shown. Determine the reactions from the supports at B and C.

14.84 Sand falls onto a chute at the rate of 100 kg/s with a velocity $\mathbf{v}_D = 4.5$ m/s $\downarrow$ and leaves the chute with a velocity $\mathbf{v}_C = 3$ m/s $\leftarrow$ as shown. The combined weight force of the chute and the sand in it is $W = 2$ kN $\downarrow$ acting through the point G. Determine the reactions from the supports at B and C.

Fig. P14.84   **Fig. P14.85**

14.85 The slipstream of an 8-lb electric fan has a diameter of 10 in. and a velocity of 30 ft/s relative to the fan as shown. Assuming that air weighs 0.076 lb/ft^3, determine the minimum value of the distance b for which tipping of the fan will not occur. (*Hint.* The velocity of the air entering the fan may be assumed to be negligibly small.)

14.86* The helicopter shown has a mass of 4.5 Mg and produces a slipstream of 10-m diameter when it is hovering in midair. Assuming that the mass density of the air is 1.22 kg/m³, determine the speed of the air in the slipstream. (Cf. the *hint* in Prob. 14.85.)

Fig. P14.86*   **Fig. P14.87**

14.87 Each of the two nozzles shown discharges a stream of water at the rate $Q = 1$ m³/s with a speed $v_S = 30$ m/s, and the stream is deflected by a series of blades mounted on the periphery of a turbine wheel as shown. If the wheel rotates at 240 rev/min, determine (a) the input power from the streams of water, (b) the output power of the turbine, (c) the mechanical efficiency.

14.88 A jet aircraft is equipped with movable vanes as reverse thrusters to provide reverse thrust for deceleration during landing.[†] At the instant shown, the speed of the aircraft is 225 km/h, each of its engines scoops in air at the rate of 50 kg/s and discharges it with a velocity of 600 m/s relative to the reverse thrusters as shown. Determine the reverse thrust generated by each engine at this instant.

Fig. P14.88   **Fig. P14.89***

14.89* A 5-Mg jet aircraft cruising at a constant speed of 900 km/h in a level flight, as shown in Fig. (a), scoops in air at the rate of 100 kg/s and discharges it with a velocity of 800 m/s relative to the aircraft. (i) Determine the total drag due to air friction in such a flight. (ii) Assuming that the drag is proportional to the square of the speed, determine the constant speed v' with which the aircraft may climb at an angle of 10° as shown in Fig. (b) without additional power.

14.90 A jet aircraft is cruising at a constant speed of 900 km/h in a level flight as shown. Each of its three engines of equal capacities discharges air with a velocity of 600 m/s relative to the aircraft. Assuming that the drag is proportional to the square of the speed, determine the cruising speed v' of the aircraft if the engine at its tail is shut off so that only the two engines under its wings are in operation.

Fig. P14.90

[†]Reverse thrusters are also deployed to provide reverse thrust for the backward motion of the jet aircraft at the terminal of an airport.

14.91 Solve Prob. 14.89* if it is to be taken into account that the jet fuel is consumed at the rate of 1 kg/s and ejected in the exhaust with the same velocity of 800 m/s relative to the aircraft.

14.92 A chain of mass ρ_L per unit length and total length L is released from rest when $y = L$, where y is the distance between the floor and the top end of the chain as shown. Determine as a function of y the magnitude of the reaction from the floor as the chain is falling. (*Hint.* The portion of the chain which is still off the floor is in a state of free fall.)

Fig. P14.92   **Fig. P14.93**

14.93 The pile of chain shown has a mass ρ_L per unit length and a total length L. One end of the chain falls through a small hole in its support when the chain is released from rest at the time $t = 0$. Neglecting effects of friction, determine (a) the speed of the falling chain as a function of the length y, (b) the length y as a function of the time t, (c) the total energy loss as the last link leaves the hole.

14.94 The chain shown has a mass ρ_L per unit length and a total length L. One end of the chain is attached to a hook at H, and the other end is released from rest when $y = 0$. Determine as a function of y the magnitude of the reaction from the hook at H as the chain is falling. (*Hint.* The falling portion of the chain is in a state of free fall through a distance of $2y$ and with a speed v, while the time rate of mass added to the stationary portion of the chain is only $\frac{1}{2}\rho_L v$.)

14.95* A space shuttle is launched vertically by firing its main engines and two booster rockets as shown. The main engines consist of three identical rocket engines, each of which burns the hydrogen-oxygen liquid propellant at the rate of 1200 lb/s and ejects it with a velocity of 12.5×10^3 ft/s relative to the engines. Determine the total thrust provided by the main engines.

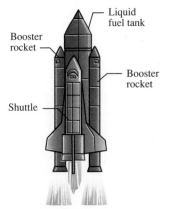

Fig. P14.95* and P14.96

14.96 A space shuttle is launched vertically by firing its main engines and two booster rockets as shown. Each booster rocket is used to provide 2×10^6 lb of thrust by burning solid propellant and ejecting it with a velocity of 9000 ft/s relative to the rocket. Determine the rate of fuel consumption by each booster rocket in tons per second.

Fig. P14.94

14.97   The rocket engine of a lunar landing module is operated in such a way that it hovers motionlessly moments before its touchdown on the surface of the moon as shown. The fuel burned in the rocket is ejected with a velocity of 2500 m/s. Knowing that the gravitational acceleration on the moon is 1.618 m/s², determine the rate of fuel consumption when the total mass of the module is 18 Mg.

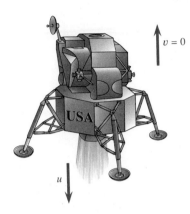

$v = 0$

u

Fig. P14.97

14.98*   A rocket of gross mass 2 Mg, including 1.8 Mg of fuel, is fired vertically from the ground at the time $t = 0$. If fuel is consumed at the rate of 25 kg/s and ejected with a velocity of 2500 m/s relative to the rocket, determine the acceleration and velocity of the rocket when (a) $t = 36$ s, (b) $t = 72$ s.

14.99   Suppose that the rocket in Prob. 14.98* is redesigned as a two-stage rocket consisting of rockets A and B as shown. Each of the rockets A and B has a gross mass of 1 Mg, including 0.9 Mg of fuel. The two-stage rocket is likewise launched vertically from the ground. When the rocket A ejects its last fuel particle, its shell is released and the rocket B is ignited. Knowing that fuel is likewise consumed at the rate of 25 kg/s and ejected with a velocity of 2500 m/s relative to the rocket, determine the speed of the rocket B when (a) the rocket A is released, (b) the last fuel particle in B is ejected.

B

A

Fig. P14.99

14.14   Concluding Remarks

The *principle of impulse and momentum* for particles is the major principle in this chapter. It has been shown to be very suitable for attacking (1) *kinetic problems where time and velocity* (among others) *are to be directly related*, (2) *problems of central-force motion*, and (3) *problems of impact*. The *drawing* of appropriate *vector-diagram equations* for the momenta and impulses in the system during the time interval considered is recommended because it provides a useful geometric aid in the solutions of such problems.

The *principle of generalized virtual work* for particles propounded in Sec. 14.10 is an extension of the principle of virtual work. This principle often allows us to obtain direct solutions of some kinetic quantities in problems of impulse and momentum related to systems of particles with one degree of freedom. The advantage of such a principle lies in the fact that many of the unknown impulse vectors can be excluded when we write the generalized virtual work equation for the momenta and impulses in the system during a constrained virtual displacement. Besides, the *conservation of energy* may be applied to provide an additional equation in the solution of a problem when the system is conservative.

Variable systems are rather common in modern technology. The motion of a variable system of particles is usually analyzed by considering the motion of *an auxiliary constant system* isolated from the variable system during a short time interval Δt. The *principle of impulse and momentum* is then applied to solve the problem.

With this chapter, we have now completed the study of major basic physical laws and principles used in elementary kinetics of particles. A thorough comprehension of these laws and principles as applied to systems of particles will provide a helpful background for the subsequent study of the kinetics of rigid bodies.

REVIEW PROBLEMS

14.100 The magnitude of a force **P** varies as shown. This force is applied to a 5-kg block at the time $t = 0$ when the block is at rest. If $\mu_s = 0.6$ and $\mu_k = 0.5$ between the block and the support, determine (a) the maximum speed attained by the block, (b) the time at which the block stops again.

14.101 A 20-g bullet is fired with a velocity $\mathbf{v}_0$ at the 2-kg block B and the 4-kg block C which are resting on a smooth surface as shown. The bullet pierces B and lodges in C. If the bullet causes B and C to start moving with velocities of 3 m/s and 2 m/s, respectively, determine the magnitude of $\mathbf{v}_0$.

Fig. P14.100

Fig. P14.101   **Fig. P14.102**

14.102 A 25-g bullet is fired with a velocity of 600 m/s at the 4-kg block B which is resting on a 3-kg plate C as shown. The bullet lodges in B at the time $t = 0$ and causes B to slide on C and C to slide on the horizontal support D. If $\mu_k = 0.6$ between B and C and friction between C and D is negligible, determine (a) the time t_1 at which B ceases to slide on C, (b) the displacement of C during the interval $0 \le t \le t_1$, (c) the displacement of B relative to C during the same interval.

14.103 Solve Prob. 14.102 if $\mu_k = 0.1$ between C and D is taken into account.

14.104* The velocity of the cylinder A shown is $v_A = 2$ m/s $\downarrow$ when the system is released at the time $t = 0$. Determine the tension in each cable and the velocity of each cylinder when $t = 5$ s.

14.105 Refer to Prob. 14.104*. Using the *principle of generalized virtual work* for particles, determine the time t_1 at which the velocities of the cylinders become zero.

14.106 A 22-lb body slides with negligible friction on a horizontal xy plane. The body is observed to pass through the origin O with a velocity $\mathbf{v}_0 = 15\mathbf{i}$ ft/s at the time $t = 0$. Following an explosion at $t = t_1$, where $0 < t_1 < 5$ s, the body separates into three fragments, A, B, and C, as shown, of weights $W_A = 9$ lb, $W_B = 8$ lb, and $W_C = 5$ lb, respectively. At $t = 5$ s, it is observed that the position vectors of B and C are $\mathbf{r}_B = 48\mathbf{i} + 36\mathbf{j}$ ft and $\mathbf{r}_C = 75\mathbf{i} - 90\mathbf{j}$ ft, and their velocities are $\mathbf{v}_B = 6\mathbf{i} + (v_B)_y\mathbf{j}$ ft/s and $\mathbf{v}_C = 15\mathbf{i} + (v_C)_y\mathbf{j}$ ft/s. Determine the position vector $\mathbf{r}_A$ and the velocity $\mathbf{v}_A$ of A at $t = 5$ s.

Fig. P14.104*

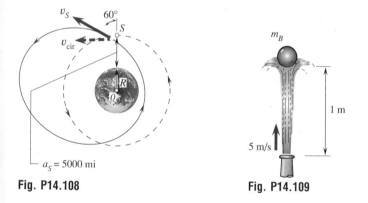

Fig. P14.106

Fig. P14.107

14.107 Two particles A and B, each having a mass of 4 kg, are moving out radially at a constant rate of $\dot{r} = 0.5$ m/s as shown. At the same time, the system is rotating freely about the vertical rod CD with an angular speed $\dot{\theta}$. At the time $t = 0$, it is known that $r = 1$ m, $\dot{\theta} = 3$ rad/s, and a torque $M_{CD} = 3t^2 + 4t$ N·m in the same sense as $\dot{\theta}$ is applied to the vertical rod CD, where t is measured in seconds. Neglecting the masses of the rods, determine the value of $\dot{\theta}$ when $r = 2$ m.

14.108 A spacecraft starts its free flight around the earth in the position S at an altitude $a_S = 5000$ mi and with a velocity $\mathbf{v}_S$ as shown. If the magnitude of $\mathbf{v}_S$ is equal to the speed v_{cir} of a satellite describing a circular orbit passing through S, determine the maximum and minimum altitudes of the spacecraft in its elliptic orbit.

Fig. P14.108

Fig. P14.109

Fig. P14.110

14.109 A vertical jet of water is issued with a velocity of 5 m/s from a nozzle of 20-mm inside diameter and supports a ball at a height of 1 m as shown. Knowing that the mass density of water is 1 Mg/m³, determine the mass m_B of the ball. (*Hint.* The diameter of the jet near the ball is larger than that near the nozzle. However, the time rate of mass flow dm/dt just below the ball is equal to that at the nozzle.)

14.110 A 200-lb cart moving freely on horizontal tracks is loaded by dropping gravel vertically into it from a stationary chute, as shown, at a constant rate of 80 lb/s. The speed of the cart is 2 ft/s when the loading begins at the time $t = 0$. Determine (a) the time t_1 at which the speed of the cart is reduced to 1 ft/s, (b) the acceleration of the cart at $t = t_1$.

14.111 An inverted pendulum formed by attaching a sphere of weight W to a light rod can rotate freely about the hinge O as shown. The system is released from the vertical position OA_1 to fall to the horizontal position OA_2 where impact with the support occurs. If the highest position to which it rebounds is OA_3, determine the coefficient of restitution between the sphere and the support.

Fig. P14.111

14.112* A chain of mass ρ_L per unit length and total length L is held at one end by a force **P** as shown. Starting with $y = L$, the chain is lowered with a constant velocity of magnitude v_0. Determine as a function of y (a) the magnitude of **P**, (b) the magnitude of the reaction from the floor as the chain is lowered.

Fig. P14.112*   **Fig. P14.113***

14.113* The jet engine at the tail of an airplane cruising at a speed of 540 mi/h scoops in air at S at the rate of 150 lb/s and discharges it at D with a velocity of 2500 ft/s relative to the engine. Determine the magnitude and line of action of the propulsive thrust generated by the engine.

Plane Kinematics of Rigid Bodies

Chapter 15

KEY CONCEPTS

- Translation, rotation, general plane motion, angular velocity, angular acceleration, Chasles' theorem, linkage equation, velocity center, space centrode, body centrode, acceleration center, and Coriolis acceleration.
- Methods of solution for kinematic quantities of rigid bodies in plane motion using one or more of the following: (1) relations between linear and angular motions, (2) equations of relative motion, (3) linkage equation for velocities, (4) velocity center, (5) linkage equation for accelerations, (6) acceleration center, and (7) parametric method.
- Plane kinematics using rotating reference frames.

The plane kinematics of rigid bodies is a study of the motion of rigid bodies as taking place in a single plane without considering the cause of motion. Traditional methods of analysis using nonrotating reference frames are covered first. Moreover, the technique using acceleration center is introduced. This technique is particularly useful in analyzing systems of rigid bodies which have nonzero angular accelerations but zero angular velocities at the instant under consideration. Later, the method of analysis using rotating reference frames is presented. This method is useful in many situations, especially in the analysis of accelerations of a system where two members rotate as well as slide at a guided connection. The Coriolis acceleration and other components of acceleration involved in such a system are interpreted and illustrated.

15.1 Types of Plane Motion of a Rigid Body

A rigid body is said to be in *motion* if the position of any particle of the body or the orientation of any line drawn on the body is changed during a finite time interval. A body is at *rest* if and only if its velocity and acceleration are both equal to zero. A rigid body is said to be in *plane motion* if each particle of the body remains at a constant distance from a fixed reference plane during the motion. Thus, the plane motion of a rigid body may be analyzed as taking place in a single plane.[†]

In general, the plane motion of a rigid body may be classified as one of the following three types:

(1) *Translation.* The motion of a rigid body is said to be a *translation* if the orientation of any line drawn on the body remains unchanged during the motion. In translation, the paths of the particles forming the body must be either parallel straight lines or congruent curves. If all paths are parallel straight lines, as in Fig. 15.1(a), the motion is called a *rectilinear translation*; if all paths are congruent curves, as in Fig. 15.1(b), the motion is called a *curvilinear translation.* In either case, the motion of the body is completely specified by the motion of any single point in the body.

(a)   (b)

Fig. 15.1

(2) *Rotation.* The motion of a rigid body is said to be a *rotation* if all particles of the body move along circles centered on a fixed axis, which is perpendicular to the plane of motion and is called the *axis of rotation.* A rotation is an angular motion. If the axis of rotation passes through the rigid body, as in Fig. 15.2(a), the particles located on the axis will not move. If the axis of rotation is located outside the rigid body, as in Fig. 15.2(b), all particles of the rigid body will move.

(a)   (b)

Fig. 15.2

[†]The kinematics of a rigid body requiring three spatial coordinates for its description is covered in Chap. 19.

(3) *General plane motion.* Any plane motion which is neither a translation nor a rotation is referred to as a *general plane motion.* When a rigid body is in general plane motion, the body may be viewed as undergoing a combined motion of a translation parallel to a fixed plane and a rotation about an axis perpendicular to the fixed plane, as in Fig. 15.3. Many problems of plane motion may be studied using nonrotating reference frames, but there are others whose study would require the use of rotating reference frames.

Fig. 15.3

Developmental Exercises

D15.1   Define the following terms related to a rigid body: (a) translation, (b) rotation, (c) general plane motion.

D15.2   Identify the type of plane motion for each of the following:

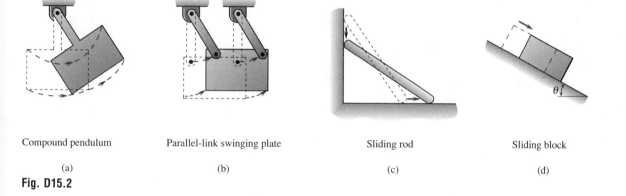

| Compound pendulum | Parallel-link swinging plate | Sliding rod | Sliding block |
| (a) | (b) | (c) | (d) |

Fig. D15.2

USE OF NONROTATING REFERENCE FRAMES

15.2   Translation

Let A and B be any two of the particles of a rigid body in either rectilinear or curvilinear translation parallel to the xy plane as shown in Fig. 15.4. Denoting respectively the position vectors of A and B by $\mathbf{r}_A$ and $\mathbf{r}_B$, and the position vector of B relative to A by $\mathbf{r}_{B/A}$, as shown, we write

$$\mathbf{r}_B = \mathbf{r}_A + \mathbf{r}_{B/A} \tag{15.1}$$

Since A and B belong to the same rigid body, the direction and magnitude of the vector $\mathbf{r}_{B/A}$ must remain unchanged during the translation of the rigid body. Thus, $\mathbf{r}_{B/A}$ is a constant vector and the increments of $\mathbf{r}_B$ and $\mathbf{r}_A$ in Eq. (15.1) are related by

$$\Delta \mathbf{r}_B = \Delta \mathbf{r}_A \tag{15.2}$$

which expresses the property that all points of a rigid body *in translation* during any time interval have the *same displacement.* By differentiating Eq. (15.1) with respect to the time t, we obtain the velocities

$$\mathbf{v}_B = \mathbf{v}_A \tag{15.3}$$

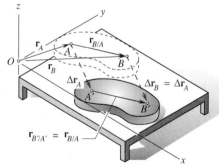

Fig. 15.4

Differentiating Eq. (15.3) with respect to t, we obtain the accelerations

$$\mathbf{a}_B = \mathbf{a}_A \qquad (15.4)$$

Equations (15.3) and (15.4) indicate that, *when a rigid body is in translation during a certain time interval, all points of the body must have the same velocity and the same acceleration at any instant during that time interval.*

Developmental Exercise

D15.3 The rod BC of the linkage rotates about the pin at C, and the ends A and B of the rod AB have the same velocity but different accelerations at the instant shown. Is the rod AB in translation at this instant?

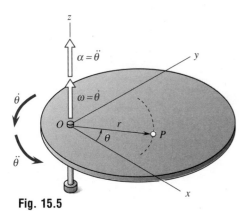

Fig. D15.3

Fig. 15.5

15.3 Rotation

Let us consider a thin rigid slab which rotates about the z axis as shown in Fig. 15.5, where the reference frame $Oxyz$ is fixed in space. Note that the angular position coordinate θ for the line segment $\overline{OP}$ on the slab serves to specify the angular position of the slab. The value of θ must be measured in radians (rad) when a circular arc length s subtended by the angle θ and a radius r is given by the relation $s = r\theta$. Otherwise, θ may also be measured in degrees (°) or revolutions (rev). We note that

$$1 \text{ rev} = 2\pi \text{ rad} = 360° \qquad (15.5)$$

The *angular velocity* of a rigid body is a vector denoted by $\boldsymbol{\omega}$ and is defined as the time rate of change of the angular position coordinate θ of the rigid body. The direction of $\boldsymbol{\omega}$ is given by the extended thumb of the right hand when the other fingers of the right hand are curled in the direction of rotation of the body. Referring to Fig. 15.5, we write

$$\boldsymbol{\omega} = \omega\mathbf{k} \qquad \omega = \frac{d\theta}{dt} = \dot{\theta} \qquad (15.6)^\dagger$$

where the scalar quantity ω is called the *angular speed* of the body.

The *angular acceleration* of a rigid body is a vector denoted by $\boldsymbol{\alpha}$ and is defined as the time rate of change of the angular velocity $\boldsymbol{\omega}$ of the body. Since $\mathbf{k}$ is a constant unit vector, we write

$$\boldsymbol{\alpha} = \frac{d\boldsymbol{\omega}}{dt} = \frac{d}{dt}(\omega\mathbf{k}) = \frac{d\omega}{dt}\mathbf{k}$$

$$\boldsymbol{\alpha} = \alpha\mathbf{k} \qquad \alpha = \frac{d\omega}{dt} = \dot{\omega} \qquad (15.7)$$

Applying the chain rule of differentiation, we write

$$\alpha = \frac{d\omega}{dt} = \frac{d\omega}{d\theta}\frac{d\theta}{dt} = \frac{d\omega}{d\theta}\omega \qquad \alpha = \omega\frac{d\omega}{d\theta} \qquad (15.8)$$

---

†Cf. the angular velocities defined by Eqs. (15.34) and (15.40).

Usually, the angular velocity ω is measured in rad/s or in rpm, while the angular acceleration α is measured in rad/s^2. In drawings, the angular velocity and angular acceleration vectors are represented either by *curved arrows* in the plane of rotation or by *hollow arrows* perpendicular to the plane of rotation, as shown in Fig. 15.5.

The orientations of ω and α of a body rotating about a fixed axis are always parallel to the axis of rotation. In this case, only the *magnitudes* and *senses* of ω and α are the nontrivial information to be described or predicted. Consequently, the rotation of a body about a fixed axis can be studied using essentially *scalar analysis*.

The scalar equations in Eqs. (15.6) through (15.8) should be recognized as mathematically analogous to Eqs. (11.3), (11.6), and (11.8), which define the rectilinear motion of particles. Thus, all relations described for the rectilinear motion of particles are applicable to the case of rotation of a rigid body about a fixed axis. In application, we may simply replace the linear kinematic quantities x, v, and a in Chap. 11 with their respective counterparts θ, ω, and α in rotation.

If the angular acceleration is a constant, $\alpha = \alpha_c$, the scalar equations in Eqs. (15.6) through (15.8) may readily be integrated to yield

$$\omega = \omega_0 + \alpha_c t \qquad (15.9)$$

$$\theta = \theta_0 + \omega_0 t + \tfrac{1}{2}\alpha_c t^2 \qquad (15.10)$$

$$\omega^2 = \omega_0^2 + 2\alpha_c(\theta - \theta_0) \qquad (15.11)$$

where t is the time variable, and ω_0 and θ_0 are the angular velocity and the angular position coordinate at $t = 0$.

EXAMPLE 15.1

A grinding wheel is rotating at the rate of 1500 rpm ⤸ when the power is cut off at the time $t = 0$. Because of friction, the wheel decelerates at a constant rate α_c and comes to rest in 500 rev. Determine α_c and the time t at which the wheel comes to rest.

Solution. The angular speed of the grinding wheel at $t = 0$ is

$$\omega_0 = 1500\,\frac{\text{rev}}{\text{min}} \cdot \frac{2\pi\ \text{rad}}{1\ \text{rev}} \cdot \frac{1\ \text{min}}{60\ \text{s}} = 50\pi\ \text{rad/s}$$

When it comes to rest, its final speed is $\omega = 0$. Its angular displacement $\Delta\theta$ from the time it starts to slow down until the time it comes to rest is

$$\Delta\theta = \theta - \theta_0 = 500\ \text{rev} \cdot \frac{2\pi\ \text{rad}}{1\ \text{rev}} = 1000\pi\ \text{rad}$$

Since the wheel slows down at a constant rate, we write

$$\omega^2 = \omega_0^2 + 2\alpha_c(\theta - \theta_0): \qquad 0^2 = (50\pi)^2 + 2\alpha_c(1000\pi)$$

$$\alpha_c = -1.25\pi \qquad \boldsymbol{\alpha_c = 3.93\ \text{rad/s}^2} \ ⤸ \ \blacktriangleleft$$

To find the time t at which the wheel comes to rest, we write

$$\omega = \omega_0 + \alpha_c t: \qquad 0 = 50\pi + (-1.25\pi)t$$

$$t = 40\ \text{s} \ \blacktriangleleft$$

Developmental Exercise

D15.4   Refer to Example 15.1. Determine (a) the number of revolutions that the grinding wheel has executed from the time it starts to slow down to the time its angular speed is equal to 25π rad/s, (b) the time elapsed during this period.

15.4   Linear and Angular Motions

A thin rigid slab which lies in the xy plane and rotates about the z axis is shown in Fig. 15.6, where the reference frame $Oxyz$ is fixed in space. In this case, all points of the slab move along concentric circles with centers all located at the origin O. As the slab in Fig. 15.6 rotates about the z axis, the point P of the slab will describe a circular path of radius $\overline{OP}$ with center at O. We here use the notations $r = \overline{OP} = \rho$ for the radius of curvature of the path of P, $\omega = \dot{\theta}$ for the angular speed of $\overline{OP}$, and $\alpha = \dot{\omega} = \ddot{\theta}$ for the time rate of change of the angular speed. Of course, the angular velocity and angular acceleration of $\overline{OP}$ are $\boldsymbol{\omega} = \omega\mathbf{k}$ and $\boldsymbol{\alpha} = \alpha\mathbf{k}$, respectively, as indicated in parts (a) and (b) of Fig. 15.6.

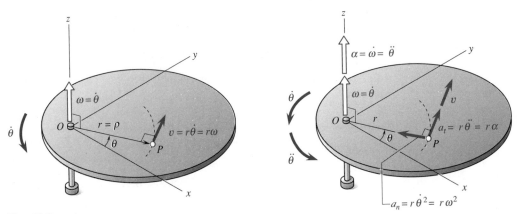

Fig. 15.6

The relations between the linear motion of the point P and the angular motion of the line segment $\overline{OP}$, which exemplifies the angular motion of the rigid slab, can be established from the relations expressed in Eqs. (11.56). Substituting the aforementioned notations into Eqs. (11.56), we write

$$v = r\omega \qquad a_t = r\alpha \qquad a_n = r\omega^2 \tag{15.12}$$

where v, a_t, and a_n are, respectively, the speed, the tangential component of acceleration, and the normal component of acceleration of the point P. The quantities v, a_t, and a_n have been illustrated in Fig. 15.6(a) and (b). Note that *the resultant acceleration* $\mathbf{a}$, which is the vector sum of $\mathbf{a}_t$ and $\mathbf{a}_n$, *is always directed toward the concave side of the path of* P.

The velocity $\mathbf{v}$ and the acceleration components $\mathbf{a}_t$ and $\mathbf{a}_n$ of P may be related to its position vector $\mathbf{r}$, the angular velocity $\boldsymbol{\omega}$ of $\mathbf{r}$, and the angular acceleration $\boldsymbol{\alpha}$ of $\mathbf{r}$ through the use of vector analysis. By the definition of the cross product of two vectors, we see that $\mathbf{v}$ is obtained by crossing $\boldsymbol{\omega}$ into $\mathbf{r}$ as shown in Fig. 15.7(a). This cross product gives precisely the correct magnitude and direction of $\mathbf{v}$. Thus, we write

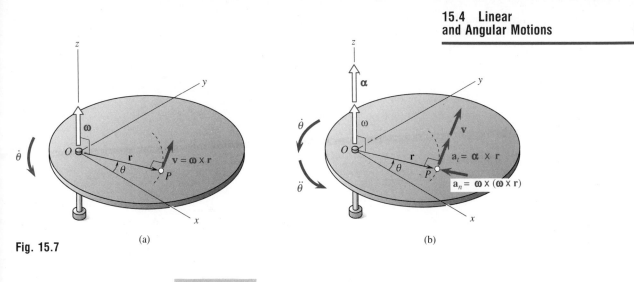

Fig. 15.7

(a)   (b)

$$\mathbf{v} = \boldsymbol{\omega} \times \mathbf{r} \qquad (15.13)$$

Note that $\mathbf{r} \times \boldsymbol{\omega} = -\mathbf{v}$. Since the acceleration $\mathbf{a}$ of P is, by definition, the time rate of change of $\mathbf{v}$, we write

$$\mathbf{a} = \dot{\mathbf{v}} = \frac{d}{dt}(\boldsymbol{\omega} \times \mathbf{r}) = \dot{\boldsymbol{\omega}} \times \mathbf{r} + \boldsymbol{\omega} \times \dot{\mathbf{r}}$$

$$= \boldsymbol{\alpha} \times \mathbf{r} + \boldsymbol{\omega} \times \mathbf{v}$$

$$= \boldsymbol{\alpha} \times \mathbf{r} + \boldsymbol{\omega} \times (\boldsymbol{\omega} \times \mathbf{r}) = \mathbf{a}_t + \mathbf{a}_n$$

where $\boldsymbol{\alpha} = \dot{\boldsymbol{\omega}}$, the angular acceleration of $\mathbf{r}$. Thus, we have

$$\mathbf{a}_t = \boldsymbol{\alpha} \times \mathbf{r} \qquad \mathbf{a}_n = \boldsymbol{\omega} \times (\boldsymbol{\omega} \times \mathbf{r}) \qquad (15.14)$$

which are shown in Fig. 15.7(b). Note that Eqs. (15.13) and (15.14) are the vector versions of Eqs. (15.12).

In plane motion of a rigid body, the directions of $\boldsymbol{\omega}$ and $\boldsymbol{\alpha}$ are parallel to each other because both of them are perpendicular to the plane of motion. In three-dimensional motion of a rigid body, the directions of $\boldsymbol{\omega}$ and $\boldsymbol{\alpha}$ are generally not parallel to each other.

EXAMPLE 15.2

A load L is raised from rest through a distance of 9 m by a cable wrapped around the drum of a hoisting system as shown. Determine the angular displacements of the gears A and B in raising the load.

Solution. From the sketch shown on next page, we see that the distance s traveled by the load L must be equal to the length of the arc CC' which corresponds to the part of the cable wound around the drum during the raising of the load. Thus, the angular displacement $\Delta\theta_B$ of the gear B in raising the load L through a distance $s = 9$ m is

$$\Delta\theta_B = \frac{s}{\overline{HC}} = \frac{9}{0.36} \qquad \Delta\theta_B = 25 \text{ rad } \circlearrowright \blacktriangleleft$$

The length of the arc DD' subtended by $\Delta\theta_B$, on the gear B, is s' $= \overline{HD} \, \Delta\theta_B$, while the length of the arc EE' subtended by the corresponding

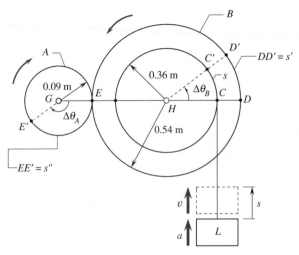

angular displacement $\Delta\theta_A$ of the gear A is $s'' = \overline{GE}\ \Delta\theta_A$. Since the lengths s' and s'' of the arcs rolled out by the gears A and B must be equal, we write

$$\overline{HD}\ \Delta\theta_B = \overline{GE}\ \Delta\theta_A \qquad 0.54(25) = 0.09\ \Delta\theta_A$$

$$\Delta\theta_A = 150 \text{ rad} \quad \blacktriangleleft$$

Developmental Exercise

D15.5   Work Example 15.2 if the radius of the gear A is equal to 180 mm.

EXAMPLE 15.3

Refer to Example 15.2. Suppose that the gear A of the hoist motor accelerates from rest at a constant rate to a speed of 150 rpm in 5 s and then remains at this speed. Determine the total time required to raise the load L from rest through the distance of 9 m.

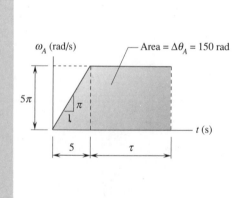

Solution. Since $\omega_A = 0$ at the time $t = 0$ and $\omega_A = 150$ rpm $= 5\pi$ rad/s at $t = 5$ s, we draw the angular velocity curve for the gear A as shown. From the solution in Example 15.2, we know that the angular displacement of the gear A in raising the load through 9 m is $\Delta\theta_A = 150$ rad. By analogy to that in Sec. 11.8, we may equate the 150-rad angular displacement to the shaded trapezoidal area as shown. We write

$$150 = \tfrac{1}{2}[(5 + \tau) + \tau](5\pi) \qquad \tau = 7.05$$

Thus, the total time required is

$$t = 5 + \tau = 5 + 7.05 \qquad t = 12.05 \text{ s} \quad \blacktriangleleft$$

Developmental Exercise

D15.6   Refer to Example 15.3. Determine the velocity of the load L when $t = 5$ s.

EXAMPLE 15.4

Refer to Example 15.2, where $\omega_A = 6\omega_B$ and $\alpha_A = 6\alpha_B$. Suppose that, at a certain instant, the gears A and B have $\omega_A = 2\pi$ rad/s $\circlearrowleft$, $\alpha_A = 6\pi$ rad/s^2 $\circlearrowleft$, and $\omega_B = \pi/3$ rad/s $\circlearrowright$, $\alpha_B = \pi$ rad/s^2 $\circlearrowright$, respectively. The point of contact between the two gears is E. For this instant, determine the acceleration $\mathbf{a}_E'$ of the particle at E on the gear A and the acceleration $\mathbf{a}_E''$ of the particle at E on the gear B.

Solution. Since no slipping occurs at E between the two gears, the tangential component $(\mathbf{a}_E)_t$ of the acceleration of the particle at E on either gear must be the same and may be computed from the motion of either gear. From the given data and the sketch as shown, we write

$$(a_E)_t = r_A\alpha_A = 0.09(6\pi) = 0.54\pi$$

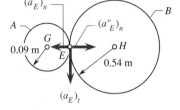

The normal components $(\mathbf{a}_E')_n$ and $(\mathbf{a}_E'')_n$ of the accelerations of the particles at E on the gears A and B, respectively, are different. We write

$$(a_E')_n = r_A\omega_A^2 = 0.09(2\pi)^2 = 0.36\pi^2$$

$$(a_E'')_n = r_B\omega_B^2 = 0.54(\pi/3)^2 = 0.06\pi^2$$

Adding vectorially the normal and tangential components, we write

$$\mathbf{a}_E' = -(a_E')_n\mathbf{i} - (a_E)_t\mathbf{j} = -0.36\pi^2\mathbf{i} - 0.54\pi\mathbf{j}$$

$$\mathbf{a}_E'' = (a_E'')_n\mathbf{i} - (a_E)_t\mathbf{j} = 0.06\pi^2\mathbf{i} - 0.54\pi\mathbf{j}$$

$$\mathbf{a}_E' = 3.94 \text{ m/s}^2 \ \angle 25.5° \qquad \mathbf{a}_E'' = 1.797 \text{ m/s}^2 \ \angle 70.8° \ \blacktriangleleft$$

We see here that $\mathbf{a}_E'$ and $\mathbf{a}_E''$ are considerably different.

Developmental Exercise

D15.7 A system consisting of a gear and a rack is shown. If the rack moves with $\mathbf{v} = 16$ in./s $\rightarrow$ and $\mathbf{a} = 120$ in./s^2 $\rightarrow$ at the given instant, determine ω, α, and $\mathbf{a}_P$ of the point P of the gear.

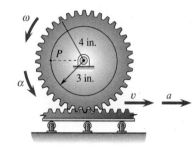

Fig. D15.7

PROBLEMS

15.1 A flywheel is attached to the shaft of a motor which has a rated speed of 2400 rpm. When the motor is turned on, it accelerates at a constant rate from rest and reaches its rated speed after executing 80 rev; and when the motor is turned off, it decelerates at a constant rate and comes to rest after executing 900 rev. Determine the time required for the flywheel (a) to reach its rated speed, (b) to come to rest.

15.2 The angular position coordinate of a swinging pendulum as shown is given by $\theta = (\pi/3)\cos 2\pi t$, where θ is measured in radians and t in seconds. Determine the magnitudes of the maximum angular velocity and the maximum angular acceleration of the pendulum.

Fig. P15.2

15.3* At the instant shown, the gear A rotates with $\omega_A = 2$ rad/s ↻ and the gear B rotates with $\alpha_B = 6$ rad/s^2 ↺. Determine $\mathbf{v}_D$, $\mathbf{v}_E$, $\mathbf{a}_D$, and $\mathbf{a}_E$ of the points D and E at this instant.

15.4 The gear A of the system shown executes 200 rev over a time interval Δt, while its angular velocity is being increased from 300 rpm ↻ to 900 rpm ↻ at a constant rate α_c. Determine α_c and Δt.

Fig. P15.3* and P15.4

Fig. P15.5 and P15.6*

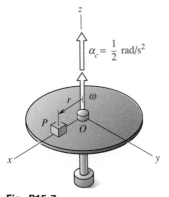

Fig. P15.7

✗15.5 A circular disk rotates about the point O as shown. At the instant shown, the point P moves with $\mathbf{v}_P = 0.9$ m/s ↑ and $\mathbf{a}_Q$ of the point Q of the disk is directed as indicated. Determine α of the disk and the magnitude of $\mathbf{a}_Q$ at this instant.

15.6* At the instant shown, $\mathbf{a}_Q$ of the point Q of the rotating disk is directed as indicated and has a magnitude of 10 m/s^2. Determine the speed v_P of the point P and α of the disk at this instant.

✗15.7 A small object P rests on a horizontal disk as shown. At the time $t = 0$, the disk accelerates at a constant rate $\alpha_c = \frac{1}{2}\mathbf{k}$ rad/s^2 from rest until $t = 5$ s, then it rotates freely without any angular acceleration. It is known that P will start to slide on the disk when its resultant acceleration exceeds 3 m/s^2. Determine the maximum radial distance r for which sliding of P on the disk will not occur.

15.8 The gears A, B, and C, of radii $r_A = 500$ mm, $r_B = 200$ mm, and $r_C = 150$ mm, respectively, all rotate about fixed axes parallel to the x axis as shown. At a certain instant, the gear A rotates with $\omega_A = 6\mathbf{i}$ rad/s and the gear C rotates with $\alpha_C = 10\mathbf{i}$ rad/s^2. Determine ω_B and α_B of the gear B at this instant.

Fig. P15.8

Fig. P15.9*

15.9* A 0.6-m radius drum carrying the load A is rigidly attached to a 0.9-m radius pulley carrying the load B as shown. At the time $t = 0$, the load B moves with a velocity of 2 m/s ↓ and a constant acceleration of 3 m/s^2 ↓. Over the time interval $0 \le t \le 2$ s, determine (a) the number of revolutions executed by the pulley, (b) the displacement of the load A.

15.10 The relation between ω and θ of a cam rotating about a fixed axis is plotted as shown. Estimate the magnitudes of $\boldsymbol{\alpha}$ of the cam when (a) $\theta = 2$ rad, (b) $\theta = 6$ rad.[†]

15.11* Refer to Prob. 15.10. Estimate the magnitudes of $\boldsymbol{\alpha}$ of the cam when (a) $\theta = 4$ rad, (b) $\theta = 8$ rad.[†]

Fig. P15.10   **Fig. P15.12**

15.12 The rack R moves with $\mathbf{a}_R = 6$ ft/s^2 ↙ on the gears A and B. At the instant shown, the point P of the gear A has an acceleration of magnitude 10 ft/s^2. Determine the magnitude of the acceleration of the point Q of the gear B at this instant.

15.13* A roll of carpet material is unrolled by pulling its end downward with a constant speed v_0 as shown, where h denotes the thickness of the carpet and r denotes the radius of the carpet roll at the instant under consideration. Determine the angular acceleration of the carpet roll in terms of the parameters v_0, h, and r. [*Hint.* Note that $v_0 = r\omega$, $\alpha = \dot{\omega}$, and $h(v_0\, dt) = 2\pi r(-dr)$ from the change in volume.]

Fig. P15.13*   **Fig. P15.14**

15.14 The roll of paper at A is being transferred to that at B by turning the paper roll at B with a constant angular speed $(\omega_B)_0$ as shown. The radii of the paper rolls at A and B are r_A and r_B, respectively, and the thickness of the paper is h. Determine α_A of the paper roll at A at the instant shown in terms of the parameters r_A, r_B, $(\omega_B)_0$, and h. [*Hint.* The time rates of change $\dot{r}_A$ and $\dot{r}_B$ are related to each other by the fact that, during the time interval dt, the volume of paper per unit width unrolled from the roll at A is equal to that taken up by the roll at B; i.e., $2\pi r_A(-dr_A) = 2\pi r_B(dr_B) = hr_B(\omega_B)_0\, dt$.]

[†]Cf. the *hint* in Prob. 13.23*.

15.15   A circular cylinder is rotating about the fixed axis AB with a constant angular speed $\omega_0 = 13$ rad/s as shown, where AB passes through the origin O and the unit vector along AB is $\boldsymbol{\lambda}_{AB} = \frac{1}{13}(-5\mathbf{i} + 12\mathbf{k})$. At the instant considered, the coordinates of the point P of the cylinder are (3, 2, 6) m. Determine $\mathbf{v}_P$ and $\mathbf{a}_P$ of P at this instant. [*Hint*. Apply Eqs. (15.13) and (15.14).]

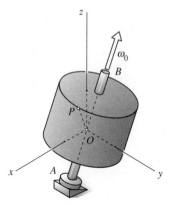

Fig. P15.15

15.16   Refer to Prob. 15.15. At the instant considered, determine (a) the shortest distance d_s between the point P and the axis AB, (b) v, a_t, a_n, and a of the point P. [*Hint*. Use Eqs. (15.12).]

15.5   General Plane Motion: Chasles' Theorem

The most general motion of a rigid body is equivalent to a translation of the body with some point of the body (or its massless extension) *plus a rotation of the body about an axis passing through that point, and the order of such translation and rotation is immaterial.* This property is referred to as *Chasles' theorem*, after the French geometer Michel Chasles (1793–1880).[†] This theorem will here be applied to the study of the general plane motion of rigid bodies.

Let a body executing a general plane motion as shown in Fig. 15.8 be considered. Over a certain time interval, two given points P and Q of the body will have moved, respectively, from P_1 to P_2 and from Q_1 to Q_2. This motion is equivalent to a translation of the body with the point P from P_1 to P_2 plus a rotation of the body about the point P from the position P_2Q_1' to the position P_2Q_2; of course, it is also equivalent to a rotation about P first plus a translation with P later as shown. The actual motion of the body is a combination of translation and rotation simultaneously.

The motion of the body in Fig. 15.8 may also be represented in Fig. 15.9. In the latter figure, the general plane motion of the body is shown to be equivalent to a translation of the body with the point Q from Q_1 to Q_2 plus a rotation of the body about the point Q from the position Q_2P_1' to the position Q_2P_2; of course, it is also equivalent to a rotation about Q first plus

[†]We accept Chasles' theorem as a more-or-less-obvious axiom. This theorem is valid in studying both the kinematics and the kinetics of rigid bodies. When this theorem is used in studying the plane kinetics of rigid bodies, the point of reference mentioned in the theorem is usually the mass center G of the rigid body, although the theorem remains valid when the point of reference is not the mass center. An illustration for the latter case is provided in D16.23. For further treatment, see Beggs, J. S., *Advanced Mechanism*, Macmillan, New York, 1966.

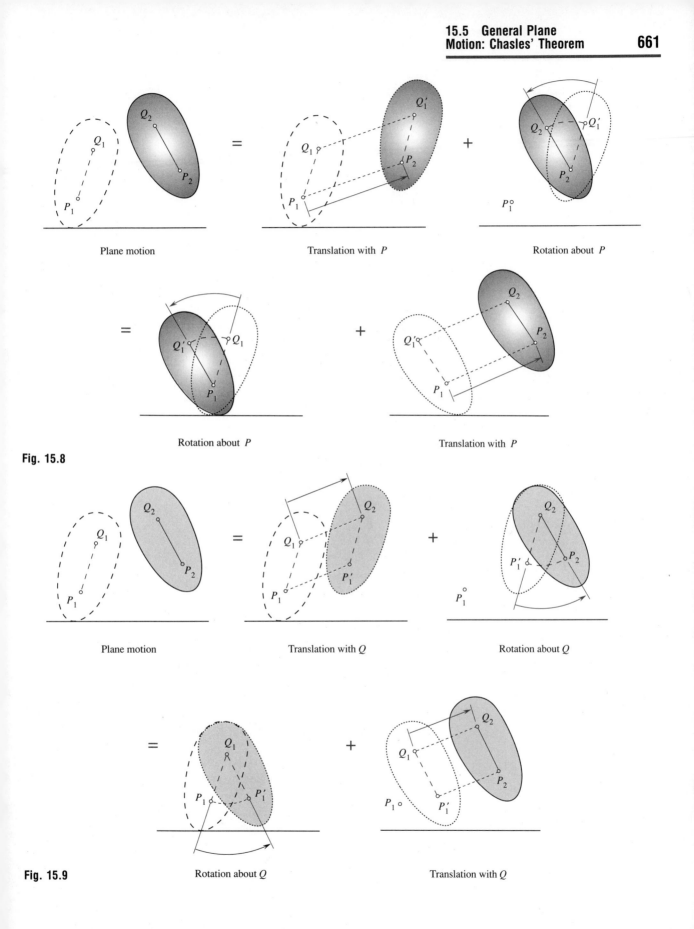

Plane motion   Translation with P Rotation about P

Rotation about P Translation with P

Fig. 15.8

Plane motion   Translation with Q Rotation about Q

Rotation about Q Translation with Q

Fig. 15.9

Plane motion   Translation with P Rotation about P

Fig. 15.10

a translation with Q later as shown. Note in Figs. 15.8 and 15.9 that the amounts of translation with different points may be unequal; the amounts of rotation about those points are always equal.

As additional examples, a sliding rod is shown in Fig. 15.10 and a rolling cylinder is shown in Fig. 15.11. Explanations for these two figures are similar to those given in the preceding paragraphs.

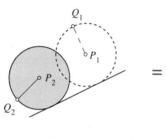

Plane motion   Translation with P Rotation about P

Fig. 15.11

Developmental Exercises

D15.8   Describe Chasles' theorem.

D15.9   Show that the motion of the sliding rod in Fig. 15.10 is also equivalent to a rotation about the end Q plus a translation with Q.

D15.10   Show that the motion of the cylinder in Fig. 15.11 is also equivalent to a translation with the point Q plus a rotation about Q.

15.6   Velocities in Relative Motion

Suppose that the points A and B are two different points of a rigid slab in plane motion with an angular velocity $\boldsymbol{\omega} = \omega \mathbf{k}$, as shown in Fig. 15.12. By Chasles' theorem, we may view the plane motion of the rigid slab as a rotation of the slab about A plus a translation of the slab with A. Such a viewpoint is depicted in the graphical equation shown. Thus, we may express the velocity $\mathbf{v}_B$ of B as the velocity $\mathbf{v}_{B/A}$ of B relative to A plus the velocity $\mathbf{v}_A$ of A; i.e.,

| Plane motion | Rotation about *A* | Translation with *A* | $\mathbf{v}_B = \mathbf{v}_{B/A} + \mathbf{v}_A$ |

Fig. 15.12

$$\mathbf{v}_B = \mathbf{v}_{B/A} + \mathbf{v}_A \qquad (15.15)^\dagger$$

which is equivalent to Eq. (11.36). Furthermore, letting $r = r_{B/A} = \overline{AB}$ in Fig. 15.12 and using the cross product of two vectors, we see that

$$v_{B/A} = r\omega \qquad \mathbf{v}_{B/A} = \boldsymbol{\omega} \times \mathbf{r} \qquad (15.16)$$

where $\mathbf{r} = \mathbf{r}_{B/A} = \overrightarrow{AB}$. Note that (1) *the velocity* $\mathbf{v}_{B/A}$ *of B relative to A is due solely to the rotational part of the motion of the slab* (i.e., the value of $\mathbf{v}_{B/A}$ vanishes whenever $\boldsymbol{\omega} = \mathbf{0}$), (2) *the direction of* $\mathbf{v}_{B/A}$ *is perpendicular to* $\overrightarrow{AB}$, and (3) *the vectors* $\boldsymbol{\omega}$, $\overrightarrow{AB}$, *and* $\mathbf{v}_{B/A}$ (in this order) *form a right-handed trihedral*. As a matter of fact, we may view $\mathbf{v}_{B/A}$ as the velocity of the tip B of the radius vector $\overrightarrow{AB}$ which rotates with the angular velocity $\boldsymbol{\omega}$ about A as a "fixed point."

For a system consisting of a single member or several pin-connected members, the equation linking the motions of the key points (e.g., the connecting pins) of the system is referred to as a *linkage equation* for the motion of the system. Such an equation ensures the compatibility of motion of the linkage with its constraint conditions. For the linkage illustrated in Fig. 15.13, we note that the pins at A and E are fixed; therefore, their velocities

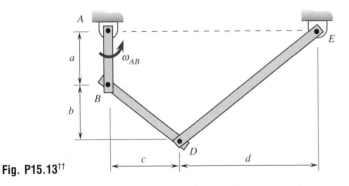

Fig. P15.13††

---

†The order of terms in Eq. (15.15) is preferable to that in Eq. (11.36) in establishing the *linkage equation* for velocities of a system, such as Eq. (15.17).

††In *mechanism*, the linkage shown is called a *four-bar linkage*, where AE is the *frame* or *ground link*, AB is the *driver* or *crank*, BD is the *coupler link*, and DE is the *follower* or *output link*.

v_A and v_E are zero. Taking these constraint conditions $v_A = v_E = 0$ into account and applying Eq. (15.15), we write

$$v_B = v_{B/A} + v_A = v_{B/A}$$

$$v_B = v_{B/D} + v_D = v_{B/D} + (v_{D/E} + v_E) = v_{B/D} + v_{D/E}$$

Thus, we have

$$v_{B/A} = v_{B/D} + v_{D/E} \qquad (15.17)$$

which is the *linkage equation* for velocities of the system shown in Fig. 15.13. Such an equation may be used as a basis for determining the linear velocities of certain points and the angular velocities of certain members in a system.

EXAMPLE 15.5

The end B of the rod AB is pulled with $v_B = 0.9$ m/s $\rightarrow$. For the position shown, determine ω of the rod and v_A of the end A of the rod.

Solution. From the constraints of the rod AB, we see that v_A of its end A must be directed downward and ω of the rod must be counterclockwise when $v_B = 0.9$ m/s $\rightarrow$ as given. Thus, the *linkage equation* for velocities of the rod may be written as

$$v_B = v_{B/A} + v_A$$

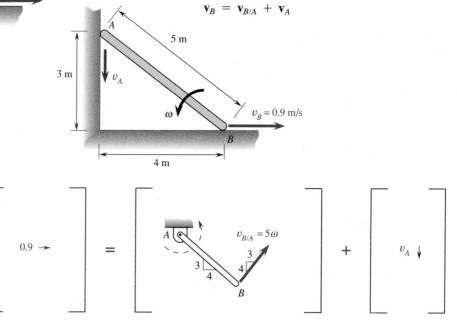

This vector equation contains two unknowns: ω and v_A, which may be solved from the two ensuing scalar equations as follows:

$$\xrightarrow{+} \Sigma V_x: \qquad 0.9 = \tfrac{3}{5}(5\omega) + 0 \qquad \omega = 0.3$$

$$+\uparrow \Sigma V_y: \qquad 0 = \tfrac{4}{5}(5\omega) - v_A \qquad v_A = 1.2$$

$$\omega = 0.3 \text{ rad/s } \circlearrowleft \qquad v_A = 1.2 \text{ m/s } \downarrow \quad \blacktriangleleft$$

Developmental Exercise

D15.11   What is a *linkage equation* for the motion of a system consisting of a single member or several pin-connected members?

EXAMPLE 15.6

Refer to the system in Fig. 15.13. If $a = b = 3$ ft, $c = 4$ ft, $d = 8$ ft, and $\omega_{AB} = 10$ rad/s $\circlearrowright$ in the position shown, determine (a) ω_{BD} and ω_{DE} of the members BD and DE, (b) $\mathbf{v}_D$ of the pin at D.

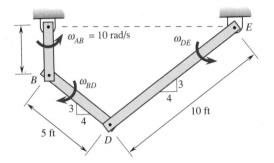

Solution.   We first indicate ω_{AB} as shown. The directions of ω_{BD} and ω_{DE} are perpendicular to the plane of motion, and their senses may initially be assumed as indicated. Since the *linkage equation* for velocities of the system has already been derived as Eq. (15.17), we here use that equation and write

$$\mathbf{v}_{B/A} = \mathbf{v}_{B/D} + \mathbf{v}_{D/E}$$

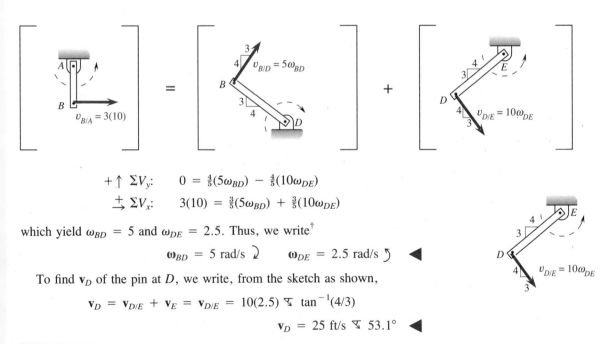

$$+\uparrow \Sigma V_y: \quad 0 = \tfrac{4}{5}(5\omega_{BD}) - \tfrac{4}{5}(10\omega_{DE})$$

$$\xrightarrow{+} \Sigma V_x: \quad 3(10) = \tfrac{3}{5}(5\omega_{BD}) + \tfrac{3}{5}(10\omega_{DE})$$

which yield $\omega_{BD} = 5$ and $\omega_{DE} = 2.5$. Thus, we write[†]

$$\omega_{BD} = 5 \text{ rad/s} \circlearrowright \qquad \omega_{DE} = 2.5 \text{ rad/s} \circlearrowright \quad \blacktriangleleft$$

To find $\mathbf{v}_D$ of the pin at D, we write, from the sketch as shown,

$$\mathbf{v}_D = \mathbf{v}_{D/E} + \mathbf{v}_E = \mathbf{v}_{D/E} = 10(2.5) \ \diagdown \ \tan^{-1}(4/3)$$

$$\mathbf{v}_D = 25 \text{ ft/s} \ \diagdown \ 53.1° \quad \blacktriangleleft$$

[†]If either of ω_{BD} and ω_{DE} were found to be negative, its initially assumed sense would have to be reversed in reporting the answers.

Developmental Exercise

D15.12 Suppose that A and B are two different points of a rigid slab in plane motion. (a) What is the angular velocity of the rigid slab if $\mathbf{v}_{B/A} = \mathbf{0}$? (b) Define $\mathbf{v}_{B/A}$ with a cross product.

EXAMPLE 15.7

The gear A of the system shown rotates with $\mathbf{\omega}_A = 180$ rpm $\circlearrowright$, and the connecting arm CD rotates with $\mathbf{\omega}_{CD} = 60$ rpm $\circlearrowleft$. Determine $\mathbf{\omega}_B$ of the gear B.

Solution. We note that

$$\omega_A = 180 \text{ rpm } \circlearrowright = 6\pi \text{ rad/s } \circlearrowright$$
$$\omega_{CD} = 60 \text{ rpm } \circlearrowleft = 2\pi \text{ rad/s } \circlearrowleft$$

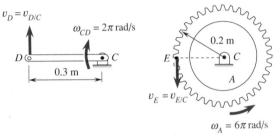

Seeing that the point C is fixed, we write

$$\mathbf{v}_D = \mathbf{v}_{D/C} + \mathbf{v}_C = \mathbf{v}_{D/C} = 0.3(2\pi) \uparrow = 0.6\pi \uparrow$$
$$\mathbf{v}_E = \mathbf{v}_{E/C} + \mathbf{v}_C = \mathbf{v}_{E/C} = 0.2(6\pi) \downarrow = 1.2\pi \downarrow$$

Since E is the point of contact between the gears A and B, the velocity of the point E on these two gears must be the same. From the velocities of D and E on the gear B as shown, we readily see that $\mathbf{\omega}_B$ of the gear B must be clockwise as indicated. We write

$$\mathbf{v}_D = \mathbf{v}_{D/E} + \mathbf{v}_E$$

$$\begin{bmatrix} \uparrow \; v_D = 0.6\pi \\ \circ D \end{bmatrix} = \begin{bmatrix} \uparrow \; v_{D/E} = 0.1\omega_B \\ D \; \diagdown \; E \end{bmatrix} + \begin{bmatrix} E \circ \\ \downarrow \; v_E = 1.2\pi \end{bmatrix}$$

$$+\uparrow \Sigma V_y: \qquad 0.6\pi = 0.1\omega_B - 1.2\pi$$

$$\omega_B = 18\pi \text{ rad/s} = 540 \text{ rpm} \qquad \omega_B = 540 \text{ rpm } \circlearrowleft \blacktriangleleft$$

Developmental Exercise

D15.13 Refer to Example 15.7. If $\mathbf{\omega}_{CD} = 60$ rpm $\circlearrowleft$ and the gear B is in curvilinear translation, determine $\mathbf{\omega}_A$.

PROBLEMS

15.17 The end B of the rod AB slides with $\mathbf{v}_B = 2.8$ m/s $\rightarrow$. For the position shown, determine $\boldsymbol{\omega}$ of the rod and $\mathbf{v}_A$ of the end A.

Fig. P15.17

Fig. P15.18*

15.18* A gear rolls on a stationary rack as shown. If the center A of the gear moves with $\mathbf{v}_A = 1.5$ m/s $\rightarrow$, determine the velocities of the points B, C, D, and E of the gear.

15.19 A cable is wrapped around a drum which is mounted on a gear that rolls on a stationary rack as shown. The end D of the cable is pulled with $\mathbf{v}_D = 0.9$ m/s $\rightarrow$. For the position shown, determine $\mathbf{v}_A$ of the center A and $\mathbf{v}_B$ of the point B of the gear.

Fig. P15.19

Fig. P15.20

Fig. P15.22

15.20 A triangular plate moves in the xy plane. For the position shown, the corners P and Q move with $\mathbf{v}_P = 4.2\mathbf{i} + 0.9\mathbf{j}$ m/s and $\mathbf{v}_Q = 1.2\mathbf{i} + (v_Q)_y\mathbf{j}$ m/s, respectively. Determine $\boldsymbol{\omega}$ of the plate and $(v_Q)_y$ for this position.

15.21* Refer to Prob. 15.20. For the position shown, (a) determine $\mathbf{v}_R$ of the corner R, (b) locate the point $C(x, y)$ of the plate where $\mathbf{v}_C = \mathbf{0}$.

15.22 The crank OB of an internal-combustion engine rotates with $\boldsymbol{\omega}_{OB} = 1200$ rpm $\circlearrowleft$. For the position shown, determine $\boldsymbol{\omega}_{AB}$ of the connecting rod AB and $\mathbf{v}_P$ of the piston P.

15.23 The planetary gear train shown consists of a sun gear S which is fixed, an annular gear A which is driven with $\boldsymbol{\omega}_A = 1800$ rpm $\circlearrowright$, three identical planet gears P, Q, and R, and a carrier C which is connected to the centers of the planet gears. For $a = 60$ mm and $b = 40$ mm, determine $\boldsymbol{\omega}_P$ of the planet gear P and $\boldsymbol{\omega}_C$ of the carrier C.

Fig. P15.23

15.24* Refer to the planetary gear train in Prob. 15.23. If $a/b = 2$, determine the ratio of ω_C of the carrier C to ω_A of the annular gear A.

15.25 The end A of the rack AB moves with a constant velocity of 8.75 in./s $\rightarrow$. For the position shown, determine ω_{AB} of AB and ω_O of the gear.

Fig. P15.25

Fig. P15.26

$\times$**15.26** The crank OA of a slider-crank mechanism rotates with a constant angular velocity of 3.6 rad/s $\circlearrowright$. Determine ω_{AB} of the connecting rod AB and v_B of the slider B when (a) $\theta = 0$, (b) $\theta = 30°$.

15.27* Refer to Prob. 15.26. Determine ω_{AB} and v_B when (a) $\theta = 90°$, (b) $\theta = 180°$.

Fig. P15.28

Fig. P15.29

15.28 and 15.29 In the position shown, the crank AB rotates with $\omega_{AB} = 10$ rad/s $\circlearrowright$. Suppose that B' is a point of the member CD coinciding with the pin at B. (a) Is the component of v_B of the pin at B in the direction perpendicular to the slot equal to $v_{B'}$? (b) Determine v_D of the point D.

15.30* and 15.31 In the position shown, the link AB rotates with $\omega_{AB} = 6$ rad/s $\circlearrowright$. Determine ω_{BD} and ω_{DE} of the links BD and DE.

Fig. P15.30*

Fig. P15.31

15.32 In the position shown, the end A of the bar AD moves with $\mathbf{v}_A = 1134$ mm/s $\rightarrow$. Determine $\mathbf{v}_B$ of the end B of the bar BD and $\boldsymbol{\omega}_{CD}$ of the bar CD.

15.33* The bar CD of the mechanism shown rotates with $\boldsymbol{\omega}_{CD} = 4.5$ rad/s $\circlearrowleft$. Determine $\boldsymbol{\omega}_{AD}$ and $\boldsymbol{\omega}_{BD}$ of the bars AD and BD.

15.34 The rod AB slides in the sleeve S of the member DE. In the position shown, the link OB rotates with $\boldsymbol{\omega}_{OB} = 3$ rad/s $\circlearrowleft$. Determine $\mathbf{v}_E$ of the point E. (*Hint.* $\boldsymbol{\omega}_{AB} = \boldsymbol{\omega}_{DE}$.)

15.35* In the position shown, the member DE containing a sleeve S rotates with $\boldsymbol{\omega}_{DE} = 4$ rad/s $\circlearrowleft$. Determine $\boldsymbol{\omega}_{OB}$ of the link OB.

15.36 In the position shown, the arm ABC of the mechanism rotates with $\boldsymbol{\omega}_{ABC} = 3$ rad/s $\circlearrowleft$. Determine $\mathbf{v}_P$ of the point P of the smaller gear.

Fig. P15.32 and P15.33*

Fig. P15.34 and P15.35*

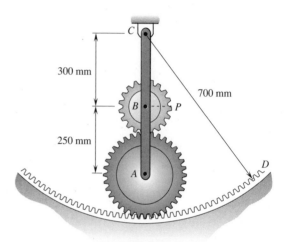

Fig. P15.36

15.7 Velocity Center

An *instantaneous center of zero velocity* of a rigid slab in plane motion is a point of the slab (or its massless extension), denoted by C as shown in Fig. 15.14, which lies in the plane of motion and has a *zero velocity* at the instant considered. The axis which passes through such a center and is perpendic-

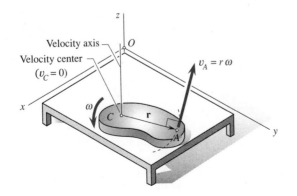

Fig. 15.14

ular to the plane of motion of the slab is called the *instantaneous axis of zero velocity* of the slab at the instant considered. For brevity, the term the *instantaneous center* (or axis) *of zero velocity* of a body will simply be referred to as the *velocity center* (or axis) of the body.[†] Naturally, the velocity of any point A of a rigid slab in plane motion is simply equal to the velocity of the point A relative to the velocity center C of the slab; i.e.,

$$\mathbf{v}_A = \mathbf{v}_{A/C} \tag{15.18}$$

Furthermore, rods, bars, links, cranks, disks, plates, gears, wheels, and bodies of other shapes may be regarded as special cases of the rigid slab in the present study.

In general, each rigid slab in a mechanism has its own velocity center, which is generally different from that of another slab. If two or more points qualify as velocity centers of a slab, the slab must have no velocity at the instant considered. However, the velocity center of a rigid slab does *not* necessarily have zero acceleration.

Fig. 15.15

Theorem of velocity center.[††] *If the velocities* $\mathbf{v}_A$ *and* $\mathbf{v}_B$ *of any two points* A *and* B *of a rigid slab in plane motion are not parallel to each other, then the velocity center* C *of the slab at the instant considered is located at the intersection of* $\overrightarrow{AC}$ *and* $\overrightarrow{BC}$, *where* $\overrightarrow{AC} \perp \mathbf{v}_A$ *and* $\overrightarrow{BC} \perp \mathbf{v}_B$. *(See Fig. 15.15.)*

Proof. To prove that the velocity $\mathbf{v}_C$ of the point C is zero when C is located as described, we first apply Chasles' theorem to write

$$\mathbf{v}_A = \mathbf{v}_{A/C} + \mathbf{v}_C \qquad \mathbf{v}_B = \mathbf{v}_{B/C} + \mathbf{v}_C \tag{15.19}$$

As $\overrightarrow{AC} \perp \mathbf{v}_A$ and $\overrightarrow{BC} \perp \mathbf{v}_B$, we infer from Eqs. (15.16) that

$$\mathbf{v}_{A/C} \parallel \mathbf{v}_A \qquad \mathbf{v}_{B/C} \parallel \mathbf{v}_B \tag{15.20}$$

For the conditions in (15.20) and Eqs. (15.19) to hold true simultaneously, we must require that

$$\mathbf{v}_C \parallel \mathbf{v}_A \qquad \mathbf{v}_C \parallel \mathbf{v}_B \tag{15.21}$$

Since $\mathbf{v}_A$ and $\mathbf{v}_B$ are, as stated, *not* parallel to each other and the conditions in (15.21) must be true, we logically conclude that $\mathbf{v}_C = \mathbf{0}$.   Q.E.D.

Corollary I. (Refer to the theorem.) *The speeds of A and B are* $v_A = \overline{CA}\omega$ *and* $v_B = \overline{CB}\omega$, *where* ω *is the angular speed of the slab at the instant considered.*

Corollary II. (Refer to the theorem.) *If* $\mathbf{v}_A$ *is not perpendicular to the line AB, but* $\mathbf{v}_A \parallel \mathbf{v}_B$, *then the velocity center* C *is located at infinity and the angular velocity of the slab is zero.*

Corollary III. (Refer to the theorem.) *If* $\mathbf{v}_A \parallel \mathbf{v}_B$ *and* $\mathbf{v}_A \perp \overline{AB}$, *then the velocity center* C *is located at the intersection of the line AB and the line joining the extremities of the vectors* $\mathbf{v}_A$ *and* $\mathbf{v}_B$ *which are drawn to scale from A and B, respectively, as shown in Fig. 15.16.*

[†]Cf. the *acceleration center* and the *acceleration axis* in Sec. 15.9.

[††]In the text, the symbols $\perp$ and $\parallel$ denote "is perpendicular to" and "is parallel to," respectively.

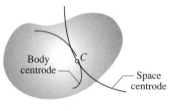

Fig. 15.16 (a) (b)

As a slab undergoes a general plane motion, the position of the velocity center C will generally change from time to time. The curve representing the locus of the velocity center in space is called the *space centrode*, and the curve representing the locus of the velocity center on the slab is called the *body centrode*. These two centrodes are illustrated in Fig. 15.17 for a general slab and in Fig. 15.18 for a wheel rolling without slipping on a rigid surface, where the *circumference* of the wheel is the body centrode and the *track* of the wheel is the space centrode. Note that these two centrodes are always tangent to each other at the velocity center C, and that, as the motion proceeds, *the body centrode appears to roll on the space centrode.*

Fig. 15.17 **Fig. 15.18**

EXAMPLE 15.8

Using *velocity center*, work Example 15.5.

Solution. From the constraint condition of the rod AB, we note that $\mathbf{v}_A$ must be directed downward when $\mathbf{v}_B$ is directed to the right as given. Thus, the velocity center C of the rod is the point of intersection of the lines AC and BC as shown, where $\overrightarrow{AC} \perp \mathbf{v}_A$ and $\overrightarrow{BC} \perp \mathbf{v}_B$. For computing velocities at the instant considered, the rod may be assumed to rotate about C as a fixed point as shown. We write

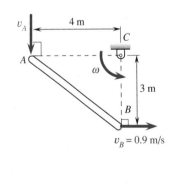

$$v_B = 0.9 = \overline{CB}\omega = 3\omega \qquad \omega = 0.3 \text{ rad/s} \ \circlearrowright \ \blacktriangleleft$$

$$v_A = \overline{CA}\omega = 4(0.3) \qquad \mathbf{v}_A = 1.2 \text{ m/s} \downarrow \ \blacktriangleleft$$

Developmental Exercises

D15.14   Define the terms: (a) velocity center, (b) space centrode, (c) body centrode.

D15.15   Describe the *theorem of velocity center* and its three corollaries.

EXAMPLE 15.9

Using *velocity center*, work Example 15.6.

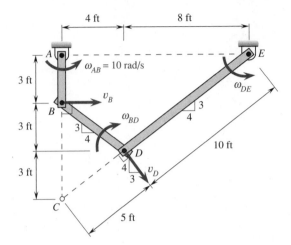

Solution. The system in Fig. 15.13 together with the given data in Example 15.6 may be drawn and indicated as shown above. Since the pins at A and E are fixed, the points A and E are the *velocity centers* of the links AB and DE, respectively. Next, we note that $\mathbf{v}_B \perp \overrightarrow{BA}$ and $\mathbf{v}_D \perp \overrightarrow{DE}$. Thus, by the theorem of velocity center, the *velocity center* of the link BD is the point of intersection C of the lines AB and DE extended as shown. For computing velocities at the instant considered, the links AB, BD, and DE may be assumed to rotate about their velocity centers A, C, and E, respectively, as fixed points as shown in Figs. (a), (b), and (c).

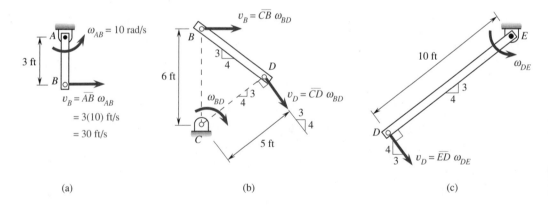

(a)   (b)   (c)

From Figs. (a) and (b), we write

$$v_B = 30 = \overline{CB}\,\omega_{BD} = 6\omega_{BD}$$

$$\omega_{BD} = 5 \text{ rads/s} \;\mathord{\circlearrowright} \quad \blacktriangleleft$$

From Figs. (b) and (c), we write

$$v_D = \overline{CD}\,\omega_{BD} = 5(5) = \overline{ED}\,\omega_{DE} = 10\omega_{DE}$$

$$\mathbf{v}_D = 25 \text{ ft/s} \;\measuredangle\; 53.1° \qquad \omega_{DE} = 2.5 \text{ rad/s} \;\mathord{\circlearrowright} \quad \blacktriangleleft$$

Developmental Exercise

D15.16 If a point P of a body in plane motion has a zero velocity, is the point P the *velocity center* of the body?

EXAMPLE 15.10

A drum of radius 120 mm is mounted on a wheel of radius 180 mm as shown, where a rope is wound around the drum. The end E of the rope is pulled with a constant velocity $\mathbf{v}_E = 0.6$ m/s $\rightarrow$, and the wheel rolls without slipping. Using *velocity center*, determine (a) $\mathbf{v}_D$ of the point D, (b) the rate at which the rope is being wound or unwound.

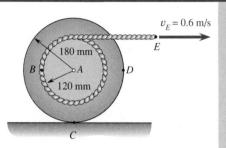

Solution. As the wheel rolls without slipping, the point C of the wheel in contact with the support has zero velocity; therefore, the point C is the *velocity center* of the body consisting of the wheel and the drum. Since the rope may be taken as inextensible, the point F on the top of the drum and the end E of the rope should have the same velocity; i.e.,

$$\mathbf{v}_F = \mathbf{v}_E = 0.6 \text{ m/s} \rightarrow$$

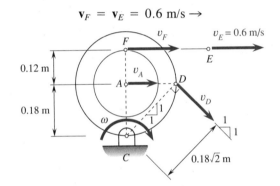

For computing velocities at the instant considered, the wheel and the drum may be assumed to rotate with $\boldsymbol{\omega}$ about the velocity center C as a fixed point as shown. We write

$$v_F = \overline{CF}\omega = 0.3\omega = 0.6 \qquad \boldsymbol{\omega} = 2 \text{ rad/s } \curvearrowleft$$

$$v_D = \overline{CD}\omega = 0.18\sqrt{2}(2)$$

$$\mathbf{v}_D = 0.509 \text{ m/s } \searrow 45° \blacktriangleleft$$

$$v_A = \overline{CA}\omega = 0.18(2) \qquad \mathbf{v}_A = 0.36 \text{ m/s} \rightarrow$$

Since $v_E > v_A$, we see that the rope is being *unwound* at the rate of

$$v_{E/A} = v_E - v_A = 0.6 - 0.36$$

$$v_{E/A} = 0.24 \text{ m/s } \blacktriangleleft$$

Developmental Exercises

D15.17 Is the acceleration $\mathbf{a}_C$ of the velocity center C of a rigid slab also equal to zero at the instant considered?

D15.18 Refer to Example 15.10. Using *velocity center*, determine $\mathbf{v}_B$.

PROBLEMS

15.37* through 15.50* Using *velocity center*, solve the following problems:

15.37* Prob. 15.17.	15.38* Prob. 15.18*.
15.39* Prob. 15.19.	15.40* Prob. 15.22.
15.41* Prob. 15.25.	15.42* Prob. 15.26.
15.43* Prob. 15.27*.	15.44* Prob. 15.30*.
15.45* Prob. 15.31.	15.46* Prob. 15.32.
15.47* Prob. 15.33*.	15.48* Prob. 15.34.
15.49* Prob. 15.35*.	15.50* Prob. 15.36.

15.51 A driving wheel of a car is shown. The car travels at 15 mi/h → while the wheel spins with ω = 300 rpm ↺ on a slippery road. Determine (a) the position of the velocity center C of the wheel, (b) $\mathbf{v}_B$ of the point B of the wheel in contact with the road.

23 in.

Fig. P15.51   **Fig. P15.52***

15.52* A rigid slab is in plane motion with an angular velocity ω as shown. Knowing the vector identity $\mathbf{P} \times (\mathbf{Q} \times \mathbf{R}) = \mathbf{Q}(\mathbf{P \cdot R}) - \mathbf{R}(\mathbf{P \cdot Q})$, show that $\overrightarrow{AC} = \omega \times \mathbf{v}_A/\omega^2$, where C is the velocity center and A is a particle of the slab.

15.53 At the instant shown, the quadrilateral plate in plane motion has an angular velocity of 3 rad/s ↻ and its corner A has a velocity $\mathbf{v}_A$ of 0.45 m/s as indicated. Determine (a) the distance $\overline{CA}$ between the velocity center C and the corner A, (b) $\mathbf{v}_B$ of the corner B, (c) $\mathbf{v}_D$ of the corner D.

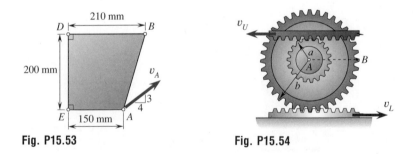

Fig. P15.53   **Fig. P15.54**

15.54 A double gear is engaged with an upper rack and a lower rack as shown. The upper rack is pulled with $\mathbf{v}_U$ = 260 mm/s ← while the lower rack is pulled with $\mathbf{v}_L$ = 100 mm/s →. If a = 60 mm and b = 120 mm, determine (a) the distance between the velocity center C and the center A of the gear, (b) ω of the gear, (c) $\mathbf{v}_B$ of the point B.

Fig. P15.55   **Fig. P15.56***

15.55 and 15.56* The end D of the string wound around the core of a yoyo is pulled with $\mathbf{v}_D$ as shown. If the yoyo rolls without slipping, determine $\boldsymbol{\omega}$ of the yoyo and $\mathbf{v}_B$ of the point B.

15.57 The end A of the slender rod AB leaning against a fixed semicylindrical surface as shown is pulled with $\mathbf{v}_A = 0.6$ m/s $\rightarrow$. If $x = 2.25$ m, determine $\boldsymbol{\omega}_{AB}$ of AB and $\mathbf{v}_B$ of the end B.

Fig. P15.57   **Fig. P15.58**

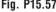**15.58** At the instant shown, the collar A slides with $\mathbf{v}_A = 1.5$ ft/s $\rightarrow$. Determine (a) $\boldsymbol{\omega}_{AB}$ of the rod AB, (b) $\mathbf{v}_D$ of the point D in contact with the roller support, (c) $\mathbf{v}_B$ of the end B.

15.59* The rod AB is mounted on sliders as shown, where the slider at A is moved with $\mathbf{v}_A = 0.24$ m/s $\rightarrow$. If $x = 0.3$ m, determine $\boldsymbol{\omega}_{AB}$ of the rod and $\mathbf{v}_B$ of the end B.

15.60 Solve Prob. 15.59* if $x = 0.4$ m.

15.61 At the instant shown, the crank OA rotates with $\boldsymbol{\omega}_{OA} = 5$ rad/s $\circlearrowright$. Determine $\mathbf{v}_B$ of collar B and $\mathbf{v}_M$ of the midpoint M of the link AB.

15.62* At the instant shown, the collar B moves with $\mathbf{v}_B = 7$ in./s $\downarrow$. Determine $\boldsymbol{\omega}_{OA}$ of the crank OA and $\mathbf{v}_M$ of the midpoint M of the link AB.

Fig. P15.59*

Fig. P15.61 and P15.62*

Fig. P15.63 and P15.64*

50 mm

50 mm

150 mm

15.63 The gear D shown rotates about O with $\boldsymbol{\omega}_D = 4$ rad/s $\circlearrowright$, while the arm AOB rotates about O with $\boldsymbol{\omega}_{AOB} = 2$ rad/s $\circlearrowleft$. Determine $\boldsymbol{\omega}_A$ of the smaller gear at A and $\boldsymbol{\omega}_B$ of the smaller gear at B.

15.64* The gear D shown rotates about O with a constant angular velocity $\boldsymbol{\omega}_D = 6$ rad/s $\circlearrowleft$. Determine the angular velocity $\boldsymbol{\omega}_{AOB}$ of the arm AOB for which the smaller gears at A and B will have curvilinear translation.

15.65 At the instant shown, the crank AB rotates with $\boldsymbol{\omega}_{AB} = 2.4$ rad/s $\circlearrowright$. Determine $\boldsymbol{\omega}_{DE}$ of the crank DE and $\mathbf{v}_F$ of the point F of the link BD.

Fig. P15.65 and P15.66

200 mm

300 mm

110 mm

320 mm 160 mm

15.66 At the instant shown, the pin at D moves with a speed of 408 mm/s. Determine (a) the angular speed of the crank AB, (b) the speed of the point F of the link BD.

15.8   Accelerations in Relative Motion

Suppose that the points A and B are two different points of a rigid slab in plane motion with an angular velocity $\boldsymbol{\omega} = \omega\mathbf{k}$ and an angular acceleration $\boldsymbol{\alpha} = \alpha\mathbf{k}$, as shown in Fig. 15.19. By Chasles' theorem, we may view the plane motion of the rigid slab as a rotation of the slab about A plus a translation of the slab with A. Such a viewpoint is depicted in the graphical equation shown. Thus, we may express the acceleration $\mathbf{a}_B$ of B in the ref-

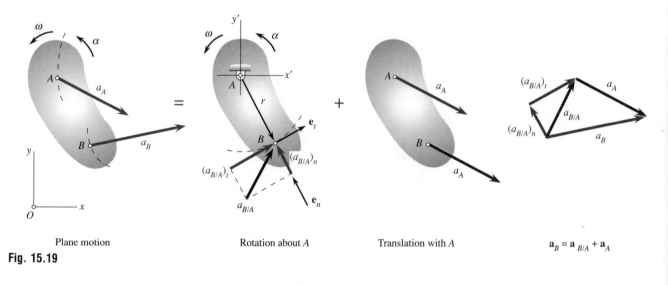

Plane motion

Rotation about A

Translation with A

$\mathbf{a}_B = \mathbf{a}_{B/A} + \mathbf{a}_A$

Fig. 15.19

erence frame Oxy as the acceleration $\mathbf{a}_{B/A}$ of B relative to A plus the acceleration $\mathbf{a}_A$ of A; i.e.,

$$\mathbf{a}_B = \mathbf{a}_{B/A} + \mathbf{a}_A \qquad (15.22)^{\dagger}$$

Recall that, more precisely, $\mathbf{a}_{B/A}$ denotes the acceleration of B in a nonrotating reference frame (e.g., $Ax'y'$) which translates with A.

Clearly, $\mathbf{a}_{B/A}$ is due solely to the rotational part of the motion of the slab. As a matter of fact, we may view $\mathbf{a}_{B/A}$ as the acceleration of the tip B of the radius vector $\overrightarrow{AB}$ which rotates with the angular velocity $\boldsymbol{\omega}$ and the angular acceleration $\boldsymbol{\alpha}$ about A as a "fixed point." Thus, $\mathbf{a}_{B/A}$ *is equivalent to the acceleration of B in circular motion about A as a fixed center, and is composed of a tangential component* $(\mathbf{a}_{B/A})_t$ *and a normal component* $(\mathbf{a}_{B/A})_n$; i.e.,

$$\mathbf{a}_{B/A} = (\mathbf{a}_{B/A})_t + (\mathbf{a}_{B/A})_n \qquad (15.23)$$

$$(\mathbf{a}_{B/A})_t = r\alpha\mathbf{e}_t \qquad (\mathbf{a}_{B/A})_n = r\omega^2\mathbf{e}_n \qquad (15.24)$$

where $r = \overline{AB}$, and $\mathbf{e}_t$ and $\mathbf{e}_n$ are the tangential and normal unit vectors pertaining to the circular arc at B, respectively, as indicated in Fig. 15.19. Since $\boldsymbol{\omega} = \omega\mathbf{k}$ and $\boldsymbol{\alpha} = \alpha\mathbf{k}$, we may use the cross product of vectors to write

$$(\mathbf{a}_{B/A})_t = \boldsymbol{\alpha} \times \mathbf{r} \qquad (\mathbf{a}_{B/A})_t = \alpha\mathbf{k} \times \mathbf{r} \qquad (15.25)$$

$$(\mathbf{a}_{B/A})_n = \boldsymbol{\omega} \times (\boldsymbol{\omega} \times \mathbf{r}) \qquad (\mathbf{a}_{B/A})_n = -\omega^2\mathbf{r} \qquad (15.26)$$

where $\mathbf{r} = \overrightarrow{AB}$. Note that (1) *the magnitude of* $(\mathbf{a}_{B/A})_t$ *is* $r\alpha$, (2) *the magnitude of* $(\mathbf{a}_{B/A})_n$ *is* $r\omega^2$, (3) *the direction of* $(\mathbf{a}_{B/A})_t$ *is perpendicular to* $\overrightarrow{AB}$, (4) *the vectors* $\boldsymbol{\alpha}$, $\overrightarrow{AB}$, *and* $(\mathbf{a}_{B/A})_t$ *(in this order) form a right-handed trihedral, and* (5) *the normal component* $(\mathbf{a}_{B/A})_n$ *is always directed from B toward A regardless of the senses of* $\boldsymbol{\omega}$ *and* $\boldsymbol{\alpha}$.

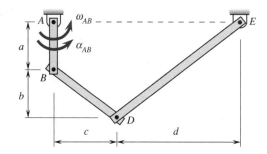

Fig. 15.20

Suppose that the link AB in Fig. 15.13 rotates with $\boldsymbol{\omega}_{AB}$ and $\boldsymbol{\alpha}_{AB}$ as shown in Fig. 15.20, where the pins at A and E are fixed. Taking the constraint conditions $\mathbf{a}_A = \mathbf{a}_E = \mathbf{0}$ into account and applying Eq. (15.22), we write

$$\mathbf{a}_B = \mathbf{a}_{B/A} + \mathbf{a}_A = \mathbf{a}_{B/A}$$

$$\mathbf{a}_B = \mathbf{a}_{B/D} + \mathbf{a}_D = \mathbf{a}_{B/D} + (\mathbf{a}_{D/E} + \mathbf{a}_E) = \mathbf{a}_{B/D} + \mathbf{a}_{D/E}$$

---

†This equation is equivalent to Eq. (11.37). However, the order of terms in Eq. (15.22) is preferable to that in Eq. (11.37) in establishing the *linkage equation* for accelerations of a system, such as Eq. (15.27).

Thus, we have

$$\mathbf{a}_{B/A} = \mathbf{a}_{B/D} + \mathbf{a}_{D/E} \qquad (15.27)^\dagger$$

which is the *linkage equation* for accelerations of the system shown in Fig. 15.20. Such an equation may be used as a basis for determining the linear and angular accelerations in a system.

EXAMPLE 15.11

The end B of the rod AB is pulled with a constant velocity $\mathbf{v}_B = 0.9$ m/s $\rightarrow$. For the position shown, determine $\boldsymbol{\alpha}$ of the rod and $\mathbf{a}_A$ of the end A of the rod.

Solution. We first need to know $\boldsymbol{\omega}$ of the rod. From the solution in Example 15.5 or 15.8, we have $\boldsymbol{\omega} = 0.3$ rad/s $\circlearrowright$, which is indicated in the sketch, where $\mathbf{a}_B = \mathbf{0}$. The constraint condition of the rod prompts us to initially assume that $\mathbf{a}_A = a_A \downarrow$ and $\boldsymbol{\alpha} = \alpha \circlearrowright$ as indicated. Thus, we write

$$\mathbf{a}_B = \mathbf{a}_{B/A} + \mathbf{a}_A$$

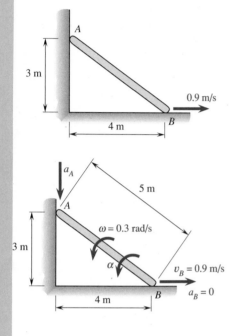

From this vector equation, we write

$$\xrightarrow{+} \Sigma V_x: \qquad 0 = \tfrac{3}{5}(5\alpha) - \tfrac{4}{5}(5)(0.3)^2 + 0 \qquad \alpha = 0.12$$

$$+\uparrow \Sigma V_y: \qquad 0 = \tfrac{4}{5}(5\alpha) + \tfrac{3}{5}(5)(0.3)^2 - a_A \qquad a_A = 0.75$$

$$\boldsymbol{\alpha} = 0.12 \text{ rad/s}^2 \circlearrowright \qquad \mathbf{a}_A = 0.75 \text{ m/s}^2 \downarrow \quad \blacktriangleleft$$

Developmental Exercise

D15.19 Determine $\boldsymbol{\alpha}$ and $\mathbf{a}_A$ in Example 15.11 if $\mathbf{a}_B = 0.6$ m/s$^2 \rightarrow$ and $\mathbf{v}_B = 0.9$ m/s $\rightarrow$.

EXAMPLE 15.12

Refer to the system in Fig. 15.20. If $a = b = 3$ ft, $c = 4$ ft, $d = 8$ ft, and the link AB rotates with $\boldsymbol{\alpha}_{AB} = 12$ rad/s$^2 \circlearrowright$ as well as $\boldsymbol{\omega}_{AB} = 10$ rad/s $\circlearrowright$ in the position shown, determine (a) $\boldsymbol{\alpha}_{BD}$ and $\boldsymbol{\alpha}_{DE}$ of the links BD and DE, (b) $\mathbf{a}_D$ of the pin at D.

---

†Cf. Eq. (15.17).

Solution. We first need to know ω_{BD} and ω_{DE} in the position shown. From the solution in Example 15.6 or 15.9, we have $\omega_{BD} = 5$ rad/s $\circlearrowright$ and $\omega_{DE} = 2.5$ rad/s $\circlearrowleft$, which are indicated in the sketch.

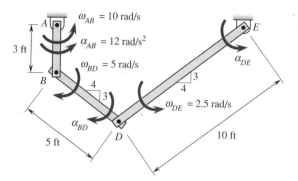

The directions of $\boldsymbol{\alpha}_{BD}$ and $\boldsymbol{\alpha}_{DE}$ of the links BD and DE may initially be assumed as indicated. Since the *linkage equation* for accelerations of the system has already been derived as Eq. (15.27), we write

$$\mathbf{a}_{B/A} = \mathbf{a}_{B/D} + \mathbf{a}_{D/E}$$

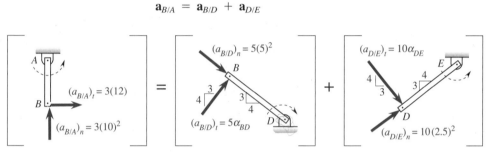

From this vector equation, we write

$$\overset{+}{\rightarrow} \Sigma V_x: \quad 3(12) = \tfrac{3}{5}(5\alpha_{BD}) + \tfrac{4}{5}(5)(5)^2$$
$$+ \tfrac{3}{5}(10\alpha_{DE}) + \tfrac{4}{5}(10)(2.5)^2$$

$$+\uparrow \Sigma V_y: \quad 3(10)^2 = \tfrac{4}{5}(5\alpha_{BD}) - \tfrac{3}{5}(5)(5)^2$$
$$- \tfrac{4}{5}(10\alpha_{DE}) + \tfrac{3}{5}(10)(2.5)^2$$

which yield $\alpha_{BD} = 23.188$ and $\alpha_{DE} = -30.594$. Since α_{DE} is negative, the true sense of $\boldsymbol{\alpha}_{DE}$ is opposite to that assumed. Thus, we write

$$\boldsymbol{\alpha}_{BD} = 23.2 \text{ rad/s}^2 \circlearrowright \qquad \boldsymbol{\alpha}_{DE} = 30.6 \text{ rad/s}^2 \circlearrowright \quad \blacktriangleleft$$

To find the acceleration of the pin at D, we write

$$\mathbf{a}_D = \mathbf{a}_{D/E} + \mathbf{a}_E = \mathbf{a}_{D/E} + \mathbf{0} = \mathbf{a}_{D/E}$$

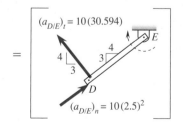

$$\mathbf{a}_D = [-\tfrac{3}{5}(10)(30.594) + \tfrac{4}{5}(10)(2.5)^2]\mathbf{i}$$
$$+ [\tfrac{4}{5}(10)(30.594) + \tfrac{3}{5}(10)(2.5)^2]\mathbf{j}$$
$$= -133.6\mathbf{i} + 282.3\mathbf{j}$$

$$\mathbf{a}_D = 312 \text{ ft/s}^2 \; \diagdown \; 64.7° \quad \blacktriangleleft$$

Developmental Exercises

D15.20 Refer to Example 15.12. Determine $\mathbf{a}_B$.

D15.21 Refer to Example 15.12. Determine the acceleration of the midpoint M of the bar DE.

EXAMPLE 15.13

Refer to Example 15.10. Suppose that the end E of the rope wound around the drum is pulled with $\mathbf{a}_E = 1.5 \text{ m/s}^2 \rightarrow$ as well as $\mathbf{v}_E = 0.6 \text{ m/s} \rightarrow$ at the instant shown, and the wheel rolls without slipping. For this instant, determine the accelerations of points A, C, and F of the wheel.

Solution. The corresponding angular velocity of the wheel has been determined in the solution of Example 15.10 as $\boldsymbol{\omega} = 2 \text{ rad/s} \;\circlearrowright$ which is indicated in the sketch. Furthermore, we may assume that the unknown angular acceleration $\boldsymbol{\alpha}$ of the wheel to be clockwise as indicated. From the constraints on the wheel, we recognize the following conditions:

(a) The center A of the wheel must move in the horizontal direction; i.e.,

$$\mathbf{a}_A = a_A \rightarrow \tag{1}$$

(b) As the wheel rolls without slipping, the point C of the wheel in contact with the support must have no tangential component of acceleration. However, C may have a normal component of acceleration; i.e.,

$$\mathbf{a}_C = a_C \uparrow \tag{2}$$

(c) The tangential component of the acceleration of the point F of the wheel should be equal to $\mathbf{a}_E$; i.e., $(\mathbf{a}_F)_x = \mathbf{a}_E = 1.5 \text{ m/s}^2 \rightarrow$. Assuming the normal component of $\mathbf{a}_F$ to be downward, we write

$$\mathbf{a}_F = [1.5 \rightarrow] + [(a_F)_y \downarrow] \tag{3}$$

Since $\mathbf{a}_F = \mathbf{a}_{F/A} + \mathbf{a}_A$, we write

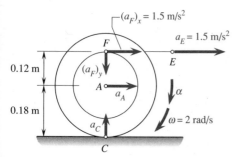

$$[1.5 \rightarrow] + [(a_F)_y \downarrow] = \begin{bmatrix} \\ \\ \end{bmatrix} + [a_A \rightarrow]$$

(with diagram showing $0.12(2)^2$ downward at F, 0.12α to the right, pivot at A)

$$+\downarrow \Sigma V_y: \qquad (a_F)_y = 0.12(2)^2 \qquad (4)$$

$$\xrightarrow{+} \Sigma V_x: \qquad 1.5 = 0.12\alpha + a_A \qquad (5)$$

Moreover, $\mathbf{a}_F = \mathbf{a}_{F/C} + \mathbf{a}_C$. Thus, we write

$$[1.5 \rightarrow] + [(a_F)_y \downarrow] = \begin{bmatrix} \\ \\ \end{bmatrix} + [a_C \uparrow]$$

(with diagram showing $0.3(2)^2$ downward at F, 0.3α to the right, pivot at C)

$$\xrightarrow{+} \Sigma V_x: \qquad 1.5 = 0.3\alpha \qquad (6)$$

$$+\downarrow \Sigma V_y: \qquad (a_F)_y = 0.3(2)^2 - a_C \qquad (7)$$

Equations (4) through (7) yield $(a_F)_y = 0.48$, $\alpha = 5$, $a_C = 0.72$, and $a_A = 0.9$. By Eqs. (1) through (3), we write

$$\mathbf{a}_A = 0.9 \text{ m/s}^2 \rightarrow \qquad \mathbf{a}_C = 0.72 \text{ m/s}^2 \uparrow \quad \blacktriangleleft$$

$$\mathbf{a}_F = 1.5\mathbf{i} - 0.48\mathbf{j} \qquad \mathbf{a}_F = 1.575 \text{ m/s}^2 \searrow 17.74° \quad \blacktriangleleft$$

REMARK. As $\mathbf{a}_C \neq 0$ and the support is stationary, we see that when a body rolls without slipping on another body, the points of contact have *equal* velocities and *equal* tangential components of acceleration, but *unequal* normal components of acceleration.

Developmental Exercises

D15.22   Refer to Example 15.13. Determine $\mathbf{a}_B$.

D15.23   Refer to Example 15.13. Determine $\mathbf{a}_D$.

PROBLEMS†

15.67   Refer to Prob. 15.17. In the position shown, the end B of the rod AB slides with $\mathbf{v}_B = 2.8$ m/s $\rightarrow$ and $\mathbf{a}_B = 0.9$ m/s^2 $\rightarrow$. For this position, determine $\boldsymbol{\alpha}$ of the rod and $\mathbf{a}_A$ of the end A.

---

†In this problem set, we include systems which are formed by taking accelerations into account in a number of previous systems.

Fig. P15.70

15.68 Refer to Prob. 15.18*. In the position shown, the center A of the gear moves with $v_A = 1.5$ m/s $\rightarrow$ and $a_A = 1.2$ m/s^2 $\rightarrow$. For this position, determine the accelerations of the points B, C, and E of the gear.

15.69* Refer to Prob. 15.19. In the position shown, the end D of the cable is pulled with $v_D = 0.9$ m/s $\rightarrow$ and $a_D = 0.63$ m/s^2 $\rightarrow$. For this position, determine the accelerations of the points A, B, and C of the gear.

15.70 A triangular plate moves in the xy plane. In the position shown, the corners P and Q have accelerations $\mathbf{a}_P = -14\mathbf{i} - 3\mathbf{j}$ m/s^2 and $\mathbf{a}_Q = 17\mathbf{i} - 11\mathbf{j}$ m/s^2, respectively. For this position, determine (a) $\boldsymbol{\omega}$ and $\boldsymbol{\alpha}$ of the plate, (b) $\mathbf{a}_R$ of the corner R.

15.71 Refer to Prob. 15.70. For the position shown, determine the coordinates of the point $Z(x, y)$ of the plate whose acceleration $\mathbf{a}_Z = \mathbf{0}$.

15.72* Refer to Prob. 15.22. In the position shown, the crank OB rotates with $\omega_{OB} = 60$ rpm and $\alpha_{OB} = 10$ rad/s^2. For this position, determine α_{AB} of the connecting rod AB and $\mathbf{a}_P$ of the position P.

15.73 Refer to Prob. 15.23. At the instant shown, the annular gear A is driven with $\omega_A = 120$ rpm and $\alpha_A = 20$ rad/s^2. For $a = 60$ mm and $b = 40$ mm, determine α_P and α_C of the planet gear P and the carrier C, respectively, and $\mathbf{a}_T$ of the particle T which belongs to the planet gear P and is in contact with the annular gear A at the instant shown.

15.74 Refer to Prob. 15.25. In the position shown, the end A of the rack AB moves with $v_A = 8.75$ in./s $\rightarrow$ and $a_A = 15$ in./s^2 $\rightarrow$. For this position, determine α_{AB} of the rack AB and α_O of the gear O.

15.75 In the position shown, the crank OA of a slider-crank mechanism rotates with $\omega_{OA} = 6$ rad/s and $\alpha_{OA} = 5$ rad/s^2. For this position, determine α_{AB} of the connecting rod AB and $\mathbf{a}_B$ of the slider B.

Fig. P15.75

Fig. P15.76*   **Fig. P15.77**

15.76* and **15.77** In the position shown, the crank AB of a slider-crank mechanism rotates with $\omega_{AB} = 4$ rad/s and $\alpha_{AB} = 10$ rad/s^2. For this position, determine α_{BD} of the connecting rod BD and $\mathbf{a}_D$ of the slider D.

15.78 and **15.79*** In the position shown, the link AB rotates with $\omega_{AB} = 1.2$ rad/s and $\alpha_{AB} = 3$ rad/s^2. For this position, determine α_{BD} of the link BD and $\mathbf{a}_D$ of the pin at D.

Fig. P15.78 and P15.80   Fig. P15.79* and P15.81*

15.80 and 15.81* In the position shown, the link AB is accelerated from rest (i.e., $\omega_{AB} = 0$) with $\alpha_{AB} = 3$ rad/s² $\circlearrowright$. For this position, determine α_{BD} of the link BD and $\mathbf{a}_D$ of the pin at D.

15.82 Refer to Prob. 15.32. The end A of the bar AD moves with a constant velocity $\mathbf{v}_A = 1134$ mm/s $\rightarrow$. For the position shown, determine (a) $\mathbf{a}_B$ of the end B of the bar BD, (b) the angular acceleration of each bar.

15.83 Refer to Prob. 15.32. In the position shown, the end A of the bar AD is accelerated from rest (i.e., $\mathbf{v}_A = 0$) with $\mathbf{a}_A = 2.268$ m/s² $\rightarrow$. For this position, determine (a) $\mathbf{a}_B$ and $\mathbf{a}_E$ of the points B and the midpoint E of AD, (b) the angular acceleration of each bar.

15.84* In the position shown, the center O of the gear moves with $\mathbf{v}_O = 19.2$ in./s $\leftarrow$ and $\mathbf{a}_O = 24$ in./s² $\rightarrow$ over the stationary rack. For this position, determine α_{AB} and α_{BC} of the links AB and BC.

Fig. P15.84* and P15.85

15.85 In the position shown, the center O of the gear is accelerated from rest (i.e., $\mathbf{v}_O = 0$) with $\mathbf{a}_O = 48$ in./s² $\leftarrow$ over the stationary rack. For this position, determine (a) α_{AB} and α_{BC} of the links AB and BC, (b) $\mathbf{a}_M$ of the midpoint M of AB.

15.86 Refer to Prob. 15.36. In the position shown, the arm ABC of the mechanism rotates with $\omega_{ABC} = 3$ rad/s $\circlearrowleft$ and $\alpha_{ABC} = 6$ rad/s² $\circlearrowright$. For this position, determine (a) α_A of the gear at A, (b) α_B of the gear at B, (c) $\mathbf{a}_P$ of the point P.

15.9   Acceleration Center

An *instantaneous center of zero acceleration* of a rigid slab in plane motion is a point of the slab (or its massless extension), denoted by Z as shown in Fig. 15.21, which lies in the plane of motion and has a *zero acceleration* at the instant considered. The axis which passes through such a center and is perpendicular to the plane of motion of the slab is called the *instantaneous axis of zero acceleration* of the slab at the instant considered. For brevity, the term the *instantaneous center* (or axis) *of zero acceleration* of a body will simply be referred to as the *acceleration center* (or axis) of the body. Naturally, the acceleration of any point A of a rigid slab in plane motion is simply equal to the acceleration of the point A relative to the acceleration center Z of the slab; i.e.,

$$\mathbf{a}_A = \mathbf{a}_{A/Z} \qquad\qquad (15.28)$$

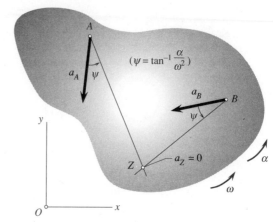

Fig. 15.21

Theorem of acceleration center.[†] *If the accelerations $\mathbf{a}_A$ and $\mathbf{a}_B$ of any two points A and B of a rigid slab, which is in plane motion with an angular velocity $\boldsymbol{\omega}$ and an angular acceleration $\boldsymbol{\alpha}$, are not parallel to each other, then the acceleration center Z of the slab at the instant considered is located at the intersection of $\overline{AZ}$ and $\overline{BZ}$, which make an angle ψ with $\mathbf{a}_A$ and $\mathbf{a}_B$, respectively, where*

$$\psi = \tan^{-1}\left(\frac{\alpha}{\omega^2}\right) \tag{15.29}$$

and the set of vectors $\mathbf{a}_A$, $\overrightarrow{AZ}$, and $\boldsymbol{\alpha}$, as well as the set of vectors $\mathbf{a}_B$, $\overrightarrow{BZ}$, and $\boldsymbol{\alpha}$, forms a right-handed trihedral. (See Fig. 15.21.)

Proof. To prove that the acceleration $\mathbf{a}_Z$ of the point Z is zero when Z is located as described, we first apply Chasles' theorem to write

$$\mathbf{a}_A = \mathbf{a}_{A/Z} + \mathbf{a}_Z \qquad \mathbf{a}_B = \mathbf{a}_{B/Z} + \mathbf{a}_Z \tag{15.30}$$

Then, as shown in Fig. 15.22, we let ψ' be the angle between $\mathbf{a}_{A/Z}$ and $\overline{AZ}$.

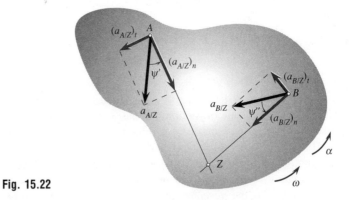

Fig. 15.22

[†]This theorem and its corollary, which follows, were presented in the authors' paper, "Teaching the Concepts of Acceleration Center and Virtual Work in Dynamics," *Engineering Education*, vol. 79, no. 3, April 1989, pp. 441–443. Furthermore, note that this theorem is mainly intended to provide a general basis for the corollary. We shall emphasize the application of the corollary.

The magnitude of the tangential and normal components of $\mathbf{a}_{A/Z}$ are

$$(a_{A/Z})_t = \overline{AZ}\alpha \qquad (a_{A/Z})_n = \overline{AZ}\omega^2$$

Thus, we have

$$\tan\psi' = \frac{(a_{A/Z})_t}{(a_{A/Z})_n} = \frac{\overline{AZ}\alpha}{\overline{AZ}\omega^2} \qquad \psi' = \tan^{-1}\left(\frac{\alpha}{\omega^2}\right)$$

Letting ψ'' be the angle between $\mathbf{a}_{B/Z}$ and $\overline{BZ}$ as shown in Fig. 15.22, we can, in a similar manner, show that $\psi'' = \tan^{-1}(\alpha/\omega^2)$. From the angle ψ as given in Eq. (15.29), we see that $\psi = \psi' = \psi''$. Therefore,

$$\mathbf{a}_{A/Z} \parallel \mathbf{a}_A \qquad \mathbf{a}_{B/Z} \parallel \mathbf{a}_B \qquad (15.31)$$

For the conditions in (15.31) and Eqs. (15.30) to hold true simultaneously, we must require that

$$\mathbf{a}_Z \parallel \mathbf{a}_A \qquad \mathbf{a}_Z \parallel \mathbf{a}_B \qquad (15.32)$$

Since $\mathbf{a}_A$ and $\mathbf{a}_B$ are, as stated, *not* parallel to each other and the conditions in (15.32) must be true, we logically conclude that $\mathbf{a}_Z = \mathbf{0}$. Q.E.D.

Corollary. (Refer to the theorem.) *If a rigid slab at an instant is accelerated to undergo a plane motion from zero angular velocity, then $\omega = 0$, $\alpha \neq 0$, $\psi = 90°$, $\overrightarrow{AZ} \perp \mathbf{a}_A$, and $\overrightarrow{BZ} \perp \mathbf{a}_B$, where $\mathbf{a}_A$ and $\mathbf{a}_B$ are accelerations of the points A and B of the slab, and Z is the acceleration center of the slab at that instant. (See Fig. 15.23.) Furthermore, for any point P of the slab at that instant, we have*

$$\boxed{a_P = \overline{ZP}\alpha} \qquad \boxed{\mathbf{a}_P = \boldsymbol{\alpha} \times \overrightarrow{ZP}} \qquad (15.33)^\dagger$$

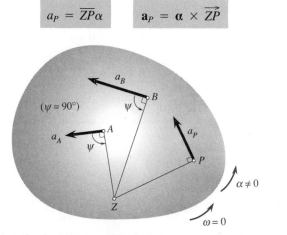

Fig. 15.23

Note that, at the instant considered, (1) any acceleration center Z must have a zero acceleration, (2) any point of a rigid slab (or its massless extension) having a zero acceleration is an acceleration center of the slab, (3) both the angular acceleration and angular velocity of a rigid slab are zero if two or more points of the slab have zero accelerations, (4) the velocity of an acceleration center is *not* necessarily zero. In fact, each rigid slab in a mechanism has its own acceleration center, which is generally different from that of another slab.

---

†This *corollary* resembles the theorem of velocity center in Sec. 15.7 and is *very useful* in cases where a rigid slab in plane motion has $\boldsymbol{\omega} = \mathbf{0}$ and $\boldsymbol{\alpha} \neq \mathbf{0}$ at the instant considered. Cf. Examples 15.14 through 15.16.

EXAMPLE 15.14

The rod AB is at rest when its end A is given $\mathbf{a}_A = 1.2$ m/s$^2 \rightarrow$ in the position as shown. For this position, determine (a) the acceleration center Z of the rod, (b) $\boldsymbol{\alpha}$ of the rod, (c) $\mathbf{a}_B$ and $\mathbf{a}_P$ of the end B and the midpoint P of the rod, respectively.

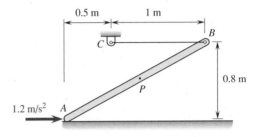

Solution. Since the rod is accelerated from rest, we see that it has $\boldsymbol{\omega} = \mathbf{0}$, $\boldsymbol{\alpha} \neq \mathbf{0}$, and $\psi = 90°$ in the position shown. From the constraint condition of the rod, we know that $\mathbf{a}_B = a_B \uparrow$. Thus, we draw the lines AZ and BZ, where $\overrightarrow{AZ} \perp \mathbf{a}_A$ and $\overrightarrow{BZ} \perp \mathbf{a}_B$, to locate the *acceleration center* at Z as shown. To be consistent with the direction of $\mathbf{a}_A$, the direction of $\boldsymbol{\alpha}$ must be counterclockwise as indicated. We write

$$a_A = \overline{ZA}\alpha \qquad 1.2 = 0.8\alpha \qquad \alpha = 1.5 \text{ rad/s}^2 \text{ ↺} \blacktriangleleft$$

$$\mathbf{a}_B = \boldsymbol{\alpha} \times \overrightarrow{ZB} = 1.5\mathbf{k} \times (1.5\mathbf{i}) = 2.25\mathbf{j}$$

$$\mathbf{a}_P = \boldsymbol{\alpha} \times \overrightarrow{ZP} = 1.5\mathbf{k} \times (0.75\mathbf{i} - 0.4\mathbf{j}) = 0.6\mathbf{i} + 1.125\mathbf{j}$$

$$\mathbf{a}_B = 2.25 \text{ m/s}^2 \uparrow \qquad \mathbf{a}_P = 1.275 \text{ m/s}^2 \, \measuredangle \, 61.9° \blacktriangleleft$$

REMARK. If the angular velocity of the rod is not zero in the position shown, then $\psi \neq 90°$ and the foregoing solution would be invalid.

Developmental Exercises

D15.24   Refer to the given data and the solution in Example 15.14. Verify that $\mathbf{a}_B = \mathbf{a}_{B/A} + \mathbf{a}_A$.

D15.25   Describe the *theorem of acceleration center* and its corollary.

EXAMPLE 15.15

The pin-connected bars are at rest when the bar OA is given $\alpha_{OA} = 2$ rad/s^2 ↻ in the position shown. For this position, determine (a) the acceleration center Z of the bar AB, (b) $\mathbf{a}_P$ of the midpoint P of the bar AB.

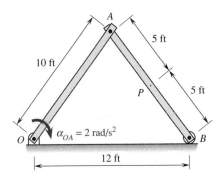

Solution.   Clearly, the point O is the acceleration center of the bar OA because $\mathbf{a}_O = \mathbf{0}$. As the system is accelerated from rest, we have $\omega_{OA} = \omega_{AB} = \mathbf{0}$. Thus, we write

$$\mathbf{a}_A = \mathbf{a}_{A/O} =$$

$$= 20 \text{ ft/s}^2 \searrow 36.9°$$

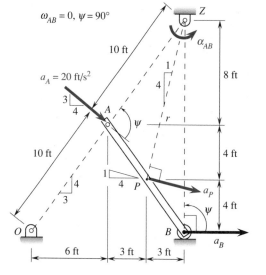

The constraint condition of the bar AB requires that $\mathbf{a}_B = a_B \rightarrow$. Therefore, we locate the *acceleration center* of the bar AB at Z as shown, where $\overrightarrow{AZ} \perp \mathbf{a}_A$ and $\overrightarrow{BZ} \perp \mathbf{a}_B$; i.e.,

Z is 16 ft directly above B.   ◀

From the sketch, we write

$$\overrightarrow{ZP} = -3\mathbf{i} - 12\mathbf{j} \text{ ft}$$

$$a_A = 20 = \overline{ZA}\alpha_{AB} = 10\alpha_{AB} \qquad \alpha_{AB} = 2$$

$$\mathbf{a}_P = \boldsymbol{\alpha}_{AB} \times \overrightarrow{ZP} = 2\mathbf{k} \times (-3\mathbf{i} - 12\mathbf{j}) = 2(-3\mathbf{j} + 12\mathbf{i})$$

$$\mathbf{a}_P = 24\mathbf{i} - 6\mathbf{j}$$

$$\mathbf{a}_P = 24.7 \text{ ft/s}^2 \searrow 14.04° \quad ◀$$

REMARK. If the angular velocities of the bars are not zero in the position shown, then $\psi \neq 90°$ and the foregoing solution would be invalid.

Developmental Exercises

D15.26 Refer to Example 15.15. Determine $\mathbf{a}_B$ of the end point B.

D15.27 Is the velocity of an acceleration center equal to zero?

EXAMPLE 15.16

Using *acceleration center*, work Example 15.11.

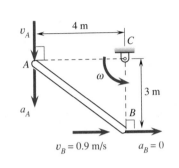

Solution. Since the end B of the rod AB is pulled with a constant velocity $\mathbf{v}_B = 0.9$ m/s $\rightarrow$, we have $\mathbf{a}_B = \mathbf{0}$ as indicated. The constraint condition of the rod requires that $\mathbf{v}_A = v_A \downarrow$ and $\mathbf{a}_A = a_A \downarrow$ as shown, where the point C is the velocity center of the rod. We have

$$v_B = 0.9 = \overline{CB}\omega = 3\omega \qquad \omega = 0.3 \text{ rad/s } \circlearrowright$$

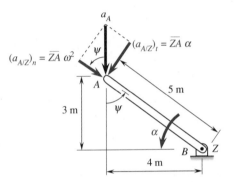

Since $\mathbf{a}_B = \mathbf{0}$, the *acceleration center* Z of the rod is, therefore, coincident with the point B. Referring to the second sketch, we write

$$\tan\psi = \frac{\alpha}{\omega^2} \qquad \frac{4}{3} = \frac{\alpha}{(0.3)^2}$$

$$a_A \cos\psi = \overline{ZA}\omega^2 \qquad a_A\left(\frac{3}{5}\right) = 5(0.3)^2$$

$$\alpha = 0.12 \text{ rad/s}^2 \circlearrowright \qquad \mathbf{a}_A = 0.75 \text{ m/s}^2 \downarrow \blacktriangleleft$$

Fig. D15.28

Developmental Exercise

D15.28 A gear of radius $r = 120$ mm is accelerated from rest with $\alpha = 2$ rad/s^2 $\circlearrowright$ as shown, where the rack R is fixed to the support. For this position, determine (a) the acceleration center Z of the gear, (b) $\mathbf{a}_B$ and $\mathbf{a}_D$.

Know

15.10 Parametric Method

We know that a one-degree-of-freedom system is a system whose position and motion can be defined by a single independent parameter or by a set of n dependent parameters subject to $n - 1$ constraint equations. For a one-degree-of-freedom mechanism moving in the xy plane, the x and y coordinates of all key points of the mechanism can be expressed, directly or indirectly, in terms of a single independent parameter; and their velocities and accelerations can be obtained by taking time derivatives of their coordinates. Of course, the angular velocity and angular acceleration of any line in the mechanism are given by the first and the second time derivatives of the angular position coordinate of that line, respectively. The procedure employing appropriate parameters to define and analyze the motion of a system is referred to as the *parametric method*.

EXAMPLE 15.17[†]

The end B of the rod AB moves with a constant velocity $\mathbf{v}_B = 0.9$ m/s $\rightarrow$. Using the *parametric method*, determine for the position shown (a) $\boldsymbol{\omega}$ and $\boldsymbol{\alpha}$ of the rod, (b) $\mathbf{v}_A$ and $\mathbf{a}_A$ of the end A.

Solution. The system under consideration is a one-degree-of-freedom system whose motion can be completely defined by the parameter θ as indicated. Let y_A and x_B be the position coordinates of the ends A and B of the rod as shown. In the position shown, we have $y_A = 3$ m, $x_B = 4$ m, $\theta = \tan^{-1}(3/4)$, $v_B = \dot{x}_B = 0.9$ m/s, and $a_B = \ddot{x}_B = 0$, where the positive (i.e., increasing) direction of θ is clockwise. We write

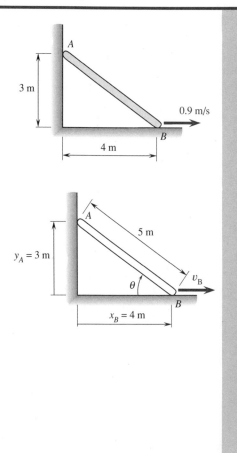

$$x_B = 5\cos\theta \qquad v_B = \dot{x}_B = -5(\sin\theta)\dot{\theta} = 0.9$$

$$\dot{\theta} = -0.3 \qquad \omega = \dot{\theta} \qquad \omega = 0.3 \text{ rad/s} \;\circlearrowright \;\blacktriangleleft$$

$$a_B = \ddot{x}_B = -5[(\cos\theta)\dot{\theta}^2 + (\sin\theta)\ddot{\theta}]$$

$$= -5[\tfrac{4}{5}(-0.3)^2 + \tfrac{3}{5}\ddot{\theta}] = 0$$

$$\ddot{\theta} = -0.12 \qquad \alpha = \ddot{\theta} \qquad \alpha = 0.12 \text{ rad/s}^2 \;\circlearrowright\;\blacktriangleleft$$

$$y_A = 5\sin\theta \qquad v_A = \dot{y}_A = 5(\cos\theta)\dot{\theta} = -1.2$$

$$a_A = \ddot{y}_A = 5[-(\sin\theta)\dot{\theta}^2 + (\cos\theta)\ddot{\theta}] = -0.75$$

Since v_A and a_A are found to be negative, the actual directions of $\mathbf{v}_A$ and $\mathbf{a}_A$ must point in the negative sense of y_A. Thus, we write

$$\mathbf{v}_A = 1.2 \text{ m/s} \downarrow \qquad \mathbf{a}_A = 0.75 \text{ m/s}^2 \downarrow \;\blacktriangleleft$$

Developmental Exercises

D15.29 Work Example 15.17 if the end B has $\mathbf{a}_B = 0.6$ m/s$^2 \rightarrow$ as well as $\mathbf{v}_B = 0.9$ m/s $\rightarrow$.

D15.30 The end B of the rod AB is pulled with a constant velocity $\mathbf{v}_B = 5.76$ ft/s $\rightarrow$ as shown. (a) Express x_B of the end B in terms of the angle θ. (b) Determine $\boldsymbol{\omega}$ and $\boldsymbol{\alpha}$ of the rod when $y_A = 12$ ft.

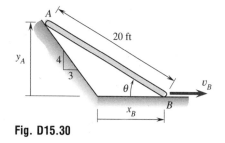

Fig. D15.30

[†]Cf. Examples 15.11 and 15.16.

EXAMPLE 15.18

A linkage moving in a plane is shown, where the crank AB rotates with a constant angular velocity $\omega_{AB} = 10$ rad/s $\circlearrowright$. Determine the angular velocities ω_{BD} and ω_{DE} of the coupler link BD and the output link DE when AB, BD, and DE all become collinear with the ground link AE.

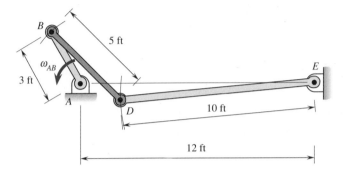

Solution. Moments before AB, BD, and DE all become collinear with the ground link AE, the configuration of AB, BD, and DE may be specified by the position coordinates θ_1, θ_2, and θ_3 as parameters as indicated in the sketch, where the positive directions of θ_1 and θ_2 are clockwise, but the positive direction of θ_3 is counterclockwise. The horizontal and vertical projections of AB, BD, and DE in this one-degree-of-freedom system require that θ_1, θ_2, and θ_3 be subject to the following two constraint equations:

$$-3\cos\theta_1 + 5\cos\theta_2 + 10\cos\theta_3 = 12 \tag{1}$$

$$3\sin\theta_1 + 10\sin\theta_3 = 5\sin\theta_2 \tag{2}$$

Since the values of θ_1, θ_2, and θ_3 under consideration are very small, we may use series expansion (cf. Sec. B3.1 of App. B) to approximate Eqs. (1) and (2) as follows:

$$-3(1 - \tfrac{1}{2}\theta_1^2) + 5(1 - \tfrac{1}{2}\theta_2^2) + 10(1 - \tfrac{1}{2}\theta_3^2) = 12 \tag{1'}$$

$$3\theta_1 + 10\theta_3 = 5\theta_2 \tag{2'}$$

where third-degree and higher terms in θ_1, θ_2 and θ_3 are neglected. Equations (1') and (2') yield

$$\theta_2 = \tfrac{1}{5}(1 \pm 2\sqrt{2})\theta_1 \qquad \theta_3 = \tfrac{1}{5}(-1 \pm \sqrt{2})\theta_1 \tag{3}$$

Differentiating with respect to the time, we write

$$\dot{\theta}_2 = \tfrac{1}{5}(1 \pm 2\sqrt{2})\dot{\theta}_1 \qquad \dot{\theta}_3 = \tfrac{1}{5}(-1 \pm \sqrt{2})\dot{\theta}_1 \qquad (4)$$

Note that $\dot{\theta}_1 = -10$ rad/s because $\omega_{AB} = 10$ rad/s $\circlearrowright$. Thus, we have $\dot{\theta}_2$ $= -7.66$ rad/s or 3.66 rad/s and $\dot{\theta}_3 = -0.828$ rad/s or 4.83 rad/s; i.e., we obtain two sets of values:

$$\omega_{BD} = 7.66 \text{ rad/s } \circlearrowright \qquad \text{and} \qquad \omega_{DE} = 0.828 \text{ rad/s } \circlearrowleft \quad \blacktriangleleft$$

or

$$\omega_{BD} = 3.66 \text{ rad/s } \circlearrowleft \qquad \text{and} \qquad \omega_{DE} = 4.83 \text{ rad/s } \circlearrowright \quad \blacktriangleleft$$

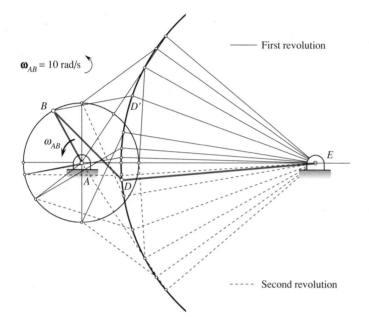

A plot showing that the configuration *ABDE* is repeated after two revolutions of *AB* about *A*

REMARK. Let us assume that *no sudden reversal* of rotation exists at the time the links become collinear. Then the *first set* of values is the solution for the case in which the configuration of *ABDE* (just before the links become collinear) is as depicted in the problem. The *second set* of values corresponds to the solution for the case in which the crank *AB* has completed one more revolution. Thus, each configuration of the entire linkage is repeated in every two revolutions of the crank *AB* as illustrated in the plot. Furthermore, it is well to point out that the problem in this example cannot be solved using the methods described in Secs. 15.6 and 15.7 because they yield only *one* nontrivial equation to relate the *two* unknowns: ω_{BD} and ω_{DE}.

Developmental Exercise

D15.31 Refer to Example 15.18. What are the values of α_{BD} and α_{DE} when *AB*, *BD*, and *DE* are collinear with *AE*?

PROBLEMS

15.87 through 15.89   At the instant shown, the end A of the rod AB is accelerated from rest with $\mathbf{a}_A$ as indicated. Using *acceleration center*, determine for this instant (a) $\boldsymbol{\alpha}$ of the rod, (b) $\mathbf{a}_B$ and $\mathbf{a}_D$ of the points B and D.

Fig. P15.87

Fig. P15.88

Fig. P15.89

Fig. P15.90

15.90   The plate $ABDE$ moves in the xy plane. In the position shown, the plate rotates with $\boldsymbol{\omega} = 2$ rad/s $\curvearrowright$, $\boldsymbol{\alpha} = 3$ rad/s^2 $\curvearrowright$, and the corner A moves with $\mathbf{a}_A = 25$ in./s^2 $\uparrow$. For this position, determine the position vector $\mathbf{r}_Z$ of the acceleration center Z of the plate.

Fig. P15.91

15.91   In the position shown, the double gear is accelerated from rest by pulling the upper rack with $\mathbf{a}_U = 132$ mm/s^2 $\leftarrow$ and the lower rack with $\mathbf{a}_L = 84$ mm/s^2 $\rightarrow$. If $a = 90$ mm and $b = 180$ mm, determine for this position (a) the distance $\overline{ZA}$ between the acceleration center Z and the center A of the gear, (b) $\boldsymbol{\alpha}$ of the gear, (c) $\mathbf{a}_B$ of the point B.

15.92*   In the position shown, the yoyo is accelerated from rest by pulling the end D of the string, which is wound around its core, with $\mathbf{a}_D = 7$ mm/s^2 $\rightarrow$. If the yoyo rolls without slipping, determine for this position $\boldsymbol{\alpha}$ of the yoyo and $\mathbf{a}_B$ of the point B.

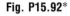

Fig. P15.92*

15.93* through 15.96* Using *acceleration center*, solve the following problems:

15.93* Prob. 15.80.   15.94* Prob. 15.81*.
15.95* Prob. 15.83.   15.96* Prob. 15.85.

15.97 In the position shown, the slider is accelerated from rest with $\mathbf{a}_D$ = 210 mm/s² ←. Using *acceleration center*, determine for this position $\boldsymbol{\alpha}_{AB}$ of the crank AB and $\mathbf{a}_E$ of the point E.

Fig. P15.97

15.98 The linkage is released from rest in the position shown, and the link OB acquires instantaneously $\boldsymbol{\alpha}_{OB} = \alpha_{OB}$ rad/s² ↺ due to gravity. Using *acceleration center*, determine for the instant of release $\mathbf{a}_D$ and $\mathbf{a}_E$ of the points D and E in terms of α_{OB} as a parameter.

Fig. P15.98   **Fig. P15.99**

15.99 The bent rod DEF is suspended by three wires AD, BD, and CE. At the instant shown, the wire BD is suddenly cut and the bent rod instantaneously accelerates with $\boldsymbol{\alpha} = \alpha$ rad/s² ↻ due to gravity. Using *acceleration center*, determine for this instant $\mathbf{a}_G$ and $\mathbf{a}_H$ of the points G and H in terms of α as a parameter.

15.100* through 15.102* The end A of the rod AB is moved to the right with a speed v_A as shown. Using the *parametric method*, determine $\boldsymbol{\omega}_{AB}$ of the rod in terms of v_A, b, and x.

Fig. P15.100* and P15.103   **Fig. P15.101 and P15.104***   **Fig. P15.102 and P15.105**

15.103 through 15.105 The end A of the rod AB is moved to the right with a constant speed v_A as shown. Using the *parametric method*, derive an expression for $\boldsymbol{\alpha}_{AB}$ of the rod in terms of v_A, b, and x.

15.106* The crank AB of the linkage shown rotates with a constant angular velocity $\omega_{AB} = 5$ rad/s $\circlearrowright$. Using the *parametric method*, determine ω_{BD} and ω_{DE} of the coupler link BD and the output link DE when AB, BD, and DE all become collinear with the ground link AE.[†]

Fig. P15.106*

15.107 A linkage is shown, where the slider D moves along a smooth circular groove of radius 6 ft as indicated. It is known that the crank AB rotates with a constant angular velocity $\omega_{AB} = 2$ rad/s $\circlearrowleft$. Using the *parametric method*, determine ω_{BD} of the link BD and $\mathbf{v}_D$ of the slider D when AB and BD become collinear with the line AO. (*Hint.* Can the line joining O and D be taken as a link?)[†]

Fig. P15.107

USE OF ROTATING REFERENCE FRAMES

15.11 Time Derivatives of Rotating Unit Vectors

We recall that, in kinematics, any reference frame may be designated as *fixed*; all others not rigidly attached to it are regarded as *moving*.[††] A *moving reference frame* may be translating, rotating, or doing both, in a *fixed reference frame*. It is true that the time rate of change of a vector in a fixed reference frame is the *same* as that in a translating reference frame. However, the time rate of change of a vector in a rotating reference frame is generally *different* from that in a fixed or translating reference frame.

Let a rigid slab S move in the XY plane of the *fixed reference frame OXYZ* as shown in Fig. 15.24, where $\mathbf{I}$, $\mathbf{J}$, and $\mathbf{K}$ are the unit vectors directed along the axes of $OXYZ$. As a fundamental step, we first consider the time

---

[†]Cf. Example 15.18.

[††]A reference frame designated as fixed in *kinetics* is usually an *inertial reference frame* in which Newton's laws are directly applicable.

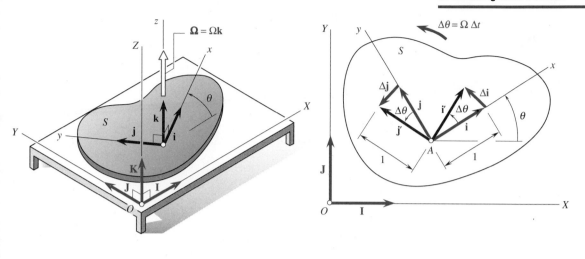

(a)   (b)

Fig. 15.24

rates of change of the unit vectors $\mathbf{i}$, $\mathbf{j}$, and $\mathbf{k}$ directed along the axes of a rectangular coordinate system $Axyz$ which is embedded in the moving rigid slab S to form a *moving reference frame Axyz*.[†] For the study of plane kinematics where $\mathbf{K} = \mathbf{k}$, let the angular velocity of $Axyz$ be

$$\mathbf{\Omega} = \Omega\mathbf{K} = \Omega\mathbf{k} \qquad (15.34)^{††}$$

and the velocity of the point A be $\mathbf{v}_A$ in $OXYZ$. Note that $\mathbf{k}$ is constant in $OXYZ$. The time rate of change of $\mathbf{k}$ in $OXYZ$ is, therefore, zero; i.e.,

$$\left(\frac{d\mathbf{k}}{dt}\right)_{OXYZ} = (\dot{\mathbf{k}})_{OXYZ} = \dot{\mathbf{k}} = \mathbf{0} \qquad (15.35)$$

NOTE: Since $\mathbf{\Omega} = \Omega\mathbf{k}$ is being used, the Z and z axes are included in the names of the reference frames. For subsequent studies, *a quantity, or the derivative of a quantity, without the name of a specific reference frame as subscripts is to be understood as one measured, or taken, in the fixed reference frame*, such as $OXYZ$.

The magnitudes of the unit vectors $\mathbf{i}$ and $\mathbf{j}$ are always equal to 1, a constant; however, their directions may change during the motion of $Axyz$ in the XY plane as shown in Fig. 15.24(a). It is readily seen that the changes of $\mathbf{i}$ and $\mathbf{j}$ are due *only* to the *rotation* of $Axyz$. Thus, the changes $\Delta\mathbf{i}$ and $\Delta\mathbf{j}$ of $\mathbf{i}$ and $\mathbf{j}$ during the time interval Δt are due to the incremental rotation $\Delta\theta$ of $Axyz$ as shown in Fig. 15.24(b). From this figure and Eq. (15.34), we derive the time rates of change of $\mathbf{i}$ and $\mathbf{j}$ as follows:

$$\lim_{\Delta t \to 0} \frac{\Delta\theta}{\Delta t} = \dot{\theta} = \Omega \qquad \mathbf{j} = \mathbf{k} \times \mathbf{i} \qquad \mathbf{i} = -\mathbf{k} \times \mathbf{j}$$

$$\Delta\mathbf{i} = \mathbf{i}' - \mathbf{i} \approx \Delta\theta\mathbf{j}$$

[†]Cf. the footnote on p. 479.

[††]Cf. the angular velocity defined by Eq. (15.40).

$$\dot{\mathbf{i}} = \lim_{\Delta t \to 0} \frac{\Delta \mathbf{i}}{\Delta t} = \lim_{\Delta t \to 0} \frac{\Delta \theta \mathbf{j}}{\Delta t} = \Omega \mathbf{j} = \Omega \mathbf{k} \times \mathbf{i}$$

$$\dot{\mathbf{i}} = \mathbf{\Omega} \times \mathbf{i} \qquad (15.36)$$

$$\Delta \mathbf{j} = \mathbf{j}' - \mathbf{j} \approx -\Delta \theta \mathbf{i}$$

$$\dot{\mathbf{j}} = \lim_{\Delta t \to 0} \frac{\Delta \mathbf{j}}{\Delta t} = \lim_{\Delta t \to 0} \frac{-\Delta \theta \mathbf{i}}{\Delta t} = -\Omega \mathbf{i} = \Omega \mathbf{k} \times \mathbf{j}$$

$$\dot{\mathbf{j}} = \mathbf{\Omega} \times \mathbf{j} \qquad (15.37)$$

Due to the fact that $\mathbf{\Omega} \times \mathbf{k} = \Omega \mathbf{k} \times \mathbf{k} = \mathbf{0}$, we may also write Eqs. (15.35) as

$$\dot{\mathbf{k}} = \mathbf{\Omega} \times \mathbf{k} = \mathbf{0} \qquad (15.38)$$

Since $\mathbf{i}$, $\mathbf{j}$, and $\mathbf{k}$ are all constant in $Axyz$, we have

$$(\dot{\mathbf{i}})_{Axyz} = (\dot{\mathbf{j}})_{Axyz} = (\dot{\mathbf{k}})_{Axyz} = \mathbf{0} \qquad (15.39)$$

If $Axyz$ were to have a general motion in $OXYZ$, its angular velocity $\mathbf{\Omega}$ in $OXYZ$ would be given by

$$\mathbf{\Omega} = (\dot{\mathbf{j}} \cdot \mathbf{k})\mathbf{i} + (\dot{\mathbf{k}} \cdot \mathbf{i})\mathbf{j} + (\dot{\mathbf{i}} \cdot \mathbf{j})\mathbf{k} \qquad (15.40)[†]$$

which agrees with Eq. (15.34) as a special case for the plane kinematics of rigid bodies. However, we shall not be concerned with the derivation of Eq. (15.40) until Sec. 19.2.

Developmental Exercise

D15.32   In the notations adopted for subsequent studies, what does each of the following represent? (a) $OXYZ$, (b) $Axyz$, (c) $\mathbf{\Omega}$, (d) $(\dot{\mathbf{i}})_{OXYZ}$, (e) $(\dot{\mathbf{i}})_{Axyz}$, (f) $\dot{\mathbf{i}}$.

15.12   Time Derivatives of a Vector in Two Reference Frames

We first note that the *time derivative of a scalar function* ϕ is $\dot{\phi}$, which is still a scalar quantity whose value is the same in any reference frame. In other words, the time derivative of a scalar in a *moving reference frame Axyz* (which is rotating, translating, or doing both) is the same as that in a *fixed reference frame OXYZ*; i.e.,

$$\dot{\phi} = (\dot{\phi})_{OXYZ} = (\dot{\phi})_{Axyz} \qquad (15.41)$$

As an example of the property expressed in Eq. (15.41), let ϕ denote the

---

[†]This general definition was set forth by the dynamicist T. R. Kane in his book *Dynamics*, Holt, Rinehart and Winston, New York, 1968, p. 21. The formulation of this definition may be found in the development leading to Eq. (19.30).

temperature at a point of a region. We readily see that the time rate of change (i.e., the time derivative) of the temperature ϕ has (and must have) the same value at a given time in all moving and fixed reference frames which are used to describe it.

Fig. 15.25

Next, let us consider a vector $\mathbf{Q}$ in the xy plane of $Axyz$ which moves in the XY plane of $OXYZ$ as shown in Fig. 15.25, where $Axyz$ has an angular velocity $\boldsymbol{\Omega} = \Omega\mathbf{k}$ and the point A has a velocity $\mathbf{v}_A$ in $OXYZ$. Resolving $\mathbf{Q}$ into components parallel to the x and y axes, we write

$$\mathbf{Q} = Q_x\mathbf{i} + Q_y\mathbf{j} \tag{15.42}$$

In following the adopted notation, we denote by $\dot{\mathbf{Q}}$ the time derivative of Q in $OXYZ$, and by $(\dot{\mathbf{Q}})_{Axyz}$ the time derivative of $\mathbf{Q}$ in $Axyz$. Taking the time derivative first in $Axyz$, then in $OXYZ$, on both sides of Eq. (15.42) and applying Eqs. (15.36), (15.37), (15.39), and (15.41), we write

$$(\dot{\mathbf{Q}})_{Axyz} = (\dot{Q}_x)_{Axyz}\mathbf{i} + (\dot{Q}_y)_{Axyz}\mathbf{j}$$
$$= (\dot{Q}_x)_{OXYZ}\mathbf{i} + (\dot{Q}_y)_{OXYZ}\mathbf{j}$$

$$\boxed{(\dot{\mathbf{Q}})_{Axyz} = \dot{Q}_x\mathbf{i} + \dot{Q}_y\mathbf{j}} \tag{15.43}$$

$$\dot{\mathbf{Q}} = \dot{Q}_x\mathbf{i} + Q_x\dot{\mathbf{i}} + \dot{Q}_y\mathbf{j} + Q_y\dot{\mathbf{j}}$$
$$= \dot{Q}_x\mathbf{i} + Q_x(\boldsymbol{\Omega} \times \mathbf{i}) + \dot{Q}_y\mathbf{j} + Q_y(\boldsymbol{\Omega} \times \mathbf{j})$$
$$= (\dot{Q}_x\mathbf{i} + \dot{Q}_y\mathbf{j}) + \boldsymbol{\Omega} \times (Q_x\mathbf{i} + Q_y\mathbf{j})$$

$$\boxed{\dot{\mathbf{Q}} = (\dot{\mathbf{Q}})_{Axyz} + \boldsymbol{\Omega} \times \mathbf{Q}} \tag{15.44}$$

Thus, we see that the time derivative of the vector $\mathbf{Q}$ taken in the *fixed reference frame OXYZ* is equal to the sum of the following two parts: (a) the time derivative of $\mathbf{Q}$ taken in the *moving reference frame Axyz*, and (b) the cross product $\boldsymbol{\Omega} \times \mathbf{Q}$, where $\boldsymbol{\Omega}$ is the *angular velocity* of $Axyz$ in $OXYZ$. If $\boldsymbol{\Omega} = \mathbf{0}$, then $Axyz$ is nonrotating and Eq. (15.44) degenerates to $\dot{\mathbf{Q}} = (\dot{\mathbf{Q}})_{Axyz}$, which is a property we used in the first part of this chapter.

Developmental Exercises

D15.33 Notations are a means for effective communication. In the notations adopted, which of the following are always true?
  (a) $\dot{Q}_x = (\dot{Q}_x)_{OXYZ}$, (b) $\dot{Q}_x = (\dot{Q}_x)_{Axyz}$, (c) $\dot{\mathbf{Q}} = (\dot{\mathbf{Q}})_{OXYZ}$, (d) $\dot{\mathbf{Q}} = (\dot{\mathbf{Q}})_{Axyz}$.

D15.34 Suppose that $\mathbf{Q}$ is a constant vector in $Axyz$, whose angular velocity in $OXYZ$ is $\mathbf{\Omega}$. Express the time derivative of $\mathbf{Q}$ taken in $OXYZ$.

15.13   Velocities in Different Reference Frames

Let us consider the motion of a particle B that moves along a path (e.g., a curved groove) C on a rigid slab S which is itself translating and rotating in the XY plane of the *fixed reference frame OXYZ* as shown in Fig. 15.26. To facilitate the study, we embed a coordinate system $Axyz$ in the slab S to form a *moving reference frame Axyz* as shown. The angular velocity of $Axyz$ in $OXYZ$ is $\mathbf{\Omega} = \Omega\mathbf{K} = \Omega\mathbf{k}$, and the velocity of the point A in $OXYZ$ is $\mathbf{v}_A$. In $OXYZ$, the positions of A and B are specified by the position vectors $\mathbf{r}_A$ and $\mathbf{r}_B$, while the position of B relative to A is specified by the position vector $\mathbf{r}_{B/A}$.

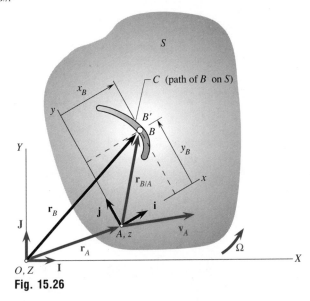

Fig. 15.26

It is well to recall in studies involving rotating reference frames that *a quantity, or the derivative of a quantity, without the name of a specific reference frame as subscripts is to be understood as one measured, or taken, in the fixed reference frame, such as OXYZ.* Besides, note that

$$\mathbf{v}_A = \dot{\mathbf{r}}_A \qquad \mathbf{v}_B = \dot{\mathbf{r}}_B \qquad \mathbf{v}_{B/A} = \dot{\mathbf{r}}_{B/A} \tag{15.45}$$

$$\mathbf{r}_{B/Axyz} = (\mathbf{r}_{B/A})_{Axyz} \qquad \mathbf{v}_{B/Axyz} = (\dot{\mathbf{r}}_{B/A})_{Axyz} \tag{15.46}$$

Suppose that B' is a point *of* the slab S, and B' *coincides* with the moving particle B at the instant considered, as shown in Fig. 15.26. It is obvious that the velocity of B' in $Axyz$, which is embedded in S, is zero; i.e.,

$$\mathbf{v}_{B'/Axyz} = (\dot{\mathbf{r}}_{B'/A})_{Axyz} = \mathbf{0} \tag{15.47}$$

The position vector of B' in $OXYZ$ is

$$\mathbf{r}_{B'} = \overrightarrow{OB'} = \mathbf{r}_A + \mathbf{r}_{B'/A} \tag{15.48}$$

Thus, we apply Eqs. (15.44) through (15.48) to write

$$\mathbf{v}_{B'} = \dot{\mathbf{r}}_{B'} = \dot{\mathbf{r}}_A + \dot{\mathbf{r}}_{B'/A}$$

$$= \mathbf{v}_A + [(\dot{\mathbf{r}}_{B'/A})_{Axyz} + \boldsymbol{\Omega} \times \mathbf{r}_{B'/A}]$$

$$= \mathbf{v}_A + (\mathbf{0} + \boldsymbol{\Omega} \times \mathbf{r}_{B'/A})$$

$$\mathbf{v}_{B'} = \boldsymbol{\Omega} \times \mathbf{r}_{B'/A} + \mathbf{v}_A \tag{15.49}$$

Since $\mathbf{r}_{B'/A} = \mathbf{r}_{B/A}$ *at the instant considered*, we find that the velocity of B' is given by

$$\boxed{\mathbf{v}_{B'} = \boldsymbol{\Omega} \times \mathbf{r}_{B/A} + \mathbf{v}_A} \tag{15.50}$$

The position vector of the particle B in $OXYZ$, as shown in Fig. 15.26, is

$$\mathbf{r}_B = \mathbf{r}_A + \mathbf{r}_{B/A} \tag{15.51}$$

Applying Eqs. (15.44) through (15.46), (15.50), and (15.51), we derive the velocity of B in $OXYZ$ as follows:

$$\mathbf{v}_B = \dot{\mathbf{r}}_B = \dot{\mathbf{r}}_A + \dot{\mathbf{r}}_{B/A}$$

$$= \mathbf{v}_A + [(\dot{\mathbf{r}}_{B/A})_{Axyz} + \boldsymbol{\Omega} \times \mathbf{r}_{B/A}]$$

$$= \mathbf{v}_{B/Axyz} + (\boldsymbol{\Omega} \times \mathbf{r}_{B/A} + \mathbf{v}_A)$$

$$\boxed{\mathbf{v}_B = \mathbf{v}_{B/Axyz} + \mathbf{v}_{B'}} \tag{15.52}^\dagger$$

$$\boxed{\mathbf{v}_B = \mathbf{v}_{B/Axyz} + \boldsymbol{\Omega} \times \mathbf{r}_{B/A} + \mathbf{v}_A} \tag{15.53}$$

We note that the velocity of B relative to A in $OXYZ$ is given by

$$\mathbf{v}_{B/A} = \dot{\mathbf{r}}_{B/A} = (\dot{\mathbf{r}}_{B/A})_{Axyz} + \boldsymbol{\Omega} \times \mathbf{r}_{B/A}$$

$$\mathbf{v}_{B/A} = \mathbf{v}_{B/Axyz} + \boldsymbol{\Omega} \times \mathbf{r}_{B/A} \tag{15.54}$$

Substitution of Eq. (15.54) into Eq. (15.53) yields

$$\mathbf{v}_B = \mathbf{v}_{B/A} + \mathbf{v}_A \tag{15.55}$$

In recapitulation of the symbols used, note that

$\mathbf{r}_{B/A} = \overrightarrow{AB} = $ position vector of B relative to A

$\mathbf{v}_A, \mathbf{v}_B = $ velocities of A and B in the *fixed* reference frame $OXYZ$

$\mathbf{v}_{B'} = $ velocity of the *coinciding point* B' in $OXYZ$

$\mathbf{v}_{B/A} = $ velocity of B relative to A in $OXYZ$

$\mathbf{v}_{B/Axyz} = $ velocity of B in the *moving* reference frame $Axyz$

$\boldsymbol{\Omega} = $ angular velocity of $Axyz$ in $OXYZ$

For motion in the XY plane, the various vectors in Eqs. (15.53) through (15.55) are illustrated in Fig. 15.27. In this figure, note that (1) the path C

[†]Note that the relation expressed in Eq. (15.52) is very *succinct*; it will provide a helpful background for understanding why the *Coriolis acceleration* in Eq. (15.64) is called a *complementary acceleration*.

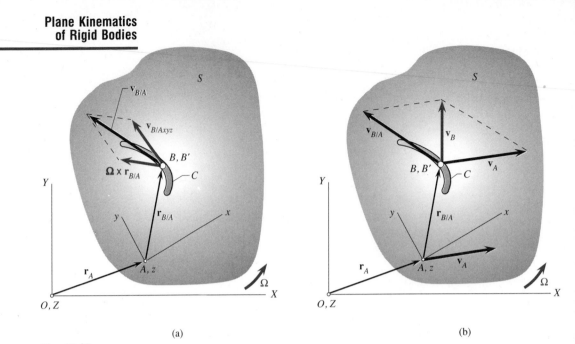

(a)

(b)

Fig. 15.27

is the curve traveled by the particle B in the moving reference frame $Axyz$, (2) the path C is generally different from the curve traveled by B in the fixed reference frame $OXYZ$, (3) $\mathbf{v}_{B/Axyz}$ is tangent to the path C, (4) $\mathbf{v}_{B/Axyz}$ as well as $\mathbf{v}_{B/A}$ is generally not perpendicular to the position vector $\mathbf{r}_{B/A}$, (5) $\mathbf{\Omega} \times \mathbf{r}_{B/A}$ is a vector perpendicular to $\mathbf{r}_{B/A}$, and (6) $\mathbf{v}_{B/A}$ is equal to $\mathbf{v}_{B/Axyz}$ only when $\mathbf{\Omega} = \mathbf{0}$. In summarizing velocities in different reference frames, we refer to Fig. 15.27 and write

$$\mathbf{v}_B = \underbrace{\mathbf{v}_{B/Axyz} + \mathbf{\Omega} \times \mathbf{r}_{B/A}} + \mathbf{v}_A \qquad (15.53)$$
(repeated)

$$\mathbf{v}_B = \underbrace{\mathbf{v}_{B/A}} + \mathbf{v}_A \qquad (15.55)$$
(repeated)

$$\mathbf{v}_B = \mathbf{v}_{B/Axyz} + \mathbf{v}_{B'} \qquad (15.52)$$
(repeated)

EXAMPLE 15.19

In the position shown, the crank BC rotates with a constant angular velocity of 4.25 rad/s $\circlearrowleft$, and its pin at B is confined to move in the slot of the link

OD. Determine for this position $\boldsymbol{\omega}_{OD}$ of the link *OD* and the velocity $\mathbf{v}_{B/OD}$
of the pin at *B* relative to the link *OD*.

Solution. To solve this problem, we let *OXYZ be fixed to the ground* and
Oxyz be embedded in the link OD, as shown. Since $\boldsymbol{\omega}_{BC} = 4.25$ rad/s
$\boldsymbol{\Omega} = -4.25\mathbf{K}$ rad/s and $\mathbf{v}_C = \mathbf{0}$, we write

$$\mathbf{v}_B = \mathbf{v}_{B/C} = \boldsymbol{\omega}_{BC} \times \mathbf{r}_{B/C} = -4.25\mathbf{K} \times (-15\mathbf{I} + 8\mathbf{J})$$

$$\mathbf{v}_B = 34\mathbf{I} + 63.75\mathbf{J} \tag{1}$$

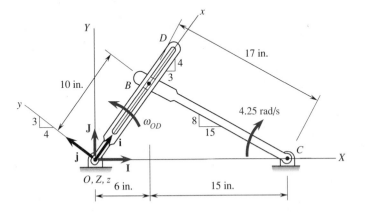

The angular velocity of *Oxyz* is $\boldsymbol{\Omega} = \omega_{OD}\mathbf{k}$. Applying Eq. (15.53), we write

$$\mathbf{v}_B = \mathbf{v}_{B/Oxyz} + \boldsymbol{\Omega} \times \mathbf{r}_{B/O} + \mathbf{v}_O$$

$$= v_{B/OD}\mathbf{i} + \omega_{OD}\mathbf{k} \times (10\mathbf{i}) + \mathbf{0}$$

$$= v_{B/OD}\mathbf{i} + 10\omega_{OD}\mathbf{j}$$

$$= v_{B/OD}(\tfrac{1}{5})(3\mathbf{I} + 4\mathbf{J}) + 10\omega_{OD}(\tfrac{1}{5})(-4\mathbf{I} + 3\mathbf{J})$$

$$\mathbf{v}_B = (\tfrac{3}{5} v_{B/OD} - 8\omega_{OD})\mathbf{I} + (\tfrac{4}{5} v_{B/OD} + 6\omega_{OD})\mathbf{J} \tag{2}$$

From Eqs. (1) and (2), we obtain two ensuing scalar equations, which yield
$v_{B/OD} = 71.4$ and $\omega_{OD} = 1.105$. Thus, we have

$$\boldsymbol{\omega}_{OD} = 1.105 \text{ rad/s} \; \circlearrowright \qquad \mathbf{v}_{B/OD} = 71.4 \text{ in./s} \quad 53.1° \; \blacktriangleleft$$

REMARK. If *B'* is a point of the link *OD* coinciding with the pin at *B*, then,
without using a rotating reference frame, we can solve for ω_{OD} by equating
the velocity of *B'* to the component of the velocity of *B* in the direction
perpendicular to the slot. On the other hand, it can be shown from the
solution in this example that

$$\mathbf{v}_B = 71.4\mathbf{i} + 11.05\mathbf{j} \text{ in./s}$$

$$\mathbf{v}_{B'} = 11.05\mathbf{j} \text{ in./s}$$

$$\mathbf{v}_{B/B'} = \mathbf{v}_B - \mathbf{v}_{B'} = 71.4\mathbf{i} \text{ in./s}$$

Thus, we see that $\mathbf{v}_{B/OD} = \mathbf{v}_{B/B'}$. (Cf. the *Remark* in Example 15.20.)

Fig. D15.37

Developmental Exercises

D15.35   Derive Eq. (2) in Example 15.19 by replacing $Oxyz$ with $B'xyz$ where B' is a point of the link OD coinciding with the pin at B.

D15.36   Both Eqs. (15.52) and (15.55) are useful formulas. Describe these two formulas in words.

D15.37   In the position shown, the crank BC rotates with a constant angular velocity $\omega_{BC} = 2$ rad/s $\circlearrowleft$. Making use of a rotating reference frame, determine ω_{OD} and $\mathbf{v}_{B/OD}$.

15.14   Accelerations in Different Reference Frames

Refer to Fig. 15.28 where a particle B moves along a path C on a rigid slab S which is itself translating and rotating in the XY plane of the *fixed reference frame OXYZ*. The coordinate system $Axyz$ is embedded in the slab S to form a *moving reference frame Axyz*, which has an angular acceleration $\dot{\boldsymbol{\Omega}} = \dot{\Omega}\mathbf{k}$ as well as an angular velocity $\boldsymbol{\Omega} = \Omega\mathbf{k}$ at the instant considered. The velocity and acceleration of the origin A of $Axyz$ are $\mathbf{v}_A$ and $\mathbf{a}_A$ in $OXYZ$, respectively.

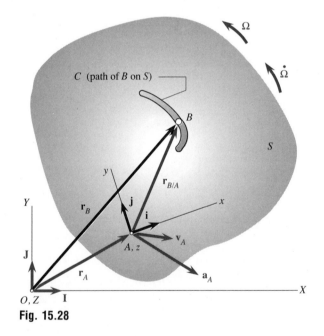

Fig. 15.28

In the ensuing study of acceleration in different reference frames, note that

$$\mathbf{a}_A = \dot{\mathbf{v}}_A \qquad \mathbf{a}_B = \dot{\mathbf{v}}_B \qquad \mathbf{a}_{B/A} = \dot{\mathbf{v}}_{B/A} \tag{15.56}$$

$$\mathbf{a}_{B/Axyz} = (\dot{\mathbf{v}}_{B/Axyz})_{Axyz} \tag{15.57}$$

Taking the time derivative in $OXYZ$ on both sides of Eq. (15.53) and applying Eqs. (15.44), (15.46), (15.56), and (15.57), we write

$$\dot{\mathbf{v}}_B = [(\dot{\mathbf{v}}_{B/Axyz})_{Axyz} + \mathbf{\Omega} \times \mathbf{v}_{B/Axyz}] + \dot{\mathbf{\Omega}} \times \mathbf{r}_{B/A}$$
$$+ \mathbf{\Omega} \times [(\dot{\mathbf{r}}_{B/A})_{Axyz} + \mathbf{\Omega} \times \mathbf{r}_{B/A}] + \dot{\mathbf{v}}_A$$

$$\mathbf{a}_B = \mathbf{a}_{B/Axyz} + \mathbf{\Omega} \times \mathbf{v}_{B/Axyz} + \dot{\mathbf{\Omega}} \times \mathbf{r}_{B/A}$$
$$+ \mathbf{\Omega} \times \mathbf{v}_{B/Axyz} + \mathbf{\Omega} \times (\mathbf{\Omega} \times \mathbf{r}_{B/A}) + \mathbf{a}_A$$

$$\mathbf{a}_B = \mathbf{a}_{B/Axyz} + \dot{\mathbf{\Omega}} \times \mathbf{r}_{B/A} + \mathbf{\Omega} \times (\mathbf{\Omega} \times \mathbf{r}_{B/A})$$
$$+ 2\mathbf{\Omega} \times \mathbf{v}_{B/Axyz} + \mathbf{a}_A \qquad (15.58)$$

Performing similar operations on Eqs. (15.54) and (15.55), we obtain

$$\mathbf{a}_{B/A} = \mathbf{a}_{B/Axyz} + \dot{\mathbf{\Omega}} \times \mathbf{r}_{B/A} + \mathbf{\Omega} \times (\mathbf{\Omega} \times \mathbf{r}_{B/A})$$
$$+ 2\mathbf{\Omega} \times \mathbf{v}_{B/Axyz} \qquad (15.59)$$

$$\mathbf{a}_B = \mathbf{a}_{B/A} + \mathbf{a}_A \qquad (15.60)$$

Note that $\mathbf{v}_{B/Axyz}$, $\mathbf{\Omega}$, and $\mathbf{r}_{B/A}$ are the same as those defined earlier, and

$\mathbf{a}_A$, $\mathbf{a}_B$ = accelerations of A and B in the *fixed* reference frame $OXYZ$

$\mathbf{a}_{B/A}$ = acceleration of B relative to A in $OXYZ$

$\mathbf{a}_{B/Axyz}$ = acceleration of B in the *moving* reference frame $Axyz$

$\dot{\mathbf{\Omega}}$ = angular acceleration of $Axyz$ in $OXYZ$

For motion in the XY plane, the various terms on the right side of Eq. (15.58) are illustrated in Fig. 15.29. Note that the vector sum of the four rust vectors at B gives the acceleration vector $\mathbf{a}_{B/A}$ of B relative to A in $OXYZ$ as defined in Eq. (15.59). Suppose that B' is a point *of* the slab S, and B' *coincides* with the moving particle B at the instant considered, as shown in Fig. 15.29. The acceleration of the *coinciding point* B' in $OXYZ$ may be derived by applying Eqs. (15.44), (15.47), (15.49), [not (15.50)], and

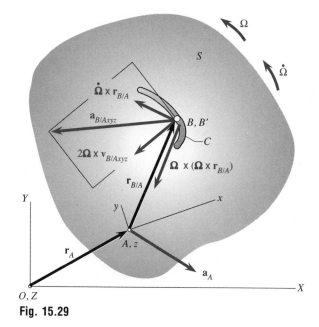

Fig. 15.29

(15.56) as follows:

$$\mathbf{a}_{B'} = \dot{\mathbf{v}}_{B'} = \dot{\mathbf{\Omega}} \times \mathbf{r}_{B'/A} + \mathbf{\Omega} \times \dot{\mathbf{r}}_{B'/A} + \dot{\mathbf{v}}_A$$
$$= \dot{\mathbf{\Omega}} \times \mathbf{r}_{B'/A} + \mathbf{\Omega} \times [(\dot{\mathbf{r}}_{B'/A})_{Axyz} + \mathbf{\Omega} \times \mathbf{r}_{B'/A}] + \mathbf{a}_A$$
$$= \dot{\mathbf{\Omega}} \times \mathbf{r}_{B'/A} + \mathbf{\Omega} \times [0 + \mathbf{\Omega} \times \mathbf{r}_{B'/A}] + \mathbf{a}_A$$

Since $\mathbf{r}_{B'/A} = \mathbf{r}_{B/A}$ *at the instant considered*, we find that the acceleration of B' in *OXYZ* is given by

$$\mathbf{a}_{B'} = \dot{\mathbf{\Omega}} \times \mathbf{r}_{B/A} + \mathbf{\Omega} \times (\mathbf{\Omega} \times \mathbf{r}_{B/A}) + \mathbf{a}_A \tag{15.61}$$

It can readily be verified that the terms $\mathbf{\Omega} \times \mathbf{r}_{B/A}$ and $\mathbf{\Omega} \times (\mathbf{\Omega} \times \mathbf{r}_{B/A})$ in Eq. (15.61) represent, respectively, the *tangential* and the *normal components* of the *acceleration* $\mathbf{a}_{B'/A}$ of the coinciding point B' relative to A in *OXYZ*. We write

$$\mathbf{a}_{B'/A} = \dot{\mathbf{\Omega}} \times \mathbf{r}_{B/A} + \mathbf{\Omega} \times (\mathbf{\Omega} \times \mathbf{r}_{B/A}) \tag{15.62}[†]$$

Substitution of Eq. (15.62) into Eq. (15.61) yields

$$\mathbf{a}_{B'} = \mathbf{a}_{B'/A} + \mathbf{a}_A \tag{15.63}$$

Furthermore, substituting Eq. (15.61) into Eq. (15.58), we obtain

$$\mathbf{a}_B = \mathbf{a}_{B/Axyz} + \mathbf{a}_{B'} + 2\mathbf{\Omega} \times \mathbf{v}_{B/Axyz} \tag{15.64}$$

Unlike Eq. (15.52), we see here that $\mathbf{a}_B \neq \mathbf{a}_{B/Axyz} + \mathbf{a}_{B'}$. The needed additional term $2\mathbf{\Omega} \times \mathbf{v}_{B/Axyz}$ is called the *complementary acceleration*[††] or, more often, the *Coriolis acceleration*, after the French mathematician and engineer Gaspard Gustave de Coriolis (1792–1843). In recapitulation, we refer to Fig. 15.29 and write

$$\mathbf{a}_B = \mathbf{a}_{B/Axyz} + \dot{\mathbf{\Omega}} \times \mathbf{r}_{B/A} + \mathbf{\Omega} \times (\mathbf{\Omega} \times \mathbf{r}_{B/A}) + 2\mathbf{\Omega} \times \mathbf{v}_{B/Axyz} + \mathbf{a}_A \tag{15.58}$$
(repeated)

$$\mathbf{a}_B = \mathbf{a}_{B/A} + \mathbf{a}_A \tag{15.60}$$
(repeated)

$$\mathbf{a}_B = \mathbf{a}_{B/Axyz} + \dot{\mathbf{\Omega}} \times \mathbf{r}_{B/A} + \mathbf{\Omega} \times (\mathbf{\Omega} \times \mathbf{r}_{B/A}) + \mathbf{a}_A + 2\mathbf{\Omega} \times \mathbf{v}_{B/Axyz}$$
$$\mathbf{a}_B = \mathbf{a}_{B/Axyz} + \mathbf{a}_{B'/A} + \mathbf{a}_A + 2\mathbf{\Omega} \times \mathbf{v}_{B/Axyz}$$
$$\mathbf{a}_B = \mathbf{a}_{B/Axyz} + \mathbf{a}_{B'} + 2\mathbf{\Omega} \times \mathbf{v}_{B/Axyz} \tag{15.64}$$
(repeated)

Developmental Exercises

D15.38 Suppose that B' is a point *of* the moving reference frame *Axyz*, and B' *coincides* with the moving particle B. (a) Is $\mathbf{v}_B = \mathbf{v}_{B/Axyz} + \mathbf{v}_{B'}$? (b) Is $\mathbf{a}_B = \mathbf{a}_{B/Axyz} + \mathbf{a}_{B'}$?

D15.39 Derive Eq. (15.62) by applying Eq. (15.44) and taking the second time derivative of $\mathbf{r}_{B'/A}$ in the fixed reference frame *OXYZ*.

[†]Cf. D15.39.

[††]Cf. Eqs. (15.22) and (15.52).

15.15   Interpretations for Coriolis Acceleration

Let us now focus our attention on some possible interpretations for the Coriolis acceleration $2\mathbf{\Omega} \times \mathbf{v}_{B/Axyz}$ as manifested in Eq. (15.64). We recall that the coordinate system $Axyz$ is embedded in the slab S to form a *moving reference frame Axyz*, whose angular velocity in the *fixed reference frame OXYZ* is $\mathbf{\Omega}$. Furthermore, we suppose that B' is a point *of* the slab S, and B' *coincides* with the moving particle B at the instant considered, as shown in Fig. 15.29. The *general interpretation* for the Coriolis acceleration $2\mathbf{\Omega} \times \mathbf{v}_{B/Axyz}$ is that *it represents the difference between the acceleration $\mathbf{a}_{B/B'}$ of B relative to B' in the fixed reference frame OXYZ and the acceleration $\mathbf{a}_{B/Axyz}$ of B in the moving reference frame Axyz*; i.e.,

$$2\mathbf{\Omega} \times \mathbf{v}_{B/Axyz} = \mathbf{a}_{B/B'} - \mathbf{a}_{B/Axyz} \qquad (15.65)$$

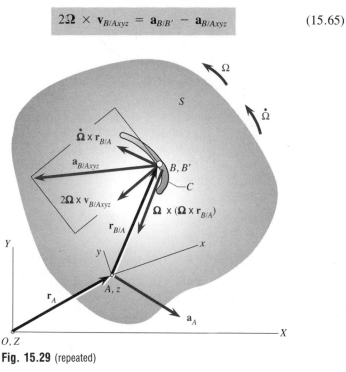

Fig. 15.29 (repeated)

To verify the truth of the above general interpretation, we first note from relative motion that

$$\mathbf{a}_{B/B'} = \mathbf{a}_B - \mathbf{a}_{B'} \qquad (15.66)$$

Rearranging Eq. (15.64) and employing Eq. (15.66), we write

$$2\mathbf{\Omega} \times \mathbf{v}_{B/Axyz} = (\mathbf{a}_B - \mathbf{a}_{B'}) - \mathbf{a}_{B/Axyz}$$

$$= \mathbf{a}_{B/B'} - \mathbf{a}_{B/Axyz}$$

which confirms the truth of the general interpretation.

In the case in which the particle B has a zero acceleration in the moving reference frame $Axyz$, the Coriolis acceleration $2\mathbf{\Omega} \times \mathbf{v}_{B/Axyz}$ lends itself to a *special interpretation* that *it simply represents the acceleration $\mathbf{a}_{B/B'}$ of B relative to B' in the fixed reference frame OXYZ*; i.e.,

$$\mathbf{a}_{B/Axyz} = \mathbf{0} \qquad 2\mathbf{\Omega} \times \mathbf{v}_{B/Axyz} = \mathbf{a}_{B/B'} \qquad (15.67)$$

The validity of the preceding special interpretation is obvious because it simply states a special case of Eq. (15.65).

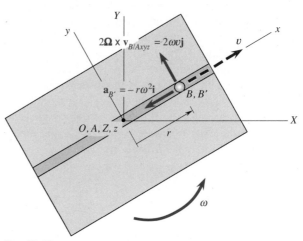

Fig. 15.30

To aid in visualizing the Coriolis acceleration in the simplest possible case in which it exists, let us consider a rotating slab with a slot in which a small particle B is confined to slide, as shown in Fig. 15.30, where $Axyz$ is embedded in the slab and $OXYZ$ is fixed to the ground. Assume that the slab rotates with a constant angular velocity $\boldsymbol{\omega} = \omega\mathbf{K}$ and the particle B moves along the slot with a constant speed v in $Axyz$. Then, referring to Fig. 15.30, we have

$$\boldsymbol{\Omega} = \omega\mathbf{K} = \omega\mathbf{k} \qquad \dot{\boldsymbol{\Omega}} = 0 \qquad \mathbf{r}_{B/A} = r\mathbf{i}$$
$$\mathbf{v}_{B/Axyz} = v\mathbf{i} \qquad \mathbf{a}_{B/Axyz} = 0 \qquad \mathbf{a}_A = 0 \qquad (15.68)$$

Substituting Eqs. (15.68) into Eq. (15.58) and simplifying, we obtain

$$\mathbf{a}_B = -r\omega^2\mathbf{i} + 2\omega v\mathbf{j} \qquad (15.69)$$

Recall that B' is a point *of* the slab, and B' *coincides* with B at the instant considered, as shown in Fig. 15.30. By Eqs. (15.61) and (15.68), we obtain

$$\mathbf{a}_{B'} = -r\omega^2\mathbf{i} \qquad (15.70)$$

$$2\boldsymbol{\Omega} \times \mathbf{v}_{B/Axyz} = 2\omega v\mathbf{j} \qquad (15.71)$$

Substituting Eqs. (15.69) and (15.70) into Eq. (15.66), we get

$$\mathbf{a}_{B/B'} = 2\omega v\mathbf{j} \qquad (15.72)$$

Equations (15.71) and (15.72) clearly demonstrate that *the Coriolis acceleration* $2\boldsymbol{\Omega} \times \mathbf{v}_{B/Axyz}$ *simply represents the acceleration* $\mathbf{a}_{B/B'}$ *of B relative to B' in the fixed reference frame $OXYZ$ when the acceleration of B in the moving reference frame $Axyz$ is* $\mathbf{a}_{B/Axyz} = 0$.

Developmental Exercise

D15.40 Give general and special interpretations for the *Coriolis acceleration* of a particle.

EXAMPLE 15.20

Refer to Example 15.19. Making use of a rotating reference frame, determine, for the position shown, α_{OD} of the link OD and $\mathbf{a}_{B/OD}$ of the pin at B relative to the link OD.

Solution. As we did in Example 15.19, we first let $OXYZ$ *be fixed to the ground* and $Oxyz$ *be embedded in the link OD*, as shown, where we have $\mathbf{k} = \mathbf{K}$. From Example 15.19, we know that the crank BC rotates with a constant angular velocity $\boldsymbol{\omega}_{BC} = -4.25\mathbf{k}$ rad/s; therefore, $\boldsymbol{\alpha}_{BC} = \mathbf{0}$. Besides, we know that, in the position shown,

$$\boldsymbol{\Omega} = \omega_{OD}\mathbf{k} = 1.105\mathbf{k} \text{ rad/s}$$

$$\mathbf{v}_{B/Oxyz} = v_{B/OD}\mathbf{i} = 71.4\mathbf{i} \text{ in./s}$$

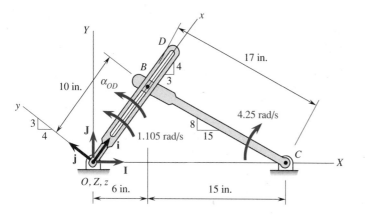

Since $\mathbf{a}_C = \mathbf{0}$ and $\boldsymbol{\alpha}_{BC} = \mathbf{0}$, we apply Eqs. (15.25) and (15.26) to write

$$\mathbf{a}_B = \mathbf{a}_{B/C} = (\mathbf{a}_{B/C})_t + (\mathbf{a}_{B/C})_n$$

$$= -\omega_{BC}^2\mathbf{r}_{B/C} = -(4.25)^2(-15\mathbf{I} + 8\mathbf{J})$$

$$\mathbf{a}_B = 270.94\mathbf{I} - 144.5\mathbf{J} \tag{1}$$

Next, we write

$$\dot{\boldsymbol{\Omega}} = \alpha_{OD}\mathbf{k} \qquad \mathbf{a}_{B/Oxyz} = \mathbf{a}_{B/OD} = a_{B/OD}\mathbf{i}$$

Since $\mathbf{a}_O = \mathbf{0}$ and $\mathbf{r}_{B/O} = 10\mathbf{i}$, we apply Eq. (15.58) to write

$$\mathbf{a}_B = \mathbf{a}_{B/Oxyz} + \dot{\boldsymbol{\Omega}} \times \mathbf{r}_{B/O} + \boldsymbol{\Omega} \times (\boldsymbol{\Omega} \times \mathbf{r}_{B/O})$$

$$+ 2\boldsymbol{\Omega} \times \mathbf{v}_{B/Oxyz} + \mathbf{a}_O$$

$$= a_{B/OD}\mathbf{i} + \alpha_{OD}\mathbf{k} \times 10\mathbf{i} + 1.105\mathbf{k} \times (1.105\mathbf{k} \times 10\mathbf{i})$$

$$+ 2(1.105\mathbf{k}) \times 71.4\mathbf{i} + \mathbf{0}$$

$$\mathbf{a}_B = (a_{B/OD} - 12.21)\mathbf{i} + (10\alpha_{OD} + 157.79)\mathbf{j} \tag{2}$$

Substituting $\mathbf{i} = \frac{1}{5}(3\mathbf{I} + 4\mathbf{J})$ and $\mathbf{j} = \frac{1}{5}(-4\mathbf{I} + 3\mathbf{J})$ into Eq. (2), we get

$$\mathbf{a}_B = (0.6a_{B/OD} - 8\alpha_{OD} - 133.56)\mathbf{I}$$

$$+ (0.8a_{B/OD} + 6\alpha_{OD} + 84.91)\mathbf{J} \tag{2'}$$

From Eqs. (1) and (2'), we obtain two ensuing scalar equations, which yield $a_{B/OD} = 59.17$ and $\alpha_{OD} = -46.12$. Thus, we have

$$\alpha_{OD} = 46.1 \text{ rad/s}^2 \;\curvearrowright \qquad \mathbf{a}_{B/OD} = 59.2 \text{ in./s}^2 \;\measuredangle\; 53.1° \;\blacktriangleleft$$

REMARK. Let B' be a point of the link OD coinciding with the pin at B. Then, from the solution in this example, it can be shown that

$$\mathbf{a}_B = 47.0\mathbf{i} - 303.4\mathbf{j} \text{ in./s}^2$$

$$\mathbf{a}_{B'} = -12.2\mathbf{i} - 461.2\mathbf{j} \text{ in./s}^2$$

$$\mathbf{a}_{B/B'} = \mathbf{a}_B - \mathbf{a}_{B'} = 59.2\mathbf{i} + 157.8\mathbf{j} \text{ in./s}^2$$

Clearly, $\mathbf{a}_{B/OD} \neq \mathbf{a}_{B/B'}$. However, by projection, we see that

$$a_{B/OD} = \mathbf{i} \cdot \mathbf{a}_{B/B'}$$

Developmental Exercises

D15.41 Derive Eq. (2) in Example 15.20 by replacing $Oxyz$ with $B'xyz$ where B' is a point of the link OD coinciding with the pin at B.

D15.42 Refer to D15.37. Determine α_{OD} and $\mathbf{a}_{B/OD}$.

EXAMPLE 15.21

The car B is rounding a circular curve while the car A is moving along a straight path with speeds and time rates of change of speed as indicated. For the instant shown, determine the velocity and acceleration of the car A in the coordinate system $Bxyz$ embedded in the car B.

Solution. Since the motion of the car A relative to the car B is to be determined, we let *BXYZ be fixed to the ground* and *Bxyz be embedded in the car B* as shown. Applying Eqs. (15.53) and (15.58) to the motion of the car A relative to the car B, we write

$$\mathbf{v}_A = \mathbf{v}_{A/Bxyz} + \mathbf{\Omega} \times \mathbf{r}_{A/B} + \mathbf{v}_B \tag{1}$$

$$\mathbf{a}_A = \mathbf{a}_{A/Bxyz} + \dot{\mathbf{\Omega}} \times \mathbf{r}_{A/B} + \mathbf{\Omega} \times (\mathbf{\Omega} \times \mathbf{r}_{A/B})$$
$$+ 2\mathbf{\Omega} \times \mathbf{v}_{A/Bxyz} + \mathbf{a}_B \tag{2}$$

where $\mathbf{v}_{A/Bxyz}$ and $\mathbf{a}_{A/Bxyz}$ are the velocity and acceleration of the car A in $Bxyz$, and $\mathbf{\Omega}$ and $\dot{\mathbf{\Omega}}$ are the angular velocity and angular acceleration of $Bxyz$ (and hence the car B) in $BXYZ$, respectively. From the given data, we write

$$\mathbf{v}_A = 25\mathbf{j} \qquad \mathbf{v}_B = 20\mathbf{j} \qquad \mathbf{r}_{A/B} = -10\mathbf{i}$$

$$\mathbf{\Omega} = -\frac{v_B}{\rho}\mathbf{k} = -\frac{20}{100}\mathbf{k} = -0.2\mathbf{k}$$

Substituting these quantities into Eq. (1), we write

$$25\mathbf{j} = \mathbf{v}_{A/Bxyz} + (-0.2\mathbf{k}) \times (-10\mathbf{i}) + 20\mathbf{j}$$

$$\mathbf{v}_{A/Bxyz} = 3\mathbf{j} \text{ m/s} \quad \blacktriangleleft$$

Furthermore, from the given data, we write

$$(\mathbf{a}_B)_t = 15\mathbf{j} \qquad (\mathbf{a}_B)_n = \frac{v_B^2}{\rho}\mathbf{i} = \frac{(20)^2}{100}\mathbf{i} = 4\mathbf{i}$$

$$\mathbf{a}_B = (\mathbf{a}_B)_t + (\mathbf{a}_B)_n = 15\mathbf{j} + 4\mathbf{i} = 4\mathbf{i} + 15\mathbf{j}$$

$$\mathbf{a}_A = -5\mathbf{j}$$

$$\dot{\mathbf{\Omega}} = -\frac{(a_B)_t}{\rho}\mathbf{k} = -\frac{15}{100}\mathbf{k} = -0.15\mathbf{k}$$

Substituting these quantities and others into Eq. (2), we write

$$-5\mathbf{j} = \mathbf{a}_{A/Bxyz} + (-0.15\mathbf{k}) \times (-10\mathbf{i})$$
$$+ (-0.2\mathbf{k}) \times [-0.2\mathbf{k} \times (-10\mathbf{i})]$$
$$+ 2(-0.2\mathbf{k}) \times (3\mathbf{j}) + (4\mathbf{i} + 15\mathbf{j})$$

$$\mathbf{a}_{A/Bxyz} = -5.6\mathbf{i} - 21.5\mathbf{j} \text{ m/s}^2 \quad \blacktriangleleft$$

Developmental Exercises

D15.43 Refer to Example 15.21. For the instant shown, determine the velocity and acceleration of the car A in a nonrotating coordinate system translating with the car B.

D15.44 Refer to Example 15.21. For the instant shown, determine the velocity and acceleration of the car B in a coordinate system embedded in the car A.

PROBLEMS

15.108 through 15.110* In the position shown, the crank BC rotates with ω_{BC} as indicated, and its pin at B slides in the slot of the member AD. Determine for this position ω_{AD} of AD and $\mathbf{v}_{B/AD}$ of the pin at B relative to AD.

Fig. P15.108 and P15.111

Fig. P15.109 and P15.112*

Fig. P15.110* and P15.113

15.111 through 15.113 In the position shown, the crank BC rotates with $\alpha_{BC} = 2$ rad/s² as well as ω_{BC} as indicated. For this position, determine α_{AD} of the member AD and $\mathbf{a}_{B/AD}$ of the pin at B relative to AD.

15.114* In this position shown, the rod BD moves downward through the fixed collar at C with $\mathbf{v}_{BD}$ as indicated, and its pin at B slides in the slot of the member AE. Determine this position ω_{AE} of AE and $\mathbf{v}_{B/AE}$ of the pin at B relative to AE.

15.115 In the position shown, the rod BD moves with $\mathbf{a}_{BD} = 25$ mm/s² ↑ as well as $\mathbf{v}_{BD}$ as indicated. For this position, determine α_{AE} of the member AE and $\mathbf{a}_{B/AE}$ of the pin at B relative to AE.

15.116 The dumper pivoted at C is operated by the hydraulic cylinder AB. In the position shown, the piston rod is being extended with a velocity of 120 mm/s and an acceleration of 23.1 mm/s² relative to the cylinder. For this position, determine ω_D and α_D of the container D.

Fig. P15.114* and P15.115

Fig. P15.116　　　　　**Fig. P15.117**

15.117 The rod DE slides freely through the pivoted collar at A. In the position shown, the pin at its end D moves with $\mathbf{v}_D = 2.89$ ft/s ↓ and $\mathbf{a}_D = 5.78$ ft/s² ↑. For this position, determine ω and α of the rod DE.

15.118 In the position shown, the crank OD rotates with a constant angular velocity $\omega_{OD} = 1.69$ rad/s ⤵. For this position, determine (a) α_{DE} of the rod DE which slides freely through the pivoted collar at A, (b) $\mathbf{a}_B$ of the point B of the rod DE coinciding with A.

Fig. P15.118

15.119* The boats A and B are rounding circular curves of radius 50 m with equal constant speeds of 10 m/s on a lake. For the instant shown, determine the velocity and acceleration of the boat B in a reference frame embedded in the boat A.

Fig. P15.119*

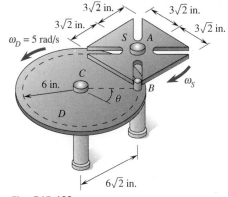

Fig. P15.120

15.120 The Geneva mechanism shown consists of a star wheel S, which rotates about A, and a driving wheel D, which rotates about C. The pin B attached to D engages one of the four radial slots of S in the position $\theta = 45°$ and turns S through 90° before disengaging itself from the slot in the position $\theta = 135°$. Note that the angular velocity of S is zero as B enters and leaves each slot of S. Thus, S produces only intermittent angular motion while D executes a constant angular velocity $\omega_D = 5$ rad/s $\circlearrowright$ as indicated. At the instant when $\theta = 60°$, determine (a) ω_S of the star wheel S, (b) the velocity of the pin B in a coordinate system embedded in S.

15.121* Refer to Prob. 15.120. At the instant when $\theta = 60°$, determine (a) α_S of the star wheel S, (b) the acceleration of the pin B in a coordinate system embedded in S.

15.16 Concluding Remarks

It has been pointed out that kinematics deals with the study of motion of bodies without considering the cause of motion, while kinetics relates the motion of bodies to the force system causing the motion; however, kinetics generally contains some kinematics. Thus, this chapter provides a useful background for the study of kinetics of rigid bodies in the subsequent chapters.

Early on in the chapter, we note that the kinematics of a rigid body in rotation about a fixed axis is mathematically similar to that of the rectilinear motion of a particle. The analytical and graphical techniques presented in Chap. 11 are therefore applicable to the solutions of problems of rigid bodies rotating about fixed axes.

The major concepts and techniques for studying the plane motion of rigid bodies using nonrotating reference frames include the following: relations between linear and angular motions, equations of relative motion, linkage equations for velocities and accelerations, velocity and acceleration centers, and the parametric method. The technique using acceleration center is particularly useful in analyzing accelerations in a system at the instant the system is *accelerated from zero angular velocity*.

The latter part of this chapter is devoted to the study of plane kinematics involving rotating reference frames. In a system where rotations of two members as well as sliding between them occur at a guided connection, the technique using rotating reference frames often brings about an effective analysis of velocities and accelerations in the system. Furthermore, this technique is shown to be useful in the kinematic analysis of particles where the particles and the reference frame are moving independently of one another or along curved paths. Although the use of rotating reference frames is challenging and mind-bending at times, its development does set the stage for the study of motions of bodies in three dimensions in Chap. 19.

REVIEW PROBLEMS

15.122   The load L is pulled up the incline from rest through a distance of 6 ft by a cable wrapped around the drum D which is attached to the gear B of a hoisting system as shown. If the gear A drives the gear B from rest with a constant angular acceleration α_A from the time $t = 0$ to $t = 4$ s, determine (a) α_A, (b) the speed of the load L when $t = 4$ s.

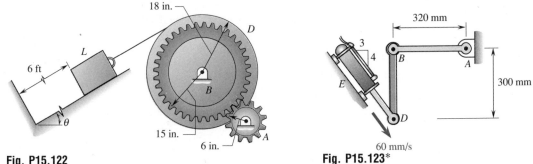

Fig. P15.122   **Fig. P15.123***

15.123*   In the position shown, the piston rod of the hydraulic cylinder at E is being extended at a constant rate of 60 mm/s. For this position, determine ω_{AB} and ω_{BD} of the links AB and BD.

15.124   Refer to Prob. 15.123*. For the position shown, determine α_{AB} and α_{BD} of the links AB and BD.

15.125   The rod BD slides freely through the fixed collar at E. In the position shown, the wheel at O is rotated with a constant angular velocity of 3 rad/s ↻. For this position, determine $\mathbf{v}_B$ and $\mathbf{a}_B$ of the point B.

Fig. P15.125

Fig. P15.126

15.126 The crank *OA* of the oil pumping unit rotates at a constant rate of 2 rad/s ⤵. The pump rod *EF* is connected by cables *DE* to the sector *D* which is welded to the walking beam *BCD*. Note that the rod hanger at *E* as well as the pump rod always moves vertically during the pumping. For the position shown, determine $\mathbf{a}_E$ of the hanger at *E* and $\boldsymbol{\alpha}_{AB}$ of the connecting rod *AB*.

15.127* In the position shown, the link *AB* rotates with a constant angular velocity of 3.2 rad/s ⤴. For this position, determine $\boldsymbol{\alpha}_{BD}$ and $\boldsymbol{\alpha}_{DE}$ of the links *BD* and *DE*.

Fig. P15.127* and P15.128   **Fig. P15.129**

15.128 In the position shown, the link *AB* is accelerated from rest with $\alpha_{AB} = 4$ rad/s² ⤴. Determine for this position (a) $\boldsymbol{\alpha}_{BD}$ and $\boldsymbol{\alpha}_{DE}$ of the links *BD* and *DE*, (b) $\mathbf{a}_G$ of the midpoint *G* of the link *BD*.

15.129 Each of the four particles *P*, *Q*, *R*, and *S* moves with a constant speed *u* relative to the slot of the rigid slab, which rotates about *O* with a constant counter-clockwise angular velocity $\boldsymbol{\omega}$ in the position shown. For this position and the directions of motion as indicated, determine the acceleration of each particle.

15.130* Solve Prob. 15.129 if the rigid slab rotates about *O* with a constant clockwise angular velocity $\boldsymbol{\omega}$ in the position shown.

15.131 The crank *OD* rotates with a constant angular velocity $\omega_{OD} = 2.5$ rad/s ⤴. When $\theta = 0$, determine $\boldsymbol{\alpha}_{DE}$ of the rod *DE*, which slides freely through the pivoted collar at *A*.

Fig. P15.131

15.132* Using the *parametric method* with θ and ϕ as dependent parameters, solve Prob. 15.131.

15.133 If the motion of the rod *OA* is as indicated, determine for the instant shown $\boldsymbol{\alpha}_{CD}$ of the rod *CD* and $\mathbf{a}_{A/CD}$ of the collar at *A* relative to the rod *CD*.

Fig. P15.133

Plane Kinetics of Rigid Bodies: Force and Acceleration

Chapter 16

KEY CONCEPTS

- Moment of inertia of a mass about an axis.
- Radius of gyration of a mass about an axis.
- Parallel-axis theorem for moments of inertia of a mass.
- Effective force-moment system on a rigid body in plane motion.
- D'Alembert's principle and the method of force and acceleration for rigid bodies in plane motion.
- Constrained general plane motion of rigid bodies.

Unlike a particle which may only be accelerated to translate by the concurrent forces acting on it, a rigid body may be accelerated to both translate and rotate by the nonconcurrent force system acting on it. We know that the innate property of a body to resist a translational acceleration is due to the *translational inertia* of the body, which is measured by its *mass*. Similarly, the innate property of a rigid body to resist a rotational acceleration is due to the *rotational inertia* of the rigid body, which is measured by its *moment of inertia*. Thus, mass moments of inertia are important characteristics of rigid bodies in kinetics; they are covered in the first part of the chapter. Later, we utilize Chasles' theorem, among other concepts, to establish the effective force-moment system acting on a rigid body in plane motion. The development leads to the property known as d'Alembert's principle, which underlies the *method of force and acceleration* for rigid bodies. Additional methods for studying plane kinetics of rigid bodies are presented in the next two chapters.

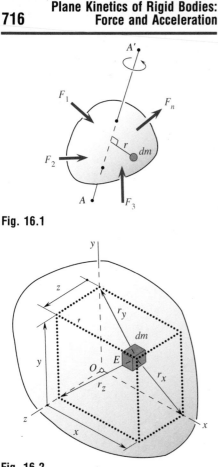

Fig. 16.1

Fig. 16.2

MASS MOMENTS OF INERTIA[†]

16.1 Moments of Inertia of a Mass

In addition to having a mass, a rigid body has a definite shape and size. Thus, under the action of nonconcurrent forces $\mathbf{F}_1$, $\mathbf{F}_2$, . . . , $\mathbf{F}_n$ as shown in Fig. 16.1, a rigid body will generally be accelerated to rotate as well as translate. The tendency of a body to resist a translational acceleration is, of course, measured by the *mass* of the body. On the other hand, the tendency of a body to resist a rotational acceleration about an axis is measured by a quantity called the *moment of inertia* of the body about that axis. This quantity, to be encountered in Sec. 16.5 and interpreted in Sec. 16.8, is an integral of the form $I = \int$ (distance)^2 d(mass) or

$$I = \int r^2 \, dm \tag{16.1}$$

where dm is a differential mass element of the body, r is the shortest distance between dm and the axis of rotation, say AA', and the integration extends over the entire body. The moment of inertia of a body defined in Eq. (16.1) is often referred to as the *mass moment of inertia*.

The differential mass element dm illustrated in Fig. 16.1 for Eq. (16.1) is of the *third order*. However, dm in Eq. (16.1) often takes the form of a *first-order* slender strip, thin ring, or thin shell if the particles in dm can be taken as equidistant from the axis of rotation, with r being the *equidistance* within the framework of first-order accuracy.

The moments of inertia of a body about the x, y, and z axes are customarily denoted by I_x, I_y, and I_z (sometimes, I_{xx}, I_{yy}, and I_{zz}), respectively. As shown in Fig. 16.2, the rectangular coordinates of the differential mass element dm at the point E are (x, y, z) and the shortest distances between dm and the x, y, and z axes are r_x, r_y, and r_z, respectively. We write

$$\begin{aligned}
I_x &= \int r_x^2 \, dm = \int (y^2 + z^2) \, dm \\
I_y &= \int r_y^2 \, dm = \int (z^2 + x^2) \, dm \\
I_z &= \int r_z^2 \, dm = \int (x^2 + y^2) \, dm
\end{aligned} \tag{16.2}$$

Note from Eqs. (16.1) and (16.2) that the moment of inertia of a body about any axis is always a *positive scalar quantity*, whose dimensions are *mass* $\times$ *(length)*^{2}, because dm is always taken as positive regardless of the position in which it is situated. The primary units for the moment of inertia of a body are kg·m^2 in the SI and slug·ft^2 (or lb·ft·s^2) in the U.S. customary system. Unless otherwise stated, it will be assumed that each body considered is homogeneous and has a constant mass density.

Developmental Exercise

D16.1   Show that 1 slug·ft^2 = 1 lb·ft·s^2 = 1.356 kg·m^2.

[†]Those who have included the *mass moments of inertia* (cf. Secs. 7.11 through 7.14) in the study of *Statics* may proceed directly to Sec. 16.5. The *mass products of inertia* are needed and presented in Chap. 19.

EXAMPLE 16.1

A slender rod of length L and mass m is shown. Determine the moments of inertia I_x, I_y, and I_z of the rod about the axes as indicated.

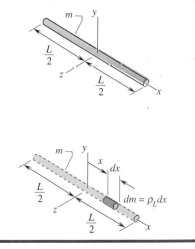

Solution. The mass density of the rod may be expressed as $\rho_L = m/L$. We note that the y and z coordinates of the mass element dm may be taken as zero throughout the rod as shown. Thus, we write

$$I_x = \int r_x^2 \, dm = \int (y^2 + z^2) \, dm = 0$$

$$I_y = \int r_y^2 \, dm = \int_{-L/2}^{L/2} x^2 \rho_L \, dx = \frac{1}{12} \rho_L L^3 = \frac{1}{12}\left(\frac{m}{L}\right)L^3$$

By symmetry, we note that $I_z = I_y$. Thus, we have

$$I_x = 0 \qquad I_y = I_z = \frac{1}{12}mL^2 \qquad \blacktriangleleft$$

EXAMPLE 16.2

A thin rectangular plate of mass m lies in the zx plane as shown. Determine I_z of the plate.

Solution. The mass density of the plate may be expressed as $\rho_A = m/(ca)$. Choosing dm to be a slender rod as shown, we write

$$I_z = \int r_z^2 \, dm = \int_{-a/2}^{a/2} x^2 \rho_A c \, dx = \frac{1}{12} \rho_A c a^3$$

$$I_z = \frac{1}{12} ma^2 \qquad \blacktriangleleft$$

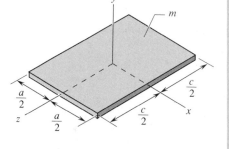

REMARK. For the thin plate in this example, we note that $y \approx 0$ throughout the plate. By Eq. (16.2), we write

$$I_x = \int z^2 \, dm$$

$$I_y = \int (z^2 + x^2) \, dm$$

$$I_z = \int x^2 \, dm$$

Thus, for such a plate, we have $I_y = I_z + I_x$.

Developmental Exercises

D16.2 Define the moment of inertia of a body about an axis.

D16.3 Determine I_x of the plate in Example 16.2.

EXAMPLE 16.3

A thin circular disk of radius r and mass m lies in the xy plane as shown. Determine I_x, I_y, and I_z of the disk.

Solution. Since the thin circular disk lies in the xy plane with $z \approx 0$, we readily deduce from Eqs. (16.2) that

$$I_z = I_x + I_y$$

By symmetry, we note that $I_x = I_y$. Thus, we have

$$I_z = 2I_x \qquad I_x = I_y = \tfrac{1}{2}I_z$$

The mass density of the disk may be expressed as $\rho_A = m/(\pi r^2)$. Referring to the sketch, we write

$$I_z = \int r_z^2\, dm = \int_0^r u^2 \rho_A(2\pi u)\, du$$

$$= \frac{1}{2}\rho_A \pi r^4 = \frac{1}{2}\left(\frac{m}{\pi r^2}\right)\pi r^4$$

$$I_z = \frac{1}{2}mr^2 \qquad I_x = I_y = \frac{1}{4}mr^2 \quad \blacktriangleleft$$

Developmental Exercise

D16.4 A circular cylinder of total mass m and radius r is shown. Applying the results obtained in Example 16.3, determine I_z of the cylinder.

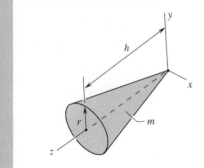

Fig. D16.4

EXAMPLE 16.4

A circular cone having a mass m, a height h, and a base of radius r is shown. Using the solution for I_z in Example 16.3 as a formula, determine I_z of the cone.

Solution. Choosing dm to be a thin disk as shown in a different perspective for better clarity, we first determine the mass density ρ and dm, in terms

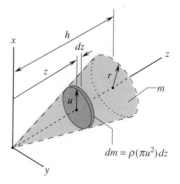

of z, as follows:

$$u = \frac{r}{h} z \qquad dm = \rho(\pi u^2) \, dz = \frac{\rho \pi r^2}{h^2} z^2 \, dz$$

$$m = \int dm = \frac{\rho \pi r^2}{h^2} \int_0^h z^2 \, dz = \frac{\rho \pi r^2 h}{3}$$

$$\rho = \frac{3m}{\pi r^2 h} \qquad dm = \frac{3m}{h^3} z^2 \, dz$$

Using the solution for I_z in Example 16.3 as a formula, we write

$$dI_z = \frac{1}{2} (dm) \, u^2 = \frac{3mr^2}{2h^5} z^4 \, dz$$

$$I_z = \int dI_z = \frac{3mr^2}{2h^5} \int_0^h z^4 \, dz = \frac{3mr^2}{2h^5} \left(\frac{1}{5} h^5 \right)$$

$$I_z = \frac{3}{10} mr^2 \quad \blacktriangleleft$$

Developmental Exercise

D16.5   For the circular cone in Example 16.4, is $I_z = I_x + I_y$?

EXAMPLE 16.5

A tetrahedron of mass density $\rho = 15$ slugs/ft^2 is bounded by the planes $x + 2y + 4z = 8$, $x = 0$, $y = 0$, and $z = 0$, where the lengths are in feet. Determine I_x of the tetrahedron.

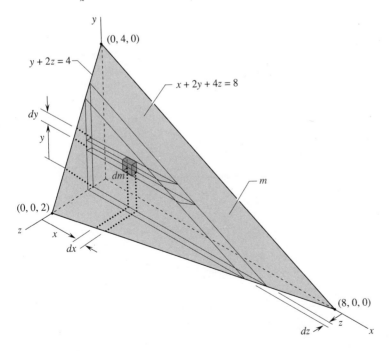

Solution. Let the differential mass element dm of the tetrahedron be chosen as sketched. We have $dm = \rho \, dV = \rho \, dx \, dy \, dz = 15 \, dx \, dy \, dz$, and

$$I_x = \int (y^2 + z^2) \, dm$$

$$= 15 \int_{z=0}^{2} \int_{y=0}^{4-2z} \int_{x=0}^{8-2y-4z} (y^2 + z^2) \, dx \, dy \, dz$$

$$= 15 \int_{z=0}^{2} \int_{y=0}^{4-2z} (y^2 + z^2)(8 - 2y - 4z) \, dy \, dz$$

$$= 20 \int_{0}^{2} (5z^4 - 28z^3 + 60z^2 - 64z + 32) \, dz$$

$$I_x = 320 \text{ slug·ft}^2 \quad \blacktriangleleft$$

Developmental Exercise

D16.6 Determine I_y of the tetrahedron in Example 16.5.

16.2 Radius of Gyration of a Mass

The *radius of gyration* of a body about an axis is a distance whose square multiplied by the mass of the body gives the moment of inertia of the body about that axis. For a body of mass m as shown in Fig. 16.3, we write

$$I_x = k_x^2 m \qquad I_y = k_y^2 m \qquad I_z = k_z^2 m \tag{16.3}$$

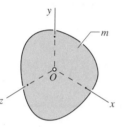

Fig. 16.3

where k_x, k_y, and k_z are the radii of gyration of the body about the x, y, and z axes, respectively.

The radius of gyration $k_{AA'}$ of a body of mass m about an axis AA' is illustrated in Fig. 16.4, where $I_{AA'}$ is the moment of inertia of the body about the axis AA'. Note that the moment of inertia of a body about an axis will remain unchanged if the entire mass of the body is concentrated at a point which is away from that axis at a *distance* equal to the *radius of gyration* of the body about that axis. Clearly, the dimension of the radius of gyration is *length*, and it is measured in meters (m) in the SI and in feet (ft) in the U.S. customary system.

Fig. 16.4

EXAMPLE 16.6

Refer to Example 16.1. Determine k_y of the rod.

Solution. From the solution in Example 16.1, we know that $I_y = \frac{1}{12} mL^2$. Thus, we write

$$\frac{1}{12} mL^2 = k_y^2 m$$

$$k_y = \frac{\sqrt{3}}{6} L = 0.289L \quad \blacktriangleleft$$

EXAMPLE 16.7

Determine k_z of the disk in Example 16.3.

Solution. From the solution in Example 16.3, we know that $I_z = \frac{1}{2} mr^2$. Thus, we write

$$\tfrac{1}{2} mr^2 = k_z^2 m$$

$$k_z = \frac{\sqrt{2}}{2} r = 0.707r \blacktriangleleft$$

Developmental Exercises

D16.7 Define the radius of gyration of a body about an axis.

D16.8 Determine k_z of the cone in Example 16.4.

D16.9 The radius of gyration of a 50-kg flywheel about its axis of rotation A is $k_A = 0.4$ m. Determine I_A of the flywheel.

D16.10 Does it make sense for the radius of gyration of any wheel about its axle to be greater than its radius? (*Hint.* Cf. Example 16.7.)

PROBLEMS

NOTE. The following problems are to be solved by *integration*.

16.1* and 16.2* Determine I_x and I_y of the slender rod of mass m as shown.

Fig. P16.1*

Fig. P16.2*

Fig. P16.3

Circular cylindrical shell

Fig. P16.4*

16.3 Determine I_x, I_y, and I_z of the thin plate of mass m as shown.

16.4* through 16.9* Determine I_z and k_z of the body of mass m as shown.

Fig. P16.5*
Circular conic shell

Fig. P16.6*
Hemispherical shell

Fig. P16.7*
Spherical shell

Fig. P16.8*
Hemisphere

Fig. P16.9*
Sphere

16.10 A body of revolution and its generating area are shown. If the specific weight is $\gamma = 322$ lb/ft^3, determine I_x and k_x of the body.

Fig. P16.10

Fig. P16.11

16.11 An ellipsoid and its generating area are shown. If the mass density is $\rho = 6$ Mg/m^3, determine I_x and k_x of the ellipsoid.

16.12* Determine I_y of the torus shown if its mass is m. (*Hint*. Choose dm to be a thin shell around the y axis. Cf. Sec. B5.2 of App. B.)

Fig. P16.12*

Fig. P16.13 and P16.15

Fig. P16.14* and P16.16

16.13 and 16.14* Determine I_x of the body shown if its mass is m.

16.15 and 16.16 Determine I_y and k_y of the body shown if its mass is m.

16.17 The mass density of the prism shown is $\rho = 4$ Mg/m^3. If $a = 0.4$ m, $b = 0.3$ m, and $c = 0.6$ m, determine I_x and k_x of the prism.

16.18* The mass density of the prism shown is $\rho = 5$ Mg/m^3. If $a = 80$ mm, $b = 6$ mm, and $c = 120$ mm, determine I_y and k_y of the prism.

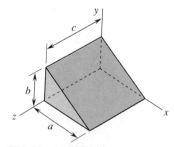

Fig. P16.17 and P16.18*

16.3   Parallel-Axis Theorem: Mass Moments of Inertia

Let the point $G(\overline{x}, \overline{y}, \overline{z})$ be the center of mass of a body of mass m as shown in Fig. 16.5 where the *central axes* x', y', and z' of the body are parallel to the fixed coordinate axes x, y, and z, respectively.[†] The coordinates of the

---

[†]The word *central* indicates something associated with a center (e.g., center of mass or center of gravity), while the word *centroidal* indicates something associated with the *centroid* of a geometric figure (e.g., line, area, or volume). Thus, *central axes* of a body are axes passing through the center of mass of the body, while *centroidal axes* of a body are axes passing through the centroid of the volume of the body. Note that the *center of mass* of a body does not necessarily coincide with the *centroid* of the volume of the body. If these two points coincide in a body, then the *centroidal axes* are also the *central axes* of the body.

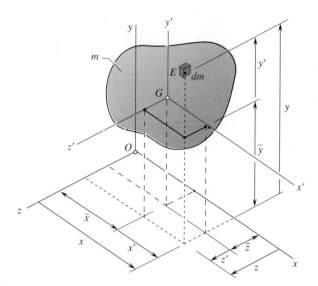

Fig. 16.5

differential mass element dm at the point E are (x, y, z) in the $Oxyz$ fixed coordinate system but are (x', y', z') in the $Gx'y'z'$ central coordinate system as indicated. Thus, we write

$$x = \bar{x} + x' \qquad y = \bar{y} + y' \qquad z = \bar{z} + z' \tag{16.4}$$

$$I_x = \int (y^2 + z^2)\, dm = \int [(\bar{y} + y')^2 + (\bar{z} + z')^2]\, dm$$

$$= \int (y'^2 + z'^2)\, dm + 2\bar{y} \int y'\, dm + 2\bar{z} \int z'\, dm + (\bar{y}^2 + \bar{z}^2) \int dm$$

$$= \bar{I}_{x'} + 2\bar{y}\bar{y}'m + 2\bar{z}\bar{z}'m + (\bar{y}^2 + \bar{z}^2)m$$

where $\bar{I}_{x'}$, represents the moment of inertia of the body about the central axis x', $\bar{y}'$ represents the distance between G and the $z'x'$ plane, and $\bar{z}'$ represents the distance between G and the $x'y'$ plane. Since G lies in the $z'x'$ and $x'y'$ planes, we see that $\bar{y}' = \bar{z}' = 0$. Thus, we write

$$I_x = \bar{I}_{x'} + m(\bar{y}^2 + \bar{z}^2) \tag{16.5}$$

and similarly,

$$I_y = \bar{I}_{y'} + m(\bar{z}^2 + \bar{x}^2)\, I_z \qquad \bar{I}_{z'} + m(\bar{x}^2 + \bar{y}^2) \tag{16.6}$$

From Fig. 16.5, we see that the sum $\bar{y}^2 + \bar{z}^2$ in Eq. (16.5) represents the square of the distance between the x and x' axes. Similarly, the sums $\bar{z}^2 + \bar{x}^2$ and $\bar{x}^2 + \bar{y}^2$ represent the squares of the distances between the y and y' axes, and the z and z' axes, respectively.

The formulas in Eqs. (16.5) and (16.6) are called the *parallel-axis theorem* for mass moments of inertia, which may be stated as follows: *The moment of inertia I of a body of mass m about any given axis is equal to the moment of inertia $\bar{I}$ of the body about a central axis parallel to the given axis plus the product md^2, where d is the distance between those two axes,* as shown in Fig. 16.6; i.e.,

$$I = \bar{I} + md^2 \tag{16.7}$$

Fig. 16.6

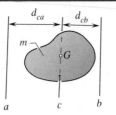

Fig. D16.11

Developmental Exercise

D16.11 If a, b, and c are three parallel axes and the axis c passes through the center of mass G of a body of mass m as shown, show that

$$I_a = I_b + m(d_{ca}^2 - d_{cb}^2)$$

EXAMPLE 16.8

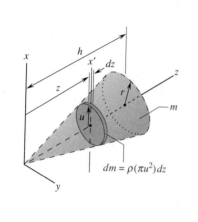

Determine I_x of the circular cone in Example 16.4.

Solution. Let dm be a thin disk and x' be a central axis of dm as shown, where the radius of the disk is $u = (r/h)\,z$ and its mass density is $\rho = 3m/(\pi^2 h)$. Applying the *parallel-axis theorem* and the solution for I_x in Example 16.3 as a formula, we write

$$dI_x = d\bar{I}_{x'} + (dm)\,z^2 = \tfrac{1}{4}\,(dm)\,u^2 + (dm)\,z^2$$

$$= (\tfrac{1}{4}u^2 + z^2)(\rho\pi u^2\,dz) = \frac{\rho\pi r^2}{4h^4}(r^2 + 4h^2)z^4\,dz$$

$$I_x = \int dI_x = \frac{\rho\pi r^2}{4h^4}(r^2 + 4h^2)\int_0^h z^4\,dz$$

$$= \frac{\rho\pi r^2 h}{20}(r^2 + 4h^2) = \frac{3m}{\pi r^2 h}\frac{\pi r^2 h}{20}(r^2 + 4h^2)$$

$$I_x = \frac{3}{20}\,m(r^2 + 4h^2) \blacktriangleleft$$

Fig. D16.13

Developmental Exercises

D16.12 Describe the *parallel-axis theorem* for mass moments of inertia.

D16.13 The circular cone shown has a mass m. Determine I_x of the cone.

16.4 Composite Bodies

A composite body has several component bodies. The moment of inertia I of a composite body about an axis is equal to the sum of the moments of inertia I_1, I_2, . . . , I_n of its component bodies about that axis; i.e.,

$$I = I_1 + I_2 + \cdots + I_n \tag{16.8}$$

For use in determining the moments of inertia of composite bodies, the moments of inertia of homogeneous bodies of some common shapes are given in Fig. 16.7, where the plates, disks, rings, and shells are understood to be thin, and m is the total mass of the body shown.

Slender rod

$$I_x = \frac{1}{3} mL^2 \sin^2\theta \qquad I_y = \frac{1}{3} mL^2 \cos^2\theta$$

$$I_z = \frac{1}{3} mL^2 \qquad \bar{I}_{x'} = \frac{1}{12} mL^2 \sin^2\theta$$

$$\bar{I}_{y'} = \frac{1}{12} mL^2 \cos^2\theta \qquad \bar{I}_{z'} = \frac{1}{12} mL^2$$

Circular ring

$$I_x = I_z = \frac{1}{2} mr^2$$

$$I_y = mr^2$$

Rectangular parallelepiped

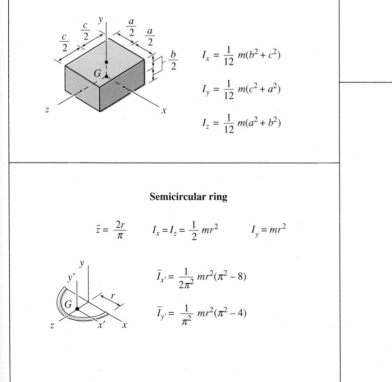

$$I_x = \frac{1}{12} m(b^2 + c^2)$$

$$I_y = \frac{1}{12} m(c^2 + a^2)$$

$$I_z = \frac{1}{12} m(a^2 + b^2)$$

Semicircular ring

$$\bar{z} = \frac{2r}{\pi} \qquad I_x = I_z = \frac{1}{2} mr^2 \qquad I_y = mr^2$$

$$\bar{I}_{x'} = \frac{1}{2\pi^2} mr^2(\pi^2 - 8)$$

$$\bar{I}_{y'} = \frac{1}{\pi^2} mr^2(\pi^2 - 4)$$

Rectangular plate

$$I_x = \frac{1}{12} mc^2 \qquad I_y = \frac{1}{12} m(c^2 + a^2) \qquad I_z = \frac{1}{12} ma^2$$

Circular disk

$$I_y = \frac{1}{2} mr^2$$

$$I_x = I_z = \frac{1}{4} mr^2$$

Circular cylinder

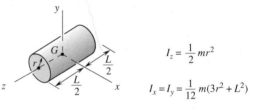

$$I_z = \frac{1}{2} mr^2$$

$$I_x = I_y = \frac{1}{12} m(3r^2 + L^2)$$

Semicircular disk

$$\bar{z} = \frac{4r}{3\pi}$$

$$I_y = \frac{1}{2} mr^2$$

$$I_x = I_z = \frac{1}{4} mr^2$$

$$\bar{I}_{x'} = \frac{1}{36\pi^2} mr^2(9\pi^2 - 64)$$

$$\bar{I}_{y'} = \frac{1}{18\pi^2} mr^2(9\pi^2 - 32)$$

Fig. 16.7(a)   Moments of inertia of bodies of common shapes.

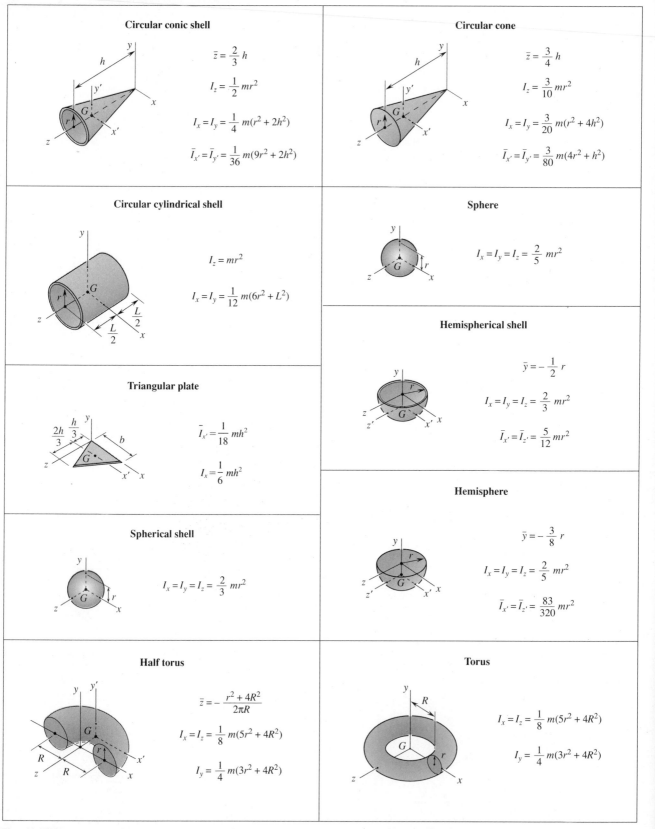

Circular conic shell

$$\bar{z} = \frac{2}{3} h$$

$$I_z = \frac{1}{2} mr^2$$

$$I_x = I_y = \frac{1}{4} m(r^2 + 2h^2)$$

$$\bar{I}_{x'} = \bar{I}_{y'} = \frac{1}{36} m(9r^2 + 2h^2)$$

Circular cone

$$\bar{z} = \frac{3}{4} h$$

$$I_z = \frac{3}{10} mr^2$$

$$I_x = I_y = \frac{3}{20} m(r^2 + 4h^2)$$

$$\bar{I}_{x'} = \bar{I}_{y'} = \frac{3}{80} m(4r^2 + h^2)$$

Circular cylindrical shell

$$I_z = mr^2$$

$$I_x = I_y = \frac{1}{12} m(6r^2 + L^2)$$

Sphere

$$I_x = I_y = I_z = \frac{2}{5} mr^2$$

Hemispherical shell

$$\bar{y} = -\frac{1}{2} r$$

$$I_x = I_y = I_z = \frac{2}{3} mr^2$$

$$\bar{I}_{x'} = \bar{I}_{z'} = \frac{5}{12} mr^2$$

Triangular plate

$$\bar{I}_{x'} = \frac{1}{18} mh^2$$

$$I_x = \frac{1}{6} mh^2$$

Spherical shell

$$I_x = I_y = I_z = \frac{2}{3} mr^2$$

Hemisphere

$$\bar{y} = -\frac{3}{8} r$$

$$I_x = I_y = I_z = \frac{2}{5} mr^2$$

$$\bar{I}_{x'} = \bar{I}_{z'} = \frac{83}{320} mr^2$$

Half torus

$$\bar{z} = -\frac{r^2 + 4R^2}{2\pi R}$$

$$I_x = I_z = \frac{1}{8} m(5r^2 + 4R^2)$$

$$I_y = \frac{1}{4} m(3r^2 + 4R^2)$$

Torus

$$I_x = I_z = \frac{1}{8} m(5r^2 + 4R^2)$$

$$I_y = \frac{1}{4} m(3r^2 + 4R^2)$$

Fig. 16.7(b) Moments of inertia of bodies of common shapes.

EXAMPLE 16.9

The mass density of the slender bent rod shown is $\rho_L = 3.75$ kg/m. Determine I_y and k_y of the rod.

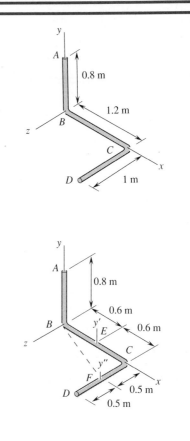

Solution. The rod may be treated as a composite body consisting of the segments AB, BC, and CD as the component bodies. Thus, we write

$$I_y = (I_y)_{AB} + (I_y)_{BC} + (I_y)_{CD}$$

Since the segment AB lies along the y axis, we have $(I_y)_{AB} = 0$. To find $(I_y)_{BC}$ and $(I_y)_{CD}$ of the segments BC and CD, we first establish their central axes y' and y'' which are parallel to the y axis. Then, we apply the *parallel-axis theorem* and the formulas for a slender rod to write

$$(I_y)_{BC} = (\bar{I}_{y'})_{BC} + m_{BC}\,\overline{BE}^2 = \tfrac{1}{12}\,m_{BC}\,\overline{BC}^2 + m_{BC}\,\overline{BE}^2$$

$$= \tfrac{1}{12}[3.75(1.2)](1.2)^2 + [3.75(1.2)](0.6)^2$$

$$= 2.16$$

$$(I_y)_{CD} = (\bar{I}_{y''})_{CD} + m_{CD}\,\overline{BF}^2$$

$$= \tfrac{1}{12}\,m_{CD}\,\overline{CD}^2 + m_{CD}(\overline{BC}^2 + \overline{CF}^2)$$

$$= \tfrac{1}{12}[3.75(1)](1)^2 + [3.75(1)][(1.2)^2 + (0.5)^2]$$

$$= 6.65$$

$$I_y = 0 + 2.16 + 6.65 \qquad\qquad I_y = 8.81 \text{ kg·m}^2 \blacktriangleleft$$

$$I_y = k_y^2 m: \qquad 8.81 = k_y^2(3.75)(0.8 + 1.2 + 1)$$

$$k_y = 0.885 \text{ m} \blacktriangleleft$$

Developmental Exercises

D16.14   Deterime I_x and k_x of the bent rod in Example 16.9.

D16.15   Determine I_z and k_z of the bent rod in Example 16.9.

EXAMPLE 16.10

A thin machine part of mass density $\rho_A = 4.83$ lbm/ft^2 is shown, where $a = 6$ in. Determine I_y of the machine part.

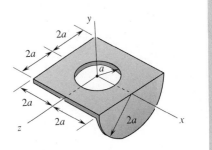

Solution. Let the machine part be a composite body consisting of a square plate of mass m_1, a semicircular plate of mass m_2, and an *empty* circular plate of mass m_3. Then, we have

$$m_1 = \rho_A(4a)^2 = 16\rho_A a^2$$

$$m_2 = \rho_A[\tfrac{1}{2}\pi(2a)^2] = 2\pi\rho_A a^2$$

$$m_3 = \rho_A(\pi a^2) = \pi\rho_A a^2$$

where $\rho_A = 0.15$ slug/ft^2 and $a = 0.5$ ft. Applying the *parallel-axis theorem* and appropriate formulas in Fig. 16.7, we write

$$I_y = (I_y)_1 + (I_y)_2 - (I_y)_3$$
$$= (I_y)_1 + [(\bar{I}_{y'})_2 + m_2 d_2^2] - (I_y)_3$$
$$= \tfrac{1}{12} m_1[(4a)^2 + (4a)^2]$$
$$+ [\tfrac{1}{4} m_2(2a)^2 + m_2(2a)^2] - \tfrac{1}{2} m_3 a^2$$
$$= \tfrac{1}{6}(256 + 57\pi)\rho_A a^4$$

$$I_y = 0.680 \text{ slug·ft}^2 \quad \blacktriangleleft$$

Developmental Exercises

D16.16 Determine I_x and k_x of the machine part in Example 16.10.

D16.17 Determine I_z and k_z of the machine part in Example 16.10.

EXAMPLE 16.11

A body consists of a rectangular parallelepiped and a circular cylinder as shown, where $a = 20$ mm and the mass density is $\rho = 6$ Mg/m^3. Determine I_z and k_z of the body.

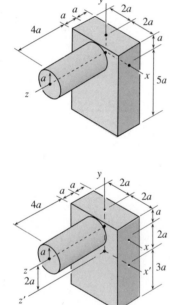

Solution. Let the component bodies be the rectangular parallelepiped of mass m_1 and the circular cylinder of mass m_2. Then, we have

$$m_1 = \rho(4a)(6a)(2a) = 48\rho a^3$$
$$m_2 = \rho(\pi a^2)(4a) = 4\pi\rho a^3$$
$$m = m_1 + m_2 = (48 + 4\pi)\rho a^3$$

where $\rho = 6000$ kg/m^3 and $a = 0.02$ m. Applying the *parallel-axis theorem* and appropriate formulas in Fig. 16.7, we write

$$I_z = (I_z)_1 + (I_z)_2 = [(\bar{I}_{z'})_1 + m_1 d_1^2] + (I_z)_2$$
$$= \{\tfrac{1}{12} m_1[(4a)^2 + (6a)^2] + m_1(2a)^2\} + \tfrac{1}{2} m_2 a^2$$
$$= (400 + 2\pi)\rho a^5 = (400 + 2\pi)(6000)(0.02)^4$$

$$I_z = 7.80 \times 10^{-3} \text{ kg·m}^2 \quad \blacktriangleleft$$

$$I_z = k_z^2 m: \quad (400 + 2\pi)\rho a^5 = k_z^2(48 + 4\pi)\rho a^3$$

$$k_z = 51.8 \text{ mm} \quad \blacktriangleleft$$

Developmental Exercises

D16.18 Determine I_x and k_x of the body in Example 16.11.

D16.19 Determine I_y and k_y of the body in Example 16.11.

PROBLEMS

NOTE: Unless otherwise specified, the formulas given in Fig. 16.7 as well as the parallel-axis theorem may be applied in solving the following problems of mass moments of inertia.

16.19 through 16.21 Determine I_x and I_y of the body made of slender rods as shown.

Fig. P16.19 and P16.22*   Fig. P16.20* and P16.23   Fig. P16.21 and P16.24

16.22* through 16.24 Determine I_z and k_z of the body made of slender rods as shown.

16.25 through 16.27 Determine I_x and I_y of the sheet metal part shown.

Fig. P16.25 and P16.28*   Fig. P16.26* and P16.29   Fig. P16.27 and P16.30

16.28* through 16.30 Determine I_z and k_z of the sheet metal part shown.

16.31 through 16.33 Determine I_y and k_y of the thin shell shown.

Truncated conic shell
Fig. P16.31 and P16.34*

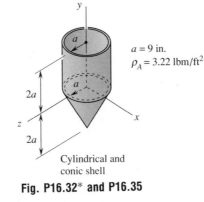

Cylindrical and conic shell
Fig. P16.32* and P16.35

Cylindrical and hemispherical shell
Fig. P16.33 and P16.36

16.34* through 16.36 Determine I_x and k_x of the thin shell shown.

16.37 through 16.39 Determine I_y and k_y of the body of revolution shown.

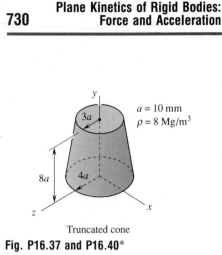

Truncated cone

Fig. P16.37 and P16.40*

Cylinder
and cone

Fig. P16.38* and P16.41

Cylinder and
hemisphere

Fig. P16.39 and P16.42

16.40* through 16.42 Determine I_x and k_x of the body of revolution shown.

16.43 A horizontal hole of radius 1 in. is drilled through a body as shown. If $\rho = 14$ slugs/ft^3, determine I_z and k_z of the body.

Fig. P16.43

Fig. P16.44

16.44 A flywheel is of the same shape and size as that obtained by revolving the area shown about the x axis one revolution, where $r_1 = 250$ mm, $r_2 = 200$ mm, $r_3 = 50$ mm, $r_4 = 25$ mm, $b_1 = 150$ mm, and $b_2 = 25$ mm. If $\rho = 7$ Mg/m^3, determine I_x and k_x of the flywheel.

16.45 A link consists of a cylinder and a half-torus as shown. It is made of an alloy with $\rho = 6$ Mg/m^3. Determine I_y and I_z of the link.

Fig. P16.45

$a = 20$ mm

16.46* Using single integration, verify the formula $I_x = \frac{1}{12}m(3r^2 + L^2)$ for the circular cylinder in Fig. 16.7.

16.47* Using single integration, verify the formula $I_x = \frac{1}{12}m(6r^2 + L^2)$ for the circular cylindrical shell in Fig. 16.7.

16.48* Using single integration, verify the formula $I_x = \frac{1}{8}m(5r^2 + 4R^2)$ for the torus in Fig. 16.7). (*Hint.* Refer to the *hint* in Prob. 16.12*.)

FORCE AND ACCELERATION

16.5 Effective Force-Moment System on a Rigid Body

A rigid body may be viewed as a continuum composed of a large number of particles which remain at fixed distances from each other. Thus, a rigid body is a special case of a system of particles. We saw in Sec. 12.6 that *the system of external forces acting on a system of particles is equipollent to the system of effective forces on the particles*. This relation may be applied to the plane motion of a rigid body of mass m under the action of the external forces $\mathbf{F}_1$, $\mathbf{F}_2$, . . . , $\mathbf{F}_n$ in an inertial reference frame Oxy, as shown in Fig. 16.8, where G is the mass center of the rigid body, $(\Delta m_i)\mathbf{a}_i$ is the effective force on the ith particle P_i having a mass Δm_i and an acceleration $\mathbf{a}_i$, $\bar{\mathbf{a}}$ is the acceleration of G, and $m\bar{\mathbf{a}}$ is the resultant effective force, which does not necessarily pass through G.

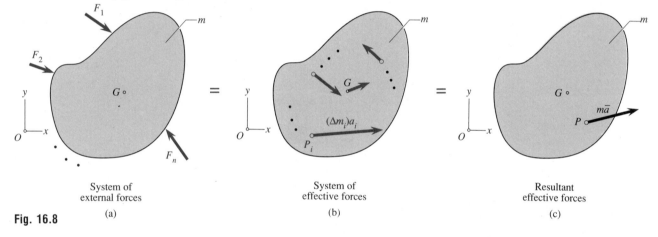

System of external forces	System of effective forces	Resultant effective forces
(a)	(b)	(c)

Fig. 16.8

Note in Fig. 16.8 that the *equal* sign $=$ (not the *equipollent* sign $\Rightarrow$) for *equivalence* is used because the particles of a rigid body are not free particles; they are interconnected at fixed distances. Applying *Chasles' theorem* in Sec. 15.5, we may treat the *plane motion* of a rigid body as being equal to the *translation* of the rigid body *with G* plus the *rotation* of the rigid body *about G*. In other words, the *system of effective forces* on a rigid body of mass m as illustrated in Fig. 16.8(b) may be treated as being equal to the vector sum of the following two systems: (1) the *translational system of effective forces* corresponding to the translation of the rigid body with G whose acceleration is $\bar{\mathbf{a}}$, and (2) the *rotational system of effective forces* corresponding to the rotation of the rigid body about G with an angular velocity $\boldsymbol{\omega}$ and an angular acceleration $\boldsymbol{\alpha}$. Such a relation is depicted in Fig. 16.9, where $Gx'y'$ is a nonrotating reference frame attached to G, and r_i' is the radial distance from G to the ith particle P_i whose mass is Δm_i and whose acceleration $\mathbf{a}_i$ is given by

$$\mathbf{a}_i = \bar{\mathbf{a}} + (r_i'\alpha\mathbf{e}_t + r_i'\omega^2\mathbf{e}_n) \qquad (16.9)$$

The vectors $\mathbf{e}_t$ and $\mathbf{e}_n$ in Eq. (16.9) and Fig. 16.9(c) are, respectively, the tangential and the normal unit vectors associated with the path of the ith particle P_i rotating about G.

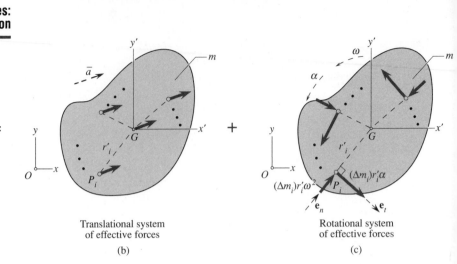

System of
effective forces

(a)

Translational system
of effective forces

(b)

Rotational system
of effective forces

(c)

Fig. 16.9

Letting $\mathbf{R}_{\text{eff,trans}}$ be the *resultant effective force* of the *translational* system of effective forces in Fig. 16.9(b), we write

$$\mathbf{R}_{\text{eff,trans}} = \sum_{i=1}^{n} (\Delta m_i)\bar{\mathbf{a}} = \bar{\mathbf{a}} \sum_{i=1}^{n} \Delta m_i = \bar{\mathbf{a}}m$$

$$\mathbf{R}_{\text{eff,trans}} = m\bar{\mathbf{a}} \tag{16.10}$$

Letting $(\mathbf{M}_G^R)_{\text{eff,trans}}$ be the *resultant effective moment* about G of the *translational* system of effective forces in Fig. 16.9(b), we write

$$(\mathbf{M}_G^R)_{\text{eff,trans}} = \sum_{i=1}^{n} \mathbf{r}_i' \times (\Delta m_i)\bar{\mathbf{a}}$$

$$= -\bar{\mathbf{a}} \times \sum_{i=1}^{n} \mathbf{r}_i' (\Delta m_i)$$

$$= -\bar{\mathbf{a}} \times \bar{\mathbf{r}}'m$$

where, by the *principle of moments* $\bar{\mathbf{r}}'$ is the position vector of G in $Gx'y'$. Since G is the origin, we see that $\bar{\mathbf{r}}' = \mathbf{0}$. Thus, we have

$$(\mathbf{M}_G^R)_{\text{eff,trans}} = \mathbf{0} \tag{16.11}$$

We find from Eqs. (16.10) and (16.11) that *the translational system of effective forces* in Fig. 16.9(b) *is equivalent to the resultant effective force* $\mathbf{R}_{\text{eff,trans}}$ *acting through* G as illustrated in Fig. 16.10.

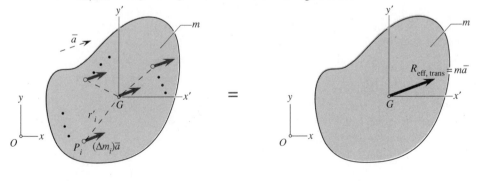

Translational system
of effective forces

Resultant of the
translational system

Fig. 16.10

Now, letting $\mathbf{R}_{\text{eff,rot}}$ be the *resultant effective force* of the *rotational* system of effective forces in Fig. 16.9(c) and applying Eq. (16.9) as well as the *principle of moments*, we write

$$\mathbf{R}_{\text{eff,rot}} = \sum_{i=1}^{n} [(\Delta m_i)r_i'\alpha\mathbf{e}_t + (\Delta m_i)r_i'\omega^2\mathbf{e}_n]$$

$$= \sum_{i=1}^{n} (\Delta m_i)(r_i'\alpha\mathbf{e}_t + r_i'\omega^2\mathbf{e}_n)$$

$$= \sum_{i=1}^{n} (\Delta m_i)(\mathbf{a}_i - \bar{\mathbf{a}})$$

$$= \sum_{i=1}^{n} \mathbf{a}_i(\Delta m_i) - \bar{\mathbf{a}} \sum_{i=1}^{n} (\Delta m_i)$$

$$= \bar{\mathbf{a}}m - \bar{\mathbf{a}}m$$

$$\mathbf{R}_{\text{eff,rot}} = \mathbf{0} \tag{16.12}$$

Letting $(\mathbf{M}_G^R)_{\text{eff,rot}}$ be the *resultant effective moment* about G of the *rotational* system of effective forces in Fig. 16.9(c) and noting that $\mathbf{e}_n \times \mathbf{e}_n = \mathbf{0}$ and $\mathbf{e}_n \times \mathbf{e}_t = -\mathbf{k}$, we write

$$(\mathbf{M}_G^R)_{\text{eff,rot}} = \sum_{i=1}^{n} (-r_i'\mathbf{e}_n) \times [(\Delta m_i)r_i'\alpha\mathbf{e}_t + (\Delta m_i)r_i'\omega^2\mathbf{e}_n]$$

$$= \alpha(-\mathbf{e}_n \times \mathbf{e}_t) \sum_{i=1}^{n} r_i'^2(\Delta m_i) + \mathbf{0}$$

$$= (\alpha\mathbf{k}) \int r'^2 \, dm = \alpha\bar{I}$$

$$(\mathbf{M}_G^R)_{\text{eff,rot}} = \bar{I}\alpha \tag{16.13}$$

where we let $n \to \infty$ in the last summation to obtain the *integral* which equals the moment of inertia $\bar{I}$ of the body about G. We find from Eqs. (16.12) and (16.13) that *the rotational system of effective forces* in Fig. 16.9(c) *is equivalent to the resultant effective moment* $(\mathbf{M}_G^R)_{\text{eff,rot}}$ *acting about* G as illustrated in Fig. 16.11.

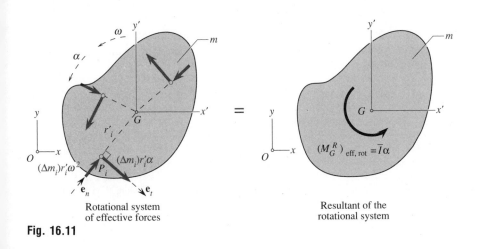

Rotational system
of effective forces

Resultant of the
rotational system

Fig. 16.11

By *Chasles' theorem* and Eqs. (16.10) and (16.12), we conclude that the *resultant effective force* acts through G and is given by

$$\mathbf{R}_{\text{eff}} = \mathbf{R}_{\text{eff,trans}} + \mathbf{R}_{\text{eff,rot}} = m\bar{\mathbf{a}} + \mathbf{0}$$

$$\boxed{\mathbf{R}_{\text{eff}} = m\bar{\mathbf{a}}} \tag{16.14}$$

Furthermore, by *Chasles' theorem* and Eqs. (16.11) and (16.13), we conclude that the coexisting *resultant effective moment* acting about G is given by

$$(\mathbf{M}_G^R)_{\text{eff}} = (\mathbf{M}_G^R)_{\text{eff, trans}} + (\mathbf{M}_G^R)_{\text{eff,rot}} = \mathbf{0} + \bar{I}\boldsymbol{\alpha}$$

$$\boxed{(\mathbf{M}_G^R)_{\text{eff}} = \bar{I}\boldsymbol{\alpha}} \tag{16.15}$$

The vectors $m\bar{\mathbf{a}}$ and $\bar{I}\boldsymbol{\alpha}$ in Eqs. (16.14) and (16.15) are respectively called the *effective force* and the *effective moment* at the mass center G of the rigid body. A diagram showing the vectors $m\bar{\mathbf{a}}$ and $\bar{I}\boldsymbol{\alpha}$ at G, or their equivalent at a different point, is referred to as the *effective-force diagram* of the rigid body, where the effective moment is dependent on the point chosen (cf. D16.20).

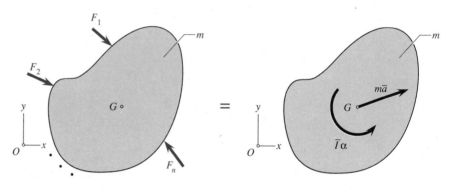

Fig. 16.12   Free–body diagram   Effective–force diagram

In general, we see that *the system of external forces* $\mathbf{F}_1, \mathbf{F}_2, \ldots, \mathbf{F}_n$ *acting on a rigid body in plane motion is equivalent to the effective force-moment system consisting of* $m\bar{\mathbf{a}}$ *acting through the mass center* G *and* $\bar{I}\boldsymbol{\alpha}$ *acting about* G *of the body*. This important property is illustrated in Fig. 16.12 and is a modern version of *d'Alembert's principle*, after the French mathematician Jean le Rond d'Alembert (1717–1783), even though d'Alembert's original statement was written in different terms for the motion of a *system of linked masses*.[†] However, d'Alembert's principle was not originally intended for the case of a *single particle*.[††]

[†]The original statement of d'Alembert's principle is seldom directly quoted; it requires interpretation for modern usage. D'Alembert essentially found that *the sum of the interaction forces in a system of connected bodies vanishes*. The modern version of his principle is really the consequence of that finding. The principle is contained in d'Alembert's *Traité de Dynamique*, which was first published in 1743 and later revised in 1758. A reprint of its second edition with an elaborate commentary and a bibliographical note by Thomas L. Hankins was published in 1968 by Johnson Reprint Corporation, New York. Despite some criticisms of d'Alembert's attempts to "geometrize" the world in his *Traité de Dynamique*, his work remains an important milestone in the history of mechanics.

[††]Cf. the first footnote on p. 511.

Developmental Exercises

D16.20 A rigid body of mass m in plane motion has an angular velocity $\boldsymbol{\omega}$, an angular acceleration $\boldsymbol{\alpha}$, and a moment of inertia $\bar{I}$ about its central axis perpendicular to the plane of motion. If the acceleration of its mass center G is $\bar{\mathbf{a}}$, give the proper effective force-moment system at G for the rigid body.

D16.21 Describe *d'Alembert's principle*.

16.6 Center of Percussion

Using the concept of equivalence in Sec. 5.1, we can move the effective force $m\bar{\mathbf{a}}$ at G in Fig. 16.12 to a certain point P plus the moment of $m\bar{\mathbf{a}}$ at G about P. If the moment of $m\bar{\mathbf{a}}$ at G about P is equal to $-\bar{I}\boldsymbol{\alpha}$, then we have reduced the effective force-moment system at G in Fig. 16.12 to an *equivalent single effective force* $m\bar{\mathbf{a}}$ *acting through the point* P, as shown in Fig. 16.13, where

$$\overrightarrow{PG} \times m\mathbf{a} = -\bar{I}\boldsymbol{\alpha} \tag{16.16}$$

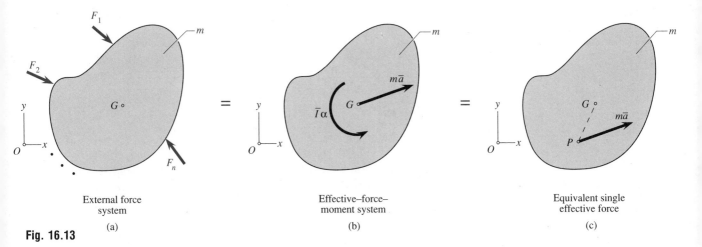

External force system
(a)

Effective–force–moment system
(b)

Equivalent single effective force
(c)

Fig. 16.13

In the case of a rigid body rotating in a plane about a fixed axis O, the point of intersection P of the *equivalent single effective force* $m\bar{\mathbf{a}}$ and the line containing the fixed point O and the mass center G of the body is called the *center of percussion* of the body. The steps which may be used to locate the center of percussion are illustrated in Fig. 16.14, where $\boldsymbol{\omega}$ is the angular velocity, $\bar{k}$ is the central radius of gyration, and the distance from the mass center G to the *center of percussion* P is

$$d = \frac{\bar{k}^2}{\bar{r}} \tag{16.17}$$

The *center of percussion* P is a unique point of the body (or its massless extension) through which the *equivalent single effective force* $m\bar{\mathbf{a}}$ must pass. If a body pivoted at a fixed point O is suddenly struck by a relatively large force $\mathbf{F}$ through its *center of percussion* P, the free-body diagram and the

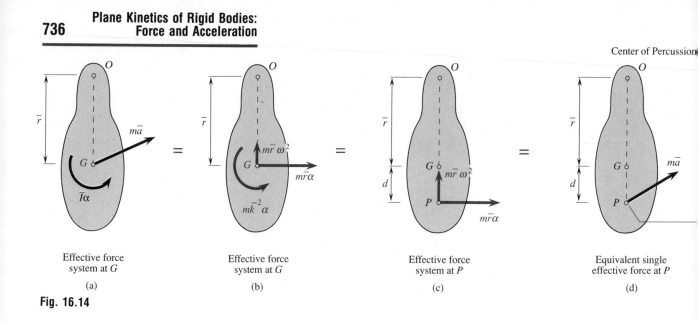

Center of Percussion

Effective force system at G	Effective force system at G	Effective force system at P	Equivalent single effective force at P
(a)	(b)	(c)	(d)

Fig. 16.14

effective-force diagram of the body may be drawn as shown in Fig. 16.15. Referring to this figure, we write

$$+ \circlearrowleft \ \Sigma M_P: \qquad -(\bar{r} + d)O_t = 0$$

Since $\bar{r} + d \neq 0$, we see that $O_t = 0$; i.e.,

$$\mathbf{O}_t = \mathbf{0} \tag{16.18}$$

This means that *any force striking a rigid body through its center of percussion will not induce any tangential component of reaction from its supporting pivot.* As an application, a baseball batter may avoid the sting in the grip by appropriately hitting the ball at the *center of percussion* (or the "sweet spot") of the bat. Furthermore, the concept of center of percussion is useful in designing impact devices and testing machines.

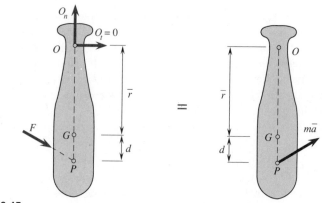

Fig. 16.15

Developmental Exercises

D16.22 If the central radius of gyration of a baseball bat AB is 7 in., determine the distance d between its *center of percussion P* and its *mass center G* as shown.

D16.23 A slender rod AB of mass m is at rest when the force $\mathbf{F}$ is applied to its lower end B as shown in the space diagram. Applying *Chasles' theorem* to the study of the rod with its hinged end A as the point of reference, we may draw the free-body diagram (FBD) and a series of effective-force diagrams (EFDs) of the rod having an angular acceleration $\boldsymbol{\alpha}$ as shown. Verify that the systems of vectors in EFD-1 through EFD-4 are equivalent to each other and have values as indicated.

Fig. D16.22

Fig. D16.23 (The point P in EFD-2 is the *center of percussion* of the rod.)

16.7 Method of Force and Acceleration

In plane kinetics of rigid bodies, an *effective-force diagram* may show *one* of the following: (a) the effective force-moment system acting at the mass center of a rigid body, (b) the effective force-moment systems acting at the various mass centers of the component parts of a composite rigid body, and (c) the effective force-moment systems acting at the various mass centers at the component parts of a set of connected rigid bodies. The *method of force and acceleration* for rigid bodies is a method of solution utilizing the *equivalence* or *equipollence* of the system of vectors in the free-body diagram of a body *to* the system of vectors in the corresponding effective-force diagram, where the body may be a simple rigid body, a composite rigid body, or a set of connected rigid bodies. Notice that the underlying principle in this method is *d'Alembert's principle*, which is illustrated in Fig. 16.12.

The equivalence depicted in Fig. 16.12 on p. 734 can be implemented algebraically by *equating* the sums of appropriate rectangular components of the vectors, as well as equating the sums of the moments of the vectors about a desirable point, as follows:

$$\Sigma V_x: \qquad \Sigma F_x = m\bar{a}_x \qquad\qquad (16.19)$$

$$\Sigma V_y: \qquad \Sigma F_y = m\bar{a}_y \qquad\qquad (16.20)$$

$$\Sigma M_G: \qquad \Sigma M_G = \bar{I}\alpha \qquad\qquad (16.21)$$

where the x and y axes can be any convenient pair of orthogonal axes in the plane of motion of the rigid body. If the moments are computed about a given point P, Eq. (16.21) is replaced by

$$\Sigma M_P: \qquad \Sigma M_P = \Sigma M_{P,\text{eff}} \qquad\qquad (16.22)$$

where the subscript "eff" stands for "effective." The point P in Eq. (16.22) is usually a propitious point, such as the point of intersection of two unknown vectors on the body. Using Eq. (16.22) and making an intelligent choice of P as the moment center, we can often reduce the computational effort involved in solving a kinetic problem. Thus, Eq. (16.22) is a powerful equation in rigid-body kinetics.

The *method of force and acceleration* for rigid bodies is best used in solving rigid-body problems where forces (including moments) and accelerations are to be directly related without initially involving displacement, velocity, and time. Furthermore, it should be noted that at most *three* independent scalar equations can be written from a vector-diagram equation for a rigid body in plane motion. For a set of connected rigid bodies, separate vector-diagram equations for the component parts of the set may sometimes be needed. In general, *the number of determinate unknowns contained in a given number of vector-diagram equations cannot exceed the total number of independent scalar equations which can be written* from (1) equating the sums of the appropriate rectangular components, as well as equating the sums of the moments of the vectors about a point, in each vector-diagram equation; (2) the laws of friction; (3) the rigid-body kinematics; and (4) the principle of virtual work in kinetics, which will be presented in the next chapter.

For a much clearer view of the situation in the problem under consideration and a much better command in setting up equations for the efficient solution of unknowns, it is recommended that both the *free-body diagram* and the *effective-force diagram* of the rigid body be drawn and equated to each other to form a *vector-diagram equation* as illustrated in Fig. 16.12. Such a geometric approach places the problem situation in good perspective. It generally makes the learning of kinetics less abstract than the approach which applies formulas directly without the aid of any diagram.

EXAMPLE 16.12

A 150-N horizontal force **P** is applied to a 30-kg file cabinet which is initially at rest on a floor as shown. If $\mu_k = 0.2$ and slipping occurs between the cabinet and the floor, determine (a) the acceleration $\bar{\mathbf{a}}$ of its mass center G, (b) the *minimum* value of h for which the cabinet will not tip while it is being moved to the right.

Solution. As the value of h is decreased to h_{min}, tipping of the cabinet about its corner at B will impend and the reaction from the floor will pass through B. Thus, we equate the *free-body diagram* to the *effective-force*

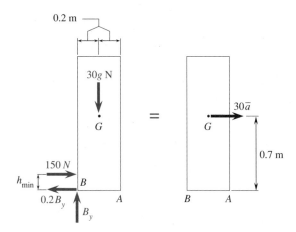

diagram for such a case as shown, where $\bar{I}\alpha = 0$ because rotation does not actually occur. We write

$$+\uparrow \Sigma V_y: \qquad B_y - 30g = 0$$

$$\xrightarrow{+} \Sigma V_x: \qquad 150 - 0.2B_y = 30\bar{a}$$

$$+\circlearrowleft \Sigma M_B: \qquad h_{min}(150) + 0.2(30g) = 0.7(30\bar{a})$$

These three equations yield $B_y = 30g$, $\bar{a} = 3.308$ m/s^2, and $h_{min} = 0.0329$ m. Thus, we have

$$\bar{\mathbf{a}} = 3.04 \text{ m/s}^2 \rightarrow \qquad h_{min} = 32.9 \text{ mm} \quad \blacktriangleleft$$

Developmental Exercise

D16.24   Refer to Example 16.12. Determine the *maximum* value of h for which the cabinet will not tip while it is being moved to the right.

EXAMPLE 16.13

The 100-kg plate *ABCD* with its mass center at *G* is held in the position shown by the wire *BE* and the links *AE* and *BF* whose masses are negligible. Determine the force in each link immediately after the wire *BE* is cut.

Solution. After the wire *BE* is cut, the plate will have a curvilinear translation, and the acceleration $\bar{\mathbf{a}}$ of *G* is equal to the acceleration $\mathbf{a}_A$ of *A* (or $\mathbf{a}_B$ of *B*) at any given instant. However, *immediately* after the wire *BE* is cut, the velocities of the plate and the links are zero; the acceleration $\bar{\mathbf{a}}$ being equal to $\mathbf{a}_A$ (or $\mathbf{a}_B$) is thus perpendicular to the link *AE* (or *BF*). Since the masses of the links *AE* and *BF* are negligible, the effective forces on these links are also negligible and the forces developed in them are taken as directed along their axes. Noting that the weight of the plate is $W = mg = 100(9.81)$ N $= 981$ N and letting $\bar{\mathbf{a}}$ be equal to $\bar{a} \ \swarrow \ 30°$, we equate

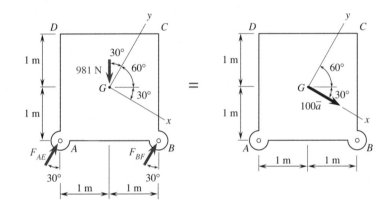

the *free-body diagram* of the plate (just after the wire is cut) to its *effective-force diagram* as shown, where the effective moment $\bar{I}\alpha$ is zero because the plate is in translation. We write

$$+\searrow \Sigma V_x: \qquad 981 \sin 30° = 100\bar{a} \qquad \bar{a} = 4.905 \text{ m/s}^2$$

$$+\ \circlearrowleft \ \Sigma M_A: \qquad 1(981) - 2(F_{BF} \cos 30°)$$

$$= 1(100\bar{a} \cos 30°) + 1(100\bar{a} \sin 30°)$$

$$+\nearrow \Sigma V_y: \qquad F_{AE} + F_{BF} - 981 \cos 30° = 0$$

$$F_{BF} = 179.5 \text{ N } C \qquad F_{AE} = 670 \text{ N } C \ \blacktriangleleft$$

Developmental Exercise

D16.25 Refer to Example 16.13. Suppose that the mass of the plate is $m > 0$. Immediately after the wire *BE* is cut, (a) does the acceleration of the plate depend on the value of *m*? (b) do the forces in the links depend on the value of *m*?

EXAMPLE 16.14

The slender bent bar *ABC* supported as shown has a mass density $\rho_L = 10$ kg/m. Determine the angular acceleration of the bent bar immediately after the wire *BD* is cut.

Solution. The bent bar *ABC* may be treated as a composite rigid body with *AB* and *BC* as its component parts. Since $\rho_L = 10$ kg/m, the masses of *AB* and *BC* are $m_{AB} = 10$ kg and $m_{BC} = 12$ kg, and their weights are $W_{AB} = 10g$ N and $W_{BC} = 12g$ N, respectively. *Immediately* after the wire *BD* is cut, the bent bar will lose balance and accelerate from rest with a counterclockwise angular acceleration α about the hinge at *C*, which is also its *acceleration center*. Thus, we equate the *free-body diagram* of the bent bar to its *effective-force diagram* as shown.

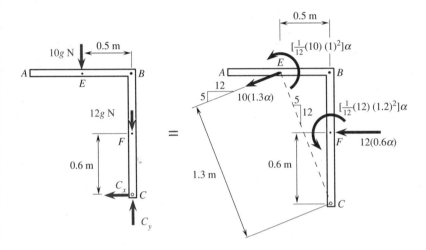

Note that the formula $\bar{I} = \frac{1}{12} mL^2$ for the central moment of inertia of a slender rod of mass *m* and length *L* is used in indicating the effective moment $\bar{I}\alpha$ about the center of mass of each of the component parts *AB* and *BC*. We write

$$+ \circlearrowleft \; \Sigma M_C: \quad 0.5(10g) = 1.3[10(1.3\alpha)]$$
$$+ \; [\tfrac{1}{12}(10)(1)^2]\alpha$$
$$+ \; 0.6[12(0.6\alpha)]$$
$$+ \; [\tfrac{1}{12}(12)(1.2)^2]\alpha$$

$$5(9.81) = \frac{70.48}{3}\alpha \qquad \alpha = 2.09 \text{ rad/s}^2 \circlearrowleft \quad \blacktriangleleft$$

Developmental Exercise

D16.26 Refer to Example 16.14. Determine the reaction from the support at *C* immediately after the wire *BD* is cut.

EXAMPLE 16.15

It is known that the cylinder B moves with $\mathbf{a}_B = 0.8$ m/s² ↓ when the system is released from rest as shown. Neglecting friction and the mass of the shaft CD, determine the moment of the inertia, $\bar{I}$ of the flywheel and its drum about the axis CD.

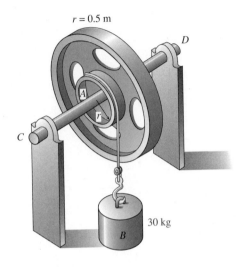

$r = 0.5$ m

Solution. The corresponding angular acceleration $\boldsymbol{\alpha}$ of the flywheel and its drum may be obtained as follows:

$$\alpha = a_B/r = 0.80/0.5 \qquad \alpha = 1.6 \text{ rad/s}^2 \text{ ↺}$$

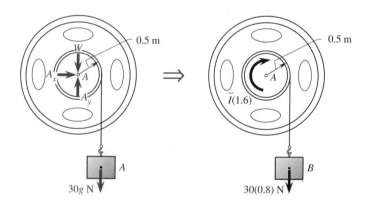

Denoting the weight of the flywheel and its drum by W, we equate the *free-body diagram* of the entire system to its *effective-force diagram* as shown, where $\bar{I}$ is the moment of inertia, and the *equipollence* sign ⇒ is used because the system is composed of connected rigid bodies. We write

$$+\text{↺} \ \Sigma M_A: \qquad 0.5(30g) = 0.5(30)(0.8) + \bar{I}(1.6)$$

$$\bar{I} = 84.5 \qquad\qquad \bar{I} = 84.5 \text{ kg·m}^2 \ \blacktriangleleft$$

Developmental Exercise

D16.27 Refer to Example 16.15. Determine the tension in the cord when the system is released from rest.

16.8 Interpretations for Mass Moments of Inertia

Referring to Fig. 16.12 and substituting $\alpha = 1 \text{ rad/s}^2$ into Eq. (16.21),

$$\Sigma M_G = \bar{I} \qquad (16.23)$$

The equation allows this interpretation: *the moment of inertia of a rigid body about its own central axis is numerically equal to the moment required to accelerate the body to rotate at the rate of* 1 rad/s^2 *about that axis.*

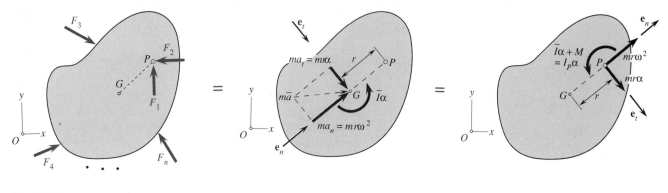

Fig. 16.16   (a)   (b)   (c)

If a rigid body of mass m rotates, at the instant considered, with an angular velocity $\boldsymbol{\omega}$ and an angular acceleration $\boldsymbol{\alpha}$ about a *fixed axis* passing through the point P, we may equate its *free-body diagram* to its *effective-force diagram* as illustrated in Fig. 16.16. Note in Fig. 16.16(b) that the effective force $m\bar{\mathbf{a}}$ is, for convenience, replaced by its tangential component $m\mathbf{a}_t = mr\alpha\mathbf{e}_t$ and its normal component $m\mathbf{a}_n = mr\omega^2\mathbf{e}_n$. These two components as well as the effective moment $\bar{I}\alpha$ may, without losing equivalence, be moved from G to P if a moment $\mathbf{M}$ equal to the sum of moments about P due to $m\mathbf{a}_t$ and $m\mathbf{a}_n$ at G is added as shown in Fig. 16.16(c). Since $m\mathbf{a}_n$ contributes no moment about P and $\mathbf{e}_n \times \mathbf{e}_t = -\mathbf{k}$, we have

$$\mathbf{M} = \overrightarrow{PG} \times m\mathbf{a}_t = -r\mathbf{e}_n \times mr\alpha\mathbf{e}_t = mr^2\alpha\mathbf{k}$$

Noting that the moment of inertia of the body about the fixed axis at P is $I_P = \bar{I} + mr^2$, we may write the effective moment of the equivalent effective force-moment system at P as

$$\Sigma \mathbf{M}_{P,\text{eff}} = \bar{I}\alpha\mathbf{k} + \mathbf{M} = (\bar{I} + mr^2)\alpha\mathbf{k} = I_P\alpha\mathbf{k}$$

$$\Sigma M_{P,\text{eff}} = I_P\alpha \qquad (16.24)$$

which is indicated in Fig. 16.16(c). Referring to this figure and Eqs. (16.22) and (16.24), we write

$$\Sigma M_P: \qquad \boxed{\Sigma M_P = I_P\alpha} \qquad (16.25)$$

Substituting $\alpha = 1 \text{ rad/s}^2$ into Eq. (16.25), we have

$$\Sigma M_P = I_P \qquad (16.26)$$

which may be interpreted as follows: *the moment of inertia of a rigid body about any axis fixed in space is numerically equal to the moment required to accelerate the body to rotate at the rate of* 1 rad/s² *about that axis.* Since the acceleration axis (cf. Sec. 15.9) has zero acceleration, a similar interpretation for the moment of inertia of a body about its acceleration axis can be shown to be valid.

Developmental Exercise

D16.28   Refer to the relations established in Fig. 16.16, where the rigid body rotates about a fixed axis passing through the point P. Show that Chasles' theorem can be applied to yield the same results as shown in Fig. 16.16(c) when the point of reference mentioned in the theorem is the fixed point P, instead of the mass center G, of the body. [*Hint.* Cf. D16.23 and note in Sec. 16.5 that $\mathbf{R}_{\text{eff,trans}} = (\mathbf{M}_P^R)_{\text{eff,trans}} = \mathbf{0}$, $\mathbf{R}_{\text{eff,rot}} = m\bar{\mathbf{a}}$, and $(\mathbf{M}_P^R)_{\text{eff,rot}} = I_P\boldsymbol{\alpha}$ in the present case.]

PROBLEMS

16.49   A plank AB of mass m is placed in a truck so that its lower end rests against a crate on the bed while its upper part rests against the top of the railing as shown. If the truck moves to the right with an acceleration $\mathbf{a}$, determine the maximum magnitude of $\mathbf{a}$ for which the plank will remain in contact with the top of the railing.

Fig. P16.49   **Fig. 16.50**

16.50   The slender bar AB of weight W is supported as shown. If the cart CD is moved to the left with an acceleration a, determine the maximum magnitude of $\mathbf{a}$ for which the wire BC will remain taut.

16.51*   Refer to Prob. 16.50. If $W = 12.88$ lb and $\mathbf{a} = 10$ ft/s² →, determine the tension in the wire BC and the reaction to the bar AB from the hinge at A.

16.52*   A crate of mass m is placed on the flat bed of a truck as shown, where the point G is the mass center of the crate. If the truck moves from rest with a forward acceleration $\mathbf{a}$ and $\mu_s = 0.35$ between the crate and the flat bed, determine the maximum magnitude of $\mathbf{a}$ for which the crate will remain unmoved relative to the truck.

16.53   Solve Prob. 16.52* if $\mu_s = 0.4$.

16.54   The forward motion of the 5000-lb forklift truck carrying a 2500-lb crate is being decelerated as shown. If $\mu_s = 0.5$ between the crate and the forklift, deter-

Fig. 16.52*

mine the maximum deceleration the truck can have without causing the crate to move relative to the forklift.

Fig. P16.54

16.55* Refer to Prob. 16.54. Neglecting the rotational effects of the wheels, determine the vertical component of the reaction from the front wheels.

16.56 Two uniform rods *AB* and *BC*, each of weight 20 lb, are welded together to form a 40-lb bent bar which is held by the wire *CE* and the links *BE* and *CD* as shown. Neglecting the masses of the links, determine $\mathbf{a}_B$ of the pin at *B* and the force in each link immediately after the wire *CE* is cut.

Fig. P16.56   **Fig. P16.57**

16.57 A 16-kg rod *AB* and a 20-kg rod *CD* are welded together to form a T bar as shown. The T bar is raised from rest at $\theta = 0$ by means of the parallel links *CE* and *DF* where a constant torque $\mathbf{M} = 360$ N·m ↻ is applied to the link *CE*. Neglecting the masses of the parallel links, determine (a) $\boldsymbol{\alpha}_{DF}$ of the link *DF* as a function of θ, (b) the force in the link *DF* when $\theta = 120°$. [*Hint.* The tangential component of the reaction from the pin at *C* to the T bar is $\mathbf{C}_t = (360/0.5)\mathbf{e}_t$ N.]

16.58 It is known that $\mu_s = 0.7$ between the road and the tires of the car as shown. Determine the shortest time in which the car can travel from rest through a distance of 100 m on a level road, assuming (a) front-wheel drive, (b) rear-wheel drive, (c) four-wheel drive.

Fig. P16.58

Fig. P16.59*

16.59* The uniform rectangular plate *ABCD* of weight *W* is suspended by two parallel wires and swings in a vertical plane. In the position shown, the tension in the wire *AE* is known to be equal to 0.1*W*. Determine the angular speed of the wire *AE* in this position.

16.60 Solve Prob. 16.59* if the tension in the wire *AE* is known to be equal to 0.05*W*.

16.61 The 1.4-Mg coal car with load is being towed by the counterweight *E* as shown. If the mass of the counterweight is 100 kg, determine the corresponding reactions from the supports to the small wheels at *A* and *B*.

Fig. P16.61

16.62* Solve Prob. 16.61 if the mass of the counterweight is 200 kg.

16.63 Each of the uniform bent bars *AB* has a weight *W* and is supported as shown. Determine the percentage of change in magnitude of the reaction from the support at *C* when the wire at *A* is suddenly cut in each case.

(a)

(b)

Fig. P16.63

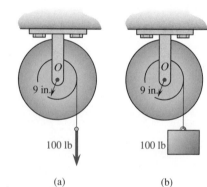

(a)

(b)

Fig. P16.64

16.64 Each of the double pulleys shown weighs 322 lb and has a central radius of gyration of 15 in. about *O*. Determine the angular acceleration in each case.

16.65 Each of the bodies shown is uniform and has a total weight *W*. Determine the angular acceleration of each body and the reaction from its support at *A* when the body is released from rest in the position shown.

(a) Square plate

(b) Square wire

(c) Circular plate

Fig. P16.65

16.66 The uniform bent bar *BCD* weighs 15 lb and is connected by a cord *AB* to a 6-lb smooth collar *A* as shown. Neglecting effects of friction, determine (a) the minimum magnitude of the constant force **F** required to cause the cord *AB* to become collinear with the segment *BC*, (b) the corresponding tension in the cord.

16.67* Solve Prob. 16.66 if $\mu_k = 0.2$ between the bent bar and the support at *C*.

16.68 The trailer shown weighs 1500 lb and is attached at *H* to a rear-bumper hitch as shown. If the car pulls the trailer with an acceleration of 10 ft/s² ← and the lightweight wheels have negligible rotational effects, determine the force exerted on the trailer by the hitch at *H*.

Fig. P16.66

Fig. P16.68

16.69* Solve Prob. 16.68 if the brakes of the car are applied, causing the trailer to decelerate at 10 ft/s² →.

16.70 The gears *A* and *B* have masses of 10 kg and 30 kg and central radii of gyration of 80 mm and 150 mm, respectively. Determine the torque **M** acting on the gear *A* if the angular acceleration of the gear *B* is 2 rad/s² ↺.

16.71* Determine the reaction from the support at *A* when the bent rod weighing 0.5 lb/ft is released from rest in the position shown.

Fig. P16.70

Fig. P16.71*   **Fig. P16.72**

16.72 Determine the reaction from the support at *A* when the plate with a mass density of 10 kg/m² is released from rest in the position shown.

16.73 A 5-Mg jet fighter aircraft decelerates at the rate of 3 m/s² by operating its thrust reverser at *C* along the runway without applying the mechanical brakes in its wheels at *A* and *B* as shown. Neglecting the aerodynamic forces on the aircraft and the rotational effects of the wheels, determine the reaction from the runway to the nose wheel at *A*.

Fig. P16.73

Fig. P16.74

16.74 A front-wheel drive car with a gross weight of 3600 lb is being tested with a dynamometer as shown. If the freely turning drum of the dynamometer has a moment of inertia $I_O = 1200$ slug·ft^2 and $\mu_s = 0.9$ between the tire and the drum, determine the maximum angular acceleration which the drum can have during the testing.

Fig. P16.75   **Fig. P16.76***

16.75 and 16.76* At the time $t = 0$, the uniform cylinder A of weight W is lowered onto the plate B which moves with a constant velocity $\mathbf{v}_0$ as shown. If $\mu_k = 0.5$ between the surfaces of contact, determine (a) the angular acceleration of the cylinder immediately after contact, (b) the time t at which the cylinder starts rolling without slipping on the plate B.

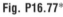
Dimensions
in millimeters
Fig. P16.77*

16.77* The moment of inertia of the 100-mm drum with flywheel as shown is 10 kg·m^2 about O, and $\mu_k = 0.4$ between the drum and the brake band. If the initial angular velocity is 600 rpm ↺ and the system is to stop in 10 revolutions, determine the force **P** which must be applied to the point D of the hand bar.[†]

16.78 Solve Prob. 16.77* if the initial angular velocity is 600 rpm ↻.

16.79 The grooved wheel weighs 128.8 lb and has a radius of gyration of 9 in. about O as shown. It is known that $\mu_k = 0.4$ between the groove and the V belt as delineated, and the wheel rotates with $\boldsymbol{\omega} = 600$ rpm ↻ when the 10-lb force **P** is applied to the point A of the lever. Determine the number of revolutions executed by the wheel during the braking.[†]

16.80* Solve Prob. 16.79 if $\boldsymbol{\omega} = 600$ rpm ↺.

Fig. P16.79

[†]Recall the belt friction formulas in Secs. 9.12 and 9.13 that $T_2 = T_1 e^{\mu_k \beta}$ for a flat belt, and $T_2 = T_1 e^{\mu_k \beta / \sin(\alpha/2)}$ for a V belt.

16.81 The central moments of inertia of the gears A and B as shown are 10 kg·m^2 and 20 kg·m^2, respectively. If the mass of the block C is $m_C = 50$ kg and the system is released from rest, determine α_A of the gear A.

Fig. P16.81

16.82* Solve Prob. 16.81 if $m_C = 100$ kg.

16.83 The 2-kg slender plank AB rests on a frictionless horizontal surface. If the end A at the instant shown has zero acceleration when a 10-N horizontal force acts on it at C as shown, determine (a) the length r, (b) $\mathbf{a}_B$ of the end B.

Fig. P16.83   **Fig. P16.84**

16.84 A 300-kg thin-walled pipe is raised from *rest* by the cables of tall overhead cranes as shown. At the instant of raising, the cable A moves with 4$\mathbf{j}$ m/s^2 and the pipe rotates with 2$\mathbf{k}$ rad/s^2. Determine the corresponding tension in each cable.

16.85* Refer to Prob. 16.84. Suppose that, at the instant of raising, the tensions in the cables A and B are 2 kN and 3 kN, respectively. Determine for that instant (a) the angular acceleration of the pipe, (b) the acceleration of each cable.

16.86* A 300-kg hemispherical shell is raised from *rest* by the cables of tall overhead cranes as shown. At the instant of raising, the accelerations of the cables A and B are 3$\mathbf{j}$ m/s^2 and 1$\mathbf{j}$ m/s^2, respectively. Determine the corresponding tension in each cable.

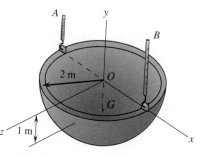

Fig. P16.86*

16.87 Refer to Prob. 16.86*. Suppose that, at the instant of raising, the tensions in the cables A and B are 1.8 kN and 1.4 kN, respectively. Determine for that instant (a) the angular acceleration of the shell, (b) the acceleration of each cable.

16.9 Constrained General Plane Motion

Fig. 16.17

A rigid body is said to be in a *constrained motion* if there exist definite relations between the components of the acceleration $\bar{\mathbf{a}}$ of the mass center G of the rigid body and its angular acceleration $\boldsymbol{\alpha}$ as well as its angular velocity $\boldsymbol{\omega}$. The rotation of a rigid body about a fixed axis is a special case of constrained plane motion, which was covered in Secs. 16.7 and 16.8. The constrained motion of rigid bodies which are in *general plane motion* is studied in this section.

An important step in the solution of a problem involving constrained general plane motion is the *kinematic analysis* of the body in the problem. For example, let us consider a slender rod AB of length L and mass m whose extremities A and B are constrained to move on the horizontal surface and the vertical wall as shown in Fig. 16.17. The rod is pushed by a force $\mathbf{P}$ applied at A and has an angular velocity $\boldsymbol{\omega}$ and an angular acceleration $\boldsymbol{\alpha}$ in the position defined by the angle θ. If effects of friction are negligible, we may draw the vector-diagram equation containing the *free-body diagram* and the *effective-force diagram* of the rod as shown in Fig. 16.18. Performing the *kinematic analysis* (cf. Secs. 15.8 through 15.10), we can express $\bar{a}_x$ and $\bar{a}_y$ in terms of α and ω for a given position defined by θ. Substituting the expressions for $\bar{a}_x$ and $\bar{a}_y$ into Fig. 16.18(b), we can reduce the total number of unknowns in the vector-diagram equation to a minimum. If the minimum number of unknowns in such an equation is three or less, the unknowns may then be determined by using the *method of force and acceleration* for rigid bodies described in Sec. 16.7.

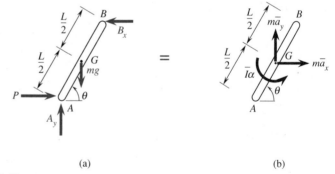

(a)   (b)

Fig. 16.18

As another example, let us consider the rolling motion of a disk (or wheel) on a rigid surface due to a horizontal force $\mathbf{P}$ applied as shown in Fig. 16.19. The vector-diagram equation containing the *free-body diagram* and *effective-force diagram* of the disk is shown in Fig. 16.20, where the

Fig. 16.19

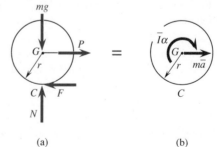

Fig. 16.20   (a)   (b)

symbols have the usual meaning. The relation between the friction force **F**
and the normal force **N**, as well as that between the acceleration $\bar{a}$ of G and
the angular acceleration α of the disk, may be summarized as follows:

1. $F \leq \mu_s N$ and $\bar{a} = r\alpha$ if it rolls without slipping.
2. $F = \mu_s N$ and $\bar{a} = r\alpha$ if it rolls with impending slipping.
3. $F = \mu_k N$ and $\bar{a} \neq r\alpha$ if it rolls and slips.

If the mass center G of the disk does not coincide with its geometric center
O, then $\bar{a}$ in the above relations is replaced by the magnitude a_O of the
acceleration $\mathbf{a}_O$ of O and $\bar{\mathbf{a}}$ may be computed from the relation $\bar{\mathbf{a}} = \mathbf{a}_{G/O} + \mathbf{a}_O$.

EXAMPLE 16.16

The 20-kg rod AB is released from rest in the position shown. Neglecting
effects of friction as well as the masses of the small wheels at A and B,
determine the angular acceleration of the rod and the reaction from its support at A immediately after release.

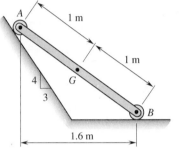

Solution.   Immediately after release, the rod has zero angular velocity,
and its *acceleration center* Z (cf. Sec. 15.9) is located as shown. Assuming
the angular acceleration α to be counterclockwise, we see that the acceleration $\bar{\mathbf{a}}$ of the mass center G is perpendicular to $\overline{ZG}$ and points to the lower
right. The magnitude of $\bar{\mathbf{a}}$ is $\bar{a} = \overline{ZG}\alpha = 0.2\sqrt{97}\,\alpha$. Thus, we draw the
vector-diagram equation containing the *free-body diagram* and the *effective-
force diagram* of the rod as shown, where the unknowns are A, B, and α.

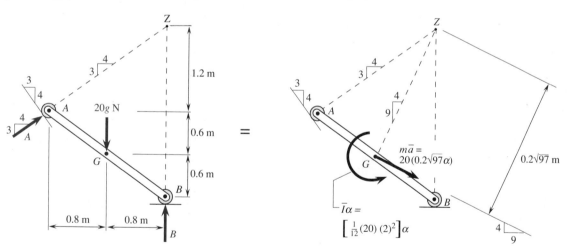

To determine $\boldsymbol{\alpha}$ and $\mathbf{A}$, we write

$$+ \circlearrowleft \ \Sigma M_Z: \qquad 0.8(20g) = [\tfrac{1}{12}(20)(2)^2]\alpha + 0.2\sqrt{97}\,[20(0.2\sqrt{97}\,\alpha)]$$

$$\alpha = 1.863 \qquad \boldsymbol{\alpha} = 1.863 \ \text{rad/s}^2 \ \circlearrowleft \ \blacktriangleleft$$

$$\xrightarrow{+} \Sigma V_x: \qquad \frac{4}{5}A = \frac{9}{\sqrt{97}}\left[20(0.2\sqrt{97}\,\alpha)\right]$$

$$\mathbf{A} = 83.8 \ \text{N} \ \measuredangle \ 36.9° \ \blacktriangleleft$$

Developmental Exercise

D16.29   Refer to Example 16.16. Determine the reaction from the support at B immediately after release.

EXAMPLE 16.17

A cord is wrapped around the outer drum of a 96.6-lb wheel and pulled vertically with a constant force of 41 lb as shown. If the wheel has a central radius of gyration of 2.5 ft and rolls without slipping on the horizontal track, determine the angular acceleration of the wheel.

Solution.   Since it rolls without slipping, we have $\bar{a} = r\alpha = 2\alpha$. Thus, we equate the *free-body diagram* of the wheel to its *effective-force diagram*

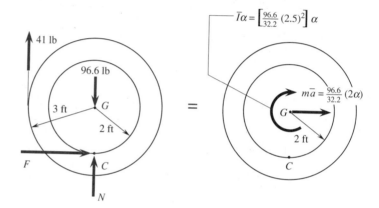

as shown, where the unknowns are N, F, and α. To determine $\boldsymbol{\alpha}$, we write

$$+ \curvearrowright \ \Sigma M_C: \qquad 3(41) = \left[\frac{96.6}{32.2}(2.5)^2\right]\alpha \ + \ 2\left[\frac{96.6}{32.2}(2\alpha)\right]$$

$$\alpha = 4 \qquad\qquad \boldsymbol{\alpha} = 4 \text{ rad/s}^2 \ \curvearrowright \ \blacktriangleleft$$

Developmental Exercise

D16.30   Refer to Example 16.17. Determine (a) the reaction components $\mathbf{F}$ and $\mathbf{N}$, (b) the minimum value of μ_s between the wheel and the track.

EXAMPLE 16.18[†]

The two slender bars OA and AB, each of length $L = 5$ ft and weight $W = 10$ lb, are released from rest in the position shown. Determine $\boldsymbol{\alpha}_{OA}$ of the bar OA immediately after release.

[†]The problem in this example is solved by using the *principle of virtual work in kinetics* in Example 17.8.

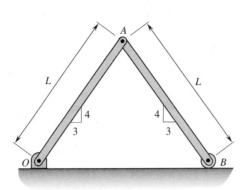

Solution. Immediately after release, the bars OA and AB have zero angular velocities and their *acceleration centers* are located at O and Z, respectively, as shown. By symmetry, we know that $\boldsymbol{\alpha}_{OA} = -\boldsymbol{\alpha}_{AB}$; so, we may assume that $\boldsymbol{\alpha}_{OA} = \alpha \, \circlearrowleft$ and $\boldsymbol{\alpha}_{AB} = \alpha \, \circlearrowright$.

In the vector-diagram equation consisting of the *free-body diagram* and the *effective-force diagram* for the entire system as shown, we note that there are four unknowns (O_x, O_y, B_y, and α) but only three independent

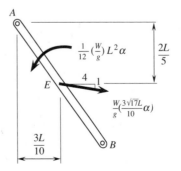

scalar equations to be written. Thus, we resort to drawing an additional vector-diagram equation for the bar AB as shown, which incurs two new unknowns (A_x and A_y) but provides three additional independent scalar equations. The six unknowns may therefore be solved from the six equations. However, to focus our effort on solving for α, we proceed as follows:

Entire system

$+\circlearrowleft \ \Sigma M_O$:

$$\frac{6L}{5} B_y - \frac{3L}{10} W - \frac{9L}{10} W$$

$$= -\frac{1}{12}\left(\frac{W}{g}\right)L^2\alpha + \frac{1}{12}\left(\frac{W}{g}\right)L^2\alpha - \frac{L}{2}\left[\frac{W}{g}\left(\frac{L}{2}\alpha\right)\right]$$

$$- \frac{2L}{5}\left\{\frac{4}{\sqrt{17}}\left[\frac{W}{g}\left(\frac{3\sqrt{17}L}{10}\alpha\right)\right]\right\} - \frac{9L}{10}\left\{\frac{1}{\sqrt{17}}\left[\frac{W}{g}\left(\frac{3\sqrt{17}L}{10}\alpha\right)\right]\right\}$$

Bar AB

$+\circlearrowleft \ \Sigma M_A$:

$$\frac{3L}{5} B_y - \frac{3L}{10} W$$

$$= \frac{1}{12}\left(\frac{W}{g}\right)L^2\alpha + \frac{2L}{5}\left\{\frac{4}{\sqrt{17}}\left[\frac{W}{g}\left(\frac{3\sqrt{17}L}{10}\alpha\right)\right]\right\}$$

$$- \frac{3L}{10}\left[\frac{1}{\sqrt{17}}\left(\frac{3\sqrt{17}L}{10}\alpha\right)\right]$$

Since $W = 10$ lb and $L = 5$ ft, the above two equations yield

$$B_y = \frac{217W}{292} = 7.43 \text{ lb} \qquad \alpha = \frac{45g}{146L} = 1.985 \text{ rad/s}^2$$

$$\alpha_{OA} = 1.985 \text{ rad/s}^2 \ \circlearrowleft \ \blacktriangleleft$$

Developmental Exercise

D16.31 Refer to Example 16.18. Determine the unknowns $\mathbf{O}_x$ and $\mathbf{O}_y$.

EXAMPLE 16.19[†]

A 40-kg block C is supported by a 20-kg cylinder A and a 30-kg cylinder B as shown, where the cylinders have the same radius $r = 100$ mm. If the cylinders roll without slipping, determine the acceleration of the block C when a 50-N force $\mathbf{P}$ is applied to it, which is initially at rest.

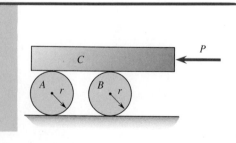

[†]The problem in this example is solved by using the *principle of virtual work in kinetics* in Example 17.9.

Solution.   Since the cylinders roll without sliding and are of the same radius $r = 0.1$ m, they must have the same angular acceleration $\boldsymbol{\alpha}$ which is related to the acceleration $\mathbf{a}_C$ of the block C by $a_C = 2r\alpha = 2(0.1)\alpha$ or $\alpha = 5a_C$. Thus, the vector-diagram equation consisting of the *free-body diagram* and the *effective-force diagram* for the entire system as well as that for each cylinder may be drawn as shown. In the three vector-diagram equations shown, there are nine unknowns: a_C, F_A, N_A, F_B, N_B, D_x, D_y, E_x, and E_y. However, our objective is mainly to find $\mathbf{a}_C$. Thus, we proceed:

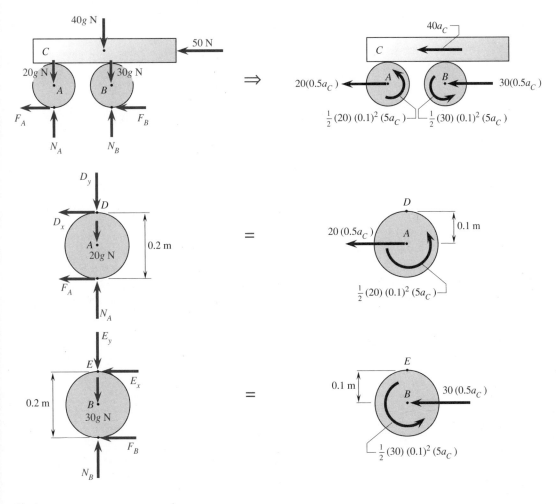

Entire system

$$\overset{+}{\rightarrow} \Sigma V_x: \qquad 50 + F_A + F_B = 40a_C + 20(0.5a_C) + 30(0.5a_C)$$

Cylinder A

$$+ \,\zeta\, \Sigma M_D: \qquad 0.2F_A = 0.1[20(0.5a_C)] - \tfrac{1}{2}(20)(0.1)^2(5a_C)$$

Cylinder B

$$+ \,\zeta\, \Sigma M_E: \qquad 0.2F_B = 0.1[30(0.5a_C)] - \tfrac{1}{2}(30)(0.1)^2(5a_C)$$

The above three equations yield

$$a_C = 0.851 \text{ m/s}^2 \qquad F_A = 2.13 \text{ N} \qquad F_B = 3.19 \text{ N}$$

$$\mathbf{a}_C = 0.851 \text{ m/s}^2 \leftarrow \blacktriangleleft$$

REMARK. Since the values of F_A and F_B are found to be positive, the directions of the friction forces $\mathbf{F}_A$ and $\mathbf{F}_B$ as indicated in the free-body diagrams are correct.

Developmental Exercises

D16.32 Refer to Example 16.19. Without knowing dimensions other than the radii of the cylinders, which of the nine unknowns in the vector-diagram equations could not be determined?

D16.33 Work Example 16.19 if the masses of the cylinders A and B are $m_A = m_B = 20$ kg.

Fig. P16.88

PROBLEMS

16.88 A small collar of weight W is securely attached at A to a hoop of weight W and radius r. The system is released from rest in the position shown and no slipping occurs. Determine the reaction from the support at B immediately after release.

16.89* and 16.90 The rod AB has a weight W and is released from rest in the position shown. Neglecting effects of friction, determine the angular acceleration of the rod immediately after release.

Fig. P16.89* and P16.91

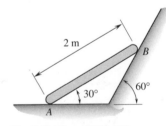

Fig. P16.90 and P16.92*

16.91 and 16.92* The rod AB has a weight W and is released from rest in the position shown. Knowing that $\mu_s = 0.1$ between all surfaces of contact, determine the normal component of reaction from the support at A immediately after release.

16.93 An aircraft lands on a level runway and covers it with a constant speed of 180 mph. The landing wheels are not turning before they touch the runway. If the length of the skid mark of the tires on the runway is 200 ft, and the normal force between the runway and the landing wheel of radius r is $8W$ where W is the weight of the wheel whose central radius of gyration is $0.7r$, determine the kinetic coefficient of friction between the tires and the runway.

Fig. P16.93

16.94* A 50-kg wheel of radius 180 mm and central radius of gyration 120 mm is released from rest on the incline as shown. Knowing that the wheel rolls without slipping, determine (a) the acceleration of the mass center G, (b) the minimum value of μ_s compatible with this motion.

Fig. P16.94*

Fig. P16.95

16.95 A 100-kg pipe rolls without slipping in a circular support of radius 2.2 m. In the position shown, the pipe has an angular speed of 10 rad/s. Determine the reaction from the support.

16.96* The pipe A and the cylinder B are in contact when they are released from rest as shown. Knowing that $r = 120$ mm and both A and B roll without slipping, determine (a) the acceleration of the center of B relative to that of A, (b) the clear distance between them after 2 s.

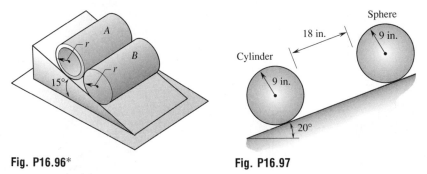

Fig. P16.96*

Fig. P16.97

16.97 The bodies as shown are released from rest at the time $t = 0$. If they roll without slipping, determine the time at which the clearance between them becomes zero.

16.98 A pipe has a cord wrapped around it at the midsection and is released from rest in the position shown. Determine the time required for the pipe to descend 2 m to the ground.

Fig. P16.98

Fig. P16.99

16.99 A cue hits a billiard ball with a horizontal force **P** at a height which is a distance d above its center O as shown. If the ball is to roll without slipping regardless of the value of μ_s between the ball and the table, determine the value of d in terms of the radius r of the ball.

16.100 A hoop is spinning at 10 rad/s when it is set on a horizontal floor as shown. If $\mu_s = 0.1$ between the hoop and the floor, determine the time during which the hoop will both roll and slip on the floor.

Fig. P16.100   **Fig. P16.101***

16.101* A half-section of pipe of mass 10 kg is released from rest in the position shown. If the section rolls without slipping, determine the acceleration of the point B and the reaction from the support at C immediately after release.

16.102 and 16.103* A cord is wrapped around the hub of the spool and pulled with a constant force **P** as shown. The spool weighs 36 lb and has a central radius of gyration of 4 in. about G. If $P = 13$ lb, $r = 3$ in., $R = 6$ in., and the spool rolls without slipping, determine (a) the acceleration of G, (b) the minimum value of μ_s between the spool and the floor.

Fig. P16.102 and P16.104*   **Fig. P16.103* and P16.105**

16.104* and 16.105 A cord is wrapped around the hub of the spool and pulled with a constant force **P** as shown. The spool has a mass of 10 kg and a radius of gyration of 150 mm about its mass center G. If $P = 40$ N, $r = 100$ mm, $R = 200$ mm, and $\mu_s \approx \mu_k \approx 0.25$ between the spool and the floor, determine (a) whether the spool slips, (b) the acceleration of G.

16.106 The door on the passenger's side of the car is unintentionally left slightly open when the driver backs up the car from rest with a constant backward acceleration of $\mathbf{a} = 5$ ft/s$^2 \rightarrow$. The door has a weight $W = 80$ lb, and its center of gravity is located at a distance $\bar{r} = 20$ in. from the hinge axis. If the central radius of gyration of the door is $\bar{k} = 12$ in., determine the angular velocity ω of the door immediately before it becomes wide open (i.e., $\theta = 90°$). [*Hint.* Cf. Eq. (15.8).]

16.107* Refer to Prob. 16.106. Determine the angular velocity ω of the door as it passes through the position $\theta = 45°$.

Fig. P16.106

16.108 A 10-kg disk has a radius of gyration of 120 mm about its mass center G. The disk rolls without slipping and has an angular velocity $\boldsymbol{\omega} = 5$ rad/s in the position shown, where $b = 60$ mm and $r = 200$ mm. Determine the corresponding angular acceleration of the disk and the reaction from the support at C.

Fig. P16.108   **Fig. P16.109***

16.109* A pipe rests on the horizontal bed of a truck as shown. If the truck moves forward with a constant acceleration **a** and the pipe rolls without slipping, determine (a) the acceleration of the mass center of the pipe, (b) the distance which the truck will move before the pipe rolls off the truck.

16.110 Solve Prob. 16.109* for the case in which the pipe is replaced (i) by a cylinder, (ii) by a sphere.

16.111 The 20-kg rod AB is initially at rest. Neglecting effects of friction, determine the tension in the cable BC at the instant a 200-N force **P** is applied as shown.

16.112* Solve Prob. 16.111 if $\mu_s = 0.2$ between the rod and the horizontal surface.

16.113 The motion of the 10-kg slender rod ABC is guided by two rollers of negligible mass. A force **P** is applied to hold the rod in equilibrium as shown. Determine the angular acceleration of the rod and the reactions from its supports immediately after the force **P** is removed.

Fig. P16.111

Fig. P16.113 and P16.114*   **Fig. P16.115**

16.114* In the position shown, the force **P** causes the 10-kg slender rod to move from rest with an angular acceleration of 5 rad/s^2, where the masses of the rollers at A and B are negligible. Determine the corresponding magnitude of **P**.

16.115 The connecting rod AB weighs 10 lb and the gear C is stationary. The gear D weighs 20 lb and has a central radius of gyration of 6 in. Determine $\mathbf{a}_A$ of the point A immediately after the system is released from rest in the position shown.

Fig. P16.116* and P16.117

16.116* Each of the two slender bars AB and BC weighs 20 lb and the roller at C has a negligible mass. A force **P** is applied at C to hold the system in equilibrium as shown where $L = 2.6$ ft. Determine α_{BC} of the bar BC and the reaction from the support at C immediately after the force **P** is removed.

16.117 Each of the two slender bars AB and BC has a mass of 10 kg and the roller at C has a negligible mass. In the position shown, the point C moves with a *constant* velocity of 1.2 m/s $\leftarrow$. If $L = 1.3$ m, determine the corresponding magnitude of the horizontal force **P** acting at C.

16.118 The 50-kg carriage C is supported by the 20-kg disk A and the 15-kg disk B as shown. Knowing that the disks roll without slipping, determine the acceleration of the carriage when the system is released from rest.

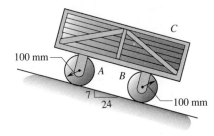

Fig. P16.118

16.119* Solve Prob. 16.118 if both disks A and B have the same mass of 20 kg.

16.120 and 16.121* The system shown consists of a section of pipe A weighing 15 lb and a cart B weighing 10 lb. If the system is released from rest and no slipping occurs between the pipe and the cart, determine (a) the angular acceleration of the pipe, (b) the acceleration of the cart, immediately after release.

Fig. P16.120

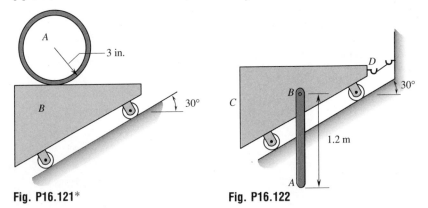

Fig. P16.121*

Fig. P16.122

16.122 A 10-kg rod AB is hinged to a 15-kg cart C as shown. If the cord at D is cut, determine (a) the angular acceleration of the rod, (b) the acceleration of the cart, immediately after cutting.

16.10 Concluding Remarks

Similar to the moment of inertia of an area, the *moment of inertia of a body* about a given axis may be thought of as the integral sum of the products of the differential masses of the body and the squares of their individual distances from the given axis. The moment of inertia of a body is also referred to as the *mass moment of inertia*. The definitions and the parallel-axis theorem related to moments of inertia of a body, as presented in the first part of this chapter, serve as a timely prerequisite to the study of the plane kinetics of rigid bodies.

Unlike the moment of inertia of an area (which does not lend itself to a simple interpretation), the moment of inertia of a body about its own central axis or an axis fixed in space can be considered as being numerically equal to *the moment required to accelerate the body to rotate at the rate of* 1 rad/s^2 *about that axis*. Physically, the *moment of inertia* of a body about a central or fixed axis is *a measure of the rotational inertia* of the body about that axis, while the *mass* of a body is *a measure of the translational inertia* of the body.

The second part of this chapter presents the *method of force and acceleration* for rigid bodies in plane motion, which is a direct extension of that presented in Chap. 12 for particles. The central concept in this method is that *the system of forces* (including moments, if any) *in the free-body diagram of a body is either equivalent or equipollent to the effective force-moment system(s)*, at the mass center of the body or mass centers of its component parts, *in the corresponding effective-force diagram of the body*. Those two systems of vectors are equivalent if the body is a single rigid body; they are equipollent if the body is composed of a set of connected rigid bodies. In any case, those two systems have the *same resultant force* and the *same resultant moment* about any given point; and these conditions generally yield up to three independent scalar equations, such as Eqs. (16.19), (16.20), and (16.21) or (16.22), which are then used to solve for the unknowns in the problem considered.

Note that the drawing of the vector-diagram equation consisting of the *free-body diagram* and the *effective-force diagram* is an effective approach in learning plane kinetics of rigid bodies using the method of force and acceleration. Without drawing such a vector-diagram equation, we would have to give up the freedom of choice of moment center accorded by Eq. (16.22), which is a more general and powerful equation than Eq. (16.21). In fact, a dexterity in the application of Eq. (16.22) is a benchmark of excellence in one's understanding of the *method of force and acceleration* for rigid bodies in plane motion.

REVIEW PROBLEMS

16.123   A slender rod weighing 1.61 lb/ft is used to form a composite body as shown. Determine the moment of inertia I_x and the radius of gyration k_x of the body.

16.124*   Refer to Prob. 16.123. Determine I_y and k_y of the body.

Fig. P16.123

16.125   A sheet of metal is cut and bent into a machine component as shown. If its mass density is $\rho_A = 16$ kg/m^2, determine I_x and k_x of the component.

16.126   Refer to Prob. 16.125. Determine I_y and k_y of the component.

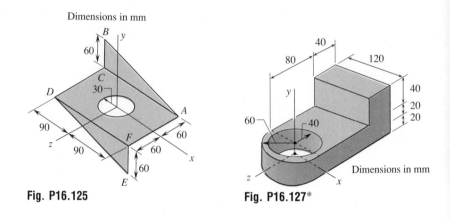

Fig. P16.125   **Fig. P16.127***

16.127*   The mass density of the steel machine element is $\rho = 7.85$ Mg/m^3. Determine I_x and k_x of the element.

16.128   Refer to Prob. 16.127*. Determine I_y and k_y of the element.

16.129   Two uniform rods AB and BC, weighing 1 lb/ft, are welded together to form a single rigid body which is held by three wires as shown. Determine the angular acceleration of the rigid body and the tension in the wires AD and BF immediately after the wire BE is cut.

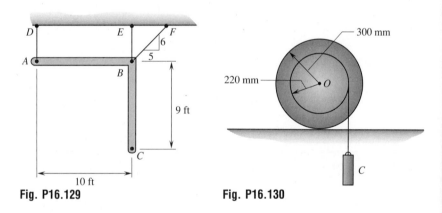

Fig. P16.129   **Fig. P16.130**

16.130   A 10-kg cylinder C is suspended by a cord which is wrapped around the drum of a wheel as shown. The wheel with the drum has a total mass of 15 kg and a central radius of gyration $\bar{k}_O = 200$ mm. If the system is released from rest and the wheel rolls without slipping, determine the acceleration of the cylinder.

16.131*   The system shown is obtained by attaching a rigid guide AOB of negligible mass to a hinge at the point O of the system described in Prob. 16.130. Neglecting effects of friction between the cylinder C and the rigid guide and knowing that the wheel will roll without slipping, determine the acceleration of the cylinder when the system is released.

Fig. P16.131*

16.132* A 50-kg wooden block C is supported by a 20-kg cylinder A and a 15-kg cylinder B as shown. If no slipping occurs, determine the acceleration of the block when the system is released from rest.

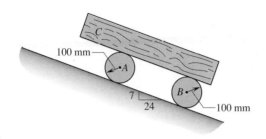

Fig. P16.132*

16.133 Solve Prob. 16.132* if both cylinders A and B have the same mass of 20 kg.

16.134 Each of the two bars OA and AB weighs 13 lb. The bars are released from rest in the position shown. Determine the angular acceleration of each bar and the reactions from the supports at O and B immediately after release.

Fig. P16.134   **Fig. P16.135**

16.135 A 10-kg block B rests on rollers of negligible mass and is in contact with a 2-kg cylinder C as shown. It is known that $\mu_k \approx \mu_s = 0.5$ between B and C as well as between C and its support A. Determine the range of values of the magnitude of the horizontal force $\mathbf{P}$ which will cause the cylinder to roll without slipping.

16.136 Refer to Prob. 16.135. Determine the magnitude of the horizontal force $\mathbf{P}$ for which the cylinder will roll with an angular acceleration of 20 rad/s² ↺ on the support without sliding.

16.137 The 50-lb carriage D is supported by a 20-lb cylinder A, a 25-lb cylinder B, and a 30-lb cylinder C as shown, where the cylinders have the same radius of 4 in. If the cylinders roll without slipping when the 10-lb force acts on it, determine $\mathbf{a}_D$ of the carriage D.

Fig. P16.137

16.138   The rods AB and BC weigh 10 lb and 20 lb, respectively. In the position shown, the end A of AB is acted on by a horizontal force $\mathbf{P}$ and moves with $\mathbf{v}_A = 5$ ft/s $\rightarrow$ and $\mathbf{a}_A = 10$ ft/s^2 $\rightarrow$. Neglecting effects of friction, determine the magnitude of $\mathbf{P}$ in this position.

Fig. P16.138   **Fig. P16.139***

16.139*   The rods AB and BC have masses of 4 kg and 2 kg, respectively. In the position shown, a moment $\mathbf{M}$ acts on the rod BC and the end A of the rod AB moves with $\mathbf{v}_A = 4.8$ m/s $\rightarrow$ and $\mathbf{a}_A = 2$ m/s^2 $\rightarrow$. Neglecting effects of friction, determine the magnitude of $\mathbf{M}$ in this position.

Plane Kinetics of Rigid Bodies: Work and Energy

Chapter 17

KEY CONCEPTS

- Work of a force sliding on a rigid body, work of a couple acting on a rigid body, kinetic energy of a rigid body in plane motion, total mechanical energy of a rigid body, and power input to a rigid body by a force or by a moment.
- Principle of work and energy for rigid bodies.
- Principle of conservation of energy for rigid bodies.
- Principle of virtual work in kinetics for studying forces and accelerations in one-degree-of-freedom systems of rigid bodies.

We saw in Chap. 13 that the principle of work and energy and the principle of conservation of energy were useful in studies involving velocity changes of particles resulting from the cumulative effect of forces which move them through specific distances. By those principles, it is possible to determine the change in velocity by analyzing the work done and the change in kinetic energy for a general system, or by analyzing just the initial and final total mechanical energies for a conservative system. In these analyses, the acceleration and the time need not be explicitly considered. Such advantages still hold true when we extend those principles to rigid bodies in this chapter. Moreover, the principle of virtual work in kinetics for particles in Sec. 13.13 is extended to rigid bodies in Sec. 17.7. It is shown in Examples 17.8 and 17.9 that the virtual work principle can effectively *complement* the method of force and acceleration for systems of rigid bodies with one degree of freedom.

WORK AND KINETIC ENERGY

17.1 Work of a Force Acting on a Rigid Body

The work of a force **F** *acting on a rigid body* is defined to be the sum of the works done by **F** to the various particles of the rigid body. Because of similarities and close relations, several of the concepts and equations in Chap. 13 may be applied to compute the work of a force acting on a rigid body. For example, the work $U_{1\to2}$ of a force $\mathbf{F}$ acting on a particle of a rigid body during the displacement of the particle from the point A_1 along a path C to the point A_2 may be computed from the equation

$$U_{1\to2} = \int_C \mathbf{F} \cdot d\mathbf{r} \qquad (17.1)$$

where $d\mathbf{r}$ is the differential displacement *of the particle*[†] (moving with the rigid body) in an inertial reference frame. If $\mathbf{F}$ is a conservative force, the value of $U_{1\to2}$ is dependent on the coordinates of A_1 and A_2, not on those of the path C. In this case, we write

$$U_{1\to2} = \int_{A_1}^{A_2} \mathbf{F} \cdot d\mathbf{r} \qquad (17.2)$$

If the force $\mathbf{F}$ is a *constant force* as shown in Fig. 17.1, we write

$$U_{1\to2} = \int_C \mathbf{F} \cdot d\mathbf{r} = \mathbf{F} \cdot \int_C d\mathbf{r} = \mathbf{F} \cdot \mathbf{q} = Fq\cos\psi$$

$$U_{1\to2} = Fq_\| \qquad (17.3)$$

where $\mathbf{q}$ is the displacement of the particle acted on by $\mathbf{F}$ from A_1 to A_2, $q_\| = q\cos\psi$ is the scalar component of $\mathbf{q}$ parallel to $\mathbf{F}$, and ψ is the angle between the positive directions of $\mathbf{F}$ and $\mathbf{q}$ as indicated in Fig. 17.1. Since $d\mathbf{r} = (d\mathbf{r}/dt)dt = \mathbf{v}\,dt$, we may write Eq. (17.1) as follows:

$$U_{1\to2} = \int_{t_1}^{t_2} \mathbf{F} \cdot \mathbf{v}\,dt \qquad (17.4)$$

where $\mathbf{v}$ is the velocity *of the particle* on which $\mathbf{F}$ acts, t is the time variable, and t_1 and t_2 are beginning and ending times of the interval during which $\mathbf{F}$ acts on the particle. Clearly, *no work* is done by a force to a rigid body which remains stationary during the action of the force.

Fig. 17.1

(a)   (b)   (c)

Fig. 17.2

At times, a force may act on *more than one particle* of a rigid body during a certain time interval. This is illustrated in Fig. 17.2, where a stationary rod BC is pressed against a spinning grinder of radius r, which is mounted at A. When the grinder rotates through an angle θ, the friction

---

[†]The *pitfall* of defining $d\mathbf{r}$ in Eq. (17.1) as the differential displacement *of the point of application* of the force $\mathbf{F}$ is underscored in the last paragraph of this section, where $\mathbf{F}$ does not continually act on the same particle of a rigid body. Furthermore, note that the dimensions and units of the work of a force have been discussed in Sec. 13.1.

force **F** will have acted on *a series of particles* lying along the arc BB' on the rim of the grinder. Assume that the normal force $\mathbf{N}_B$ and the friction force **F** are *constant* forces and the unit vector in the direction of **F** is $\boldsymbol{\lambda}_F$. Applying the definition for the work of a force acting on a rigid body and referring to Fig. 17.2(c), we may compute *the work $U_{1\rightarrow 2}$ of* **F** *acting on the grinder during its rotation through* θ as follows:

$$U_{1\rightarrow 2} = \int_C \mathbf{F} \cdot d\mathbf{r} = \int_C F\,\boldsymbol{\lambda}_F \cdot d\mathbf{r}$$

$$= \int_C F(-r\,d\theta) = -Fr \int_0^\theta d\theta$$

$$U_{1\rightarrow 2} = -Fr\theta \qquad (17.5)$$

The negative sign in Eq. (17.5) is due to the fact that $\boldsymbol{\lambda}_F$ and $d\mathbf{r}$ are opposite in direction. The work of the normal force $\mathbf{N}_B$ acting on the grinder is clearly zero because $\mathbf{N}_B$ is always perpendicular to $d\mathbf{r}$. On the other hand, the friction force $\mathbf{F}'$ as well as the normal force $\mathbf{N}_B'$ exerted by the grinder on the stationary rod does no work to the rod because the rod remains stationary. ($\mathbf{F}'$ and $\mathbf{N}_B'$ are not shown in Fig. 17.2.) For that matter, note that *no work is done by any force during the interval in which the force acts at a stationary point or a velocity center of any body.*

Furthermore, it is to be pointed out that $d\mathbf{r}$ in Eq. (17.1) should *not* be taken as the differential displacement *of the point of application of* **F** which acts on a series of particles of a rigid body. For instance, the friction force **F** exerted by a writing chalk on a stationary chalkboard does *no* work to the chalkboard. If $d\mathbf{r}$ in Eq. (17.1) were taken as the differential displacement of the point of application of **F**, one would *erroneously* assert that "a positive work was done by the friction force **F** to the stationary chalkboard."

Developmental Exercises

D17.1   Define the work of a force acting on a rigid body.

D17.2   A rigid wheel *rolls without slipping* down a fixed incline. (a) Does the friction force at the point of contact do any work to the wheel? (b) Does the friction force do any work to the fixed incline? [*Hint.* Cf. Eq. (17.4).]

D17.3   Do D17.2 if the rigid wheel *slips* on the fixed incline.

17.2 Work of a Couple Acting on a Rigid Body

The forces acting on a rigid body in *plane motion* are generally nonconcurrent coplanar forces where some of them may pair up to form couples. Each couple may be represented by a pair of noncollinear parallel forces such as **F** and $-$**F** shown in Fig. 17.3. Since the displacements of all particles of a rigid body in translation must be equal, *the work of a couple acting on a rigid body in translation is always zero.* This is obviously true because the total work done by **F** and $-$**F** to the rigid body moving through the common displacement $d\mathbf{r}$ is always zero; i.e.,

$$dU = \mathbf{F} \cdot d\mathbf{r} + (-\mathbf{F}) \cdot d\mathbf{r} = 0 \qquad (17.6)$$

Fig. 17.3

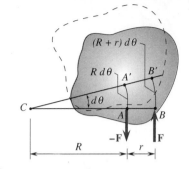

Fig. 17.4

Now, consider the case in which a rigid body is acted on by a couple of moment **M** and, at the same time, undergoes an infinitesimal angular displacement $\overrightarrow{d\theta}$ which is parallel to **M**. Let the couple be represented by the forces **F** and $-$**F** which are tangent to the arcs of rotation at the points A and B as shown in Fig. 17.4. By Eq. (17.3), the work of the couple acting on the rigid body during the angular displacement $\overrightarrow{d\theta}$ is

$$dU = |-\mathbf{F}| \, (-R \, d\theta) + |\mathbf{F}| \, (R + r) \, d\theta = Fr \, d\theta$$

Since the product Fr is equal to the magnitude of the moment **M**, we write

$$dU = M \, d\theta = \mathbf{M} \cdot \overrightarrow{d\theta} \tag{17.7}$$

Thus, *the work $U_{1\to2}$ of a couple of moment* **M** *acting on a rigid body rotating from the angular position θ_1 to the angular position θ_2 is*

$$U_{1\to2} = \int_{\theta_1}^{\theta_2} \mathbf{M} \cdot \overrightarrow{d\theta} \tag{17.8}$$

If the moment **M** is *constant*, the work $U_{1\to2}$ of **M** acting on a rigid body during its finite angular displacement $\Delta\theta$, which is expressed in radians and is in the same direction as that of **M**, is

$$U_{1\to2} = M \, \Delta\theta \tag{17.9}$$

Naturally, $U_{1\to2}$ is negative if M and $\Delta\theta$ have opposite senses. The work of a couple is a scalar whose dimensions are *force $\times$ length* and whose primary units are *joules* (1 joule = 1 J = 1 N·m) in the SI and *foot-pounds* (ft·lb) in the U.S. customary system.

Since $\overrightarrow{d\theta} = (\overrightarrow{d\theta}/dt) \, dt = \boldsymbol{\omega} \, dt$, we may write Eq. (17.8) as follows:

$$U_{1\to2} = \int_{t_1}^{t_2} \mathbf{M} \cdot \boldsymbol{\omega} \, dt \tag{17.10}$$

where $\boldsymbol{\omega}$ is the angular velocity of the rigid body on which the couple acts, t is the time variable, and t_1 and t_2 are beginning and ending times of the interval during which the couple acts on the rigid body. Clearly, *no* work is done by a couple to the rigid body which remains stationary or in translation (where $\boldsymbol{\omega} = \mathbf{0}$) during the action of the couple.

Developmental Exercises

D17.4 Define the work of a couple acting on a rigid body.

D17.5 A couple of moment $\mathbf{M} = 10$ N·m $\circlearrowleft$ acts on a rigid body during its rotation through an angle $\Delta\theta$. Determine the work of the couple if (a) $\Delta\theta = 1.5$ rad $\circlearrowleft$, (b) $\Delta\theta = 1.5$ rad $\circlearrowright$, (c) $\Delta\theta = 30°$ $\circlearrowright$.

17.3 Kinetic Energy of a Rigid Body in Plane Motion

For the present study, let Oxy be an inertial reference frame and $\mathbf{v}_i$ be the velocity, in Oxy, of the ith particle P_i of mass Δm_i in a system of n particles which form a rigid body of mass m and moves in the xy plane, as shown in

Fig. 17.5. We write

$$\mathbf{v}_i = \bar{\mathbf{v}} + \mathbf{v}_i' \qquad (17.11)$$

where $\bar{\mathbf{v}}$ is the velocity of the mass center G of the system in Oxy, and $\mathbf{v}_i'$ is the velocity of P_i in the nonrotating reference frame $Gx'y'$ which translates with G. By Eq. (13.30), we may write the kinetic energy T of the system of particles forming the rigid body in plane motion as

$$T = \frac{1}{2} m\bar{v}^2 + \frac{1}{2} \sum_{i=1}^{n} (\Delta m_i)v_i'^2 \qquad (17.12)$$

Fig. 17.5

Note that the speed of the particle in $Gx'y'$ is $v_i' = r_i'\omega$, where r_i' is the distance between P_i and G, and ω is the angular speed of the rigid body in Oxy at the instant considered. We have

$$\sum_{i=1}^{n} (\Delta m_i)v_i'^2 = \sum_{i=1}^{n} (\Delta m_i)(r_i'\omega)^2 = \omega^2 \sum_{i=1}^{n} r_i'^2 (\Delta m_i)$$

$$= \omega^2 \int r'^2 \, dm = \omega^2 \bar{I}$$

where $\bar{I}$ is the central moment of inertia of the body, taken about the perpendicular axis through G. Therefore, Eq. (17.12) may be written as

$$T = \tfrac{1}{2} m\bar{v}^2 + \tfrac{1}{2} \bar{I}\omega^2 \qquad (17.13)$$

where the term $\tfrac{1}{2} m\bar{v}^2$ represents the *translational kinetic energy* of the rigid body translating with G, and the term $\tfrac{1}{2}\bar{I}\omega^2$ represents the *rotational kinetic energy* of the rigid body rotating about G.

For computing velocities and hence the kinetic energy at a given instant, a rigid body in *general plane motion* may conveniently be taken as rotating with an angular velocity $\boldsymbol{\omega}$ about its *velocity axis* (i.e., instantaneous axis of zero velocity) as a "fixed axis" at that instant. By definition, the velocity axis is perpendicular to the plane of motion and passes through the *velocity center C* of the rigid body as shown in Fig. 17.6,[†] where the speed v_i of the ith particle P_i a distance r_i from C is given by $v_i = r_i\omega$. Denoting the mass of the ith particle P_i as Δm_i and applying the definition of moment of inertia, we write the kinetic energy of the rigid body as

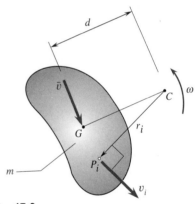

Fig. 17.6

$$T = \frac{1}{2} \sum_{i=1}^{n} (\Delta m_i)(r_i\omega)^2 = \frac{1}{2} \omega^2 \sum_{i=1}^{n} r_i^2 (\Delta m_i)$$

$$= \frac{1}{2} \omega^2 \int r^2 \, dm = \frac{1}{2} \omega^2 I_C$$

$$T = \tfrac{1}{2} I_C \omega^2 \qquad (17.14)$$

where I_C is the moment of inertia of the rigid body about its velocity axis at C at the instant considered. Of course, *Eq. (17.14) is still valid if C happens to be a fixed point about which the rigid body rotates*. Note that

$$I_C = \bar{I} + md^2 \qquad (17.15)^{\dagger\dagger}$$

[†]Cf. Sec. 15.7.
[††]Cf. Sec. 16.3.

where d is the distance between G and $\overline{C}$ in Fig. 17.6. Substituting Eq. (17.15) into Eq. (17.14) and noting that $\overline{v} = \omega d$, we readily obtain Eq. (17.13). Clearly, Eq. (17.14) is equivalent to, but simpler in form than, Eq. (17.13).

Developmental Exercises

D17.6 Applying *Chasles' theorem* in Sec. 15.5 to the general plane motion of a rigid body with its *mass center G* as the point of reference, derive the formula given by Eq. (17.13).

D17.7 Suppose that the rigid body in Fig. 17.6 is nearly in translation; i.e., $\omega \approx 0$ and $d \approx \infty$. In this case, show that Eq. (17.14) yields $T \approx \frac{1}{2} m \overline{v}^2$.

17.4 Principle of Work and Energy

In Secs. 13.6 and 13.7, the principle of work and energy was first developed for a particle and then for a system of particles. Recalling that a rigid body may be regarded as a system of a large number of particles which are at fixed distances from each other, we see that Eq. (13.27) for a system of particles is also applicable to a rigid body. Thus, the *principle of work and energy* for a rigid body is also of the form

$$T_1 + U_{1 \to 2} = T_2 \qquad (17.16)$$

where $U_{1 \to 2}$ denotes the work done by all forces acting on the various particles of the rigid body during its displacement from the position 1 to the position 2, and T_1 and T_2 denote the total kinetic energies of the particles of the rigid body in the positions 1 and 2, respectively. Since the total work done by the internal forces holding together the various particles of a rigid body is zero, the term $U_{1 \to 2}$ in Eq. (17.16) may be taken as representing *the work done by the external forces* acting on the rigid body during the displacement considered. The kinetic energy terms T_1 and T_2 in Eq. (17.16) are to be computed for the rigid body in the system using Eq. (17.14) or (17.13).

When a problem involves a *system of rigid bodies*, each rigid body in the system may be considered separately, and Eq. (17.16) may be applied to each individual rigid body. However, *the principle of work and energy in the form of Eq. (17.16) is also applicable to the entire system of rigid bodies*, where T_1 and T_2 represent, respectively, the arithmetic sums of the kinetic energies of the rigid bodies forming the system in the positions 1 and 2, and $U_{1 \to 2}$ represents the work done by all the forces acting on the various rigid bodies, whether these forces are *internal* or *external* to the system as a whole. If the rigid bodies in a system are pin-connected, in mesh with one another as gears, or connected by cordlike elements, then the total work done by the internal forces is zero, and $U_{1 \to 2}$ reduces to the total work done by the *external forces* acting on the system.

The *principle of work and energy* for rigid bodies is particularly well adapted to the solution of kinetic problems of rigid bodies where *displacement* and *velocity* (among others) are to be directly related without initially involving acceleration and time.[†] An effective visual aid in computing the work done $U_{1\rightarrow2}$ in applying this principle is, of course, the appropriate *free-body diagrams*. Note that the procedure which uses the principle of work and energy to analyze and solve appropriate kinetic problems is often referred to as the *method of work and energy*.

EXAMPLE 17.1

The system shown is released from rest, where the 150-kg flywheel with a drum has a central radius of gyration $\bar{k}_O = 0.3$ m and the spring with a modulus $k = 1$ kN/m is unstretched. Determine the angular velocity of the flywheel when the end A of the cord is displaced by the amount $\mathbf{q} = 0.6$ m $\downarrow$ due to the constant force $\mathbf{P} = 100$ N $\downarrow$.

Solution.   Since the problem directly relates velocity to displacement, it may be solved by applying the *principle of work and energy*. In the initial position, the kinetic energy is $T_1 = 0$. Let ω be the angular velocity of the flywheel when the end A of the cord is displaced 0.6 m $\downarrow$. In this final position, the kinetic energy is $T_2 = \frac{1}{2}\bar{I}\omega^2 = \frac{1}{2}m\bar{k}^2\omega^2$. During the motion, only the force $\mathbf{P}$ and the spring force $\mathbf{F}_s$ will do work because the point O remains unmoved. When the end A of the cord is displaced 0.6 m $\downarrow$, the corresponding angular displacement of the flywheel is

$$\Delta\theta = \frac{0.6}{0.4} \qquad \Delta\theta = 1.5 \text{ rad } \circlearrowright$$

and the spring is stretched by the amount

$$x = r\,\Delta\theta = 0.2(1.5) \qquad x = 0.3 \text{ m}$$

Thus, the work done is $U_{1\rightarrow2} = Pq - \frac{1}{2}kx^2$. Applying $T_1 + U_{1\rightarrow2} = T_2$ to the flywheel and noting that $m = 150$ kg, $\bar{k}_O = 0.3$ m, $P = 100$ N, $q = 0.6$ m, $k = 1000$ N/m, and $x = 0.3$ m, we write

$$0 + [100(0.6) - \tfrac{1}{2}(1000)(0.3)^2] = \tfrac{1}{2}(150)(0.3)^2\omega^2$$

$$\omega = 1.4907 \qquad \boldsymbol{\omega} = 1.491 \text{ rad/s } \circlearrowright \quad \blacktriangleleft$$

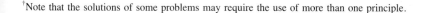

[†]Note that the solutions of some problems may require the use of more than one principle.

Developmental Exercises

D17.8 Describe the *principle of work and energy* for (a) a rigid body, (b) a system of rigid bodies.

D17.9 Refer to Example 17.1. Determine (a) the maximum displacement of the end *A* of the cord due to the constant force **P** = 100 N ↓, (b) the angular acceleration of the flywheel when the displacement of the end *A* of the cord is maximum. (*Hint.* A constant force is suddenly or instantaneously applied. The system will exhibit oscillatory rotation about *O* after it is released from rest.)

EXAMPLE 17.2

The mass center *G* of a 48.3-lb eccentric wheel of radius $r = 24$ in. is located at a distance $e = 7$ in. from its geometric center *A* as shown. As the wheel rolls without sliding, its angular velocity is observed to vary from $\omega_1 = 2$ rad/s ↻ in the position 1 to $\omega_2 = 3.5$ rad/s ↻ in the position 2. Determine the central radius of gyration $\bar{k}$ of the wheel.

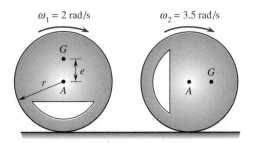

Solution. The central radius of gyration $\bar{k}$ is a parameter contained in the expression of the kinetic energy of the wheel. Thus, the *principle of work and energy* may be applied to solve for $\bar{k}$. In the position 1, the velocity center is at C_1 and

$$T_1 = \frac{1}{2} I_{C_1}\omega_1^2 = \frac{1}{2}(\bar{I} + md_1^2)\omega_1^2 = \frac{1}{2}(m\bar{k}^2 + md_1^2)\omega_1^2$$

$$= \frac{1}{2}m(\bar{k}^2 + d_1^2)\omega_1^2 = \frac{1}{2}\left(\frac{48.3}{32.2}\right)\left[\bar{k}^2 + \left(\frac{31}{12}\right)^2\right](2)^2$$

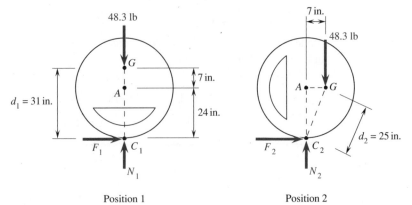

Position 1   Position 2

In the position 2, the velocity center is at C_2 and

$$T_2 = \frac{1}{2} I_{C_2} \omega_2^2 = \frac{1}{2} (\bar{I} + md_2^2)\omega_2^2 = \frac{1}{2} (m\bar{k}^2 + md_2^2)\omega_2^2$$

$$T_2 = \frac{1}{2} m(\bar{k}^2 + d_2^2)\omega_2^2 = \frac{1}{2} \left(\frac{48.3}{32.2}\right)\left[\bar{k}^2 + \left(\frac{25}{12}\right)^2\right](3.5)^2$$

Recalling that *no work is done by any force which acts at a velocity center of any body*, we see that, during the motion, the friction and normal forces do no work to the wheel because they always act on the wheel at its velocity center. Thus, only the weight force does work. We have

$$U_{1\to2} = We = 48.3\left(\frac{7}{12}\right)$$

Substituting the above expressions into $T_1 + U_{1\to2} = T_2$, we write

$$\frac{1}{2}\left(\frac{48.3}{32.2}\right)\left[\bar{k}^2 + \left(\frac{31}{12}\right)^2\right](2)^2 + 48.3\left(\frac{7}{12}\right)$$

$$= \frac{1}{2}\left(\frac{48.3}{32.2}\right)\left[\bar{k}^2 + \left(\frac{25}{12}\right)^2\right](3.5)^2$$

$$\bar{k} = 1.15956 \text{ ft} \qquad \bar{k} = 13.91 \text{ in.} \quad \blacktriangleleft$$

Developmental Exercises

D17.10   Work Example 17.2 if $\omega_2 = 3.4$ rad/s $\curvearrowright$. Note the change in $\bar{k}$.

D17.11   Refer to Example 17.2. Show that the same values for the kinetic energies T_1 and T_2 can be obtained by using Eq. (17.13).

EXAMPLE 17.3

A 10-kg block B is suspended by a cord which is wrapped around the drum of a 100-kg flywheel A as shown. The flywheel is known to have a central radius of gyration $\bar{k}_O = 0.5$ m and the system is released from rest. Neglecting effects of friction, determine $\mathbf{v}_B$ of the block B after it has moved 2 m $\downarrow$.

Solution.   This problem directly relates displacement and velocity, but not acceleration and time. Thus, the *principle of work and energy* may be applied to obtain the solution. Let the *entire system* be considered and its *free-body diagram* as well as the positions 1 and 2 be sketched as shown. Since the system is released from rest, we have $T_1 = 0$. During the motion, the forces at O do no work; only the weight force of the block B does work. For the entire system, we write

$$U_{1\to2} = W_B q_B = 10(9.81)(2) \qquad U_{1\to2} = 196.2 \text{ J}$$

In the position 2, the block B translates with $\mathbf{v}_B = v_B \downarrow$, while the flywheel rotates about the fixed axis at O with $\omega_A = (v_B/0.6)$ $\curvearrowright$. Therefore, for the

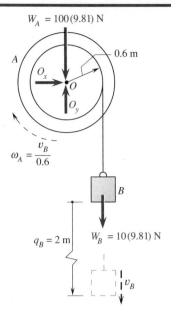

entire system in this position, we write

$$T_2 = (T_2)_A + (T_2)_B = \tfrac{1}{2}(I_O)_A\, \omega_A^2 + \tfrac{1}{2} m_B v_B^2$$

$$= \tfrac{1}{2}\left[100(0.5)^2 \right]\left(\frac{v_B}{0.6} \right)^2 + \tfrac{1}{2}(10)v_B^2 = 39.72 v_B^2$$

Substituting the above expressions into $T_1 + U_{1\to2} = T_2$, we write

$$0 + 196.2 = 39.72 v_B^2 \qquad v_B = 2.22 \text{ m/s}$$

$$\mathbf{v}_B = 2.22 \text{ m/s} \downarrow \quad \blacktriangleleft$$

Alternative Solution. This problem may also be solved by considering the *block B* and the *flywheel A* separately. Denoting the tension in the cord by F, we may draw their *free-body diagrams* and the positions 1 and 2 as

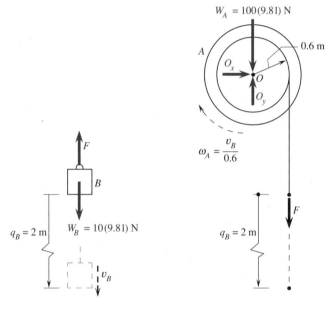

Block B   Flywheel A

shown. Applying $T_1 + U_{1\to2} = T_2$ to the *block B*, we write

$$0 + [10(9.81)(2) + F(-2)] = \tfrac{1}{2}(10)v_B^2 \qquad (1)$$

Applying $T_1 + U_{1\to2} = T_2$ to the *flywheel A*, we write

$$0 + F(2) = \frac{1}{2}\left[100(0.5)^2 \right]\left(\frac{v_B}{0.6} \right)^2 \qquad (2)$$

Solving Eqs. (1) and (2), we obtain $v_B = 2.22$ m/s and $F = 85.8$ N. Thus,

$$\mathbf{v}_B = 2.22 \text{ m/s} \downarrow \quad \blacktriangleleft$$

Developmental Exercise

D17.12 Work Example 17.3 if $\bar{k}_O = 0.4$ m. Note the change in $\mathbf{v}_B$.

EXAMPLE 17.4

A 10-kg block B is suspended by a cord which passes around a frictionless pulley of negligible mass at D and is wrapped around the drum of a 100-kg wheel A as shown. The wheel with the drum has a central radius of gyration $\bar{k}_O = 0.5$ m and the system is released from rest. Knowing that the wheel rolls without slipping, determine $\mathbf{v}_B$ of the block B after it has moved 2 m $\downarrow$.

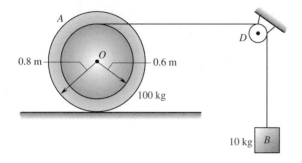

Solution. Clearly, the *principle of work and energy* may be applied to obtain the solution. Let the *entire system* be considered and its free-body diagram as well as the positions 1 and 2 be sketched as shown. Since the system is released from rest, we have $T_1 = 0$. During the motion, the *velocity center* of the wheel is located at the point of contact between the wheel and the horizontal support, the displacement of the mass center of the wheel is perpendicular to the weight force of the wheel, and the center of the pulley never moves. We recall that *no work is done by any force which acts at a stationary point or a velocity center of any body*. Thus, the weight force of the block B is the only force which does work. For the entire system, we write

$$U_{1 \to 2} = W_B q_B = 10(9.81)(2) \qquad U_{1 \to 2} = 196.2 \text{ J}$$

In the position 2, the block B translates with $\mathbf{v}_B = v_B \downarrow$, while the wheel rotates with $\boldsymbol{\omega}_A = (v_B/1.4) \, \rotatebox{0}{↺}$ about its velocity center at C_2. For the entire system, we write

$$T_2 = (T_2)_A + (T_2)_B = \tfrac{1}{2}(I_{C_2})_A \omega_A^2 + \tfrac{1}{2} m_B v_B^2$$

$$= \tfrac{1}{2} m_A(\bar{k}_O^2 + d_2^2)\omega_A^2 + \tfrac{1}{2} m_B v_B^2$$

$$= \tfrac{1}{2}(100)\left[(0.5)^2 + (0.8)^2\right]\left(\frac{v_B}{1.4}\right)^2 + \tfrac{1}{2}(10)v_B^2$$

$$= 27.7 v_B^2$$

Substituting the above expressions into $T_1 + U_{1\to2} = T_2$, we write

$$0 + 196.2 = 27.7 v_B^2 \qquad v_B = 2.66$$

$$\mathbf{v}_B = 2.66 \text{ m/s} \downarrow \quad \blacktriangleleft$$

Developmental Exercises

D17.13 Refer to Example 17.4. What is the displacement of the center of the wheel A when the block B has moved 2 m $\downarrow$?

D17.14 Refer to Example 17.4. Determine the tension in the cord by considering the block B and the wheel A separately.

EXAMPLE 17.5

A 15-lb gear B of central radius of gyration $\bar{k} = 4.5$ in. is in mesh with a stationary gear D as shown. The system is at rest when the 20-lb·ft torque is applied to the 10-lb bar OA. Determine $\boldsymbol{\omega}_{OA}$ of the bar OA after it is rotated to the horizontal position.

Solution. We may apply the *principle of work and energy* to obtain the solution. Let the *entire system* be considered and its free-body diagram as well as the positions 1 and 2 be sketched as shown. Since the system starts from rest, we have $T_1 = 0$. During the motion, the *velocity center* of the bar OA is at O, while the *velocity center* of the gear B is at the point of contact between the gear B and the stationary gear D. Thus, we have[†]

[†]Recall that no work is done by any force acting at a velocity center.

$$U_{1\to2} = 20\left(\frac{\pi}{2}\right) + 10(-0.5) + 15(-1) \qquad U_{1\to2} = 11.416 \text{ ft·lb}$$

In the position 2, the bar OA rotates with $\boldsymbol{\omega}_{OA} = \omega_{OA}$ ↺ and the speed of the point A is $v_A = \overline{OA}\,\omega_{OA} = 1\omega_{OA} = \overline{C_2A}\,\omega_B = 0.5\omega_B$. In other words, the gear B rotates with $\boldsymbol{\omega}_B = 2\omega_{OA}$ ↻ about its velocity center C_2. We write

$$T_2 = (T_2)_{OA} + (T_2)_B = \frac{1}{2}(I_O)_{OA}\omega_{OA}^2 + \frac{1}{2}(I_{C_2})_B\omega_B^2$$

$$= \frac{1}{2}\left[\frac{1}{12}\left(\frac{10}{32.2}\right)(1)^2 + \frac{10}{32.2}(0.5)^2\right]\omega_{OA}^2$$

$$+ \frac{1}{2}\left(\frac{15}{32.2}\right)\left[\left(\frac{4.5}{12}\right)^2 + (0.5)^2\right](2\omega_{OA})^2$$

$$= 0.4157\omega_{OA}^2$$

Substituting the above expressions into $T_1 + U_{1\to2} = T_2$, we write

$$0 + 11.416 = 0.4157\omega_{OA}^2 \qquad \omega_{OA} = 5.24$$

$$\boldsymbol{\omega}_{OA} = 5.24 \text{ rad/s} \text{ ↺} \quad \blacktriangleleft$$

Developmental Exercise

D17.15 Refer to Example 17.5. Determine $\boldsymbol{\omega}_{OA}$ of the bar OA after it is rotated from the vertical position through 45°.

PROBLEMS

17.1 A bolt is tightened by applying a couple as shown. The wheel has a mass of 20 kg and a central radius of gyration of 280 mm. If the wheel is rotated by the couple through one-quarter revolution and friction is negligible, determine the angular velocity of the wheel.

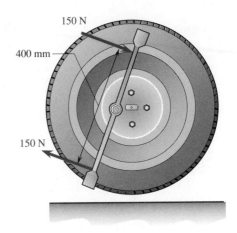

Fig. P17.1

17.2 and 17.3* The system shown is released from rest, where the 300-lb flywheel has a central radius of gyration $\bar{k}_O = 1.6$ ft. Determine the angular velocity of the flywheel after it has rotated 1 revolution.

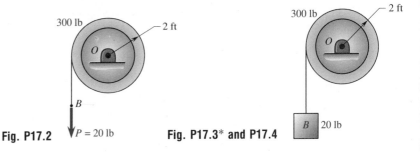

Fig. P17.2   **Fig. P17.3* and P17.4**

17.4   A hoisting engine has a flywheel, as shown, which weighs 300 lb and has a central radius of gyration $\bar{k}_O = 1.5$ ft. If the power is cut off when the flywheel is rotating at 10 rad/s $\curvearrowright$, determine the number of revolutions that the flywheel executes before stopping occurs.

17.5   A 50-kg gear of central radius of gyration $\bar{k}_O = 300$ mm is rolled up a rack from rest by a 200-N force **P** applied to a cord which is wrapped around the drum of the gear as shown, where $r_1 = 200$ mm and $r_2 = 400$ mm. Determine the speed of its center O after it has traveled 2 m up the rack.

17.6*   A 100-lb gear of central radius of gyration $\bar{k}_O = 0.75$ ft is rolled up a rack from rest by a constant force **P** applied to a cord which is wrapped around the drum of the gear as shown, where $r_1 = 0.5$ ft and $r_2 = 1$ ft. It is known that the angular velocity of the gear is 10 rad/s $\downarrow$ after its center O has traveled 4 ft up the rack. Determine the magnitude of **P**.

17.7*   A 60-g yo-yo of central radius of gyration $\bar{k}_O = 18$ mm is released from rest as shown. Assume that the string is wound around the central peg such that the mean radius at which it unravels is $r = 6$ mm. Determine the distance it must descend in order to attain an angular speed of 600 rpm.

17.8   The driver and the jeep have a total mass of 750 kg, excluding the four wheels. Each wheel, with an all-terrain tire mounted on it, has a mass of 30 kg and a central radius of gyration of 0.24 m. The driver sets the jeep in neutral gear and lets it coast from rest down the slope for 50 m before applying the brakes. Neglecting the effects of friction and assuming that the wheels roll without slipping, determine the maximum speed attained by the jeep in km/h.

Fig. P17.5 and P17.6*

Fig. P17.7*

Fig. P17.8

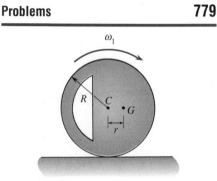

Fig. P17.9

17.9 The mass center of a 16.1-lb wheel of radius $R = 15$ in. is at G as shown. The central radius of gyration of the wheel is $\bar{k} = 9$ in. and $r = 8$ in. If the wheel rolls without slipping and has an angular velocity of $\omega_1 = 5$ rad/s $\circlearrowleft$ in the position shown, determine its angular velocity ω_2 in the next position where G is directly *below* its center C.

17.10* Refer to Prob. 17.9. Determine the angular velocity ω_2 of the wheel in the next position where G is directly *above* its center C.

17.11 A collar C of mass m_C is securely attached to a slender rod AB of mass m_{AB} as shown. The rod with the collar is released from rest to rotate downward from $\theta = 0$ to $\theta = 180°$. If $m_{AB}/m_C = 4$, determine the ratio r/L for which the angular velocity of the rod is maximum when $\theta = 180°$.

17.12* A collar C of weight W_C is securely attached to a slender rod AB of weight W_{AB} as shown. The rod with the collar is released from rest to rotate downward from $\theta = 0$ to $\theta = 180°$. If $W_{AB}/W_C = 4$, $r = 3$ ft, and $L = 4$ ft, determine the angular velocity of the rod when $\theta = 180°$.

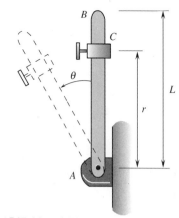

Fig. P17.11 and P17.12*

17.13 Three identical rectangular plate glasses, each of mass 24 kg, are mounted on a vertical axle to form a three-door revolving door, which is initially at rest. A man passes through a door by pushing with a 50-N force **F** which is constantly directed perpendicular to the plane of the door as shown. If the angular velocity of the door is increased to 1.5 rad/s after the door is rotated through 90°, determine the resisting moment due to friction in the door.

Fig. P17.13

Fig. P17.14 and P17.15*

17.14   The 10-kg rod shown is given an angular velocity of 6 rad/s ↻ in the vertical position where $\theta = 0$. It hits the spring at B and is stopped after compressing it by 0.1 m. Determine the modulus of the spring.

17.15*   The 10-kg rod shown is given an angular velocity of 6 rad/s ↻ in the vertical position where $\theta = 0$. The spring at B has a modulus $k = 100$ kN/m. Determine the angular velocity of the rod as it passes through the position where $\theta = 90°$.

17.16   The disk shown has a mass of 20 kg and is attached to a spring which has a free length of 500 mm and a modulus of 600 N/m. If the disk is released from rest and rolls without slipping, determine the maximum speed attained by its center O.

Fig. P17.16   **Fig. P17.17***

17.17*   The disk D of weight W and radius r is released from rest to press on the belt, which moves with a constant velocity **v** as shown. If the coefficient of kinetic friction between the disk and the belt is μ_k, determine the number of revolutions executed by the disk before its slipping on the belt ceases. (*Hint.* The answer is a function of r, v, μ_k, θ, and g.)

17.18   Solve Prob. 17.17* if $r = 0.1$ m, $v = 10$ m/s, $\mu_k = 0.2$, $\theta = 20°$, and $g = 9.81$ m/s².

17.19   Solve Prob. 17.17* if the direction of **v** is reversed.

17.20   The driver and the disabled tractor have a total weight of 1500 lb, excluding the four wheels. Each of the two front wheels has a radius $r = 10$ in., a weight of 30 lb, and a central radius of gyration of 7 in. The two rear wheels are much larger; each of them has a radius $R = 20$ in., a weight of 120 lb, and a central radius of gyration of 15 in. The driver sets the tractor in neutral gear and lets it be towed from rest by a constant force **P** $= 100$ lb ← on a level surface until the tractor reaches a speed of 30 mph. Neglecting effects of friction and assuming that the wheels roll without slipping, determine the distance through which the tractor is towed.

Fig. P17.20

17.21   The 15-lb wheel A of central radius of gyration $\bar{k} = 5$ in. is released from rest to press on the disk B as shown, where B is driven by a motor to rotate at a constant angular velocity **ω** $= 600$ rpm ↻. If $\mu_k = 0.2$ between A and B, determine the number of revolutions executed by the wheel before its slipping on the disk ceases.

17.22* Solve Prob. 17.21 if ω = 600 rpm ↺.

17.23 The 100-kg wheel shown has a central radius of gyration $\bar{k}_O$ = 150 mm and is initially rotating with ω = 300 rpm ↻ when the hydraulic cylinder CD is activated to exert a constant force $\mathbf{F}_{CD}$ on the lever AC for braking at B, where μ_k = 0.25. If the wheel is to stop in 3 revolutions, determine the magnitude of $\mathbf{F}_{CD}$.

Fig. P17.21

Fig. P17.23   **Fig. P17.24**

17.24 The gears A and B shown have masses 5 kg and 12 kg, and central radii of gyration 80 mm and 120 mm, respectively. A constant torque $\mathbf{M}$ = 10 N·m ↻ is acting on the gear A. Determine the number of revolutions executed by the gear A during the interval in which the angular velocity of the gear B is increased from 300 rpm ↺ to 600 rpm ↺.

17.25* Solve Prob. 17.24 if the angular velocity of the gear B is increased from 150 rpm ↺ to 450 rpm ↺.

17.26 The flywheel at A weighs 40 lb and has a central radius of gyration of 1.2 ft. The system is released from rest in the position shown. After the 30-lb block B has moved 3 ft, the 15-lb collar C is lifted from the block B which continues to move. Determine the speed of the block B just before it strikes the bottom of the pit.

Fig. P17.26   **Fig. P17.27**

17.27 Each of the two slender rods has a mass m and length L. If the system is released from rest in the position shown, determine the angular velocity of each rod as (a) the end D passes through the x axis, (b) the rod AB becomes vertical.

CONSERVATION OF ENERGY AND SPECIAL TOPIC

17.5 Conservation of Energy

We note that Eq. (13.44) developed in Sec. 13.12 is applicable to either a single particle or a system of particles. Along the same line of reasoning in Sec. 17.4, we readily see that Eq. (13.44) is also applicable to one or more rigid bodies. Thus, the *principle of conservation of energy* for any conservative system consisting of one or more rigid bodies is also of the form

$$T_1 + V_1 = T_2 + V_2 \qquad (17.17)$$

where T_1 and T_2 denote the total kinetic energies of the system in the positions 1 and 2, and V_1 and V_2 denote the total potential energies of the system in the positions 1 and 2, respectively. The sum of the kinetic energy T and the potential energy V of a rigid body is called the *total mechanical energy* of the rigid body. Naturally, Eq. (17.17) is a special case of Eq. (17.16) and is applicable to only *conservative systems*, where *displacement* and *velocity* (among others) are to be directly related. Note that bodies in conservative systems are subjected to only conservative forces and nonworking forces, if any.

The potential energies of a rigid body may include gravitational, elastic, and applied potential energies. For computing the gravitational potential energy of a rigid body, the weight force of the body is taken to act at the mass center of the body; i.e., the rigid body is taken as a particle located at the mass center of the body (cf. Sec. 13.9). For computing the elastic and applied potential energies of a rigid body, Eqs. (13.36), (13.38), (13.39), and (13.41) may appropriately be used (cf. Secs. 13.10 and 13.11).

EXAMPLE 17.6

Using the *principle of conservation of energy*, work Example 17.1.

Solution. For convenience, the space diagram and the free-body diagram of the system are repeated as shown. Referring to Example 17.1, we note that the spring is unstretched in the initial position and the applied force **P** = 100 N $\downarrow$ is a *constant force*. Therefore, the system is a *conservative system*; the initial position of the system may be chosen as the reference datum for the *gravitational*, the *elastic*, and the *applied* potential energies.[†] In the initial position, the flywheel is at rest; so, we have $T_1 = 0$ and $V_1 = (V_g)_1 + (V_e)_1 + (V_a)_1 = 0 + 0 + 0 = 0$. When the end A of the cord is displaced $q = 0.6$ m $\downarrow$, the angular velocity of the flywheel is ω and the spring is stretched by the amount $x = 0.3$ m, as shown in the solution in Example 17.1. Thus, in the final position, we have $T_2 = \frac{1}{2}\bar{I}\omega^2 = \frac{1}{2}m\bar{k}_O\omega^2$ and $V_2 = (V_g)_2 + (V_e)_2 + (V_a)_2 = 0 + \frac{1}{2}kx^2 + (-Pq)$. Applying the *principle of conservation of energy*, $T_1 + V_1 = T_2 + V_2$, to the flywheel and noting that $m = 150$ kg, $\bar{k}_O = 0.3$ m, $P = 100$ N,

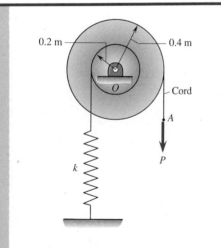

[†]Cf. Secs. 13.9 through 13.11.

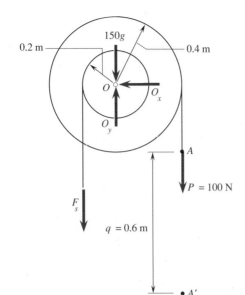

$q = 0.6$ m, $k = 1000$ N/m, and $x = 0.3$ m, we write

$$0 + 0 = \tfrac{1}{2}(150)(0.3)^2\omega^2$$

$$+ \left[0 + \tfrac{1}{2}(1000)(0.3)^2 - 100(0.6) \right]$$

$$\omega = 1.4907 \qquad \omega = 1.491 \text{ rad/s } \circlearrowright \ \blacktriangleleft$$

Note that the result obtained here is identical with that obtained in Example 17.1.

REMARK. We have here stressed that an *applied constant force* is a *conservative force*. As a matter of fact, any force $\mathbf{F}$ satisfying the condition $\nabla \times \mathbf{F} = \mathbf{0}$ (cf. Sec. 13.8) is a conservative force. This concept will help in setting the stage for more advanced studies.

Developmental Exercise

D17.16 Describe the *principle of conservation of energy* for a conservative system of rigid bodies.

EXAMPLE 17.7

A 5-lb disk A is connected to a 10-lb disk B by a cord. At the instant shown, the disk B is rotating at $(\omega_B)_1 = 15$ rad/s $\circlearrowright$. Determine how far the disk A will drop before the angular velocity of the disk B is increased to $(\omega_B)_2 = 30$ rad/s $\circlearrowright$.

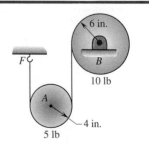

Solution. This is a conservative system to which the *principle of conservation of energy* is applicable. Suppose that the disk A will drop a distance h and the reference datum is chosen as indicated in the sketch, where b is the initial distance between the two disks. As the disk A moves, it simply

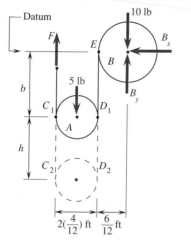

rolls without slipping down the left portion of the cord which serves as a stationary track; so, the velocity centers of the disk A in the positions 1 and 2 are at C_1 and C_2 as indicated. Furthermore, note that the extreme right point of the disk A and the extreme left point of the disk B must always have the same downward speed. We write

$$2r_A\omega_A = r_B\omega_B \qquad 2(\tfrac{4}{12})\omega_A = \tfrac{6}{12}\omega_B \qquad \omega_A = 0.75\omega_B$$

$$(\omega_B)_1 = 15 \text{ rad/s} \qquad (\omega_B)_2 = 30 \text{ rad/s}$$

$$(\omega_A)_1 = 0.75(15) \text{ rad/s} = 11.25 \text{ rad/s}$$

$$(\omega_A)_2 = 0.75(30) \text{ rad/s} = 22.5 \text{ rad/s}$$

In the position 1, the kinetic and potential energies are

$$T_1 = (T_1)_A + (T_1)_B = \frac{1}{2}(I_{C_1})_A(\omega_A)_1^2 + \frac{1}{2}\bar{I}_B(\omega_B)_1^2$$

$$= \frac{1}{2}\left[\frac{1}{2}\left(\frac{5}{32.2}\right)\left(\frac{4}{12}\right)^2 + \frac{5}{32.2}\left(\frac{4}{12}\right)^2\right](11.25)^2$$

$$+ \frac{1}{2}\left[\frac{1}{2}\left(\frac{10}{32.2}\right)\left(\frac{6}{12}\right)^2\right](15)^2$$

$$= 6.005 \text{ (ft·lb)}$$

$$V_1 = (V_1)_A + (V_1)_B = -5b + 0 = -5b$$

In the position 2, the kinetic and potential energies are

$$T_2 = (T_2)_A + (T_2)_B = \frac{1}{2}(I_{C_2})_A(\omega_A)_2^2 + \frac{1}{2}\bar{I}_B(\omega_B)_2^2$$

$$= \frac{1}{2}\left[\frac{1}{2}\left(\frac{5}{32.2}\right)\left(\frac{4}{12}\right)^2 + \frac{5}{32.2}\left(\frac{4}{12}\right)^2\right](22.5)^2$$

$$+ \frac{1}{2}\left[\frac{1}{2}\left(\frac{10}{32.2}\right)\left(\frac{6}{12}\right)^2\right](30)^2$$

$$= 24.020 \text{ (ft·lb)}$$

$$V_2 = (V_2)_A + (V_2)_B = -5(b + h) + 0 = -5(b + h)$$

Substituting the above expressions into $T_1 + V_1 = T_2 + V_2$, we write

$$6.005 - 5b = 24.020 - 5(b + h) \qquad h = 3.60 \text{ ft} \blacktriangleleft$$

Developmental Exercises

D17.17 Using the *principle of conservation of energy*, work Example 17.2.

D17.18 Using the *principle of conservation of energy*, work Example 17.3.

17.6 Power

The *power* input to a rigid body is defined as the sum of the time rates of doing work to the various particles of the rigid body at the instant considered. In the case of a force $\mathbf{F}$ acting on a rigid body, the power P input by $\mathbf{F}$ to the rigid body is

$$P = \mathbf{F} \cdot \mathbf{v} \tag{17.18}$$

where $\mathbf{v}$ is the velocity *of the particle* (of the rigid body) acted on by $\mathbf{F}$ at the instant considered. In the case of a couple of moment $\mathbf{M}$ acting on a rigid body, the power P input by the couple to the rigid body is

$$P = \mathbf{M} \cdot \boldsymbol{\omega} \tag{17.19}$$

where $\boldsymbol{\omega}$ is the angular velocity of the rigid body at the instant considered. In plane motion, we have $\mathbf{M} \parallel \boldsymbol{\omega}$. Thus, Eq. (17.19) gives

$$P = M\omega \tag{17.20}$$

For units of power, refer to Sec. 13.2.

Developmental Exercises

D17.19   Refer to the system in Fig. 17.2. Assume that $N_B = 20$ N, $\mu_k = 0.3$, $r = 100$ mm, and the grinder is spinning at 600 rpm ↻. Determine the power input by (a) the friction force $\mathbf{F}$ at B to the grinder, (b) the reactive friction force $\mathbf{F}'$ at B to the rod.

D17.20   A steel shaft transmits 10 hp at 300 rpm. What is the torque transmitted by the shaft?

*17.7   Virtual Work in Kinetics: Force and Acceleration

We recall in Sec. 13.13 that the *principle of virtual work in kinetics* may be applied to solve a kinetic problem involving a one-degree-of-freedom system of particles where *force* and *acceleration* (among others) are to be directly related. Along the same line of reasoning in Sec. 17.4, it follows that *during a constrained virtual displacement of a set of connected rigid bodies with one degree of freedom, the total virtual work δU of all the external and internal force systems acting on the various rigid bodies in the set must be equal to the total virtual work δU_{eff} of all the effective force-moment systems on the various rigid bodies in the set*; i.e.,

$$\delta U = \delta U_{\text{eff}} \tag{17.21}$$

If the set of rigid bodies under consideration is connected by *frictionless pins*, *inextensible cords*, or *links*, the total virtual work of the internal force system is zero, and δU in Eq. (17.21) reduces to the *total virtual work of the external force system only*. The property expressed in Eq. (17.21) is referred to as the *principle of virtual work in kinetics* for rigid bodies.[†] This principle complements the method of force and acceleration for rigid bodies in Sec. 16.7.

EXAMPLE 17.8

Using the *principle of virtual work in kinetics*, solve for α_{OA} in Example 16.18.

Solution.[†] For convenience, the space diagram of the system is repeated as shown, where each bar is of length $L = 5$ ft and weight $W = 10$ lb, and the system is released from rest. We are to determine α_{OA} of the bar OA immediately after release.

As explained in the solution in Example 16.18, we may, by symmetry, assume that $\alpha_{OA} = \alpha$ ↺ and $\alpha_{AB} = \alpha$ ↻ and draw the vector-diagram equation consisting of the *free-body diagram* and the *effective-force diagram* for the entire system as shown. As the system is released and undergoes a *constrained virtual displacement*, the bar OA will have rotated about its acceleration center O to the position OA', while the bar AB will have rotated about its acceleration center Z to the position $A'B'$ as shown separately.[††] Since the triangles AOA' and AZA' are congruent, we see that the measures

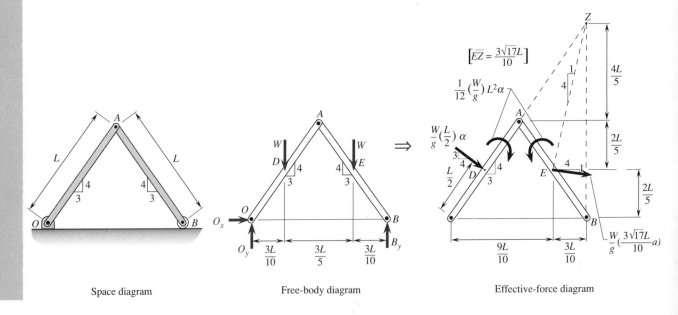

Space diagram   Free-body diagram   Effective-force diagram

[†]This principle and the solution in Example 17.8 were presented in the authors' paper, "Teaching the Concepts of Acceleration Center and Virtual Work in Dynamics," *Engineering Education*, vol. 79, no. 3, April 1989, pp. 441–443.
[††]The points O and Z are the displacement centers of the bars OA and OB, respectively. Cf. Sec. 10.6.

of the two infinitesimal rotations of OA and AB are equal in magnitude, which is denoted by $\delta\theta$. Since the set of rigid bodies is connected by frictionless pins, we see that δU in Eq. (17.21) is equal to the total virtual work of the external force system only. Referring to the diagrams shown and applying the *principle of virtual work in kinetics*, we write

$$\delta U: W\left(\frac{3L}{10}\,\delta\theta\right)(2) = \frac{1}{12}\left(\frac{W}{g}\right)L^2\alpha(\delta\theta)(2) + \frac{W}{g}\left(\frac{L}{2}\,\alpha\right)\left(\frac{L}{2}\,\delta\theta\right)$$

$$+ \frac{W}{g}\left(\frac{3\sqrt{17}\,L}{10}\,\alpha\right)\left(\frac{3\sqrt{17}\,L}{10}\,\delta\theta\right)$$

where the reaction components O_x, O_y, and B_y do no work, the weight forces do equal works, and the effective moments also do equal works. Solving for α, we write

$$\alpha = \frac{45g}{146L} = \frac{45(32.2)}{146(5)} \qquad \alpha_{OA} = 1.985 \text{ rad/s}^2 \; \circlearrowleft \; \blacktriangleleft$$

Constrained virtual displacement

REMARK. By comparing the solution in this example with that in Example 16.18, we readily note that, in order to find the magnitude α of the angular acceleration of the bar OA in the system, the *method of force and acceleration* requires the solution of at least *two* simultaneous equations, while the *principle of virtual work in kinetics* requires the solution of only *one* equation. The ability to *uncouple an unknown quantity* in some seemingly complex one-degree-of-freedom systems is, therefore, a salient feature of the principle of virtual work in kinetics. Clearly, the advantage of this principle resides in the fact that *the unknown constraining forces in the free-body diagram are nonworking forces during the constrained virtual displacement of the system.*

Developmental Exercise

D17.21 Describe the *principle of virtual work in kinetics* for rigid bodies.

EXAMPLE 17.9

Using the *principle of virtual work in kinetics*, solve for $\mathbf{a}_C$ in Example 16.19.

Solution. For convenience, the space diagram of the system is repeated as shown, where a 40-kg block C is supported by a 20-kg cylinder A and a 30-kg cylinder B of the same radius $r = 100$ mm. We are to find $\mathbf{a}_C$ of the block C if a 50-N force $\mathbf{P}$ is applied to it and the cylinders roll without slipping.

As explained in the solution of Example 16.19, we may draw the vector-diagram equation consisting of the *free-body diagram* and the *effective-force diagram* for the entire system as shown. The constrained virtual displacement of the block C is $\overrightarrow{\delta x}_C = \delta x_C \leftarrow$, which is related to the constrained

Space diagram

Free–body diagram   Effective–force diagram

angular virtual displacement $\overrightarrow{\delta\theta}$ of the cylinders A and B by $\delta x_C = 2r\,\delta\theta = 2(0.1)\,\delta\theta$ or $\delta\theta = 5\delta x_C$. The constrained virtual displacements of the centers of the cylinders are given by

$$\delta x_A = \delta x_B = r\delta\theta = 0.1(5\delta x_C) = 0.5\delta x_C$$

Thus, the *constrained virtual displacement* of the system may be indicated as shown. Referring to the diagrams shown and applying the *principle of virtual work in kinetics*, we write

$$\delta U: \qquad 50\,\delta x_C = 40a_C(\delta x_C) + 20(0.5a_C)(0.5\,\delta x_C)$$
$$+ 30(0.5a_C)(0.5\,\delta x_C)$$
$$+ \tfrac{1}{2}(20)(0.1)^2(5a_C)(5\,\delta x_C)$$
$$+ \tfrac{1}{2}(30)(0.1)^2(5a_C)(5\,\delta x_C)$$

where the weight forces and the reactions from the supports do no work during the constrained virtual displacement. Thus, we have

$$50 = 58.75a_C \qquad a_C = 0.851 \text{ m/s}^2 \leftarrow \quad \blacktriangleleft$$

Constrained
virtual displacement

REMARK. By comparing the solution in this example with that in Example 16.19, we readily note that, in order to find a_C of the block C in the system, the *method of force and acceleration* requires the solution of at least *three* simultaneous equations, while the *principle of virtual work in kinetics* requires the solution of only *one* equation. The ability to *uncouple an unknown quantity* in some seemingly complex one-degree-of-freedom systems is, again, demonstrated in this example by the principle of virtual work in kinetics.

Developmental Exercise

D17.22 Using the *principle of virtual work in kinetics*, solve for $\mathbf{a}_C$ in Example 16.19 if $r = 200$ mm. Note the change in $\mathbf{a}_C$.

PROBLEMS

17.28* through 17.45* Using the *principle of conservation of energy*, solve the following problems:

17.28* Prob. 17.1.[†]	17.29* Prob. 17.2.
17.30* Prob. 17.3*.	17.31* Prob. 17.4.
17.32* Prob. 17.5*.	17.33* Prob. 17.6.
17.34* Prob. 17.7*.	17.35* Prob. 17.8.
17.36* Prob. 17.9.	17.37* Prob. 17.10*.
17.38* Prob. 17.14.	17.39* Prob. 17.15*.
17.40* Prob. 17.16.	17.41* Prob. 17.20.
17.42* Prob. 17.24.	17.43* Prob. 17.25*.
17.44* Prob. 17.26.	17.45* Prob. 17.27.

17.46 A *cylinder* and a *sphere* of the same mass m and the same radius r are released from rest on an incline and roll without slipping to descend by $h = 1$ m in elevation as shown. Determine the velocity of the center O of each body after the descent.

Fig. P17.46

17.47* Solve Prob. 17.46 if the bodies are replaced by a *pipe* and a *spherical shell*.

17.48 A 10-lb *cylinder* of radius 6 in. is released from rest in the position shown and rolls without slipping on the curved surface of radius 30 in. Determine (a) the speed of its center O when it reaches the bottom of the surface in terms of θ, (b) the value of θ for which the magnitude of the vertical reaction from the bottom of the surface is 20 lb.

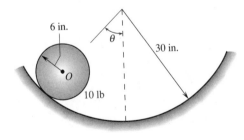

Fig. P17.48

17.49* Solve Prob. 17.48 if the cylinder is replaced by a *sphere* of the same radius and weight.

17.50 The 10-kg rod AB is released from rest when θ is essentially zero. If the modulus of the spring is $k = 1$ kN/m, determine $\mathbf{v}_A$ of the end A when (a) $\theta = 30°$, (b) $\theta = 70°$.

Fig. P17.50

[†]A couple of constant moment $\mathbf{M}$ is a conservative load. The corresponding potential energy is $V_a = -M\,\Delta\theta$, where $\Delta\theta$ is the angular displacement of the body.

17.51* The 10-kg rod *AB* is released from rest in the position shown, where the spring of modulus $k = 1$ kN/m has been compressed 0.4 m. Determine $\mathbf{v}_B$ of the end *B* when the rod passes through the vertical position.

Fig. P17.51*   **Fig. P17.52**

17.52   A constant torque $\mathbf{M} = 10$ lb·ft $\circlearrowright$ is applied to the 20-lb rod *OA* which is initially at rest as shown. If the spring has a free length of 1 ft and a modulus $k = 2$ lb/ft, determine $\boldsymbol{\omega}_{OA}$ as the rod *OA* passes through the horizontal position.[†]

17.53   The connecting rod *OA* weighs 10 lb and the gear *C* is stationary. The gear *B* weighs 20 lb and has a central radius of gyration of 6 in. If the system is released from rest in the position shown, determine $\boldsymbol{\omega}_{OA}$ of the rod *OA* after it becomes vertical.

Fig. P17.53

17.54* Solve Prob. 17.53 if the connecting rod *OA* and the gear *B* have the same weight of 20 lb.

17.55* The two slender bars *OA* and *AB*, each of mass 4 kg, are released from rest in the position shown, where $\theta = 60°$. If the modulus of the spring is $k = 400$ N/m, determine $\mathbf{v}_B$ of the end *B* as the bar *OA* passes through the position where (a) $\theta = 45°$, (b) $\theta = 30°$.

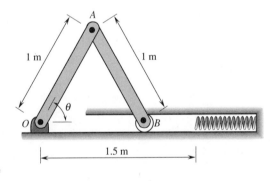

Fig. P17.55*

17.56   Solve Prob. 17.55* if the mass of each bar is 6 kg.

17.57   A 20-lb collar and a 40-lb disk are connected by a 30-lb rod *AB* as shown. It is known that the disk rolls without slipping and the modulus of the spring is 10 kips/ft. If the mechanism is released from rest when $\theta = 30°$, determine (a) the velocity of the collar when $\theta = 90°$, (b) the maximum deflection of the spring.

Fig. P17.57

[†]Cf. the footnote on p. 789.

17.58   A 40-kg sheave A of radius $r = 350$ mm and central radius of gyration $\bar{k} = 250$ mm carries a 160-kg load B and is suspended by a cable and a spring of modulus $k = 1.2$ kN/m. If the system is released from rest in the position shown, where the spring is stretched 100 mm, determine $\mathbf{v}_B$ of the load B after it has dropped 200 mm.

Fig. P17.58

Fig. P17.59

17.59   The two uniform rods OA and AB, each of mass 4 kg, are released from rest in the position shown. Determine $\mathbf{v}_B$ of the end B when (a) the rod OA is horizontal, (b) the ends B and O have the same elevation.

17.60   A 10-lb slender rod is released from rest when θ is slightly greater than zero. Determine the angle θ at which its bottom end A starts to lift off the ground.

Fig. P17.60

Fig. P17.61*

17.61*   A *hemisphere* of mass m and radius r is released from rest in the position shown. If the hemisphere rolls without slipping, determine (a) its maximum angular speed, (b) the maximum reaction from the support.

17.62   Solve Prob. 17.61* if the hemisphere is replaced by a *semicircular disk* of mass m and radius r.

17.63*   An oblong body of weight W is released from rest when θ is slightly greater than zero. Show that the angle θ at which its bottom end A starts to lift off the ground is given by the equation

$$3\cos^2\theta - 2\cos\theta + (\bar{k}/d)^2 = 0$$

where $\bar{k}$ is the central radius of gyration of the body and d is the distance between its mass center G and its bottom end A.

Fig. P17.63*

17.64   The 5-Mg dump of a dump truck is tilted at a constant rate of 5°/s about its hinge support at A by operating a hydraulic cylinder which acts at B as shown. If the gate is stuck during the tilting of the dump, determine for the hydraulic cylinder (a) the power required when $\theta = 30°$, (b) the maximum power required. (*Hint.* Does the solution depend on the values of h and L?)

Fig. P17.64

17.65*   After power shutoff, the angular speed of the 1000-kg flywheel shown is reduced from 6000 rpm to 3000 rpm in a 2-min interval. If the central radius of gyration of the flywheel is 400 mm, determine the average power supplied by the flywheel to the system during that interval.

17.66   An electric motor is delivering 5 kW at 1800 rpm to the wheel shown. If the maximum tension in the belt is 60 N, determine the minimum value of μ_s between the belt and the drum of the wheel. (*Hint.* $T_2 = T_1 e^{\mu_s \beta}$.)

Fig. P17.65* and P17.66

17.67   A 4-kg disk is supported by a cord which is connected to a spring as shown. The disk is released from the position where the spring is unstretched. If its center G attains a velocity of 1 m/s ↓ after descending 0.5 m, determine the modulus k of the spring.

Fig. P17.67

Fig. P17.68

17.68   A gear train is formed by four gears and three shafts as shown. If a power of 10 kW is steadily transmitted from the motor at M to the load at E and the motor run at 45 Hz, determine the magnitudes of the torques in the shafts MA, BC, and DE.

17.69   A section of *pipe* of radius r and mass m sits on a horizontal plate as shown. If the plate is pulled to the right from rest with a constant acceleration **a** and the pipe rolls on it without slipping, determine the total work done by the friction force to the pipe when the pipe has rotated one revolution. (*Hint.* Use the method of force and acceleration as well as the principle of work and energy.)

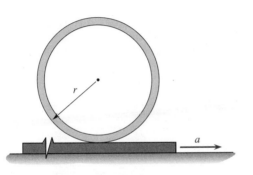

Fig. P17.69

17.70* Solve Prob. 17.69 if $\mathbf{a} = 0.5$ m/s$^2 \rightarrow$ and the section of pipe is replaced by a 2-kg *solid cylinder* of radius 100 mm.

17.71* Solve Prob. 17.69 if $\mathbf{a} = 2$ ft/s$^2 \rightarrow$ and the section of pipe is replaced by a 16.1-lb *sphere* of radius 6 in.

NOTE: Solve the following problems using the *principle of virtual work in kinetics* (cf. Sec. 17.7).

Fig. P17.72   **Fig. P17.73***

Fig. P17.74

17.72 through 17.74   Each of the two bars *OA* and *AB* has a mass of 4 kg. The system is released from rest in the position shown. Determine α_{AB} of the bar *AB* immediately after release.

17.75 and 17.76*   The 50-lb carriage *D* is supported by a 20-lb cylinder *A*, a 25-lb cylinder *B*, and a 30-lb cylinder *C* as shown, where the cylinders have the same radius of 4 in. If the cylinders roll without slipping, determine $\mathbf{a}_D$ of the carriage *D* when the 10-lb force acts on it.

Fig. P17.75   **Fig. P17.76***

17.77* Solve Prob. 17.75 if all cylinders have the same weight of 20 lb.

17.78   Solve Prob. 17.76* if all cylinders have the same weight of 20 lb.

17.79   The 50-kg carriage E is supported by a 20-kg cylinder A, a 25-kg cylinder B, a 30-kg cylinder C, and a 35-kg cylinder D as shown, where the cylinders have the same radius of 100 mm. If the cylinders roll without slipping, determine $\mathbf{a}_E$ of the carriage E when the 20-N force acts on it.

Fig. P17.79

17.80*   Solve Prob. 17.79 if all cylinders have the same mass of 20 kg.

17.8   Concluding Remarks

This chapter is clearly a direct extension of Chap. 13. It covers three principles related to work and energy for studying the motion of rigid bodies. They are (1) the *principle of work and energy*, which is applicable to both conservative and nonconservative systems; (2) the *principle of conservation of energy*, which is applicable to conservative systems only; and (3) the *principle of virtual work in kinetics*, which complements the *method of force and acceleration* presented in Chap. 16.

We note that the principle of work and energy and the principle of conservation of energy are particularly well adapted to the study of the plane kinetics problems of rigid bodies where *velocity* and *displacement* (among others) are to be directly related without initially involving acceleration and time. On the other hand, the *principle of virtual work in kinetics* for rigid bodies is particularly well adapted to the study of the plane kinetics of systems of rigid bodies with one degree of freedom, where *force* and *acceleration* (among others) are to be directly related. As demonstrated in Examples 17.8 and 17.9, this principle provides a procedure which enables us to *uncouple an unknown quantity* in a seemingly complex one-degree-of-freedom system. Thus, the principle of virtual work in kinetics is an effective complement to the method of force and acceleration for one-degree-of-freedom systems of connected rigid bodies.

Fig. P17.81

REVIEW PROBLEMS

17.81   The 20-lb rod shown is given an angular velocity of 3 rad/s $\circlearrowright$ in the vertical position where $\theta = 0$. Determine the angular velocity of the rod when (a) $\theta = 45°$, (b) $\theta = 90°$.

17.82*   The ends of a 10-kg rod AB are constrained to move along the slots shown. Its end B is suspended by the spring BC which has a free length $L = 1$ m. The rod is released from rest when $\theta = 0$ and the maximum deflection of the end B is found to be $\delta = 0.5$ m. Determine the modulus k of the spring.

Fig. P17.82*

Fig. P17.83

17.83   A 2-kg rod AB is welded to a 4-kg disk as shown. If the system is released from rest, determine the maximum angular speed of the disk.

17.84 and 17.85*   Four slender rods, each of length 500 mm and mass 2 kg, are welded together to form a square frame which is supported at its corner A as shown. If the frame is released from rest when θ is slightly less than 45°, determine $\mathbf{v}_B$ of its corner B when the rod AB becomes horizontal. (*Hint*. In Prob. 17.85*, $\bar{\mathbf{v}} = \bar{v} \downarrow$ when $\theta = 0$.)

Fig. P17.84   **Fig. P17.85***

Fig. P17.86

17.86   A section of *pipe* of mass m and radius r is placed at a sharp corner A and is released from rest when θ is slightly greater than zero. Assuming that no slipping occurs at A, determine the angle θ at which the pipe begins to leap off the corner A.

17.87   Solve Prob. 17.86 if the pipe is replaced by a *circular cylinder* of mass m and radius r.

17.88*   Solve Prob. 17.86 if the pipe is replaced by a *sphere* of mass m and radius r.

17.89*   An oblong body of weight W is placed at a sharp corner A and is released from rest when θ is slightly greater than zero. Assuming that no slipping occurs at A, show that the angle θ at which the body begins to leap off the corner A is given by

$$\cos \theta = \frac{2d^2}{\bar{k}^2 + 3d^2}$$

where $\bar{k}$ is the central radius of gyration of the body and d is the distance between its mass center G and its bottom end at A.

Fig. P17.89*

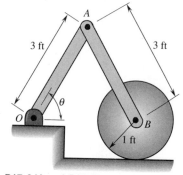

D

10 kg

A

100 mm

B

2 kg

Fig. P17.90

17.90 A 10-kg disk A is connected to a 2-kg block B by a cord which passes around a fixed drum D as shown, where $\mu_s = 0.25$ and $\mu_k = 0.2$ between the cord and the drum. If the system is released from rest, determine $\mathbf{v}_B$ of the block B after it has moved 1 m ↑.

17.91* Solve Prob. 17.90 if $\mu_k = 0.1$.

17.92 The driver and the disabled tractor have a total mass of 800 kg, excluding the four wheels. Each of the two front wheels has a radius $r = 250$ mm, a mass of 15 kg, and a central radius of gyration of 180 mm. The two rear wheels are much larger; each of them has a radius $R = 500$ mm, a weight of 60 kg, and a central radius of gyration of 380 mm. The tractor is set in neutral gear and is pulled from rest by a constant force $\mathbf{P} = 600$ N ←. Neglecting effects of friction and assuming that the wheels roll without slipping, determine the velocity of the tractor when it has been moved 10 m ←.

P

R

r

Fig. P17.92 and 17.93*

17.93* Refer to Prob. 17.92. Using the *principle of virtual work in kinetics*, determine the acceleration of the tractor when it is being pulled by the constant force $\mathbf{P} = 600$ N ←.

17.94* Each of the bars OA and AB weighs 20 lb and the disk at B weighs 30 lb. The system shown is released from rest in the position where $\theta = 60°$. If the disk rolls without slipping, determine $\boldsymbol{\omega}_{OA}$ when $\theta = 30°$.

17.95 Each of the bars OA and AB weighs 20 lb and the disk at B weighs 30 lb. The system shown is released from rest in the position where $\theta = 60°$. The disk is known to roll without slipping. Using the *principle of virtual work in kinetics*, determine the angular acceleration of the bar OA immediately after release.

A

3 ft

3 ft

θ

O

B

1 ft

Fig. P17.94* and P17.95

17.96* through 17.100* Using the *principle of virtual work in kinetics*, solve the following problems:

17.96* Prob. 16.118.   17.97* Prob. 16.119*.
17.98* Prob. 16.132.   17.99* Prob. 16.133.
17.100* Prob. 16.137.

Plane Kinetics of Rigid Bodies: Impulse and Momentum

Chapter 18

KEY CONCEPTS

- Linear impulse and angular impulse about a fixed point of a force acting on a rigid body, angular impulse of a couple, linear momentum and angular momentum of a rigid body in plane motion, and coefficient of restitution for eccentric impact.
- Principle of impulse and momentum for rigid bodies.
- Conservation of momentum for rigid bodies.
- Eccentric impact between rigid bodies.
- Principle of generalized virtual work for studying impulses and momenta in one-degree-of-freedom systems of rigid bodies.

It was shown in Chap. 14 that the principle of impulse and momentum was useful in studies involving velocity changes of particles resulting from the cumulative effect of forces which act on them over a specific interval of time. By this principle, it is possible to determine the change in velocity by analyzing the impulses exerted on the system and the change in momentum of the system, without first determining the acceleration and then integrating it over the time interval of motion. Such advantages still hold true when we extend the principle of impulse and momentum to rigid bodies in this chapter. Furthermore, the principle of generalized virtual work for particles in Sec. 14.10 is extended to rigid bodies in Sec. 18.7. It is shown in Examples 18.10 and 18.11 that the principle of generalized virtual work can effectively complement the principle of impulse and momentum for systems of rigid bodies with one degree of freedom.

IMPULSE AND MOMENTUM

18.1 Linear and Angular Impulses on a Rigid Body

The *linear impulse* $\mathbf{Imp}_{1\to2}$ of a force $\mathbf{F}$ acting on a rigid body during the time interval $t_1 \leq t \leq t_2$ is defined as the sum of the linear impulses exerted by $\mathbf{F}$ on the various particles of the rigid body during that interval.[†] We write

$$\mathbf{Imp}_{1\to2} = \int_{t_1}^{t_2} \mathbf{F}\, dt \tag{18.1}$$

The *linear impulse* of a force is often simply referred to as the *impulse* of the force. If the force $\mathbf{F}$ remains *constant* during the interval $\Delta t = t_2 - t_1$, we readily obtain

$$\mathbf{Imp}_{1\to2} = \mathbf{F}\, \Delta t \tag{18.2}$$

Similarly, the *angular impulse* $(\mathbf{Aimp}_A)_{1\to2}$ of a force $\mathbf{F}$ acting on a rigid body about a fixed point A during the time interval $t_1 \leq t \leq t_2$ is defined as the sum of the angular impulses exerted by the moment $\mathbf{M}_A$ of $\mathbf{F}$ about the point A during that interval. We write

$$(\mathbf{Aimp}_A)_{1\to2} = \int_{t_1}^{t_2} \mathbf{M}_A\, dt \tag{18.3}$$

If $\mathbf{M}_A$ remains *constant* during the interval $\Delta t = t_2 - t_1$, we get

$$(\mathbf{Aimp}_A)_{1\to2} = \mathbf{M}_A\, \Delta t \tag{18.4}$$

Note that the moment of a couple acting on a rigid body about a point is a free vector. The *angular impulse* $(\mathbf{Aimp}_A)_{1\to2}$ of a *couple* acting on a rigid body about a fixed point A during the time interval $t_1 \leq t \leq t_2$ is the same as that about any other point during that interval and is defined as a free vector equal to the integral of the torque $\mathbf{M}$ of the couple times the differential time dt, evaluated over that interval. We write

$$(\mathbf{Aimp}_A)_{1\to2} = \int_{t_1}^{t_2} \mathbf{M}\, dt \tag{18.5}$$

If $\mathbf{M}$ remains *constant* during the interval $\Delta t = t_2 - t_1$, we obtain

$$(\mathbf{Aimp}_A)_{1\to2} = \mathbf{M}\, \Delta t \tag{18.6}$$

Note that the angular impulse of a couple about an axis (instead of a point) is dependent on the direction of the axis.

Developmental Exercises

D18.1   A 300-N force $\mathbf{F}$ pulls a cord which is wrapped around a disk mounted at A and, at the same time, a moment $\mathbf{M}$ of 50 N·m is applied to the disk as shown. If they act for an interval of 2 s, determine (a) the linear impulse of $\mathbf{F}$, (b) the angular impulse of $\mathbf{F}$ about A, (c) the angular impulse of $\mathbf{M}$ about A.

D18.2   Do D18.1 if $F = 300t$ N and $M = 50t$ N·m, where t is the time in seconds.

D18.3   Define the *linear impulse* and the *angular impulse about a fixed point* of a force acting on a rigid body.

Fig. D18.1

[†]Cf. the footnote on p. 595.

18.2   Momentum of a Rigid Body in Plane Motion

The vector sum of the momenta of the various particles of a rigid body is called the *resultant momentum of the rigid body*, which does not necessarily pass through the mass center G of the rigid body. However, by *Chasles' theorem* in Sec. 15.5, we may treat the *system of momenta* of the particles of a rigid body in plane motion as being equal to the sum of (1) the *translational system of momenta* corresponding to the translation of the rigid body with G whose velocity is $\bar{\mathbf{v}}$ in an inertial reference frame Oxy, and (2) the *rotational system of momenta* corresponding to the rotation of the rigid body about G with an angular velocity $\boldsymbol{\omega}$ in Oxy. Such a relation is depicted in Fig. 18.1, where $Gx'y'$ is a nonrotating reference frame attached to G, and r_i' is the radial distance from G to the ith particle P_i whose mass is Δm_i and whose velocity $\mathbf{v}_i$ in Oxy is given by

$$\mathbf{v}_i = \bar{\mathbf{v}} + (r_i'\omega)\mathbf{e}_t \qquad (18.7)$$

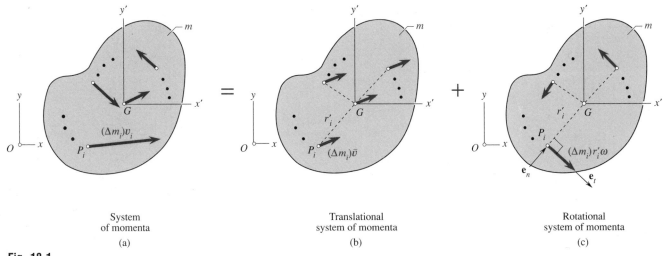

System
of momenta
(a)

Translational
system of momenta
(b)

Rotational
system of momenta
(c)

Fig. 18.1

Note that the vectors $\mathbf{e}_t$ and $\mathbf{e}_n$ in Eq. (18.7) and Fig. 18.1 are, respectively, the tangential and the normal unit vectors associated with the path of P_i rotating about G, and the total mass of the rigid body is m.

Letting $\mathbf{L}_{\text{trans}}$ be the *resultant linear momentum* of the *translational* system of momenta in Fig. 18.1(b), we write

$$\mathbf{L}_{\text{trans}} = \sum_{i=1}^{n} (\Delta m_i)\bar{\mathbf{v}} = \bar{\mathbf{v}} \sum_{i=1}^{n} (\Delta m_i) = \bar{\mathbf{v}}m$$

$$\mathbf{L}_{\text{trans}} = m\bar{\mathbf{v}} \qquad (18.8)$$

Letting $(\mathbf{H}_G)_{\text{trans}}$ be the *resultant angular momentum* about G of the *translational* system of momenta in Fig. 18.1(b) and applying the *principle of moments*, we write

$$(\mathbf{H}_G)_{\text{trans}} = \sum_{i=1}^{n} \mathbf{r}_i' \times (\Delta m_i)\bar{\mathbf{v}} = -\bar{\mathbf{v}} \times \sum_{i=1}^{n} \mathbf{r}_i' (\Delta m_i)$$

$$= -\bar{\mathbf{v}} \times \bar{\mathbf{r}}'m$$

where $\bar{\mathbf{r}}'$ is the position vector of G in $Gx'y'$. Clearly, $\bar{\mathbf{r}}' = \mathbf{0}$. Thus, we have

$$(\mathbf{H}_G)_{\text{trans}} = \mathbf{0} \qquad (18.9)$$

From Eqs. (18.8) and (18.9), we find that *the translational system of momenta in Fig. 18.1(b) is equivalent to the resultant linear momentum* $\mathbf{L}_{\text{trans}}$ *acting through* G as illustrated in Fig. 18.2.

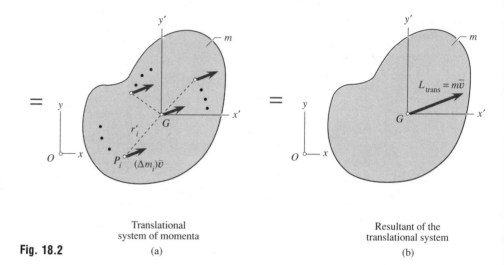

Translational system of momenta	Resultant of the translational system
(a)	(b)

Fig. 18.2

Now, letting $\mathbf{L}_{\text{rot}}$ be the *resultant linear momentum* of the *rotational* system of momenta in Fig. 18.1(c) and applying Eq. (18.7) as well as the *principle of moments*, we write

$$\mathbf{L}_{\text{rot}} = \sum_{i=1}^{n} (\Delta m_i)(r_i'\omega)\mathbf{e}_t = \sum_{i=1}^{n} (\Delta m_i)(\mathbf{v}_i - \bar{\mathbf{v}})$$

$$= \sum_{i=1}^{n} (\Delta m_i)\mathbf{v}_i - \bar{\mathbf{v}} \sum_{i=1}^{n} (\Delta m_i) = m\bar{\mathbf{v}} - \bar{\mathbf{v}}m$$

$$\mathbf{L}_{\text{rot}} = \mathbf{0} \qquad (18.10)$$

Letting $(\mathbf{H}_G)_{\text{rot}}$ be the *resultant angular momentum* about G of the *rotational* system of momenta in Fig. 18.1(c) and noting that $\mathbf{e}_n \times \mathbf{e}_t = -\mathbf{k}$, we write

$$(\mathbf{H}_G)_{\text{rot}} = \sum_{i=1}^{n} (-r_i'\mathbf{e}_n) \times (\Delta m_i)(r_i'\omega)\mathbf{e}_t$$

$$= -\omega(\mathbf{e}_n \times \mathbf{e}_t) \sum_{i=1}^{n} r_i'^2 (\Delta m_i)$$

$$= (\omega\mathbf{k}) \int r'^2 \, dm = \omega\bar{I}$$

$$(\mathbf{H}_G)_{\text{rot}} = \bar{I}\boldsymbol{\omega} \qquad (18.11)$$

where $\bar{I}$ is the moment of inertia of the rigid body about the central axis perpendicular to the plane of motion. From Eqs. (18.10) and (18.11), we find that the *rotational system of momenta* in Fig. 18.1(c) *is equivalent to the resultant angular momentum* $(\mathbf{H}_G)_{rot}$ *acting about G* as illustrated in Fig. 18.3.

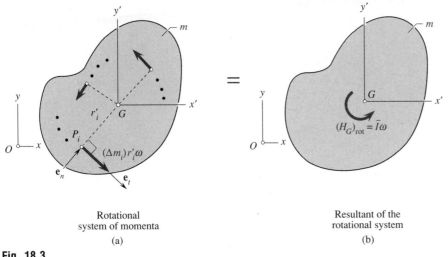

Rotational
system of momenta
(a)

Resultant of the
rotational system
(b)

Fig. 18.3

By *Chasles' theorem* and Eqs. (18.8) and (18.10), we conclude that the *resultant linear momentum* acts through G and is given by

$$\mathbf{L} = \mathbf{L}_{trans} + \mathbf{L}_{rot} = m\bar{\mathbf{v}} + \mathbf{0}$$

$$\boxed{\mathbf{L} = m\bar{\mathbf{v}}} \qquad (18.12)$$

Moreover, by *Chasles' theorem* and Eqs. (18.9) and (18.11), we conclude that the coexisting *resultant angular momentum* acting about G is given by

$$\mathbf{H}_G = (\mathbf{H}_G)_{trans} + (\mathbf{H}_G)_{rot} = \mathbf{0} + \bar{I}\omega$$

$$\boxed{\mathbf{H}_G = \bar{I}\omega} \qquad (18.13)$$

We see from Eqs. (18.12) and (18.13) that *the system of momenta of the particles of a rigid body in plane motion is equivalent to the resultant system of momenta consisting of the resultant linear momentum* $m\bar{\mathbf{v}}$ *acting through the mass center G and the resultant angular momentum* $\bar{I}\omega$ *acting about G.* This important relation is illustrated in Fig. 18.4. Of course, $m\bar{\mathbf{v}}$ and $\bar{I}\omega$ at G may be replaced by a *single equivalent linear momentum* $m\bar{\mathbf{v}}$ acting through an appropriate point.

In the case of a rigid body rotating in a plane about a fixed axis at O, the *resultant system of momenta* in Fig. 18.4(b) can be represented as shown in Fig. 18.5(a), where $m\bar{\mathbf{v}} \perp \overrightarrow{OG}$. Using the concept of equivalent systems, we can replace the system in Fig. 18.5(a) with an *equivalent single linear momentum* $m\bar{\mathbf{v}}$ acting through the point P as shown in Fig. 18.5(b). The distance d in this figure can readily be shown to be given by

$$d = \frac{\bar{k}^2}{\bar{r}} \qquad (18.14)$$

System
of momenta

Resultant system
of momenta

(a)

(b)

Fig. 18.4

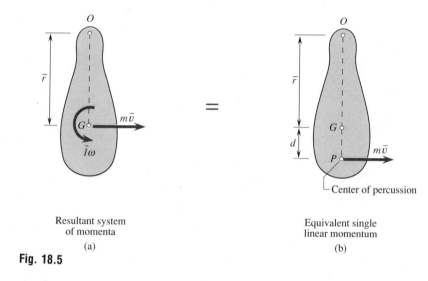

Resultant system
of momenta

Equivalent single
linear momentum

(a)

(b)

Fig. 18.5

where $\bar{k}$ is the central radius of gyration of the rigid body. Referring to Figs. 16.14 and 18.5 and noting that Eqs. (16.17) and (18.14) are identical, we see that the point P in Fig. 18.5(b) is coincident with the *center of percussion* of a rigid body rotating about a fixed pivot at O, as described in Sec. 16.6. Furthermore, it is well to note that $m\bar{v}$ is a vector measured in units of kg·m/s or slug·ft/s, while $\bar{I}\omega$ is a vector measured in units of kg·m^2/s or slug·ft^2/s.[†]

[†]The resultant angular momentum $\bar{I}\omega$ has been called a *momentum couple* or *rotational momentum* by others. (Cf. the footnote on p. 90.)

Developmental Exercises

D18.4   A 10-kg disk of radius $r = 200$ mm rotates with $\boldsymbol{\omega} = 300$ rpm $\circlearrowright$ about a fixed axle through G as shown. Determine its (a) resultant linear momentum $\mathbf{L}$, (b) resultant angular momentum $\mathbf{H}_G$.

D18.5   The mass center G of a 20-lb cylinder of radius $r = 9$ in. moves with $\bar{\mathbf{v}} = 4$ ft/s $\rightarrow$. If the cylinder rolls without slipping, determine its (a) resultant linear momentum $\mathbf{L}$, (b) resultant angular momentum $\mathbf{H}_G$.

Fig. D18.4   Fig. D18.5

18.3   Principle of Impulse and Momentum

In Secs. 14.2 and 14.4, the principle of impulse and momentum was first developed for a particle and then for a system of particles. As a rigid body may be regarded as a system of a large number of particles which are at fixed distances from each other, we readily see that Eq. (14.12) for a system of particles is also applicable to a rigid body. Using the result derived in Sec. 18.2 and expressed in Fig. 18.4, we may state and illustrate the *principle of impulse and momentum* for a rigid body as follows:

The *resultant* system of *momenta* of a rigid body at the time t_1	+	The *linear and angular impulses* exerted on the rigid body from t_1 to t_2	=	The *resultant* system of *momenta* of the rigid body at the time t_2	(18.15)

Note in Eq. (18.15) that the *resultant system of momenta* of a rigid body consists of the *resultant linear momentum* $m\bar{\mathbf{v}}$ acting through the mass center G and the *resultant angular momentum* $\bar{I}\boldsymbol{\omega}$ acting about G, as shown on the next page in Fig. 18.6. The *principle of impulse and momentum* for a rigid body, as expressed in Eq. (18.15) and illustrated in Fig. 18.6, is particularly well adapted to the solution of plane kinetic problems of rigid bodies where *time* and *velocity* (among others) are to be directly related without initially involving acceleration and displacement.

Developmental Exercise

D18.6   Describe the *principle of impulse and momentum* for a rigid body.

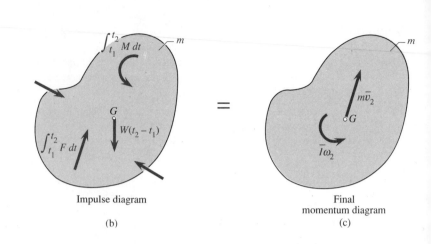

Fig. 18.6

Initial
momentum diagram
(a)

Impulse diagram
(b)

Final
momentum diagram
(c)

EXAMPLE 18.1

A flywheel with a drum of radius $r = 150$ mm has a mass of 10 kg and is rotating with $\boldsymbol{\omega}_1 = 40$ rad/s $\circlearrowright$ at the time $t = 0$ when it is acted on by a moment $\mathbf{M} = 3$ N·m $\circlearrowleft$ and a variable force $\mathbf{P} = 30t$ N $\rightarrow$ applied at the end of a cord which is wrapped around the drum as shown, where t is measured in seconds. If the central radius of gyration is $\bar{k}_A = 160$ mm, determine the angular velocity $\boldsymbol{\omega}_2$ of the flywheel at the time $t = 2$ s.

Solution. For this flywheel, we have $\mathbf{L} = m\bar{\mathbf{v}} = \mathbf{0}$ and $\mathbf{H}_G = \bar{I}\boldsymbol{\omega} = m\bar{k}_A^2\boldsymbol{\omega}$. Thus, we apply the *principle of impulse and momentum* to draw a vector-diagram equation involving momentum and impulse diagrams of the system as shown, where the impulses of $\mathbf{P}$ and the reaction component $\mathbf{A}_x$ are represented by *integrals* because both $\mathbf{P}$ and $\mathbf{A}_x$ are variable during $0 \leq t \leq 2$ s. We have

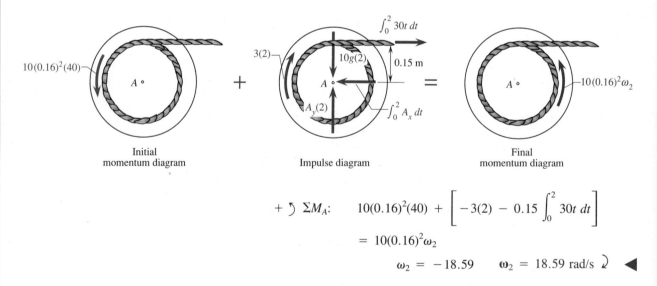

Initial
momentum diagram

Impulse diagram

Final
momentum diagram

$$+ \circlearrowright \ \Sigma M_A: \quad 10(0.16)^2(40) + \left[-3(2) - 0.15 \int_0^2 30t \ dt \right]$$

$$= 10(0.16)^2\omega_2$$

$$\omega_2 = -18.59 \qquad \omega_2 = 18.59 \text{ rad/s } \circlearrowleft \ \blacktriangleleft$$

REMARK. The actual direction of $\boldsymbol{\omega}_2$ is opposite to that assumed in the final momentum diagram because ω_2 is found to be negative in the solution.

Developmental Exercise

D18.7   Refer to Example 18.1. If the wheel comes to a stop at $t = \tau$, determine τ.

EXAMPLE 18.2

A wheel of radius $r = 500$ mm and mass $m = 100$ kg has a pure rotation with an angular velocity $\boldsymbol{\omega}_1 = 50$ rad/s $\circlearrowright$ at the time $t = 0$ when it is placed on an incline as shown, where $\theta = 15°$ and $\mu_k = 0.3$ between the wheel and the incline. If the central radius of gyration of the wheel is $\bar{k} = 400$ mm, determine (a) the time t_2 at which the wheel will start rolling without slipping, (b) the angular speed ω_2 of the wheel at the time t_2.

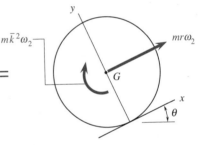

Solution.   We need to consider the motion of the wheel from the time $t = t_1 = 0$ when it is placed on the incline to the time $t = t_2$ when it starts rolling without slipping. Since *time* and *velocity* are to be directly related, we apply the *principle of impulse and momentum* to draw a vector-diagram equation for the system as shown, where $\mathbf{N}$ is the normal force and $\mathbf{F}_k = \mu_k N \nearrow$ is the kinetic friction force.

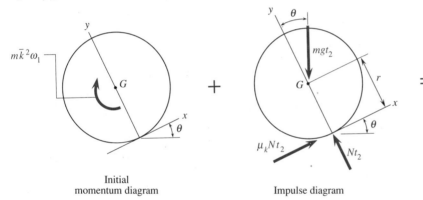

| Initial | Impulse diagram | Final |
| momentum diagram | | momentum diagram |

Referring to the vector-diagram equation, we write

$$+\nwarrow \Sigma V_y:\quad 0 + (Nt_2 - mgt_2 \cos\theta) = 0 \tag{1}$$

$$+\nearrow \Sigma V_x:\quad 0 + (\mu_k Nt_2 - mgt_2 \sin\theta) = mr\omega_2 \tag{2}$$

$$+\circlearrowright \Sigma M_G:\quad m\bar{k}^2\omega_1 - r\mu_k Nt_2 = m\bar{k}^2\omega_2 \tag{3}$$

From the given data and equations above, we find that $N = mg \cos\theta$ and

$$t_2 = \frac{\bar{k}^2 r \omega_1}{g[\mu_k(r^2 + \bar{k}^2)\cos\theta - \bar{k}^2 \sin\theta]}$$

$$t_2 = 5.27 \text{ s} \quad \blacktriangleleft$$

$$\omega_2 = \frac{\bar{k}^2 (\mu_k \cos\theta - \sin\theta)\omega_1}{\mu_k (r^2 + \bar{k}^2)\cos\theta - \bar{k}^2 \sin\theta}$$

$$\omega_2 = 3.20 \text{ rad/s} \quad \blacktriangleleft$$

Developmental Exercises

D18.8   Refer to Example 18.2. Suppose that the incline became horizontal (i.e., $\theta = 0$). (a) Would t_2 become bigger? (b) Would ω_2 become bigger?

D18.9   Work Example 18.2 if $\mu_k = 0.35$.

18.4   System of Rigid Bodies

For a *system of rigid bodies*, the principle of impulse and momentum may be applied to each rigid body separately. However, it is sometimes desirable to apply that principle to the system as a whole (cf. Examples 18.3 and 18.4). In the latter case, the momentum and impulse diagrams are drawn for the entire system of rigid bodies. The initial and final momentum diagrams should include the *resultant linear momentum* $m\bar{\mathbf{v}}$ and the *resultant angular momentum* $\bar{I}\boldsymbol{\omega}$, as appropriate, *for each component part of the system*. Linear and angular impulses of forces and moments *internal* to the system considered should be *omitted* in the impulse diagram, since they cancel out by occurring in collinear pairs equal in magnitude and opposite in sense.

In the present case, the *principle of impulse and momentum* may be stated as follows: *For a system of rigid bodies, the system of vectors in the initial momentum diagram plus the system of vectors in the impulse diagram during a time interval is equipollent to the system of vectors in the final momentum diagram.* In an abbreviated form, the *principle of impulse and momentum* for any case may, therefore, be written in terms of *equipollence*[†] as

$$\textbf{Syst Momenta}_1 + \textbf{Syst Ext Imp}_{1 \to 2} \Rightarrow \textbf{Syst Momenta}_2 \quad (18.16)^{\dagger\dagger}$$

In applying this principle, care should be taken *not* to add directly linear and angular momenta for the same reason that force and moment cannot be added directly. Furthermore, a *consistent system of kinetic units* (cf. Sec. 12.2) should be used in writing each term in an equation.

EXAMPLE 18.3

The flywheel of an engine has a drum of radius 0.4 m, a mass of 300 kg, and a central radius of gyration $\bar{k}_O = 0.5$ m. It is hoisting a 100-kg crate C as shown with an angular velocity $\omega_1 = 120$ rpm $\circlearrowright$ at the time $t = 0$ when the power is cut off. The system stops momentarily at the time $t = t_2$. Determine t_2.

0.4 m

O

300 kg

100 kg C

[†]Note that if two systems of vectors are equivalent, they are also equipollent; but the converse is not always true.

[††]Recall that the symbol $\Rightarrow$ indicates *equipollence*.

Solution. Let the motion of the flywheel and the crate as a system be considered from the time $t = t_1 = 0$ to the time $t = t_2$. Noting that $\omega_1 = 120$ rpm $\circlearrowright = 4\pi$ rad/s $\circlearrowright$, $\omega_2 = 0$, and $\bar{I} = m\bar{k}_O^2$, we apply the *principle of impulse and momentum* to draw a vector-diagram equation for the system as shown. Referring to the vector-diagram equation, we write

$$+ \circlearrowright \Sigma M_O: \quad \{300(0.5)^2(4\pi) + 0.4(100)[0.4(4\pi)]\} - 0.4(100gt_2) = 0$$

$$t_2 = 9.1\pi/g \qquad\qquad t_2 = 2.91 \text{ s} \blacktriangleleft$$

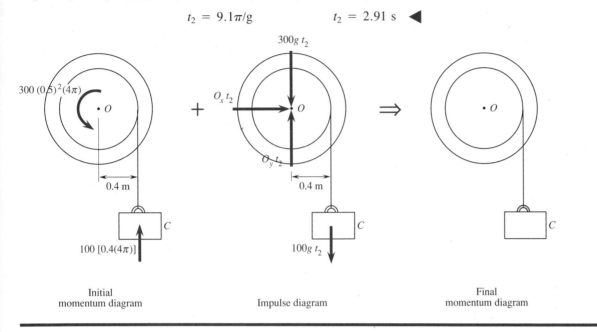

| Initial momentum diagram | Impulse diagram | Final momentum diagram |

Developmental Exercises

D18.10   Describe the *principle of impulse and momentum* for a system of rigid bodies.

D18.11   Refer to Example 18.3. Determine $\mathbf{v}_C$ of the crate C 2 s after the power is cut off.

EXAMPLE 18.4[†]

A 40-kg block C is supported by a 20-kg cylinder A and a 30-kg cylinder B as shown, where the cylinders have the same radius $r = 100$ mm. The system is at rest at the time $t = 0$ when a 50-N force $\mathbf{P}$ is applied to the block C. If the cylinders roll without slipping, determine the velocity $\mathbf{v}_C$ of the block C at the time $t = 2$ s.

Solution. For the system as described, the cylinders must have the same angular velocity ω which is related to the velocity $\mathbf{v}_C$ of the block C by $v_C = 2r\omega = 2(0.1)\omega$ or $\omega = 5v_C$. Noting that the velocity of the center of

[†]For comparative study, the problem in this example is solved by using the *principle of generalized virtual work* in Example 18.10.

either cylinder is $\bar{\mathbf{v}} = r\omega \leftarrow = 0.5v_C \leftarrow$, we apply the *principle of impulse and momentum* to draw the following vector-diagram equations:

Entire system

 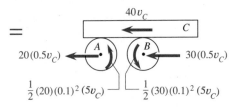

| Initial
momentum diagram | Impulse diagram | Final
momentum diagram |

Cylinder A

 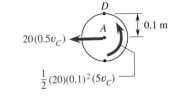

| Initial
momentum diagram | Impulse diagram | Final
momentum diagram |

Cylinder B

| Initial
momentum diagram | Impulse diagram | Final
momentum diagram |

Referring to the preceding vector-diagram equations, we write

Entire system

$$\overset{+}{\leftarrow} \Sigma V_x: \quad 0 + [50(2) + F_A(2) + F_B(2)]$$
$$= 40v_C + 20(0.5v_C) + 30(0.5v_C) \quad (1)$$

Cylinder A

$$+\,\circlearrowright\, \Sigma M_D: \quad 0 + 0.2[F_A(2)] = 0.1[20(0.5v_C)]$$
$$-\tfrac{1}{2}(20)(0.1)^2(5v_C) \quad (2)$$

Cylinder B

$$+\,\circlearrowright\, \Sigma M_E: \quad 0 + 0.2[F_B(2)] = 0.1[30(0.5v_C)]$$
$$-\tfrac{1}{2}(30)(0.1)^2(5v_C) \quad (3)$$

Solution of Eqs. (1) through (3) yields

$$v_C = 1.702 \text{ m/s} \qquad F_A = 2.13 \text{ N} \qquad F_B = 3.19 \text{ N}$$

$$\mathbf{v}_C = 1.702 \text{ m/s} \leftarrow \quad \blacktriangleleft$$

REMARK: The acceleration of the block C under the same condition was determined in Example 16.19 to be $\mathbf{a}_C = 0.851 \text{ m/s}^2 \leftarrow$. Since $\mathbf{a}_C$ is constant we readily see in 2 s that $v_C = a_C t = 0.851(2) \text{ m/s} = 1.702 \text{ m/s}$, which is what we have obtained here.

Developmental Exercise

D18.12 Work Example 18.4 if the masses of the cylinders A and B are $m_A = m_B = 30$ kg.

18.5 Conservation of Momentum

If the resultant force and the resultant moment of the external forces acting on a rigid body or a system of rigid bodies are equal to zero, Eq. (18.16) reduces to

$$\textbf{Syst Momenta}_1 \Rightarrow \textbf{Syst Momenta}_2 \qquad (18.17)$$

which expresses the *conservation of momentum*. In this case, *the system of the initial momenta is equipollent to the system of the final momenta*, and it follows that the linear momentum of the system in any direction as well as the angular momentum of the system about any point is conserved.

However, there are cases in which the resultant external force is not zero, yet in which *the resultant external moment about a given point or a given axis is zero*. In such cases, the resultant linear momentum is not conserved, yet *the resultant angular momentum about that given point or axis is conserved*. Such cases occur whenever the lines of action of the external forces (a) all pass through a given point or axis, (b) either intersect a given axis or are parallel to the given axis.

Problems involving *conservation of momentum* may be solved by applying the *principle of impulse and momentum* as before. It is usually beneficial to draw the vector-diagram equation for the impulses and momenta involved in the body or system of bodies during the interval under consideration. Scalar equations are written by appropriately summing and equating the components, or the moments, of the vector quantities in the vector-diagram equation. If the system is conservative, the principle of conservation of energy can also be applied to provide an additional scalar equation. The unknowns are then solved from the scalar equations.

Developmental Exercise

D18.13 Describe the cases in which the angular momentum of a system about a given point or a given axis is conserved, while the linear momentum of the system is not conserved.

EXAMPLE 18.5

The disk A with its attached shaft and motor armature have a mass of 20 kg and a central radius of gyration of 80 mm. The motor is mounted on the turntable B. The motor housing and the turntable B with its shaft have a mass of 30 kg and a radius of gyration of 240 mm about the y axis. Before the motor is turned on, the entire assembly is rotating with $\omega_1 = 5\mathbf{j}$ rad/s about the y axis. If the operating velocity of the motor is $\omega_{MA/B} = 600\mathbf{j}$ rpm relative to the turntable B and friction is negligible, determine the new angular velocity ω_2 of the turntable B after the motor is turned on.

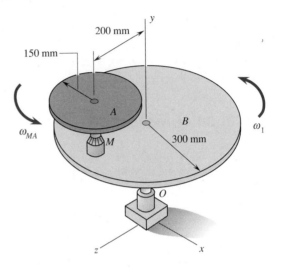

Solution. Let the entire assembly be considered as a system. Then, the action and reaction in the motor are internal forces to the system. The weight forces are all parallel to the y axis. Thus, the angular momentum of the system about the y axis is conserved. Note that

$$\omega_{MA/B} = 600\mathbf{j} \text{ rpm} = 20\pi\mathbf{j}$$

rad/s and the final angular velocity of the motor armature and the disk A in the inertial reference frame $Oxyz$ is

$$\omega_{MA} = \omega_{MA/B} + \omega_B = (20\pi + \omega_2)\mathbf{j} \text{ rad/s}$$

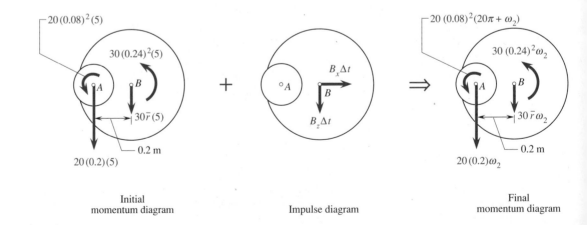

Initial momentum diagram	Impulse diagram	Final momentum diagram

Accordingly, we apply the *principle of impulse and momentum* to draw a vector-diagram equation as shown, where the impulses parallel to the y axis are not indicated and $\bar{r}$ represents the distance from the y axis to the center of mass of the composite body consisting of the motor housing and the turntable with its shaft. (We do not need to know $\bar{r}$ in the solution.) Referring to the vector-diagram equation, we write

$$\curvearrowright \Sigma M_B: \quad 30(0.24)^2(5) + 20(0.08)^2(5)$$

$$+ 0.2[20(0.2)(5)] + 0$$

$$= 30(0.24)^2\omega_2 + 20(0.08)^2(20\pi + \omega_2)$$

$$+ 0.2[20(0.2)\omega_2]$$

$$\omega_2 = 1.972 \qquad \boldsymbol{\omega}_2 = 1.972\mathbf{j} \text{ rad/s} \quad \blacktriangleleft$$

Developmental Exercises

D18.14 Refer to the vector-diagram equation in the solution in Example 18.5. Consider the composite body consisting of the motor housing and the turntable B with its shaft. Verify that the *resultant system of momenta* of this composite body in the initial momentum diagram is equivalent to a linear momentum acting through B and an angular momentum acting about B as indicated.

D18.15 Refer to Example 18.5. Determine the minimum operating velocity of the motor $\boldsymbol{\omega}_{MA/B}$ relative to the turntable B for which the turntable will reverse its direction of rotation after the motor is turned on.

EXAMPLE 18.6

A 15-lb solid sphere of radius 4 in. is mounted at B on the horizontal slender rod AC, which rotates freely about the vertical spindle with $\boldsymbol{\omega}_1 = 10\mathbf{j}$ rad/s. The cord holding the sphere is suddenly cut, and the sphere starts to slide outward. Neglecting the masses and the moments of inertia of the rod and spindle about the y axis, determine the angular velocity $\boldsymbol{\omega}_2$ of the rod when the sphere reaches the position C as shown.

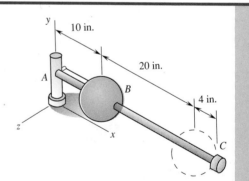

Solution. Since the spindle is free to rotate about the y axis, the reaction from the support to the spindle is composed of a force $\mathbf{A} = A_x\mathbf{i} + A_y\mathbf{j} + A_z\mathbf{k}$ and a moment $\mathbf{M} = M_x\mathbf{i} + M_z\mathbf{k}$. Thus, we apply the *principle of impulse and momentum* to draw a vector-diagram equation as shown on the next page, where the impulses parallel to the y axis are not indicated. Referring to the vector-diagram equation, we write

$$+ \curvearrowright \Sigma M_y: \quad \left\{\frac{2}{5}\left(\frac{15}{32.2}\right)\left(\frac{4}{12}\right)^2(10) + \frac{10}{12}\left[\frac{15}{32.2}\left(\frac{10}{12}\right)(10)\right]\right\} + 0$$

$$= \frac{2}{5}\left(\frac{15}{32.2}\right)\left(\frac{4}{12}\right)^2\omega_2 + \frac{30}{12}\left[\frac{15}{32.2}\left(\frac{30}{12}\right)\omega_2\right]$$

$$\omega_2 = 1.174 \qquad \boldsymbol{\omega}_2 = 1.174\mathbf{j} \text{ rad/s} \quad \blacktriangleleft$$

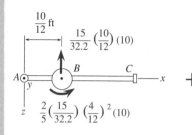

$$\frac{10}{12} \text{ ft}$$

$$\frac{15}{32.2}\left(\frac{10}{12}\right)(10)$$

$$\frac{2}{5}\left(\frac{15}{32.2}\right)\left(\frac{4}{12}\right)^2 (10)$$

Initial
momentum diagram

$+$

$\int A_z dt$

$\int A_x dt$ $\int M_x dt$

$\int M_z dt$

Impulse diagram

$=$

$$\frac{15}{32.2}\left(\frac{30}{12}\right)\omega_2$$

$$\frac{30}{12}\text{ ft}$$

$$\frac{2}{5}\left(\frac{15}{32.2}\right)\left(\frac{4}{12}\right)^2 \omega_2$$

Final
momentum diagram

REMARK. In the system considered, the angular impulse about the y axis is equal to zero at all times. Thus, the angular momentum of the system about the y axis is conserved. However, note that the direction of the linear momentum vector in the final momentum diagram is dependent on the orientation of the rod AC, which is unknown at t_2. This means that we are unable to sum and equate the components of the linear momentum vectors in the vector-diagram equation shown. We can do only what we have done with that vector-diagram equation in the above solution.

Developmental Exercise

D18.16 Work Example 18.6 if the rod weighs 3 lb and the moment of inertia of the spindle about the y axis is negligible.

PROBLEMS

18.1 A bolt is tightened by applying a couple as shown. The wheel has a mass of 20 kg and a central radius of gyration of 280 mm. If the wheel is rotated by the couple for 0.5 s and friction is negligible, determine the angular velocity of the wheel.

18.2 and 18.3* The system shown is released from rest, where the 300-lb flywheel has a central radius of gyration $\bar{k}_O = 1.6$ ft. Determine the angular velocity of the flywheel 3 s after the release.

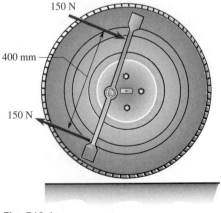

150 N

400 mm

150 N

Fig. P18.1

2 ft

O

B

$P = 20$ lb

Fig. P18.2

2 ft

O

B 20 lb

Fig. P18.3*

18.4 Solve Prob. 18.2 if $P = 20t$ lb, where t is the time measured in seconds.

18.5 A 20-kg cylinder of radius $r = 150$ mm is rotating with $\omega_1 = 12$ rad/s $\circlearrowright$ in a corner when a moment $\mathbf{M} = 10$ N·m $\circlearrowright$ is applied to it at the time $t = 0$. If $\mu_k = 0.1$ between all surfaces of contact, determine the angular velocity ω_2 of the cylinder when $t = 3$ s.

18.6* Solve Prob. 18.5 if $\mathbf{M} = 10t$ N·m $\circlearrowright$, where t is the time measured in seconds.

18.7* A bowling ball of mass m and radius r is released with a linear velocity $\mathbf{v}_1$ and no angular velocity as shown. Denoting the central radius of gyration of the ball by $\bar{k}$ and the coefficient of kinetic friction between the ball and the floor by μ_k, determine the time t_2 at which the ball begins rolling without slipping.

Fig. P18.5

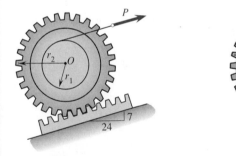

Fig. P18.7*

18.8 Solve Prob. 18.7* if $r = 110$ mm, $v_1 = 8$ m/s, $\bar{k} = 70$ mm, and $\mu_k = 0.2$.

18.9 and 18.10 A 50-kg gear of central radius of gyration $\bar{k}_O = 300$ mm is rolled up a rack from rest at the time $t = 0$ by a 200-N force $\mathbf{P}$ applied to a cord which is wrapped around the drum of the gear as shown, where $r_1 = 200$ mm and $r_2 = 400$ mm. Determine the speed of its center O when $t = 3$ s.

Fig. P18.9 and P18.11 **Fig. P18.10 and P18.12***

Fig. P18.13*

18.11 and 18.12* A 100-lb gear of central radius of gyration $\bar{k}_O = 9$ in. is rolled up a rack from rest at the time $t = 0$ by a constant force $\mathbf{P}$ applied to a cord which is wrapped around the drum of the gear as shown, where $r_1 = 6$ in. and $r_2 = 12$ in. It is known that the angular velocity of the gear is 10 rad/s $\circlearrowright$ when $t = 3$ s. Determine the magnitude of $\mathbf{P}$.

18.13* A 60-g yo-yo of central radius of gyration $\bar{k}_O = 18$ mm is released from rest at the time $t = 0$ as shown. Assume that the string is wound around the central peg such that the mean radius at which it unravels is $r = 6$ mm. Determine the time t_2 when it attains an angular speed of 600 rpm.

18.14 Constant tensions of 45 lb and 50 lb are applied to the two sides of the cable to hoist a load B of 55 lb as shown, where the pulley A weighs 30 lb and has a central radius of gyration of 7.5 in. At the time $t = 0$, the load B moves with $\mathbf{v}_1 = 4$ ft/s $\downarrow$ and the pulley A rotates with $\omega_1 = 20$ rad/s $\circlearrowright$. Determine the velocity $\mathbf{v}_2$ of the load and the angular velocity ω_2 of the pulley at the time $t = 3$ s.

Fig. P18.14

18.15 A cord is wrapped around the drum of a 60-kg wheel which has a central radius of gyration of 150 mm. If $\mu_s = 0.55$ and $\mu_k = 0.5$ between the wheel and the incline, as shown, and the system is released from rest at the time $t = 0$, determine the speed of the mass center G when $t = 4$ s.

18.16 The 100-kg wheel shown has a central radius of gyration $\bar{k}_O = 150$ mm and is rotating with $\omega = 300$ rpm $\circlearrowright$ when the hydraulic cylinder CD is activated to exert a constant force $\mathbf{F}_{CD}$ on the lever AC for braking at B, where $\mu_k = 0.25$. If the wheel is to stop in 3 s, determine the magnitude of $\mathbf{F}_{CD}$.

Fig. P18.15

Fig. P18.16   Fig. P18.17

18.17 The 15-lb wheel A has a central radius of gyration $\bar{k} = 5$ in. and is released from rest at the time $t = t_1 = 0$ to press on the disk B, which is driven by a motor to rotate at a constant angular velocity $\omega = 600$ rpm $\circlearrowright$. If $\mu_k = 0.2$ between A and B, determine the time t_2 at which slipping between them ceases.

18.18* Solve Prob. 18.17 if $\omega = 600$ rpm $\circlearrowleft$.

18.19* The gears A and B shown have masses 5 kg and 12 kg and central radii of gyration 80 mm and 120 mm, respectively. A constant torque $\mathbf{M} = 10$ N·m $\circlearrowright$ is acting on the gear A. Determine the time interval Δt during which the angular velocity of the gear B is increased from 300 rpm $\circlearrowleft$ to 600 rpm $\circlearrowleft$.

Fig. P18.19*

18.20 Solve Prob. 18.19* if the angular velocity of the gear B is increased from 150 rpm $\circlearrowleft$ to 450 rpm $\circlearrowleft$.

18.21* The square tray A with its attached shaft weighs 20 lb and has a central radius of gyration of 14 in. The tray rotates freely with $\omega_1 = 180\mathbf{j}$ rpm about the y axis as shown. The 15-lb disk B is at rest when it is dropped into the tray. Determine (a) the final angular velocity ω_2 of the tray with the disk, (b) the percentage of loss of the kinetic energy of the system.

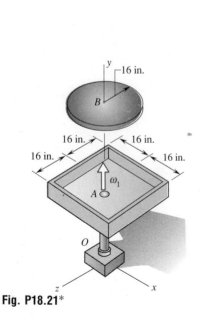

Fig. P18.21*

18.22 The 2-kg slender rod BC is hinged to the arm AB as shown. The arm with its attached shaft has a mass of 5 kg and a radius of gyration of 250 mm about the y axis. With the rod latched in the vertical position, the system rotates freely about the y axis with $\boldsymbol{\omega}_1 = 120\mathbf{j}$ rpm. If the latch is released to let the rod lie in the horizontal position, determine the new angular velocity $\boldsymbol{\omega}_2$ of the system.

18.23 The arm AB with its attached shaft OA weighs 30 lb and has a radius of gyration of 8.6 in. about the y axis. The disk C with its attached shaft BC weighs 20 lb and has a radius of gyration of 8.4 in. about BC. Initially, the disk with its shaft is rotating with $\boldsymbol{\omega}_1 = 300\mathbf{j}$ rpm while the arm with its shaft is at rest as shown. If the pin at D springs out to bear against the rotating disk to bring it to a stop relative to the arm AB and its shaft OA, determine the final angular velocity $\boldsymbol{\omega}_2$ with which the entire system will rotate freely about the y axis.

Fig. P18.22

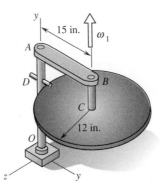

Fig. P18.23

18.24 In the assembly shown, the arm CD and the motor housing M have a mass of 30 kg and a radius of gyration of 500 mm about the y axis, the disk A and its attached shaft and motor armature have a mass of 25 kg and a radius of gyration of 200 mm about the axis of its shaft, and the 10-kg disk B has a radius of gyration of 100 mm about the axis of its shaft. The motor at M drives the disk A which turns the disk B without slipping. The entire assembly is at rest before the motor is turned on. Determine the angular speed of the arm CD when the motor is turned on and is operating at a speed of 600 rpm relative to the arm CD.

Fig. P18.24

18.25* A 10-kg block B is suspended by a cord which passes around a frictionless pulley of negligible mass at D and is wrapped around the drum of a 100-kg wheel as shown. The wheel with the drum has a central radius of gyration $\bar{k}_O = 0.5$ m and the system is released from rest. Knowing that the wheel rolls without slipping, determine $\mathbf{v}_B$ of the block B 2 s after the release.

Fig. P18.25*

Fig. P18.26

18.26 The 30-kg carriage C is supported by the 25-kg and 20-kg cylindrical wheels A and B as shown, where the wheels have the same radius of 100 mm. If the system is pushed from rest by the 20-N force and the wheels roll without slipping, determine $\mathbf{v}_C$ of the carriage C 4 s later.

18.27* Solve Prob. 18.26 if both wheels have the same mass of 20 kg.

18.28 The double pulley shown has a mass $m = 10$ kg and a central radius of gyration $\bar{k}_O = 150$ mm. The inner and outer pulleys are rigidly attached and the center O is guided by a smooth pin at O to move along the inclined slot. The pulley is at rest when a 40-N force $\mathbf{P}$ is applied to the cord A at the time $t = 0$. Determine $\mathbf{v}_O$ of the center O when $t = 2$ s.

Fig. P18.28

Fig. P18.29*

Dimensions in millimeters

Fig. P18.30

18.29* The 5-lb cylinder A and the 20-lb cylinder B are connected by a belt as shown. If the system is released from rest, determine (a) the angular velocity of the cylinder B after 4 s, (b) the tension in the portion of the belt connecting A and B.

18.30 Let m and k with appropriate subscripts denote the *mass* and the *radius of gyration* of a certain part in the gear train as shown. It is known that the gear A and the motor armature at M with their attached shaft have $m_{AM} = 18$ kg and $k_{AM} = 50$ mm, the gears B and C with their attached shaft have $m_{BC} = 20$ kg and $k_{BC} = 40$ mm, while the gear D and a working part at E with their attached shaft have $m_{DE} = 25$ kg and $k_{DE} = 45$ mm. If the motor M exerts a constant torque of 60 N·m on the shaft of the gear A for 5 s, determine the angular speed reached by the gear D.

18.31 A 10-kg disk A is connected to a 2-kg block B as shown, where $\mu_k = 0.2$ between the cord and the drum at D. If the system is released from rest, determine $\mathbf{v}_B$ of the block B after 4 s.

Fig. P18.31

Fig. P18.32

18.32 A system consists of a 10-kg rod AB and a 20-kg cart C as shown. The system is released from rest when $\theta = 60°$. Determine the velocity of the cart C when (a) $\theta = 30°$, (b) $\theta = 0$.

18.33 A 20-lb solid sphere B of radius 5 in. is mounted to slide on the 4-lb slender rod AC as shown. The sphere is held by a cord and is connected to the end C of the rod by a spring of modulus $k = 60$ lb/ft. At the instant shown, $r = 10$ in., the spring is neutral, and the system is rotating freely about the vertical spindle with $\boldsymbol{\omega}_1 = 15\mathbf{j}$ rad/s. If the cord is suddenly cut and the moment of inertia of the spindle about the y axis is negligible, determine $\boldsymbol{\omega}_2$ of the system when $r = 20$ in.

18.34* Solve Prob. 18.33 if $\boldsymbol{\omega}_1 = 30\mathbf{j}$ rad/s.

Fig. P18.33

IMPULSIVE MOTION AND SPECIAL TOPIC

18.6 Impulsive Motion of Rigid Bodies

We recall that an *impulsive motion* of a body is the motion of the body resulting from the action of an unbalanced *impulsive force*, which is a relatively large force acting on the body for a relatively short time interval and producing a significant change in the momentum of the body. Problems involving impulsive motion are generally amenable to solution by the *principle of impulse and momentum*. In such problems, the time interval considered is very short, *the bodies in the system may be assumed to occupy the same position during the time the impulsive forces act.* Furthermore, *nonimpulsive forces* (e.g., weight forces) *in juxtaposition with impulsive forces may usually be neglected.* Thus, the linear and angular impulses in problems involving impulsive motion of rigid bodies may readily be computed.

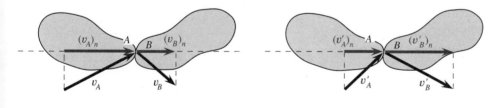

(a) Just before impact   (b) Just after impact

Fig. 18.7

In impulsive motions resulting from *impact between two rigid bodies*, the *coefficient of restitution*, denoted by e, is often employed in the study if the bodies rebound after the impact. Let the points of contact between certain two colliding bodies be A and B as shown in Fig. 18.7, where the *line of impact* is a straight line normal to the contacting surfaces during the impact, $\mathbf{v}_A$ and $\mathbf{v}_B$ are the velocities of A and B just before the impact, $\mathbf{v}'_A$ and $\mathbf{v}'_B$ are the velocities of A and B just after the impact, and $(\mathbf{v}_A)_n$, $(\mathbf{v}_B)_n$, $(\mathbf{v}'_A)_n$, and $(\mathbf{v}'_B)_n$ are the normal components of $\mathbf{v}_A$, $\mathbf{v}_B$, $\mathbf{v}'_A$, and $\mathbf{v}'_B$ directed along the line of impact. These normal components are related to the *coefficient of restitution e*, as defined in Eq. (14.34), for the *eccentric impact* as follows:

$$(v'_B)_n - (v'_A)_n = e[(v_A)_n - (v_B)_n] \qquad (18.18)$$

The impact is called a *perfectly plastic impact* if $e = 0$, and a *perfectly elastic impact* if $e = 1$. When the line of impact passes through the centers of mass of the two colliding bodies, the impact becomes a *central impact*, which was studied in Secs. 14.8 and 14.9. Note that the comments on the values of e in the footnote on p. 622 apply to eccentric impacts between rigid bodies as well.

EXAMPLE 18.7

A 10-lb small sphere is dropped from a height $h = 5$ ft and strikes the 30-lb slender plank AB with a perfectly plastic impact at S as shown. Determine the angular velocity of the plank just after the impact.

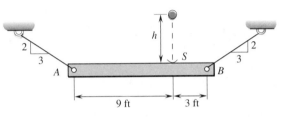

Solution. By the *principle of work and energy*, it can readily be determined that the velocity of the sphere just before it strikes the plank is $\mathbf{v}_S = \sqrt{2gh} \downarrow = \sqrt{322}$ ft/s $\downarrow$. Immediately after a perfectly plastic im-

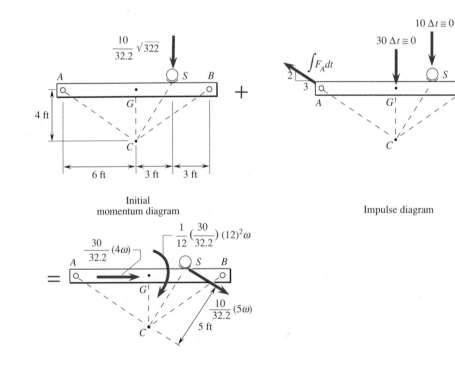

Initial
momentum diagram

Impulse diagram

Final
momentum diagram

pact, the plank and the sphere will rotate together as a rigid body with an angular velocity $\boldsymbol{\omega} = \omega \downarrow$ about their *velocity center C*. Note that the *weight forces* of the plank and the sphere in juxtaposition with the impulsive reactions $\mathbf{F}_A$ and $\mathbf{F}_B$ from the inextensible cords are *nonimpulsive forces*, which exert negligible impulses on the plank during the short time interval of impact. Thus, applying the *principle of impulse and momentum* to the sphere and the plank as a system, we have

$$+\,\downarrow\; \Sigma M_C: \qquad 3\left(\frac{10}{32.2}\,\sqrt{322}\right) + 0 = \frac{1}{12}\left(\frac{30}{32.2}\right)(12)^2\omega$$

$$+\, 4\left[\frac{30}{32.2}\,(4\omega)\right]$$

$$+\, 5\left[\frac{10}{32.2}\,(5\omega)\right]$$

$$\omega = 0.494 \qquad \boldsymbol{\omega} = 0.494 \text{ rad/s} \downarrow \quad \blacktriangleleft$$

Developmental Exercises

D18.17   Refer to Example 18.7. Verify that $\mathbf{v}_S = \sqrt{322}$ ft/s $\downarrow$.

D18.18   Refer to Example 18.7. Determine the loss of kinetic energy of the system during the impact.

EXAMPLE 18.8

The 20-kg wheel shown has a central radius of gyration $\bar{k} = 0.25$ m and rolls without slipping on the steps. If its angular velocity on the step AB is $\omega_1 = 10$ rad/s $\downarrow$ and its impact with the corner C is perfectly plastic, determine its angular velocity (a) immediately after the impact, (b) on the step CD.

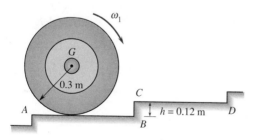

Solution.   Let the angular velocity of the wheel immediately after the impact be $\boldsymbol{\omega}_C$. The *weight force* of the wheel in juxtaposition with the impulsive reaction components $\mathbf{C}_x$ and $\mathbf{C}_y$ from the corner C is a *nonimpulsive force*, which exerts a negligible impulse on the wheel. Thus, applying the *principle of impulse and momentum* to the wheel, we have

Initial momentum diagram	Impulse diagram	Final momentum diagram

$$+ \circlearrowleft \ \Sigma M_C: \quad \{20(0.25)^2(10) + 0.18[20(0.3)(10)]\} + 0$$

$$= 20(0.25)^2\omega_C + 0.3[20(0.3\omega_C)]$$

$$\omega_C = 7.64 \qquad \omega_C = 7.64 \text{ rad/s} \ \circlearrowright \ \blacktriangleleft$$

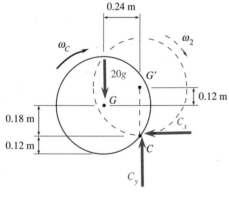

After impacting with the corner C, the wheel rises 0.12 m from the step AB to the step CD, while its angular velocity changes from $\omega_C = 7.64$ rad/s $\circlearrowright$ to ω_2. During this upward motion, the wheel simply rotates about C, its mass center moves from G to G' as shown, and the system is conservative. Note that the moment of inertia I_C of the wheel about C is

$$I_C = m\bar{k}^2 + md^2$$

$$= 20(0.25)^2 + 20(0.3)^2 \text{ kg·m}^2$$

$$= 3.05 \text{ kg·m}^2$$

Choosing the reference datum to be at G and applying the *principle of conservation of energy* we write

$$T_1 + V_1 = T_2 + V_2: \quad \tfrac{1}{2}I_C\omega_C^2 + 0 = \tfrac{1}{2}I_C\omega_2^2 + mgh$$

$$\tfrac{1}{2}(3.05)(7.64)^2 + 0 = \tfrac{1}{2}(3.05)\omega_2^2 + 20(9.81)(0.12)$$

$$\omega_2 = 6.55 \qquad \omega_2 = 6.55 \text{ rad/s} \ \circlearrowright \ \blacktriangleleft$$

Developmental Exercises

D18.19   Refer to Example 18.8. Determine the loss of kinetic energy of the wheel *during the impact*.

D18.20   Refer to Example 18.8. Determine the minimum magnitude of ω_1 with which the wheel will roll over the corner C to the step CD.

EXAMPLE 18.9[†]

Two slender rods of equal masses $m = 10$ kg may swing freely from the
pivots at A and C as shown. The rod AB is released from rest in the horizontal position. When AB swings to the vertical position, its small knob at
E hits the small knob at D on the rod CD which was at rest. If $d = 0.3$ m,
$L = 1.2$ m, and the coefficient of restitution between the knobs at D and E
on the rods is $e = 0.5$, determine the angular velocities ω'_{AB} and ω'_{CD} of the
rods AB and CD immediately after the impact.

Solution. By the *principle of work and energy*, it can readily be determined that the angular velocity of the rod AB just before the impact takes
place is $\omega_{AB} = \sqrt{3g/L}$ ↻ $= 4.952$ rad/s ↻. For simplicity in the solution,
let the impulse of an impulsive force be equal to the product of the *average*
impulsive force and the time interval Δt. Then, we apply the *principle of
impulse and momentum* to draw the following vector-diagram equations:

Rod AB

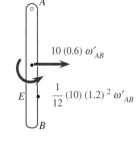

| Initial | Impulse diagram | Final |
| momentum diagram | | momentum diagram |

Rod CD

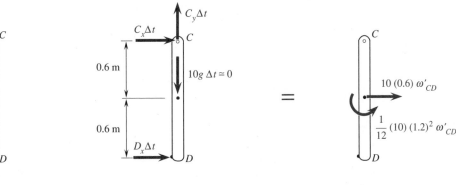

| Initial | Impulse diagram | Final |
| momentum diagram | | momentum diagram |

[†]For comparative study, the problem in this example is solved by using the *principle of gen-
eralized virtual work* in Example 18.11.

Referring to the vector-diagram equation for the rod AB, we write

$$+ \circlearrowleft \ \Sigma M_A: \quad \tfrac{1}{12} (10)(1.2)^2 (4.952) + 0.6[10(0.6)(4.952)]$$

$$- 0.9(D_x \, \Delta t)$$

$$= \tfrac{1}{12} (10)(1.2)^2 \omega'_{AB} + 0.6[10(0.6)\omega'_{AB}] \tag{1}$$

Referring to the vector-diagram equation for the rod CD, we write

$$+ \circlearrowleft \ \Sigma M_C: \quad 0 + 1.2(D_x \, \Delta t)$$

$$= \tfrac{1}{12} (10)(1.2)^2 \omega'_{CD} + 0.6[10(0.6)\omega'_{CD}] \tag{2}$$

The velocities of the points of contact E and D just before and just after the impact, in the direction of the line of impact, are $(v_E)_n = 0.9(4.952)$ m/s $\rightarrow$, $(v_D)_n = 0$, $(v'_E)_n = 0.9\omega'_{AB} \rightarrow$, and $(v'_D)_n = 1.2\omega'_{CD} \rightarrow$. Applying Eq. (18.18), which defines the *coefficient of restitution e* for the eccentric impact, we write

$$(v'_D)_n - (v'_E)_n = e[(v_E)_n - (v_D)_n]$$

$$1.2\omega'_{CD} - 0.9\omega'_{AB} = 0.5[0.9(4.952) - 0] \tag{3}$$

Solution of Eqs. (1), (2), and (3) yields

$$\omega'_{AB} = 2.28 \qquad \omega'_{CD} = 3.57 \qquad D_x \, \Delta t = 14.26$$

$$\boldsymbol{\omega}'_{AB} = 2.28 \text{ rad/s } \circlearrowleft \qquad \boldsymbol{\omega}'_{CD} = 3.57 \text{ rad/s } \circlearrowleft \quad \blacktriangleleft$$

Developmental Exercises

D18.21 Refer to Example 18.9. Verify that $\boldsymbol{\omega}_{AB} = 4.952$ rad/s $\circlearrowleft$.

D18.22 Work Example 18.9 if $d = 0.6$ m.

D18.23 Work Example 18.9 if the impact is *perfectly elastic*.

★18.7 Generalized Virtual Work: Impulse and Momentum

We recall from Sec. 14.10 that the *generalized virtual work $\delta U'$ of a linear vector* $\mathbf{L}$ (e.g., a linear impulse or linear momentum vector) acting on or associated with a particle during a virtual displacement $\delta \mathbf{r}$ of the particle is defined as

$$\delta U' = \mathbf{L} \cdot \delta \mathbf{r} \tag{18.19}$$

In a similar manner, the *generalized virtual work $\delta U'$ of an angular vector* $\mathbf{A}$ (e.g., an angular impulse or angular momentum vector) acting on or associated with a rigid body during an angular virtual displacement $\overrightarrow{\delta \theta}$ of the rigid body is defined as

$$\delta U' = \mathbf{A} \cdot \overrightarrow{\delta \theta} \tag{18.20}$$

By equivalence and equipollence in the *principle of impulse and momentum* in Secs. 18.3 and 18.4 and the nature of *constrained virtual displace-*

ment for a one-degree-of-freedom system in Sec. 13.13, it follows logically that, *during a constrained virtual displacement of a one-degree-of-freedom system of rigid bodies, the generalized virtual work of all the momentum vectors in the initial momentum diagram of the system plus the generalized virtual work of all the impulse vectors in the impulse diagram of the system must be equal to the generalized virtual work of all the momentum vectors in the final momentum diagram of the system.* This property is referred to as the *principle of generalized virtual work* for rigid bodies, which complements the traditional *principle of impulse and momentum* for rigid bodies.

EXAMPLE 18.10

Using the *principle of generalized virtual work*, do Example 18.4.

Solution.   For convenience, the space diagram of the system is repeated as shown, where a 40-kg block C is supported by a 20-kg cylinder A and a 30-kg cylinder B of the same radius $r = 100$ mm. The system is at rest at the time $t = 0$ when a 50-N force $\mathbf{P}$ is applied to the block C. The cylinders are assumed to roll without slipping, and we are to find $\mathbf{v}_C$ of the block C at $t = 2$ s.

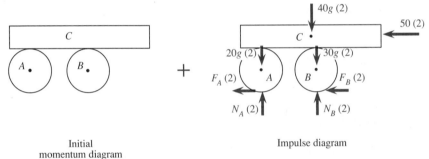

Initial
momentum diagram

Impulse diagram

Final
momentum diagram

As explained in the solution in Example 18.4, we first draw the vector-diagram equation for the entire system as shown. The constrained virtual displacement of the block C is $\overrightarrow{\delta x_C} = \delta x_C \leftarrow$, which is related to the constrained angular virtual displacement $\delta\theta$ of the cylinders A and B by $\delta x_C = 2r\delta\theta = 2(0.1)\delta\theta$ or $\delta\theta = 5\delta x_C$. The constrained virtual displacements of the centers of the cylinders are given by $\delta x_A = \delta x_B = r\,\delta\theta = 0.1(5\delta x_C) = 0.5\,\delta x_C$. Thus, the constrained virtual displacement of the system may be indicated as shown. Referring to the diagrams shown and applying the *principle of generalized virtual work*, we

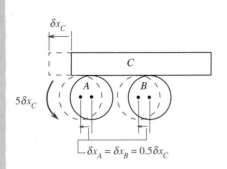

Constrained
virtual displacement

write

$$\delta U': \qquad 0 + 50(2)\, \delta x_C = 40 v_C(\delta x_C) + 20(0.5 v_C)(0.5\, \delta x_C)$$

$$+ 30(0.5 v_C)(0.5\, \delta x_C)$$

$$+ \tfrac{1}{2}(20)(0.1)^2(5 v_C)(5\, \delta x_C)$$

$$+ \tfrac{1}{2}(30)(0.1)^2(5 v_C)(5\, \delta x_C)$$

where the impulses of the weight forces and the reactions from the support to the cylinders do no generalized virtual work during the constrained virtual displacement. Thus, we have

$$100 = 58.75 v_C \qquad v_C = 1.702 \text{ m/s} \leftarrow \quad \blacktriangleleft$$

REMARK. By comparing the solution in this example with that in Example 18.4, we readily note that, in order to find v_C of the block C in the system, the traditional *principle of impulse and momentum* requires the solution of at least *three* simultaneous equations, while the *principle of generalized virtual work* requires the solution of only *one* equation. The ability to *uncouple an unknown quantity* in some seemingly complex one-degree-of-freedom systems is, therefore, a salient feature of the principle of generalized virtual work. Clearly, the advantage of this principle resides in the fact that *the impulses of the unknown constraining forces in the impulse diagram do not do any generalized virtual work during the constrained virtual displacement of the system.*

Developmental Exercise

D18.24   Describe the *principle of generalized virtual work* for rigid bodies.

EXAMPLE 18.11

Using the *principle of generalized virtual work*, work Example 18.9.

Solution. For convenience, the space diagram of the system is repeated here as shown, where the 10-kg rod AB is released from rest in the horizontal position to hit the 10-kg rod CD which is initially at rest. It is given that $d = 0.3$ m, $L = 1.2$ m, and $e = 0.5$ between the knobs at D and E. We are to find ω'_{AB} and ω'_{CD} of the rods AB and CD just after the impact.

As indicated in the solution in Example 18.9, the angular velocity of the rod AB just before the impact takes place is $\omega_{AB} = 4.952$ rad/s $\curvearrowright$. Thus, we draw the vector-diagram equation for the entire system as shown. During the impact, the system behaves as a one-degree-of-freedom system, whose constrained virtual displacement may be drawn as indicated. Referring to the diagrams drawn and applying the *principle of generalized virtual work*, we write

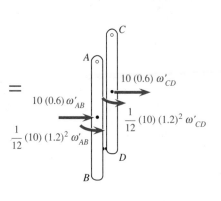

| Initial momentum diagram | Impulse diagram | Final momentum diagram |

$\delta U'$: $[10(0.6)(4.952)(0.8\ \delta\theta)$

$+ \frac{1}{12}(10)(1.2)^2(4.952)(\frac{4}{3}\ \delta\theta)] + 0$

$= 10(0.6)\omega'_{AB}(0.8\ \delta\theta) + \frac{1}{12}(10)(1.2)^2\ \omega'_{AB}\ (\frac{4}{3}\ \delta\theta)$

$+ 10(0.6)\omega'_{CD}(0.6\ \delta\theta) + \frac{1}{12}(10)(1.2)^2\omega'_{CD}(\delta\theta)$

which can be simplified and written as

$$4\omega'_{AB} + 3\omega'_{CD} = 4(4.952) \qquad (1)$$

Applying Eq. (18.18), which defines e, we write

$$(v'_D)_n - (v'_E)_n = e[(v_E)_n - (v_D)_n]$$

$$1.2\omega'_{CD} - 0.9\omega'_{AB} = 0.5[0.9(4.952) - 0] \qquad (2)$$

Solution of Eqs. (1) and (2) yields

$$\omega'_{AB} = 2.28 \qquad \omega'_{CD} = 3.57$$

$$\boldsymbol{\omega'_{AB}} = 2.28 \text{ rad/s} \quad \text{⌒} \qquad \boldsymbol{\omega'_{CD}} = 3.57 \text{ rad/s} \quad \text{⌒} \quad \blacktriangleleft$$

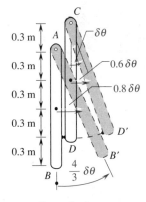

Constrained virtual displacement

REMARK. By comparing the solution in this example with that in Example 18.9, we readily note that Eq. (2) here and Eq. (3) in Example 18.9 are identical. Furthermore, by eliminating $D_x\ \Delta t$ from Eqs. (1) and (2) in Example 18.9 and simplifying, we obtain Eq. (1) in this example. Thus, the *principle of generalized virtual work* reduces the number of equations by *one* in solving the same problem in Example 18.9.

PROBLEMS

18.35 and 18.36   A 30-g bullet is fired with a velocity of 500 m/s into a 10-kg wooden rod AB suspended as shown. If $h = 0.75$ m and the bullet becomes embedded in 1 ms, determine (a) the angular velocity of the rod immediately after the impact, (b) the average impulsive reaction from the support at A.

Fig. P18.35

Fig. P18.36

18.37* Refer to Prob. 18.35. Determine (a) the distance h for which the average impulsive reaction from the support at A is zero, (b) the corresponding angular velocity of the rod immediately after the impact.

18.38* Solve Prob. 18.36 if $h = 0.5$ m.

18.39 A hammer exerts an impulse of 100 lb·s on a 300-lb sphere pivoted as shown. If $\theta = 60°$ and the impact lasts for 1.5 ms, determine (a) the angular velocity of the sphere immediately after the impact, (b) the average impulsive reaction from the support at A.

18.40 Solve Prob. 18.39 if $\theta = 120°$.

18.41 A player hits a tennis ball in a special way so that the ball has a velocity $\mathbf{v}$ and an angular velocity $\boldsymbol{\omega}$ just before it lands on the surface of the court as shown, where $\theta = 75°$ and $v = 10$ m/s. Assuming that the tennis ball can be taken as a spherical shell of diameter 65 mm and the ball does not skid during the impact, determine the minimum magnitude of $\boldsymbol{\omega}$ for which the ball will *not* rebound forward to the left.

Fig. P18.39

Fig. P18.41

Fig. P18.42

Fig. P18.43

18.42 The rod AB is falling with a velocity of 10 m/s ↓ when the wire BC becomes taut. If the impact is perfectly plastic, determine the angular velocity of the rod and the velocity of its mass center G immediately after the impact.

18.43 A 6-lb rod AB and an 8-lb rod BC are welded together to form a single 14-lb rigid body which is falling with a velocity of 88 ft/s ↓ when the wires BD and CE simultaneously become taut. If the impact is perfectly plastic and is completed in 1 ms, determine (a) the angular velocity of the rigid body immediately after the impact, (b) the average impulsive reaction from each wire.

18.44 A 0.1-kg ball B is thrown with a velocity of 30 m/s at a 10-kg rod AC as shown. If the coefficient of restitution between the ball and the rod is $e = 0.8$, determine the velocity of the ball immediately after the impact.

Fig. P18.44

18.45* A 4-oz ball B is thrown with a velocity of 100 ft/s at a 20-lb rod AC as shown. If the coefficient of restitution between the ball and the rod is $e = 0.7$, determine the angular velocity of the rod immediately after the impact.

Fig. P18.45*   **Fig. P18.46**

18.46 A uniformly loaded square crate slides with a velocity $\mathbf{v}_1$ when it strikes a small rigid obstruction at A as shown, where $\theta = 20°$. If the impact at A is perfectly plastic, determine the minimum magnitude of $\mathbf{v}_1$ for which the crate will rotate over the obstruction.

18.47* Solve Prob. 18.46 if $\theta = 10°$.

18.48 Solve Prob. 18.46 if $\theta = 0$.

18.49 A section of pipe rolls without slipping down an incline which has an obstruction at A as shown, where $\theta = 20°$. If the velocity of the center of the pipe just before the impact at A is $\mathbf{v}_1$ and the impact is perfectly plastic, determine the minimum magnitude of $\mathbf{v}_1$ for which the pipe will roll over the obstruction.

18.50* Solve Prob. 18.49 if $\theta = 10°$.

Fig. P18.49

18.51 A uniform slender rod is cut into four pieces which are then welded to form a rigid rectangular frame $ABCD$. The frame is released from rest in the position shown, and the floor is sufficiently rough to prevent the frame from slipping at the corner A. If the impact between the corner D and the floor is perfectly plastic, determine the smallest ratio of b/h for which the frame will not rebound in any way. (*Hint*. The answer is independent of the angle ϕ_0.)[†]

Fig. P18.51 and P18.52*

18.52* The rigid frame $ABCD$ is released from rest in the position shown, where the floor is sufficiently rough to prevent the frame from slipping at the corner A and the frame strikes the floor with an angular velocity ω_1 ⤵. If $b = h$ and the impact between the corner D and the floor is perfectly plastic, determine (a) the angular velocity ω_2 of the frame immediately after the impact, (b) the percentage of the kinetic energy of the frame lost during the impact.

[†]The solution of this problem leads to a cubic equation. (Cf. App. E for the solution of a nonlinear equation.)

Fig. P18.53

Fig. P18.58

18.53 The 10-kg slender rod AB is released from rest in the position shown and hits the edge C, where $b = 0.6$ m. If the impact at C is perfectly elastic and is completed in 1 ms, determine (a) the angular velocity of the rod immediately after the impact, (b) the average impulsive reactions from the edge C and the hinge at A.

18.54* Solve Prob. 18.53 if $b = 0.75$ m.

18.55* Solve Prob. 18.53 if $b = 0.9$ m.

18.56 Solve Prob. 18.53 if the coefficient of restitution between the rod and the edge at C is $e = 0.5$.

18.57* Refer to Prob. 18.53. Determine the length b for which the average impulsive reaction from the hinge at A is equal to zero.

18.58 The pendulum AB of an impact testing machine weighs 80 lb and has a radius of gyration about its mass center G of 10 in. The pendulum is designed so that the horizontal component of the reaction from the pivot at A is essentially zero during its impact from the specimen at C as shown. Determine (a) the length $\bar{r}$, (b) the reaction from the pivot at A during the impact if the pendulum is released in the position shown.

18.59 A 5-kg sphere S is dropped from a height $h = 2$ m onto the end A of the 10-kg rod AB as shown, where $b = 1.8$ m. If the impact at A is perfectly elastic, determine immediately after the impact (a) the angular velocity of the rod, (b) the velocity of the sphere.

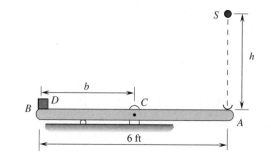

Fig. P18.59

18.60* Solve Prob. 18.59 if $b = 1.2$ m.

18.61* Solve Prob. 18.59 if $h = 4$ m.

18.62 A 10-lb sphere S is dropped from a height $h = 5$ ft onto the end A of the 20-lb rod AB as shown, where $b = 3$ ft and the 5-lb block D is not tied to the end B of the rod. If the impact at A is perfectly plastic, determine the height to which the block D will rise.

Fig. P18.62

18.63* Solve Prob. 18.62 if $b = 4$ ft.

18.64 A 45-g bullet is fired with a velocity of 650 m/s into a 12-kg wooden plank *AB* suspended by wires as shown. Determine the velocity of the bullet immediately after it becomes embedded in the plank.

Fig. P18.64

Fig. P18.65

18.65 The slender rod *AB* is dropped from rest in the position shown. The end *A* hits the corner *C* before the end *B* hits the corner *D* which is slightly lower than the corner *C*. If the impacts at *C* and *D* are perfectly plastic, determine the angular velocity of the rod immediately after it hits (a) the support *C*, (b) the support *D*.

18.66* A 20-Mg lunar landing module with mass center at *G* has a downward velocity of $\mathbf{v} = 2.5$ m/s ↓ when one of its four legs becomes the first part of the module to hit the surface of the moon with a perfectly plastic impact. The module has a radius of gyration of 2 m about *G*, which is 3 m directly above the center of the square formed by the four feet as corners. If the diagonal of that square is 8 m, determine the angular speed of the module immediately after the first impact.

18.67 Solve Prob. 18.66* if $\mathbf{v} = 1.5$ m/s ↓.

18.68* and **18.69** The rods *AB* and *CD* have equal weights $W = 20$ lb, and the rod *AB* is released from rest to hit the rod *CD* which is in equilibrium. If the coefficient of restitution between them is $e = 0.6$, determine their angular velocities immediately after the impact.

Fig. P18.66*

Fig. P18.68*

Fig. P18.69

18.70 Solve Prob. 18.68* if $e = 0.8$.

18.71* Solve Prob. 18.69 if $e = 0.8$.

18.72 and 18.73* The 50-lb carriage *D* is supported by a 20-lb cylinder *A*, a 25-lb cylinder *B*, and a 30-lb cylinder *C* as shown, where the cylinders have the same radius of 4 in. If the cylinders roll without slipping, determine the velocity $\mathbf{v}_D$ attained by the carriage *D* after the 10-lb force has been applied to move it from rest for 1.2 s.

Fig. P18.72   **Fig. P18.73***

18.74* Solve Prob. 18.72 if all cylinders have the same weight of 20 lb.

18.75 Solve Prob. 18.73* if all cylinders have the same weight of 20 lb.

18.76 The 50-kg carriage E is supported by a 20-kg cylinder A, a 25-kg cylinder B, a 30-kg cylinder C, and a 35-kg cylinder D as shown, where the cylinders have the same radius of 100 mm. If the cylinders roll without slipping, determine the velocity $\mathbf{v}_E$ attained by the carriage E after the 20-N force has been applied to move it from rest for 1.2 s.

Fig. P18.76

18.77* Solve Prob. 18.76 if all cylinders have the same mass of 20 kg.

18.78* through 18.83* Using the *principle of generalized virtual work*, solve the following problems:

18.78* Prob. 18.72.	18.79* Prob. 18.73*.
18.80* Prob. 18.74*.	18.81* Prob. 18.75.
18.82* Prob. 18.76.	18.83* Prob. 18.77*.

18.8 Concluding Remarks

Clearly, this chapter is a direct extension of Chap. 14. We recognize that, unlike a particle which only translates in space, a rigid body may rotate about its central axis and translate in space. By Chasles' theorem, the *momentum* of a rigid body in plane motion may be described by the *linear momentum* $m\mathbf{v}$ acting through the mass center G of the rigid body and the *angular momentum* $\bar{I}\boldsymbol{\omega}$ acting about G. The abilities to properly represent the angular impulse of a moment and the momentum of a rigid body in plane motion are the additional important skills to be developed when one progresses from Chap. 14 to Chap. 18.

The *principle of impulse and momentum* for rigid bodies is very useful in attacking (1) kinetic problems of rigid bodies in plane motion where *time* and *velocity* (among others) are to be directly related, and (2) problems of *eccentric impact* between rigid bodies. The *drawing* of appropriate *vector-diagram equations* for the momenta and impulses in the system during the time interval considered is recommended because it provides valuable geometric aid in the solutions of such problems.

The *principle of generalized virtual work* for rigid bodies is especially well adapted to the study of the plane kinetics of systems of rigid bodies with one degree of freedom, where impulse and momentum are involved.

As demonstrated in Example 18.10, this principle provides a procedure which enables us to *uncouple an unknown quantity* in a seemingly complex one-degree-of-freedom system. In problems of *eccentric impact*, this principle may help reduce the number of coupled simultaneous equations. Therefore, it is an effective complement to the principle of impulse and momentum.

REVIEW PROBLEMS

18.84 A 10-kg cylinder B is suspended by a cord which is wrapped around the drum of a 100-kg flywheel A as shown. The flywheel has a central radius of gyration $\bar{k}_O = 0.5$ m and the system is released from rest at the time $t = 0$. Neglecting effects of friction, determine the velocity of the cylinder when $t = 4$ s.

| Fig. P18.84 | Fig. P18.85* | Fig. P18.86* |

18.85* A 100-lb gear of central radius of gyration $\bar{k}_O = 0.75$ ft is rolled up a rack from rest at the time $t = 0$ by the force $\mathbf{P} = 2(t + 15)$ lb ↗ which is applied to a cord wrapped around the drum of the gear and remains parallel to the rack as shown, where t is measured in seconds, $r_1 = 0.5$ ft, and $r_2 = 1$ ft. Determine the angular velocity of the gear when $t = 5$ s.

18.86* The disk D of radius r is released from rest at the time $t = 0$ to press on the belt, which moves with a constant velocity $\mathbf{v}$ as shown. If the coefficient of kinetic friction between the disk and the belt is μ_k, determine the time t_2 at which slipping of the disk on the belt ceases. (*Hint.* The answer is a function of v, μ_k, θ, and g.)

18.87 Solve Prob. 18.86* if $v = 10$ m/s, $\mu_k = 0.2$, and $\theta = 20°$.

18.88 The gears A and B shown have masses 5 kg and 12 kg, and central radii of gyration 80 mm and 120 mm, respectively. The gear A is rotated from rest at the time $t = 0$ by a moment $\mathbf{M} = 12t$ N·m ↺ , where t is measured in seconds. Determine the time t at which the angular velocity of the gear B is 300 rpm ↻ .

18.89* Solve Prob. 18.88 if $\mathbf{M} = 12t^2$ N·m ↺ .

18.90 A solid sphere rolls without slipping with a horizontal velocity $\mathbf{v}_1 = 15$ m/s → when it hits the incline shown, where $\theta = 25°$. If the sphere does not rebound, determine (a) the new velocity $\mathbf{v}_2$ of the sphere as it starts up the incline, (b) the percentage of loss of kinetic energy of the sphere due to impact.

18.91* Solve Prob. 18.90 if $\mathbf{v}_1 = 30$ ft/s → and $\theta = 15°$.

Fig. P18.88

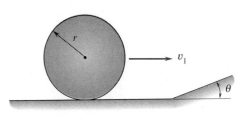

Fig. P18.90

18.92 A 120-lb acrobat stands at the end A of a 100-lb rigid plank ACB pivoted at C when a 150-lb weight W is dropped from a height $h_B = 10$ ft to hit the end B of the plank. If the impact at B is perfectly plastic and the acrobat stands completely rigid, determine the height h_A to which the acrobat at A will rise.

Fig. P18.92

18.93* Solve Prob. 18.92 if $h_B = 15$ ft.

18.94 A slender rod BC is connected by a cord to the support at A and is released from rest in the position shown. If the impact between the end C and the corner D is perfectly plastic, determine the angular velocity of the rod immediately after the impact.

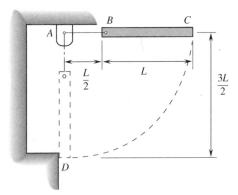

Fig. P18.94

18.95* Solve Prob. 18.94 if the impact is perfectly elastic.

18.96 Solve Prob. 18.94 if the coefficient of restitution between the end C and the corner D is $e = 0.6$.

18.97 The 10-kg rod AB rotates with $\omega = 4$ rad/s $\circlearrowright$ when it strikes the corner C as shown. If the coefficient of restitution between the rod and the corner C is $e = 0.5$, determine the maximum angular displacement of the rod after the impact.

18.98* Solve Prob. 18.97 if $e = 0.8$.

18.99 The 20-lb slender rod is released from rest in the position shown, where friction between the knobs A and B of the rod and their supports is negligible during

Fig. P18.97

the sliding. If impacts between the floor and the knobs of the rod are perfectly plastic and are completed in 0.1 s, determine the average impulsive reactions from the floor to the knobs of the rod.

18.100 A section of pipe weighing 64.4 lb rolls without slipping with $\omega_1 = 10$ rad/s $\circlearrowleft$ before it crosses a 14-in. gap as shown. Assuming the impact to be perfectly plastic, determine the new angular velocity ω_2 of the pipe after it has crossed the gap to the other side.

Fig. P18.99

Fig. P18.100

18.101* Refer to Prob. 18.100. Determine the minimum magnitude of ω_1 with which the pipe will cross the gap to the other side.

18.102* The 50-kg carriage C is supported by a 20-kg disk A and a 15-kg disk B as shown. Knowing that the disks roll without slipping, determine the velocity of the carriage 5 s after it is released from rest.

Fig. P18.102*   **Fig. P18.104**

18.103 Solve Prob. 18.102* if both disks A and B have the same mass of 20 kg.

18.104 A 50-kg wooden block C is supported by a 20-kg cylinder A and a 15-kg cylinder B as shown. Knowing that the cylinders roll without slipping, determine the velocity of the block 1.2 s after the system is released from rest.

18.105* Solve Prob. 18.104 if both cylinders A and B have the same mass of 20 kg.

Chapter 19

Motion of Rigid Bodies in Three Dimensions

KEY CONCEPTS

- Relation between the time derivatives of a vector taken in two different reference frames.
- Velocities and accelerations in different reference frames.
- Addition theorems for angular velocities and accelerations.
- Momentum and kinetic energy of a rigid body in general motion.
- Equations of motion for a rigid body in three-dimensional space, Euler's equations of motion, and Eulerian angles.
- Gyroscope, spin, nutation, precession, and steady precession.

The motion of rigid bodies in three-dimensional space is treated in this chapter. The fundamental relation between the time derivatives of a vector taken in two different reference frames is emphasized early in the chapter. The addition theorem for angular accelerations is shown to be different in form from that for angular velocities in general motion. Formulas for linear velocities and accelerations in different reference frames are given. The study of the kinetics of rigid bodies in space is based on the fundamental equations $\Sigma\mathbf{F} = \dot{\mathbf{L}}$ for translational motion, $\Sigma\mathbf{M}_G = \dot{\mathbf{H}}_G$ and $\Sigma\mathbf{M}_O = \dot{\mathbf{H}}_O$ for rotational motion, and their specialized forms for specific situations. In particular, Euler's equations of motion are derived, and the usual steps to be taken prior to applying Euler's equations of motion are outlined at the end of Sec. 19.8. Finally, we cover the gyroscopic motion and the cases of steady precession of axisymmetrical bodies either under a torque or in torque-free systems.

KINEMATICS OF RIGID BODIES IN SPACE

★19.1 Time Derivatives of a Vector in Two Reference Frames

Let $A\xi\eta\zeta$ be a reference frame *translating* in the fixed reference frame $OXYZ$, where the axes of $A\xi\eta\zeta$ remain parallel to the corresponding axes of $OXYZ$, as shown in Fig. 19.1. Then, the unit vectors $\mathbf{i}$, $\mathbf{j}$, $\mathbf{k}$ along the axes of $A\xi\eta\zeta$ are respectively equal to the unit vectors $\mathbf{I}$, $\mathbf{J}$, $\mathbf{K}$ along the axes of

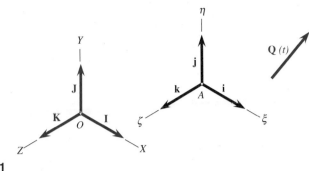

Fig. 19.1

OXYZ, and the scalar components (Q_ξ, Q_η, Q_ζ) of any vector **Q** in $A\xi\eta\zeta$ are correspondingly equal to its scalar components (Q_X, Q_Y, Q_Z) in *OXYZ* at any given instant. This means that the mathematical descriptions of **Q** in the *fixed reference frame OXYZ* and in the *translating reference frame $A\xi\eta\zeta$* are identical, and the time derivatives of **Q** in *OXYZ* and in $A\xi\eta\zeta$ must be equal; i.e.,

$$\left(\frac{d\mathbf{Q}}{dt}\right)_{OXYZ} = \left(\frac{d\mathbf{Q}}{dt}\right)_{A\xi\eta\zeta} \qquad (19.1)^\dagger$$

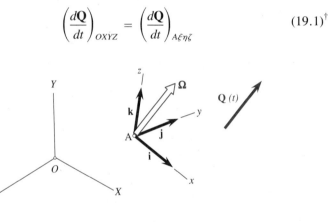

Fig. 19.2

Suppose that *Axyz* is a reference frame moving arbitrarily (e.g., both translating and rotating) in the fixed reference frame *OXYZ* as shown in Fig. 19.2, where **Ω** is the *angular velocity* of *Axyz* in *OXYZ*. As *Axyz* moves in *OXYZ*, the time rates of change of the unit vectors **i**, **j**, **k** in *Axyz* are zero because they are constant in both magnitude and direction in *Axyz*; i.e.,

$$(\dot{\mathbf{i}})_{Axyz} = (\dot{\mathbf{j}})_{Axyz} = (\dot{\mathbf{k}})_{Axyz} = 0 \qquad (19.2)$$

On the other hand, the time rates of change of the unit vectors **i**, **j**, **k** in *OXYZ* are generally not zero because their directions change from time to time as *Axyz* moves in *OXYZ*, even though each of them has a constant magnitude of 1 unit. As shown in Fig. 19.3, the unit vector **i** is changed to **i′** during the time interval Δt, and $\Delta\mathbf{i}$ represents the change in **i** due to the rotation **Ω** Δt of *Axyz*. For $\Delta t \approx 0$, we note from Fig. 19.3 that the magnitude and direction of $\Delta\mathbf{i}$ are given by the *cross product*

$$\Delta\mathbf{i} = (\mathbf{\Omega}\,\Delta t) \times \mathbf{i} \qquad (19.3)$$

Fig. 19.3

†Vectors in kinematics are *free vectors*. A free vector is characterized only by its *magnitude* and its *direction*, but not its line of action.

Thus, the time derivative of **i** in *OXYZ* is

$$(\dot{\mathbf{i}})_{OXYZ} = \lim_{\Delta t \to 0} \frac{\Delta \mathbf{i}}{\Delta t} = \boldsymbol{\Omega} \times \mathbf{i}$$

NOTE: A quantity, or the derivative of a quantity, without the name of a specific reference frame as subscripts is to be understood as one measured, or taken, in the fixed reference frame, such as *OXYZ*.

We write

$$\dot{\mathbf{i}} = \boldsymbol{\Omega} \times \mathbf{i} \tag{19.4}$$

Similarly, we can show that

$$\dot{\mathbf{j}} = \boldsymbol{\Omega} \times \mathbf{j} \qquad \dot{\mathbf{k}} = \boldsymbol{\Omega} \times \mathbf{k} \tag{19.5}$$

For a vector **Q** in *Axyz* as shown in Fig. 19.2, we write

$$\mathbf{Q} = Q_x \mathbf{i} + Q_y \mathbf{j} + Q_z \mathbf{k} \tag{19.6}$$

Taking the time derivative in *Axyz* on both sides of Eq. (19.6) and applying Eqs. (19.2) and (15.41), we write

$$(\dot{\mathbf{Q}})_{Axyz} = (\dot{Q}_x)_{Axyz}\mathbf{i} + (\dot{Q}_y)_{Axyz}\mathbf{j} + (\dot{Q}_z)_{Axyz}\mathbf{k}$$

$$(\dot{Q}_x)_{Axyz} = (\dot{Q}_x)_{OXYZ} = \dot{Q}_x$$

$$(\dot{Q}_y)_{Axyz} = (\dot{Q}_y)_{OXYZ} = \dot{Q}_y$$

$$(\dot{Q}_z)_{Axyz} = (\dot{Q}_z)_{OXYZ} = \dot{Q}_z$$

$$(\dot{\mathbf{Q}})_{Axyz} = \dot{Q}_x\mathbf{i} + \dot{Q}_y\mathbf{j} + \dot{Q}_z\mathbf{k} \tag{19.7}$$

Taking the time derivative in *OXYZ* on both sides of Eq. (19.6) and applying Eqs. (15.41), (19.4), (19.5), and (19.7), we write

$$\dot{\mathbf{Q}} = (\dot{Q}_x\mathbf{i} + \dot{Q}_y\mathbf{j} + \dot{Q}_z\mathbf{k}) + Q_x\dot{\mathbf{i}} + Q_y\dot{\mathbf{j}} + Q_z\dot{\mathbf{k}}$$

$$= (\dot{\mathbf{Q}})_{Axyz} + Q_x(\boldsymbol{\Omega} \times \mathbf{i}) + Q_y(\boldsymbol{\Omega} \times \mathbf{j}) + Q_z(\boldsymbol{\Omega} \times \mathbf{k})$$

$$= (\dot{\mathbf{Q}})_{Axyz} + \boldsymbol{\Omega} \times (Q_x\mathbf{i} + Q_y\mathbf{j} + Q_z\mathbf{k})$$

$$\dot{\mathbf{Q}} = (\dot{\mathbf{Q}})_{Axyz} + \boldsymbol{\Omega} \times \mathbf{Q} \tag{19.8}[†]$$

Equation (19.8) is an important equation in this chapter; it permits us easily to calculate the time derivative of a vector in one reference frame if that vector is expressed in another. We see from Eq. (19.8) that the time derivative of any vector **Q** taken in the *fixed reference frame OXYZ* is equal to the sum of the following two parts: (a) the time derivative of **Q** taken in the *moving reference frame Axyz*, and (b) the cross product $\boldsymbol{\Omega} \times \mathbf{Q}$, where $\boldsymbol{\Omega}$ is the *angular velocity* of *Axyz* in *OXYZ*.

[†]Cf. Eq. (15.44).

Developmental Exercises

D19.1 In the notation adopted, is the time derivative $\dot{Q}$ taken in the *moving reference frame Axyz* or the *fixed reference frame OXYZ?*

D19.2 Suppose that Q is a *constant* vector in *Axyz*, whose angular velocity in *OXYZ* is Ω. Express the first time derivative $\dot{Q}$ and the second time derivative $\ddot{Q}$ taken in *OXYZ*.

★19.2 Velocities in Different Reference Frames

In certain practical situations, a particle may move relative to two reference frames (or bodies) of interest to us. For example, a pin may be confined to move in a slot of a crank which is itself in motion relative to the ground. The general situation may be represented in Fig. 19.4, where a particle B moves along a path C in the reference frame *Axyz*, which is itself moving (i.e., translating and rotating) in the fixed reference frame *OXYZ*. For our study, we let the angular velocity of *Axyz* in *OXYZ* be Ω. The positions of A and B in *OXYZ* are specified by the position vectors r_A and r_B, while the position of B relative to A is specified by the position vector $r_{B/A}$.

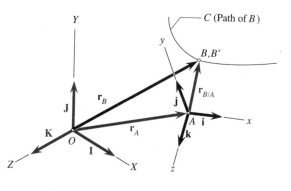

Fig. 19.4

I. LINEAR VELOCITIES IN *OXYZ* AND *Axyz*.

The analytical expression of $r_{B/A}$ may be written in terms of the unit vectors i, j, and k in *Axyz*, or the unit vectors I, J, and K in *OXYZ*, as shown in Fig. 19.4. Letting (x_B, y_B, z_B) be the coordinates of B in *Axyz*, we write

$$r_{B/A} = x_B i + y_B j + z_B k \tag{19.9}$$

$$v_{B/Axyz} = (\dot{r}_{B/A})_{Axyz} = \dot{x}_B i + \dot{y}_B j + \dot{z}_B k \tag{19.10}$$

Referring to Fig. 19.4, we write

$$r_B = r_A + r_{B/A}$$

$$\boxed{r_B = r_{B/A} + r_A} \tag{19.11}$$

Taking the time derivative in *OXYZ* on both sides of Eq. (19.11) and applying Eq. (19.8), we write

$$\dot{\mathbf{r}}_B = \dot{\mathbf{r}}_{B/A} + \dot{\mathbf{r}}_A = [(\dot{\mathbf{r}}_{B/A})_{Axyz} + \boldsymbol{\Omega} \times \mathbf{r}_{B/A}] + \dot{\mathbf{r}}_A$$

$$\mathbf{v}_B = \mathbf{v}_{B/Axyz} + \boldsymbol{\Omega} \times \mathbf{r}_{B/A} + \mathbf{v}_A \tag{19.12}$$

where

$$\mathbf{r}_{B/A} = \overrightarrow{AB} = \text{position vector of } B \text{ relative to } A$$
$$\mathbf{v}_A, \mathbf{v}_B = \text{velocities of } A \text{ and } B \text{ in the } fixed \text{ reference frame } OXYZ$$
$$\mathbf{v}_{B/Axyz} = \text{velocity of } B \text{ in the } moving \text{ reference frame } Axyz$$
$$\boldsymbol{\Omega} = \text{angular velocity of } Axyz \text{ in } OXYZ$$

Suppose that B' is a point *of* the moving reference frame $Axyz$, and B' *coincides* with the moving particle B at the instant considered, as shown in Fig. 19.4. Following similar steps as those used in deriving Eqs. (15.50) and (15.52), we can show that the velocity of B' in $OXYZ$ is

$$\mathbf{v}_{B'} = \boldsymbol{\Omega} \times \mathbf{r}_{B/A} + \mathbf{v}_A \tag{19.13}$$

and the velocity of the particle B in $OXYZ$ can be expressed as

$$\mathbf{v}_B = \mathbf{v}_{B/Axyz} + \mathbf{v}_{B'} \tag{19.14}$$

II. ANGULAR VELOCITY $\boldsymbol{\Omega}$ OF *Axyz* IN *OXYZ*.

For defining the orientation of $Axyz$ in $OXYZ$, as shown in Fig. 19.4, let (a_{11}, a_{12}, a_{13}), (a_{21}, a_{22}, a_{23}), and (a_{31}, a_{32}, a_{33}) be the sets of *direction cosines* of the unit vectors $\mathbf{i}$, $\mathbf{j}$, and $\mathbf{k}$ with respect to the fixed reference frame $OXYZ$, respectively. Then, we have

$$\mathbf{i} = a_{11}\mathbf{I} + a_{12}\mathbf{J} + a_{13}\mathbf{K} \tag{19.15}$$

$$\mathbf{j} = a_{21}\mathbf{I} + a_{22}\mathbf{J} + a_{23}\mathbf{K} \tag{19.16}$$

$$\mathbf{k} = a_{31}\mathbf{I} + a_{32}\mathbf{J} + a_{33}\mathbf{K} \tag{19.17}$$

where $\mathbf{I}$, $\mathbf{J}$, and $\mathbf{K}$ are the unit vectors directed along the axes of $OXYZ$. The direction cosines a_{mn} (m, $n = 1, 2, 3$) are functions of the time t as $Axyz$ moves in $OXYZ$. Therefore, the time derivatives of $\mathbf{i}$, $\mathbf{j}$, and $\mathbf{k}$ taken in $OXYZ$ are

$$\dot{\mathbf{i}} = \dot{a}_{11}\mathbf{I} + \dot{a}_{12}\mathbf{J} + \dot{a}_{13}\mathbf{K} \tag{19.18}$$

$$\dot{\mathbf{j}} = \dot{a}_{21}\mathbf{I} + \dot{a}_{22}\mathbf{J} + \dot{a}_{23}\mathbf{K} \tag{19.19}$$

$$\dot{\mathbf{k}} = \dot{a}_{31}\mathbf{I} + \dot{a}_{32}\mathbf{J} + \dot{a}_{33}\mathbf{K} \tag{19.20}$$

Next, let the *angular velocity* $\boldsymbol{\Omega}$ of $Axyz$ in $OXYZ$ be written as

$$\boldsymbol{\Omega} = \Omega_x\mathbf{i} + \Omega_y\mathbf{j} + \Omega_z\mathbf{k} \tag{19.21}$$

Then, Eqs. (19.6), (19.8), and (19.21) yield

$$\dot{\mathbf{Q}} = (\dot{\mathbf{Q}})_{Axyz} + (\Omega_x\mathbf{i} + \Omega_y\mathbf{j} + \Omega_z\mathbf{k}) \times (Q_x\mathbf{i} + Q_y\mathbf{j} + Q_z\mathbf{k})$$

$$\dot{\mathbf{Q}} = (\dot{\mathbf{Q}})_{Axyz} + [Q_x(-\Omega_y\mathbf{k} + \Omega_z\mathbf{j}) + Q_y(-\Omega_z\mathbf{i} + \Omega_x\mathbf{k})$$
$$+ Q_z(-\Omega_x\mathbf{j} + \Omega_y\mathbf{i})] \tag{19.22}$$

Taking the time derivative on both sides of Eq. (19.6) in *OXYZ* and using Eq. (19.7), we write

$$\dot{\mathbf{Q}} = (\dot{Q}_x\mathbf{i} + \dot{Q}_y\mathbf{j} + \dot{Q}_z\mathbf{k}) + (Q_x\dot{\mathbf{i}} + Q_y\dot{\mathbf{j}} + Q_z\dot{\mathbf{k}})$$

$$\dot{\mathbf{Q}} = (\dot{\mathbf{Q}})_{Axyz} + (Q_x\dot{\mathbf{i}} + Q_y\dot{\mathbf{j}} + Q_z\dot{\mathbf{k}}) \tag{19.23}$$

Comparing Eq. (19.23) with Eq. (19.22), we see that

$$\dot{\mathbf{i}} = -\Omega_y\mathbf{k} + \Omega_z\mathbf{j} \tag{19.24}$$

$$\dot{\mathbf{j}} = -\Omega_z\mathbf{i} + \Omega_x\mathbf{k} \tag{19.25}$$

$$\dot{\mathbf{k}} = -\Omega_x\mathbf{j} + \Omega_y\mathbf{i} \tag{19.26}$$

By dot products, we find from Eqs. (19.24) through (19.26) that

$$\Omega_x = \dot{\mathbf{j}} \cdot \mathbf{k} = -\dot{\mathbf{k}} \cdot \mathbf{j} \tag{19.27}$$

$$\Omega_y = \dot{\mathbf{k}} \cdot \mathbf{i} = -\dot{\mathbf{i}} \cdot \mathbf{k} \tag{19.28}$$

$$\Omega_z = \dot{\mathbf{i}} \cdot \mathbf{j} = -\dot{\mathbf{j}} \cdot \mathbf{i} \tag{19.29}$$

Substituting Eqs. (19.27) through (19.29) into Eq. (19.21), we arrive at an equation defining the *angular velocity* $\mathbf{\Omega}$ of *Axyz* in *OXYZ* as follows:

$$\mathbf{\Omega} = (\dot{\mathbf{j}} \cdot \mathbf{k})\mathbf{i} + (\dot{\mathbf{k}} \cdot \mathbf{i})\mathbf{j} + (\dot{\mathbf{i}} \cdot \mathbf{j})\mathbf{k} \tag{19.30}^{\dagger}$$

Thus, $\mathbf{\Omega}$ can be found when the time-dependent direction cosines a_{mn} (m, $n = 1, 2, 3$) in Eqs. (19.15) through (19.17) are known.

III. ADDITION THEOREM FOR ANGULAR VELOCITIES.

For a system consisting of several movable rigid bodies, it can be advantageous to use multiple reference frames, such as those shown in Fig. 19.5, where *OXYZ* continues to represent the *fixed reference frame*, and *Axyz*, *Bx'y'z'*, and *Cx"y"z"* are reference frames embedded in rigid bodies *A*, *B*, and *C*, respectively. The relations among the angular velocities of the various rigid bodies

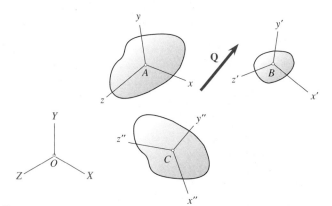

Fig. 19.5

---

†Cf. the footnote on p. 696.

with respect to each other are frequently utilized in the analysis of systems of rigid bodies moving in three dimensions. Suppose that we let $\omega_{A/B}$ denote *the angular velocity of the rigid body A* (and hence the reference frame *Axyz*) *with respect to the rigid body B* (and hence the reference frame *Bx'y'z'*), or more briefly *the angular velocity of A in B*. Similarly, let $\omega_{B/A}$, $\omega_{C/A}$, and $\omega_{C/B}$ denote the angular velocities of *B* in *A*, *C* in *A*, and *C* in *B*, respectively. Then, for any vector **Q** as shown in Fig. 19.5, we apply Eq. (19.8) to write

$$(\dot{\mathbf{Q}})_{Bx'y'z'} = (\dot{\mathbf{Q}})_{Axyz} + \omega_{A/B} \times \mathbf{Q} \tag{19.31}$$

$$(\dot{\mathbf{Q}})_{Axyz} = (\dot{\mathbf{Q}})_{Bx'y'z'} + \omega_{B/A} \times \mathbf{Q} \tag{19.32}$$

Adding Eqs. (19.31) and (19.32) and simplifying, we obtain

$$(\omega_{A/B} + \omega_{B/A}) \times \mathbf{Q} = \mathbf{0} \tag{19.33}$$

Since Eq. (19.33) must hold true for any vector **Q**, we conclude that

$$\omega_{A/B} = -\omega_{B/A} \tag{19.34}$$

Furthermore, referring to Fig. 19.5 and applying Eq. (19.8), we write

$$(\dot{\mathbf{Q}})_{Bx'y'z'} = (\dot{\mathbf{Q}})_{Cx''y''z''} + \omega_{C/B} \times \mathbf{Q} \tag{19.35}$$

$$(\dot{\mathbf{Q}})_{Cx''y''z''} = (\dot{\mathbf{Q}})_{Axyz} + \omega_{A/C} \times \mathbf{Q} \tag{19.36}$$

Subtraction of Eqs. (19.35) and (19.36) from Eq. (19.31) leads to

$$(\omega_{A/B} - \omega_{A/C} - \omega_{C/B}) \times \mathbf{Q} = \mathbf{0} \tag{19.37}$$

Since Eq. (19.37) must hold true for any vector **Q**, we conclude that

$$\omega_{A/B} = \omega_{A/C} + \omega_{C/B} \tag{19.38}$$

This property can be extended to *any number* of rigid bodies (or reference frames). For instance, we write

$$\omega_{A/O} = \omega_{A/B} + \omega_{B/O} = \omega_{A/B} + (\omega_{B/C} + \omega_{C/O})$$

$$\omega_{A/O} = \omega_{A/B} + \omega_{B/C} + \omega_{C/O} \tag{19.39}$$

where the subscript *O* refers to the fixed reference frame *OXYZ*. Following our adopted convention of notations for quantities measured in the fixed reference frame, we may write Eq. (19.39) as

$$\omega_A = \omega_{A/B} + \omega_{B/C} + \omega_C \tag{19.40}$$

The relation expressed in Eq. (19.38) or (19.40) is called the *addition theorem for angular velocities*. The validity of this theorem may seem intuitively obvious to the reader, but it will be shown in Sec. 19.3 that a similar relation does *not* hold true for angular accelerations in *general motion*.[†]

[†]Cf. Eqs. (19.48) and (19.50).

EXAMPLE 19.1

A connecting rod AB is attached to a disk by a ball-and-socket joint at A and to a collar at B by a clevis as shown. The disk is mounted on a vertical shaft at O and rotates in a horizontal plane at a constant rate of $\omega_1 = 29\mathbf{J}$ rad/s. The collar at B slides freely on the vertical rod CD, which is fixed at C. For the position shown, determine (a) the velocity of the collar at B, (b) the angular velocity of the rod AB.

Solution. We first let the reference frame $Axyz$ *be embedded in the rod AB* as indicated in the separate sketch, where the reference frame $OXYZ$ is *fixed on the ground*. Since the point A is attached to the disk and the collar at B slides freely on the vertical rod CD, we have

$$\mathbf{v}_A = \boldsymbol{\omega}_1 \times \mathbf{r}_{A/E} = 29\mathbf{J} \times 0.8\mathbf{K} = 23.2\mathbf{I}$$

$$\mathbf{v}_B = v_B\mathbf{J} \qquad \mathbf{r}_{B/A} = 2\mathbf{I} + 2.5\mathbf{J} - 0.8\mathbf{K}$$

The angular velocity $\boldsymbol{\omega}_{AB}$ of the rod AB is the same as the angular velocity $\boldsymbol{\Omega}$ of $Axyz$ in $OXYZ$. Expressing in rectangular components, we write

$$\boldsymbol{\omega}_{AB} = \boldsymbol{\Omega} = \Omega_X\mathbf{I} + \Omega_Y\mathbf{J} + \Omega_Z\mathbf{K}$$

Since B does not move in $Axyz$, we see that $\mathbf{v}_{B/Axyz} = \mathbf{0}$. Applying Eq. (19.12), we write

$$\mathbf{v}_B = \mathbf{v}_{B/Axyz} + \boldsymbol{\Omega} \times \mathbf{r}_{B/A} + \mathbf{v}_A$$

$$v_B\mathbf{J} = \mathbf{0} + \begin{vmatrix} \mathbf{I} & \mathbf{J} & \mathbf{K} \\ \Omega_X & \Omega_Y & \Omega_Z \\ 2 & 2.5 & -0.8 \end{vmatrix} + 23.2\mathbf{I}$$

$$v_B\mathbf{J} = (-0.8\Omega_Y - 2.5\Omega_Z + 23.2)\mathbf{I} + (0.8\Omega_X + 2\Omega_Z)\mathbf{J}$$

$$+ (2.5\Omega_X - 2\Omega_Y)\mathbf{K}$$

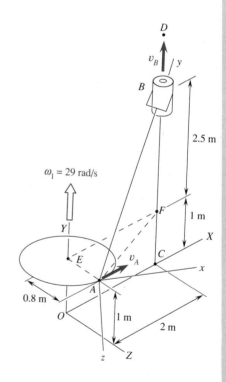

Equating the coefficients of the unit vectors, we obtain

$$0 = -0.8\Omega_Y - 2.5\Omega_Z + 23.2 \tag{1}$$

$$v_B = 0.8\Omega_X + 2\Omega_Z \tag{2}$$

$$0 = 2.5\Omega_X - 2\Omega_Y \tag{3}$$

The ball-and-socket joint at A and the collar-clevis connection at B allow the rod AB to rotate about the vertical rod CD and also about an axis perpendicular to the plane containing AB and CD. However, the rod AB *cannot* rotate about any axis (e.g., AF) which is perpendicular to CD and lies in the plane containing AB and CD. Thus, we see that $\boldsymbol{\Omega} \perp \overrightarrow{AF}$, or $\boldsymbol{\Omega} \cdot \overrightarrow{AF} = 0$. Since $\overrightarrow{AF} = 2\mathbf{I} - 0.8\mathbf{K}$, we write

$$(\Omega_X\mathbf{I} + \Omega_Y\mathbf{J} + \Omega_Z\mathbf{K}) \cdot (2\mathbf{I} - 0.8\mathbf{K}) = 0$$

$$2\Omega_X - 0.8\Omega_Z = 0 \tag{4}$$

Equations (1) through (4) yield

$$v_B = 18.56 \qquad \Omega_X = 3.2 \qquad \Omega_Y = 4 \qquad \Omega_Z = 8$$

We obtain

$$\mathbf{v}_B = 18.56\mathbf{J} \text{ m/s} \blacktriangleleft$$

$$\boldsymbol{\omega}_{AB} = 3.2\mathbf{I} + 4\mathbf{J} + 8\mathbf{K} \text{ rad/s} \blacktriangleleft$$

REMARK. If the clevis at B is replaced with a ball-and-socket joint, then the rod AB is free to rotate about its axis AB. In this case, Eq. (4) becomes invalid and Eqs. (1) through (3) yield $\Omega_X = 0.8\Omega_Y$, $\Omega_Z = 9.28 - 0.32\Omega_Y$, and $v_B = 18.56$ m/s. Thus, v_B is determinate even though Ω_X, Ω_Y, and Ω_Z are indeterminate in the modified system.

Developmental Exercises

D19.3 Work Example 19.1 for the instant at which the joint A passes through the position (0.8, 1, 0) m.

D19.4 Work Example 19.1 for the instant at which the joint A passes through the position (0, 1, -0.8) m.

D19.5 Work Example 19.1 for the instant at which the joint A passes through the position (-0.8, 1, 0) m.

EXAMPLE 19.2

The universal joint at U permits the transmission of power between the two shafts A and B which are out of alignment by an angle ψ as shown. The legs of the crosspiece C at U are perpendicular to each other, and the orientation of the yoke of A is defined by the angle θ which is measured from the vertical axis at U. If the angular speeds of the shafts A and B are ω_A and ω_B, respectively, determine the ratio ω_B/ω_A and interpret the result.

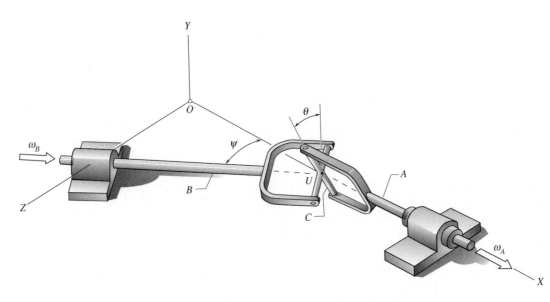

Solution. Let the unit vectors along the legs of the crosspiece C be $\mathbf{e}_1$ and $\mathbf{e}_2$, respectively, and the unit vector along the shaft B be $\boldsymbol{\lambda}$ as shown.

Then, we have

$$\mathbf{e}_1 = \cos\theta\,\mathbf{J} + \sin\theta\,\mathbf{K} \qquad \boldsymbol{\lambda} = \cos\psi\,\mathbf{I} - \sin\psi\,\mathbf{K}$$

Since $\mathbf{e}_2$ is perpendicular to both $\mathbf{e}_1$ and $\boldsymbol{\lambda}$, we have

$$\mathbf{e}_2 = \frac{\mathbf{e}_1 \times \boldsymbol{\lambda}}{|\mathbf{e}_1 \times \boldsymbol{\lambda}|}$$

$$= \frac{1}{G}\,(-\cos\theta\,\sin\psi\,\mathbf{I} + \sin\theta\,\cos\psi\,\mathbf{J} - \cos\theta\,\cos\psi\,\mathbf{K})$$

where

$$G = (\cos^2\theta + \sin^2\theta\,\cos^2\psi)^{1/2}$$

Applying the *addition theorem for angular velocities*, we write

$$\boldsymbol{\omega}_B = \boldsymbol{\omega}_{B/C} + \boldsymbol{\omega}_{C/A} + \boldsymbol{\omega}_A \qquad (1)$$

Referring to the system shown, we write

$$\boldsymbol{\omega}_B = \omega_B\boldsymbol{\lambda} = \omega_B(\cos\psi\,\mathbf{I} - \sin\psi\,\mathbf{K})$$

$$\boldsymbol{\omega}_{B/C} = \omega_{B/C}\,\mathbf{e}_2 = \frac{\omega_{B/C}}{G}\,(-\cos\theta\,\sin\psi\,\mathbf{I} + \sin\theta\,\cos\psi\,\mathbf{J}$$
$$- \cos\theta\,\cos\psi\,\mathbf{K})$$

$$\boldsymbol{\omega}_{C/A} = \omega_{C/A}\mathbf{e}_1 = \omega_{C/A}(\cos\theta\,\mathbf{J} + \sin\theta\,\mathbf{K})$$

$$\boldsymbol{\omega}_A = \omega_A\mathbf{I}$$

Substituting the above expressions into Eq. (1) and equating the coefficients of the unit vectors, we have

$$\omega_B\cos\psi = -\frac{1}{G}\,\omega_{B/C}\cos\theta\,\sin\psi + \omega_A \qquad (2)$$

$$0 = \frac{1}{G}\,\omega_{B/C}\sin\theta\,\cos\psi + \omega_{C/A}\cos\theta \qquad (3)$$

$$-\omega_B\sin\psi = -\frac{1}{G}\,\omega_{B/C}\cos\theta\,\cos\psi + \omega_{C/A}\sin\theta \qquad (4)$$

Eliminating $\omega_{C/A}$ in Eqs. (3) and (4), we get

$$\omega_{B/C} = G\omega_B \tan\psi \cos\theta \qquad (5)$$

Substituting Eq. (5) into Eq. (2) and simplifying, we obtain

$$\frac{\omega_B}{\omega_A} = \frac{\cos\psi}{1 - \sin^2\psi \sin^2\theta} \quad \blacktriangleleft$$

When the misalignment angle ψ is equal to 0, the value of the ratio ω_B/ω_A becomes 1, or $\omega_B = \omega_A$, as expected. To aid in visualizing the result, we may make use of the digital root finder in App. E to plot the variation of this ratio versus θ for the case of $\psi = 60°$ as shown, where $Y = \omega_B/\omega_A$, $X = \theta$ (in degrees), and $0 \le \theta \le 180°$.

```
XI = 0      DX = 6      XF = 180      R = 2.000001907348633

-R                                      0                                    R
!--:----!----:----!----:----!----:----!----:----!----:----!----:----!----:--!Y
                                    ! 0        *
                                    ! 6        *
                                    ! 12       *
                                    ! 18       *
                                    ! 24        *
                                    ! 30         *
                                    ! 36          *
                                    ! 42           *
                                    ! 48            *
                                    ! 54             *
                                    ! 60              *
                                    ! 66               *
                                    ! 72                *
                                    ! 78                 *
                                    ! 84                   *
                                    ! 90                    *
                                    ! 96                   *
                                    ! 102                *
                                    ! 108               *
                                    ! 114             *
                                    ! 120            *
                                    ! 126           *
                                    ! 132          *
                                    ! 138         *
                                    ! 144        *
                                    ! 150        *
                                    ! 156       *
                                    ! 162      *
                                    ! 168      *
                                    ! 174      *
                                    ! 180      *
                                    X

1000 PI = 3.141593: PSI = 60*PI/180: THETA = X*PI/180
1001 Y = COS(PSI)/(1 - (SIN(PSI)*SIN(THETA))^2)
```

We can interpret from the same plot for the result that, for a given misalignment angle ψ in the range $0 < \psi < 90°$, the shaft whose yoke comes nearly or exactly perpendicular to the plane containing the axes of the two shafts must be turning more rapidly than the other shaft. However, the ratio ω_B/ω_A becomes zero and the system becomes inoperable when $\psi = 90°$. This is the *gimbal lock*.

Developmental Exercises

D19.6　Refer to Example 19.2. Making use of the digital root finder in App. E, plot the ratio ω_B/ω_A versus θ for $0 \le \theta \le 360°$ and (a) $\psi = 65°$, (b) $\psi = 70°$.

D19.7　Suppose that the time rates of change of the unit vectors **i**, **j**, and **k** along the axes of $Axyz$, as taken in $OXYZ$, are known. Define the angular velocity $\boldsymbol{\Omega}$ of $Axyz$ in $OXYZ$.

D19.8　Describe the *addition theorem for angular velocities*.

★19.3   Accelerations in Different Reference Frames

I. LINEAR ACCELERATIONS IN *OXYZ* AND *Axyz*.

We recall that the velocity $\mathbf{v}_B$ of a particle B, as shown in Fig. 19.4, is given by Eq. (19.12). Since the acceleration is the time derivative of the velocity, we take the time derivative in $OXYZ$ on both sides of Eq. (19.12) and apply Eq. (19.8) to write

$$\dot{\mathbf{v}}_B = [(\dot{\mathbf{v}}_{B/Axyz})_{Axyz} + \boldsymbol{\Omega} \times \mathbf{v}_{B/Axyz}] + \dot{\boldsymbol{\Omega}} \times \mathbf{r}_{B/A}$$
$$+ \boldsymbol{\Omega} \times [(\dot{\mathbf{r}}_{B/A})_{Axyz} + \boldsymbol{\Omega} \times \mathbf{r}_{B/A}] + \dot{\mathbf{v}}_A$$

$$\mathbf{a}_B = \mathbf{a}_{B/Axyz} + \boldsymbol{\Omega} \times \mathbf{v}_{B/Axyz} + \dot{\boldsymbol{\Omega}} \times \mathbf{r}_{B/A}$$
$$+ \boldsymbol{\Omega} \times \mathbf{v}_{B/Axyz} + \boldsymbol{\Omega} \times (\boldsymbol{\Omega} \times \mathbf{r}_{B/A}) + \mathbf{a}_A$$

$$\boxed{\mathbf{a}_B = \mathbf{a}_{B/Axyz} + \dot{\boldsymbol{\Omega}} \times \mathbf{r}_{B/A} + \boldsymbol{\Omega} \times (\boldsymbol{\Omega} \times \mathbf{r}_{B/A}) \\ + 2\boldsymbol{\Omega} \times \mathbf{v}_{B/Axyz} + \mathbf{a}_A} \qquad (19.41)$$

Fig. 19.4 (repeated)

Suppose that B' is a point *of* the moving reference frame $Axyz$, and B' *coincides* with the moving particle B at the instant considered, as shown in Fig. 19.4. Following similar steps as those used in deriving Eqs. (15.61) and (15.64), we can show that the acceleration of B' in $OXYZ$ is

$$\boxed{\mathbf{a}_{B'} = \dot{\boldsymbol{\Omega}} \times \mathbf{r}_{B/A} + \boldsymbol{\Omega} \times (\boldsymbol{\Omega} \times \mathbf{r}_{B/A}) + \mathbf{a}_A} \qquad (19.42)$$

and the acceleration of the particle B in $OXYZ$ can be expressed as

$$\boxed{\mathbf{a}_B = \mathbf{a}_{B/Axyz} + \mathbf{a}_{B'} + 2\boldsymbol{\Omega} \times \mathbf{v}_{B/Axyz}} \qquad (19.43)$$

Note in Eqs. (19.41) through (19.43) that $\mathbf{v}_{B/Axyz}$, $\boldsymbol{\Omega}$, and $\mathbf{r}_{B/A}$ are the same as those defined in Sec. 19.2; and

$\mathbf{a}_A$, $\mathbf{a}_B$ = accelerations of A and B in the *fixed* reference frame $OXYZ$

$\mathbf{a}_{B'}$ = acceleration of the *coinciding point* B' in $OXYZ$

$\mathbf{a}_{B/Axyz}$ = acceleration of B in the *moving* reference frame $Axyz$

$\dot{\boldsymbol{\Omega}}$ = angular acceleration of $Axyz$ in $OXYZ$

Furthermore, note that $\mathbf{\Omega}$ is usually not perpendicular to $\mathbf{v}_{B/Axyz}$ and the magnitude of the Coriolis (or complementary) acceleration $2\mathbf{\Omega} \times \mathbf{v}_{B/Axyz}$ is generally *not* equal to $2\Omega v_{B/Axyz}$, as it was the case for the plane motion of a particle.

II. ADDITION THEOREM FOR ANGULAR ACCELERATIONS.

For a system consisting of several movable rigid bodies and reference frames as shown in Fig. 19.5, let $\mathbf{\alpha}_{A/B}$ denote *the angular acceleration of the rigid body A with respect to the rigid body B*, or more briefly *the angular acceleration of A in B*. Similarly, let $\mathbf{\alpha}_{A/C}$ and $\mathbf{\alpha}_{C/B}$ denote the angular accelerations of A in C, and C in B, respectively. Note in general motion that the *angular acceleration* of a rigid body in a reference frame is equal to the time derivative (taken in that reference frame) of the angular velocity (measured in that same reference frame) of the rigid body. Thus, we have

$$\mathbf{\alpha}_{A/B} = (\dot{\mathbf{\omega}}_{A/B})_{Bx'y'z'} \tag{19.44}$$

$$\mathbf{\alpha}_{A/C} = (\dot{\mathbf{\omega}}_{A/C})_{Cx''y''z''} \tag{19.45}$$

$$\mathbf{\alpha}_{C/B} = (\dot{\mathbf{\omega}}_{C/B})_{Bx'y'z'} \tag{19.46}$$

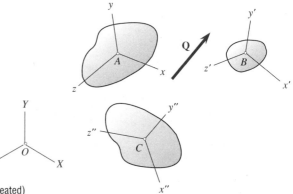

Fig. 19.5 (repeated)

Each term in the above equations involves *one* rigid body in *one* reference frame at a time and has a physical meaning. However, the quantity $(\dot{\mathbf{\omega}}_{A/C})_{Bx'y'z'}$ involves *one* rigid body in *two* reference frames and does not lend itself to a simple physical interpretation. By Eqs. (19.8) and (19.45) and the properties of cross products (cf. Sec. 4.8), we write

$$(\dot{\mathbf{\omega}}_{A/C})_{Bx'y'z'} = (\dot{\mathbf{\omega}}_{A/C})_{Cx''y''z''} + \mathbf{\omega}_{C/B} \times \mathbf{\omega}_{A/C}$$

$$(\dot{\mathbf{\omega}}_{A/C})_{Bx'y'z'} = \mathbf{\alpha}_{A/C} + \mathbf{\omega}_{C/B} \times \mathbf{\omega}_{A/C} \tag{19.47}$$

Taking the time derivative in $Bx'y'z'$ on both sides of Eq. (19.38) and applying Eqs. (19.44) through (19.47), we write

$$(\dot{\mathbf{\omega}}_{A/B})_{Bx'y'z'} = (\dot{\mathbf{\omega}}_{A/C})_{Bx'y'z'} + (\dot{\mathbf{\omega}}_{C/B})_{Bx'y'z'}$$

$$\mathbf{\alpha}_{A/B} = (\mathbf{\alpha}_{A/C} + \mathbf{\omega}_{C/B} \times \mathbf{\omega}_{A/C}) + \mathbf{\alpha}_{C/B}$$

$$\boxed{\mathbf{\alpha}_{A/B} = \mathbf{\alpha}_{A/C} + \mathbf{\alpha}_{C/B} + \mathbf{\omega}_{C/B} \times \mathbf{\omega}_{A/C}} \tag{19.48}^\dagger$$

[†] Cf. Eq. (19.38).

The property expressed in Eq. (19.48) can be extended to *any number* of rigid bodies (or reference frames). For instance, we write

$$\alpha_{A/O} = \alpha_{A/B} + \alpha_{B/O} + \omega_{B/O} \times \omega_{A/B}$$

$$= \alpha_{A/B} + (\alpha_{B/C} + \alpha_{C/O} + \omega_{C/O} \times \omega_{B/C})$$

$$+ \omega_{B/O} \times \omega_{A/B}$$

$$\alpha_{A/O} = \alpha_{A/B} + \alpha_{B/C} + \alpha_{C/O} + \omega_{B/O} \times \omega_{A/B}$$

$$+ \omega_{C/O} \times \omega_{B/C} \qquad (19.49)$$

where the subscript O refers to the fixed reference frame $OXYZ$. Following our adopted convention of notations for quantities measured in the fixed reference frame, we may write Eq. (19.49) as

$$\alpha_A = \alpha_{A/B} + \alpha_{B/C} + \alpha_C + \omega_B \times \omega_{A/B} + \omega_C \times \omega_{B/C} \quad (19.50)^\dagger$$

The relation expressed in Eq. (19.48) or (19.50) is called the *addition theorem for angular accelerations*. Note that the cross-product terms in Eqs. (19.48) and (19.50) will vanish for plane motion, where the angular velocities are parallel to each other. Thus, the addition theorems for angular velocities and angular accelerations are of *different* form in general motion, but they become of the *same* form in plane motion.

EXAMPLE 19.3

At the instant shown, the bead B slides down a semicircular bent rod R with a constant speed $v_{B/R} = 10$ ft/s relative to the rod, while the rod is rotated about the X axis at the rate $\omega_R = 3\mathbf{I}$ rad/s and $\alpha_R = 4\mathbf{I}$ rad/s^2. Determine the velocity and acceleration of the bead relative to the ground.

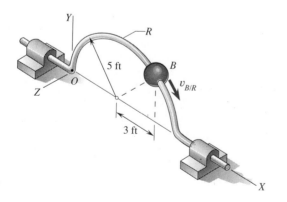

Solution. We first let the reference frame $Axyz$ be embedded in the bent rod R as shown, where the reference frame $OXYZ$ is fixed on the ground. Then, the angular velocity and angular acceleration of $Axyz$ in $OXYZ$, as shown in Fig. (a), are

$$\mathbf{\Omega} = \omega_R = 3\mathbf{I} \text{ rad/s} \qquad \dot{\mathbf{\Omega}} = \alpha_R = 4\mathbf{I} \text{ rad/s}^2$$

---

†Cf. Eq. (19.40).

(a)   (b)

At the instant shown, we see from Fig. (b) that

$$\mathbf{r}_{B/A} = 3\mathbf{I} + 4\mathbf{J} \text{ ft}$$

$$\mathbf{v}_{B/Axyz} = \mathbf{v}_{B/R} = 10\left[\frac{1}{5}(4\mathbf{I} - 3\mathbf{J})\right] \text{ ft/s}$$

$$= 8\mathbf{I} - 6\mathbf{J} \text{ ft/s}$$

$$\mathbf{a}_{B/Axyz} = \mathbf{a}_n = \frac{10^2}{5}\left[\frac{1}{5}(-3\mathbf{I} - 4\mathbf{J})\right] \text{ ft/s}^2$$

$$= -12\mathbf{I} - 16\mathbf{J} \text{ ft/s}^2$$

Since the point A does not move, we have $\mathbf{v}_A = \mathbf{a}_A = \mathbf{0}$. By Eq. (19.12), the velocity of B in $OXYZ$ is

$$\mathbf{v}_B = \mathbf{v}_{B/Axyz} + \mathbf{\Omega} \times \mathbf{r}_{B/A} + \mathbf{v}_A$$

$$= (8\mathbf{I} - 6\mathbf{J}) + 3\mathbf{I} \times (3\mathbf{I} + 4\mathbf{J}) + \mathbf{0}$$

$$\mathbf{v}_B = 8\mathbf{I} - 6\mathbf{J} + 12\mathbf{K} \text{ ft/s} \blacktriangleleft$$

By Eq. (19.41), the acceleration of B in $OXYZ$ is

$$\mathbf{a}_B = \mathbf{a}_{B/Axyz} + \dot{\mathbf{\Omega}} \times \mathbf{r}_{B/A} + \mathbf{\Omega} \times (\mathbf{\Omega} \times \mathbf{r}_{B/A})$$

$$+ 2\mathbf{\Omega} \times \mathbf{v}_{B/Axyz} + \mathbf{a}_A$$

$$= (-12\mathbf{I} - 16\mathbf{J}) + 4\mathbf{I} \times (3\mathbf{I} + 4\mathbf{J})$$

$$+ 3\mathbf{I} \times [3\mathbf{I} \times (3\mathbf{I} + 4\mathbf{J})]$$

$$+ 2(3\mathbf{I}) \times (8\mathbf{I} - 6\mathbf{J}) + \mathbf{0}$$

$$\mathbf{a}_B = -12\mathbf{I} - 52\mathbf{J} - 20\mathbf{K} \text{ ft/s}^2 \blacktriangleleft$$

Developmental Exercises

D19.9   Refer to Example 19.3. Suppose that B' is a point *of* the bent rod, and B' *coincides* with the moving bead B at the instant considered. Determine the acceleration of B' in $OXYZ$ for the instant shown.

D19.10   Refer to Example 19.3. Determine the Coriolis acceleration of the bead B for the instant shown.

EXAMPLE 19.4

At the instant shown, the disk D spins in the yoke about the pin at A with an angular velocity $\boldsymbol{\omega}_1$ and an angular acceleration $\boldsymbol{\alpha}_1$ relative to the arm C, while the arm C rotates about the vertical shaft with an angular velocity $\boldsymbol{\omega}_2$ and an angular acceleration $\boldsymbol{\alpha}_2$ relative to the ground. For this instant, determine (a) the velocity $\mathbf{v}_B$ and acceleration $\mathbf{a}_B$ of the point B of the disk, (b) the angular velocity $\boldsymbol{\omega}_D$ and angular acceleration $\boldsymbol{\alpha}_D$ of the disk.

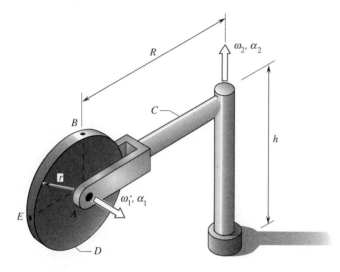

Solution. We first let the reference frame *Axyz be embedded in the arm C* (not the disk D) as shown, where the reference frame *OXYZ is fixed on the ground*. Then, the angular velocity and angular acceleration of *Axyz* in *OXYZ* are

$$\boldsymbol{\Omega} = \omega_2\mathbf{J} \qquad \dot{\boldsymbol{\Omega}} = \alpha_2\mathbf{J}$$

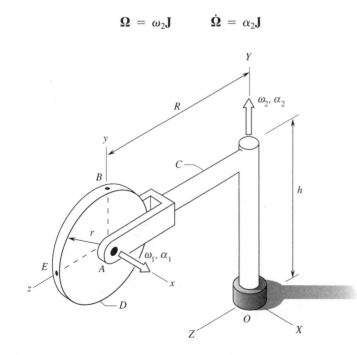

Since the point B describes a circular motion as the disk D spins about A in $Axyz$, we readily find for the instant shown that

$$\mathbf{r}_{B/A} = r\mathbf{J} \qquad \mathbf{v}_{B/Axyz} = r\omega_1\mathbf{K}$$

$$\mathbf{a}_{B/Axyz} = -r\omega_1^2\mathbf{J} + r\alpha_1\mathbf{K}$$

$$\mathbf{v}_A = R\omega_2\mathbf{I} \qquad \mathbf{a}_A = R\alpha_2\mathbf{I} - R\omega_2^2\mathbf{K}$$

By Eq. (19.12), the velocity of B in $OXYZ$ is

$$\mathbf{v}_B = \mathbf{v}_{B/Axyz} + \boldsymbol{\Omega} \times \mathbf{r}_{B/A} + \mathbf{v}_A$$

$$= r\omega_1\mathbf{K} + \omega_2\mathbf{J} \times r\mathbf{J} + R\omega_2\mathbf{I}$$

$$\mathbf{v}_B = R\omega_2\mathbf{I} + r\omega_1\mathbf{K} \qquad \blacktriangleleft$$

By Eq. (19.41), the acceleration of B in $OXYZ$ is

$$\mathbf{a}_B = \mathbf{a}_{B/Axyz} + \dot{\boldsymbol{\Omega}} \times \mathbf{r}_{B/A} + \boldsymbol{\Omega} \times (\boldsymbol{\Omega} \times \mathbf{r}_{B/A})$$

$$+ 2\boldsymbol{\Omega} \times \mathbf{v}_{B/Axyz} + \mathbf{a}_A$$

$$= (-r\omega_1^2\mathbf{J} + r\alpha_1\mathbf{K}) + \alpha_2\mathbf{J} \times r\mathbf{J} + \omega_2\mathbf{J} \times (\omega_2\mathbf{J} \times r\mathbf{J})$$

$$+ 2\omega_2\mathbf{J} \times r\omega_1\mathbf{K} + (R\alpha_2\mathbf{I} - R\omega_2^2\mathbf{K})$$

$$\mathbf{a}_B = (2r\omega_1\omega_2 + R\alpha_2)\mathbf{I} - r\omega_1^2\mathbf{J} + (r\alpha_1 - R\omega_2^2)\mathbf{K} \qquad \blacktriangleleft$$

The angular velocity of the disk D relative to the arm C is $\boldsymbol{\omega}_{D/C} = \omega_1\mathbf{I}$, while the angular velocity of the arm C in $OXYZ$ is $\boldsymbol{\omega}_C = \omega_2\mathbf{J}$. By the *addition theorem for angular velocities* [cf. Eq. (19.38)], the angular velocity of the disk D in $OXYZ$ is

$$\boldsymbol{\omega}_D = \boldsymbol{\omega}_{D/C} + \boldsymbol{\omega}_C \qquad \boldsymbol{\omega}_D = \omega_1\mathbf{I} + \omega_2\mathbf{J} \qquad \blacktriangleleft$$

The angular acceleration of the disk D relative to the arm C is $\boldsymbol{\alpha}_{D/C} = \alpha_1\mathbf{I}$, while the angular acceleration of the arm C in $OXYZ$ is $\boldsymbol{\alpha}_C = \alpha_2\mathbf{J}$. By the *addition theorem for angular accelerations* [cf. Eq. (19.48)], the angular acceleration of the disk D in $OXYZ$ is

$$\boldsymbol{\alpha}_D = \boldsymbol{\alpha}_{D/C} + \boldsymbol{\alpha}_C + \boldsymbol{\omega}_C \times \boldsymbol{\omega}_{D/C}$$

$$= \alpha_1\mathbf{I} + \alpha_2\mathbf{J} + \omega_2\mathbf{J} \times \omega_1\mathbf{I}$$

$$\boldsymbol{\alpha}_D = \alpha_1\mathbf{I} + \alpha_2\mathbf{J} - \omega_1\omega_2\mathbf{K} \qquad \blacktriangleleft$$

Developmental Exercises

D19.11   Refer to Example 19.4. Determine the velocity $\mathbf{v}_E$ and the acceleration $\mathbf{a}_E$ of the point E on the disk in $OXYZ$ for the instant shown.

D19.12   Describe the *addition theorem for angular accelerations*.

EXAMPLE 19.5

A right circular cone C rolls without slipping on the ZX plane as shown, where $r = 7$ in. and $L = 25$ in. At the instant shown, the axis OB of the cone rotates about the Y axis at the rate $\boldsymbol{\omega}_1 = 3\mathbf{J}$ rad/s and $\boldsymbol{\alpha}_1 = 4\mathbf{J}$ rad/s^2. Determine (a) the rate of spin of the cone about its axis BO and the angular velocity of the cone in $OXYZ$, (b) the angular acceleration of the cone in $OXYZ$.

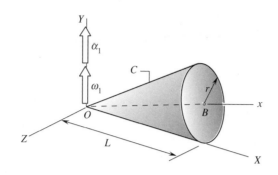

Solution. We first let $Axyz$ be a *moving reference frame* whose x axis is always collinear with the axis BO of the rolling cone as shown, and whose angular velocity and angular acceleration in the *fixed reference frame* $OXYZ$ are

$$\boldsymbol{\Omega} = \boldsymbol{\omega}_A = \boldsymbol{\omega}_1 = \omega_1\mathbf{J} \qquad \omega_1 = 3 \text{ rad/s}$$

$$\dot{\boldsymbol{\Omega}} = \boldsymbol{\alpha}_A = \boldsymbol{\alpha}_1 = \alpha_1\mathbf{J} \qquad \alpha_1 = 4 \text{ rad/s}^2$$

$$\sin\beta = 7/25$$
$$\tan\beta = 7/24$$

Since the cone C rolls without slipping, its angular velocity $\boldsymbol{\omega}_C$ in $OXYZ$ must be directed along its line of contact with the ZX plane. By inspection, we have

$$\boldsymbol{\omega}_C = -\omega_C\mathbf{I} \tag{1}$$

The *rate of spin* of the cone C about its axis BO is equivalent to the angular velocity of the cone C in $Axyz$ and may be denoted by $\boldsymbol{\omega}_{C/Axyz}$, or more briefly $\boldsymbol{\omega}_{C/A}$. Noting that $\boldsymbol{\omega}_{C/A}$ is directed along $\overrightarrow{BO}$, we write

$$\boldsymbol{\omega}_{C/A} = \omega_{C/A}\boldsymbol{\lambda}_{BO} = \omega_{C/A}(-\cos\beta\,\mathbf{I} - \sin\beta\,\mathbf{J})$$

By the *addition theorem for angular velocities* [cf. Eq. (19.38)], the angular velocity of the cone C in $OXYZ$ is

$$\boldsymbol{\omega}_C = \boldsymbol{\omega}_{C/A} + \boldsymbol{\omega}_A = \omega_{C/A}(-\cos\beta\,\mathbf{I} - \sin\beta\,\mathbf{J}) + \omega_1\mathbf{J}$$

$$\boldsymbol{\omega}_C = -\omega_{C/A}\cos\beta\,\mathbf{I} + (\omega_1 - \omega_{C/A}\sin\beta)\mathbf{J} \tag{2}$$

Comparing Eqs. (1) and (2), we find that

$$\omega_C = \omega_{C/A} \cos\beta$$

$$0 = \omega_1 - \omega_{C/A} \sin\beta$$

These two scalar equations and the given data yield

$$\omega_{C/A} = \frac{\omega_1}{\sin\beta} = \frac{3}{7/25} \qquad \omega_{C/A} = 10.71\lambda_{BO} \text{ rad/s} \quad \blacktriangleleft$$

$$\omega_C = \frac{\omega_1}{\tan\beta} = \frac{3}{7/24} \qquad \omega_C = -10.29\mathbf{I} \text{ rad/s} \quad \blacktriangleleft$$

The angular acceleration of the cone C in $Axyz$ is $\boldsymbol{\alpha}_{C/A}$, which is directed along $\overrightarrow{BO}$. We write

$$\boldsymbol{\alpha}_{C/A} = \alpha_{C/A}(-\cos\beta\,\mathbf{I} - \sin\beta\,\mathbf{J})$$

By the *addition theorem for angular accelerations* [cf. Eq. (19.48)], the angular acceleration of the cone C in $OXYZ$ is

$$\boldsymbol{\alpha}_C = \boldsymbol{\alpha}_{C/A} + \boldsymbol{\alpha}_A + \boldsymbol{\omega}_A \times \boldsymbol{\omega}_{C/A}$$

$$= \alpha_{C/A}(-\cos\beta\,\mathbf{I} - \sin\beta\,\mathbf{J}) + \alpha_1\mathbf{J}$$

$$+ \omega_1\mathbf{J} \times \frac{\omega_1}{\sin\beta}(-\cos\beta\,\mathbf{I} - \sin\beta\,\mathbf{J})$$

$$= -\alpha_{C/A}\cos\beta\,\mathbf{I} + (\alpha_1 - \alpha_{C/A}\sin\beta)\mathbf{J} + \frac{\omega_1^2}{\tan\beta}\mathbf{K}$$

Since $\boldsymbol{\omega}_C$ always lies in the ZX plane, $\boldsymbol{\alpha}_C$ must also lie in the ZX plane. Thus, we see that $\boldsymbol{\alpha}_C \perp \mathbf{J}$, or $\boldsymbol{\alpha}_C \cdot \mathbf{J} = 0$, which leads to

$$\alpha_1 - \alpha_{C/A}\sin\beta = 0 \qquad \alpha_{C/A} = \frac{\alpha_1}{\sin\beta}$$

$$\boldsymbol{\alpha}_C = -\frac{\alpha_1}{\tan\beta}\mathbf{I} + \frac{\omega_1^2}{\tan\beta}\mathbf{K} = -\frac{4}{7/24}\mathbf{I} + \frac{3^2}{7/24}\mathbf{K}$$

$$\boldsymbol{\alpha}_C = -13.71\mathbf{I} + 30.9\mathbf{K} \text{ rad/s}^2 \quad \blacktriangleleft$$

EXAMPLE 19.6

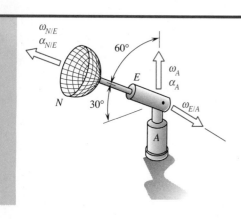

At the instant shown, the antenna N executes a *polarization* rotation about its axis in the body E at the rate of $\omega_{N/E} = 1.5$ rad/s and $\alpha_{N/E} = 2$ rad/s^2, the body E rotates about the pin in the body A to adjust the *elevation* at a constant rate $\omega_{E/A} = 1$ rad/s, while the body A rotates on the ground about the vertical axis to set the *azimuth* at the rate $\omega_A = 3$ rad/s and $\alpha_A = 2.5$ rad/s^2. For this instant, determine the angular velocity $\boldsymbol{\omega}_N$ and the angular acceleration $\boldsymbol{\alpha}_N$ of the antenna N.

Solution.   We first let the reference frame $OXYZ$ be fixed on the ground as shown, where the rigid bodies E and A will serve as frames of reference.

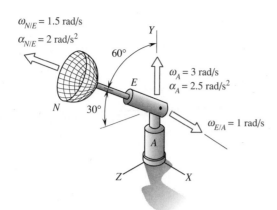

Referring to the sketch and the given data, we write

$$\boldsymbol{\omega}_{N/E} = 1.5(\cos 60° \, \mathbf{J} + \sin 60° \, \mathbf{K}) \text{ rad/s}$$

$$\boldsymbol{\alpha}_{N/E} = 2(\cos 60° \, \mathbf{J} + \sin 60° \, \mathbf{K}) \text{ rad/s}^2$$

$$\boldsymbol{\omega}_{E/A} = 1\mathbf{I} \text{ rad/s} \qquad \boldsymbol{\alpha}_{E/A} = \mathbf{0}$$

$$\boldsymbol{\omega}_A = 3\mathbf{J} \text{ rad/s} \qquad \boldsymbol{\alpha}_A = 2.5\mathbf{J} \text{ rad/s}^2$$

By the *addition theorem for angular velocities* [cf. Eq. (19.40)], the angular velocity $\boldsymbol{\omega}_N$ of the antenna N in $OXYZ$ is

$$\boldsymbol{\omega}_N = \boldsymbol{\omega}_{N/E} + \boldsymbol{\omega}_{E/A} + \boldsymbol{\omega}_A$$

$$\boldsymbol{\omega}_N = 1.5(\cos 60° \, \mathbf{J} + \sin 60° \, \mathbf{K}) + 1\mathbf{I} + 3\mathbf{J}$$

$$\boldsymbol{\omega}_N = 1\mathbf{I} + 3.75\mathbf{J} + 1.299\mathbf{K} \text{ rad/s} \quad \blacktriangleleft$$

Meanwhile, the angular velocity $\boldsymbol{\omega}_E$ of the body E in $OXYZ$ is

$$\boldsymbol{\omega}_E = \boldsymbol{\omega}_{E/A} + \boldsymbol{\omega}_A = 1\mathbf{I} + 3\mathbf{J} \text{ rad/s}$$

By the *addition theorem for angular accelerations* [cf. Eq. (19.50)], the angular acceleration $\boldsymbol{\alpha}_N$ of the antenna N in $OXYZ$ is

$$\boldsymbol{\alpha}_N = \boldsymbol{\alpha}_{N/E} + \boldsymbol{\alpha}_{E/A} + \boldsymbol{\alpha}_A + \boldsymbol{\omega}_E \times \boldsymbol{\omega}_{N/E} + \boldsymbol{\omega}_A \times \boldsymbol{\omega}_{E/A}$$

$$= 2(\cos 60° \, \mathbf{J} + \sin 60° \, \mathbf{K}) + \mathbf{0} + 2.5\mathbf{J}$$

$$+ (1\mathbf{I} + 3\mathbf{J}) \times 1.5(\cos 60° \, \mathbf{J} + \sin 60° \, \mathbf{K})$$

$$+ 3\mathbf{J} \times 1\mathbf{I}$$

$$\boldsymbol{\alpha}_N = 3.90\mathbf{I} + 2.20\mathbf{J} - 0.518\mathbf{K} \text{ rad/s}^2 \quad \blacktriangleleft$$

Developmental Exercises

D19.13 Refer to Example 19.6. For the instant shown, determine the angular velocity and angular acceleration of the antenna N relative to the body A.

D19.14 Refer to Example 19.6. For the instant shown, determine the angular velocity and angular acceleration of the body E relative to the ground.

PROBLEMS

19.1   At the instant shown, the crane turns about the vertical with a constant rate $\omega_1 = 0.25\mathbf{J}$ rad/s, while its boom, pivoted at O, is being raised with a constant rate $\omega_2 = 0.4\mathbf{K}$ rad/s relative to the cab. The length of the boom OP is 15 m. For this instant, determine (a) the velocity and acceleration of the tip P of the boom, (b) the angular velocity and angular acceleration of the boom.

Fig. P19.1

19.2   Solve Prob. 19.1 if, at the instant shown, the crane turns about the vertical at the rate $\omega_1 = 0.25\mathbf{J}$ rad/s and $\alpha_1 = 0.3\mathbf{J}$ rad/s^2, while its boom is being raised at the rate $\omega_2 = 0.4\mathbf{K}$ rad/s and $\alpha_2 = 0.5\mathbf{K}$ rad/s^2 relative to the cab.

19.3*   The bent rod rotates about the X axis with a constant rate $\omega = \pi\mathbf{I}$ rad/s. At the instant shown, the collar B moves toward C with a constant speed of 2 m/s relative to the rod. For this instant, determine the velocity and acceleration of the collar.

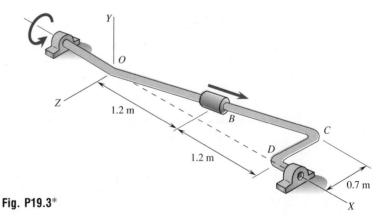

Fig. P19.3*

19.4   Solve Prob. 19.3* if, at the instant shown, the bent rod rotates at the rate $\omega = \pi\mathbf{I}$ rad/s and $\alpha = -(\pi/2)\,\mathbf{I}$ rad/s^2, while the collar B moves toward C with a speed of 2 m/s and an acceleration of 3 m/s^2, both relative to the rod.

19.5 A collar C oscillates along a rod AB as shown. At the instant considered, the turntable rotates at the rate $\boldsymbol{\omega} = 2\mathbf{J}$ rad/s and $\boldsymbol{\alpha} = 3\mathbf{J}$ rad/s^2, while the collar C is moving downward with a speed of 4 ft/s and an acceleration of 5 ft/s^2, both relative to the rod. For this instant, determine the velocity and acceleration of the collar.

19.6 The rod AB is connected by a ball-and-socket joint to the collar A and by a clevis to the collar B as shown. If $y_A = 0.032t^3$ ft, where t is the time measured in seconds, determine the velocity of the collar B and the angular velocity of the rod AB at the instant shown.

Fig. P19.5

Fig. P19.6   **Fig. P19.7***

19.7* The rod AB is connected by ball-and-socket joints to the collars A and B, which slide on the bent rod. If $y_B = 0.032t^3$ ft, where t is the time measured in seconds, determine the velocity and acceleration of the collar A at the instant shown.

19.8* The disk D rotates about the pin at A at the constant rate $\boldsymbol{\omega}_1 = -4\mathbf{I}$ rad/s relative to the frame F, while the frame F rotates about the fixed shaft at the constant rate $\boldsymbol{\omega}_2 = 8\mathbf{K}$ rad/s. For the instant shown, determine the angular velocity and the angular acceleration of the disk.

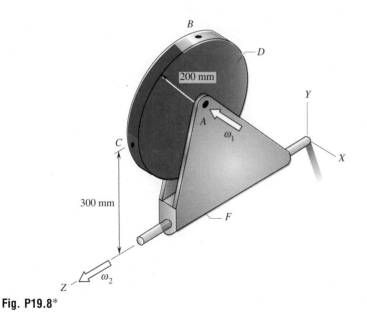

Fig. P19.8*

19.9 Refer to Prob. 19.8*. Determine the velocities and accelerations of the points B and C of the disk at the instant shown.

19.10 The disk D rotates about the pin at A at the constant rate $\boldsymbol{\omega}_1 = 6\mathbf{I}$ rad/s relative to the arm F. At the instant shown, the arm F rotates at the rate $\boldsymbol{\omega}_2 = 4\mathbf{J}$ rad/s and $\boldsymbol{\alpha}_2 = -2\mathbf{J}$ rad/s^2 relative to the ground. For this instant, determine the angular acceleration of the disk D and the accelerations of the points B and C of the disk.

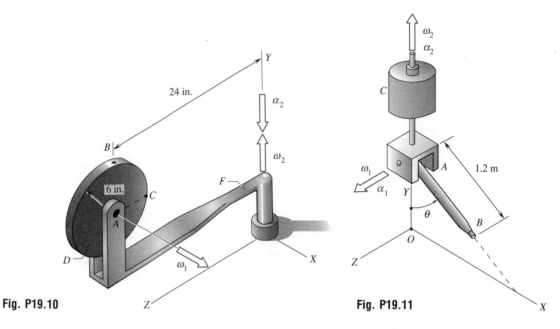

Fig. P19.10

Fig. P19.11

19.11 A 1.2-m arm AB for a remote-control mechanism, as shown, is rotated about the pin at A at the rate $\boldsymbol{\omega}_1 = 4\mathbf{K}$ rad/s and $\boldsymbol{\alpha}_1 = 2\mathbf{K}$ rad/s^2 relative to the clevis, while the clevis is turned at the rate $\boldsymbol{\omega}_2 = 5\mathbf{J}$ rad/s and $\boldsymbol{\alpha}_2 = 3\mathbf{J}$ rad/s^2 relative to the ground, where $\theta = 60°$. For this instant, determine the angular acceleration of the arm AB and the velocity and acceleration of the point B.

19.12 A small cone C rolls without slipping inside a fixed conical cavity F as shown. The cone C moves with a constant rate of spin $\omega_1 = 2$ rad/s about its axis BO. For the instant shown, determine (a) the angular velocity of its axis BO about the Y axis, (b) the angular velocity and angular acceleration of the cone C.

Fig. P19.12

Fig. P19.13

19.13 A triangular plate *OAB* is attached to a ball-and-socket at *O* as shown, where $\overline{OA}$ = 5 ft, $\overline{OB}$ = 6 ft, and $\overline{AB}$ = 7 ft. The plate is moved in such a way that the edges *OA* and *OB* always lie in the *YZ* and *ZX* planes, respectively. If the edge *OB* is rotated at a constant rate $\boldsymbol{\omega}_{OB}$ = 2**J** rad/s, determine the angular velocity of the plate *OAB* and the velocity of the corner *A* at the instant shown.

19.14* Refer to Prob. 19.13. If the edge *OA* is rotated at a constant rate $\boldsymbol{\omega}_{OA}$ = 3**I** rad/s, determine the angular velocity of the plate *OAB* and the velocity of the corner *B* at the instant shown.

19.15 At the instant shown, the disk *D* rotates about its axis relative to the motor *M* at the rate $\omega_{D/M}$ = 2 rad/s and $\alpha_{D/M}$ = 3 rad/s², the motor *M* rotates about the mounting pin relative to the bracket *B* at the rate $\omega_{M/B}$ = 1.5 rad/s and $\alpha_{M/B}$ = 2.5 rad/s², while the bracket *B* rotates about the vertical shaft relative to the ground at the rate ω_B = 4 rad/s and α_B = 5 rad/s². For this instant, determine the angular velocity $\boldsymbol{\omega}_D$ and the angular acceleration $\boldsymbol{\alpha}_D$ of the disk *D*.

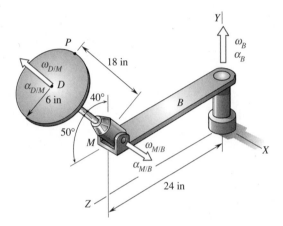

Fig. P19.15

19.16* Refer to Prob. 19.15. For the instant shown, determine the velocity and acceleration of the point *P* at the top of the disk.

KINETICS OF RIGID BODIES IN SPACE

★19.4 Moments and Products of Inertia of a Mass

The moment of inertia of a mass about a coordinate axis and the associated transfer formula as given by the parallel-axis theorem were presented in Secs. 16.1 and 16.3. To recapitulate the main points in those two sections, we refer to Fig. 16.5 and write

$$
\begin{aligned}
I_{xx} &= \int (y^2 + z^2)\, dm \\
I_{yy} &= \int (z^2 + x^2)\, dm \\
I_{zz} &= \int (x^2 + y^2)\, dm
\end{aligned}
\tag{19.51}
$$

$$
\begin{aligned}
I_{xx} &= \bar{I}_{x'x'} + m(\bar{y}^2 + \bar{z}^2) \\
I_{yy} &= \bar{I}_{y'y'} + m(\bar{z}^2 + \bar{x}^2) \\
I_{zz} &= \bar{I}_{z'z'} + m(\bar{x}^2 + \bar{y}^2)
\end{aligned}
\tag{19.52}
$$

Fig. 16.5 (repeated)

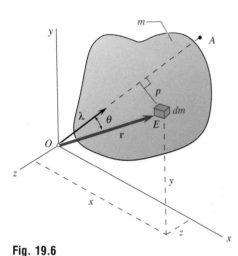

Fig. 19.6

where the mass moments of inertia about the coordinate axes x, y, z, and their parallel central axes x', y', z' are, for present purposes, denoted with double subscripts by I_{xx}, I_{yy}, I_{zz}, $\bar{I}_{x'x'}$, $\bar{I}_{y'y'}$, and $\bar{I}_{z'z'}$, respectively. The relations expressed in Eqs. (19.52) are called the *parallel-axis theorem* for mass moments of inertia.

Suppose that it is desired to determine the moment of inertia of a body about an arbitrary axis OA through the origin O as shown in Fig. 19.6. By definition, this quantity is given by

$$I_{OA} = \int p^2 \, dm \tag{19.53}$$

where p denotes the perpendicular distance from the differential mass element dm at $E(x, y, z)$ to the axis OA. We see that $p = r \sin\theta = 1 \cdot r \sin\theta = |\boldsymbol{\lambda} \times \mathbf{r}|$, where $\mathbf{r}$ is the position vector of E and $\boldsymbol{\lambda}$ is a unit vector directed along the axis OA. Note that

$$\begin{aligned}
\mathbf{r} &= x\mathbf{i} + y\mathbf{j} + z\mathbf{k} \\
\boldsymbol{\lambda} &= \lambda_x\mathbf{i} + \lambda_y\mathbf{j} + \lambda_z\mathbf{k}
\end{aligned} \tag{19.54}$$

$$p^2 = |\boldsymbol{\lambda} \times \mathbf{r}|^2 = (\boldsymbol{\lambda} \times \mathbf{r}) \cdot (\boldsymbol{\lambda} \times \mathbf{r}) \tag{19.55}$$

where the scalar components λ_x, λ_y, and λ_z of the unit vector $\boldsymbol{\lambda}$ represent the direction cosines of the axis OA in $Oxyz$. It can be shown that Eqs. (19.53) through (19.55) yield

$$\begin{aligned}
I_{OA} = {}& \lambda_x^2 \int (y^2 + z^2) \, dm + \lambda_y^2 \int (z^2 + x^2) \, dm \\
&+ \lambda_z^2 \int (x^2 + y^2) \, dm - 2\lambda_x\lambda_y \int xy \, dm \\
&- 2\lambda_y\lambda_z \int yz \, dm - 2\lambda_z\lambda_x \int zx \, dm
\end{aligned}$$

$$I_{OA} = I_{xx}\lambda_x^2 + I_{yy}\lambda_y^2 + I_{zz}\lambda_z^2 - 2I_{xy}\lambda_x\lambda_y - 2I_{yz}\lambda_y\lambda_z - 2I_{zx}\lambda_z\lambda_x \tag{19.56}$$

where

$$I_{xy} = \int xy \, dm \qquad I_{yz} = \int yz \, dm \qquad I_{zx} = \int zx \, dm \tag{19.57}$$

The quantities I_{xy}, I_{yz}, and I_{zx} defined in Eqs. (19.57) are called the *products of inertia* of the body with respect to the x and y axes, the y and z axes, and the z and x axes, respectively. Clearly, we have

$$I_{xy} = I_{yx} \qquad I_{yz} = I_{zy} \qquad I_{zx} = I_{xz} \qquad (19.58)$$

Unlike the mass moment of inertia about an axis, which is always positive, the mass products of inertia may be positive, negative, or zero. *If the mass of a body is symmetrical with respect to a coordinate plane*, such as the xy plane in Fig. 19.7(a), *then the products of inertia of the body with respect to one of the two coordinate axes in the plane of symmetry and the third coordinate axis, such as I_{yz} or I_{zx}, must be equal to zero.* Furthermore, *if the mass of a body is symmetrical about two coordinate planes*, such as the xy and yz planes in Fig. 19.7(b), *then the product of inertia of the body with respect to any two of the coordinate axes must vanish.* In other words, we have $I_{xy} = I_{yz} = I_{zx} = 0$ for the homogeneous body in Fig. 19.7(b).

We note from Fig. 16.5 that

$$x = x' + \bar{x} \qquad y = y' + \bar{y} \qquad z = z' + \bar{z} \qquad (19.59)$$

Substituting Eqs. (19.59) into Eqs. (19.57) and applying the *principle of moments* as well as the definitions of mass products of inertia, we can verify that

$$
\begin{aligned}
I_{xy} &= \bar{I}_{x'y'} + m\bar{x}\bar{y} \\
I_{yz} &= \bar{I}_{y'z'} + m\bar{y}\bar{z} \\
I_{zx} &= \bar{I}_{z'x'} + m\bar{z}\bar{x}
\end{aligned}
\qquad (19.60)
$$

The relations expressed in Eqs. (19.60) are called the *parallel-axis theorem* for mass products of inertia.

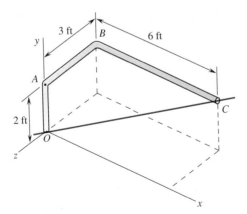

Fig. 19.7

EXAMPLE 19.7

The slender bent rod *OABC* weighs 9.66 lb/ft. Determine the moment of inertia I_{OC} of the bent rod about the axis *OC* as shown.

Solution. The mass density of the bent rod is

$$\rho_L = \frac{9.66}{32.2} \text{ slug/ft} = 0.3 \text{ slug/ft}$$

Thus, the masses of its three segments are

$$m_{OA} = 0.6 \text{ slug} \qquad m_{AB} = 0.9 \text{ slug}$$

$$m_{BC} = 1.8 \text{ slugs}$$

Referring to the bent rod as shown and applying the formula $\bar{I} = \frac{1}{12} mL^2$ (cf. Fig. 16.7 or Example 16.1) for a slender rod of mass m and length L as well as the *parallel-axis theorems*, we write

$$I_{xx} = [\tfrac{1}{12}(0.6)(2)^2 + 0.6(1)^2]$$

$$+ \{\tfrac{1}{12}(0.9)(3)^2 + 0.9[2^2 + (-1.5)^2]\}$$

$$+ \{0 + 1.8[2^2 + (-3)^2]\} \text{ slug·ft}^2$$

$$= 30.5 \text{ slug·ft}^2$$

$$I_{yy} = (0 + 0) + [\tfrac{1}{12}(0.9)(3)^2 + 0.9(1.5)^2]$$

$$+ [\tfrac{1}{12}(1.8)(6)^2 + 1.8(3^2 + 3^2)] \text{ slug·ft}^2$$

$$= 40.5 \text{ slug·ft}^2$$

$$I_{zz} = [\tfrac{1}{12}(0.6)(2)^2 + 0.6(1)^2] + [0 + 0.9(2)^2]$$

$$+ [\tfrac{1}{12}(1.8)(6)^2 + 1.8(3^2 + 2^2)] \text{ slug·ft}^2$$

$$= 33.2 \text{ slug·ft}^2$$

$$I_{xy} = (0 + 0) + (0 + 0) + [0 + 1.8(3)(2)] \text{ slug·ft}^2$$

$$= 10.8 \text{ slug·ft}^2$$

$$I_{yz} = (0 + 0) + [0 + 0.9(2)(-1.5)]$$

$$+ [0 + 1.8(2)(-3)] \text{ slug·ft}^2$$

$$= -13.5 \text{ slug·ft}^2$$

$$I_{zx} = (0 + 0) + (0 + 0)$$

$$+ [0 + 1.8(-3)(3)] \text{ slug·ft}^2$$

$$= -16.2 \text{ slug·ft}^2$$

For the axis OC, we write

$$\overrightarrow{OC} = 6\mathbf{i} + 2\mathbf{j} - 3\mathbf{k}$$

$$\boldsymbol{\lambda}_{OC} = \boldsymbol{\lambda} = \tfrac{1}{7}(6\mathbf{i} + 2\mathbf{j} - 3\mathbf{k})$$

$$\lambda_x = \tfrac{6}{7} \qquad \lambda_y = \tfrac{2}{7} \qquad \lambda_z = -\tfrac{3}{7}$$

Using the foregoing computed quantities and applying Eq. (19.56), we write

$$I_{OC} = 30.5(\tfrac{6}{7})^2 + 40.5(\tfrac{2}{7})^2 + 33.2(-\tfrac{3}{7})^2$$

$$- 2(10.8)(\tfrac{6}{7})(\tfrac{2}{7}) - 2(-13.5)(\tfrac{2}{7})(-\tfrac{3}{7})$$

$$- 2(-16.2)(-\tfrac{3}{7})(\tfrac{6}{7}) \text{ slug·ft}^2$$

$$= \frac{554.4}{49} \text{ slug·ft}^2$$

$$I_{OC} = 11.31 \text{ slug·ft}^2 \quad \blacktriangleleft$$

Developmental Exercises

D19.15 Refer to Example 19.7. Determine the moment of inertia of the bent rod about the axis (a) OB, (b) AC.

D19.16 Name the mass products of inertia which will vanish if the mass of the body is symmetrical with respect to (a) the xy plane, (b) the yz plane (c) the zx plane.

D19.17 Describe the condition under which the mass products of inertia will *all* vanish.

D19.18 Describe the *parallel-axis theorem* for mass products of inertia.

★19.5 Rotation of Axes: Principal Axes of Inertia

Let the coordinate system $Oxyz$ be rotated about its origin O so that a new coordinate system $Ox'y'z'$ is obtained as shown in Fig. 19.8, where $\boldsymbol{\lambda}_1$, $\boldsymbol{\lambda}_2$, and $\boldsymbol{\lambda}_3$ are the unit vectors directed along the x', y', and z' axes, respectively. In terms of the cartesian unit vectors $\mathbf{i}$, $\mathbf{j}$, and $\mathbf{k}$ in $Oxyz$, we may write

$$\boldsymbol{\lambda}_1 = \lambda_{11}\mathbf{i} + \lambda_{12}\mathbf{j} + \lambda_{13}\mathbf{k}$$

$$\boldsymbol{\lambda}_2 = \lambda_{21}\mathbf{i} + \lambda_{22}\mathbf{j} + \lambda_{23}\mathbf{k} \qquad (19.61)$$

$$\boldsymbol{\lambda}_3 = \lambda_{31}\mathbf{i} + \lambda_{32}\mathbf{j} + \lambda_{33}\mathbf{k}$$

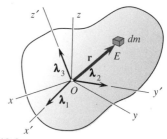

Fig. 19.8

In Fig. 19.8, the coordinates of the point E, where the differential mass element dm is located, are (x, y, z) in $Oxyz$, but are (x', y', z') in $Ox'y'z'$. Since the position vector of E is $\mathbf{r} = x\mathbf{i} + y\mathbf{j} + z\mathbf{k}$ and

$$x' = \boldsymbol{\lambda}_1 \cdot \mathbf{r} \qquad y' = \boldsymbol{\lambda}_2 \cdot \mathbf{r} \qquad z' = \boldsymbol{\lambda}_3 \cdot \mathbf{r} \qquad (19.62)$$

we readily find that

$$x' = \lambda_{11}x + \lambda_{12}y + \lambda_{13}z$$

$$y' = \lambda_{21}x + \lambda_{22}y + \lambda_{23}z \qquad (19.63)$$

$$z' = \lambda_{31}x + \lambda_{32}y + \lambda_{33}z$$

Clearly, the orientation of $Ox'y'z'$ in $Oxyz$ is completely defined by the array (or matrix) of λ_{ij} $(i, j = 1, 2, 3)$, which can be represented by

$$\mathbf{R} = \begin{bmatrix} \lambda_{11} & \lambda_{12} & \lambda_{13} \\ \lambda_{21} & \lambda_{22} & \lambda_{23} \\ \lambda_{31} & \lambda_{32} & \lambda_{33} \end{bmatrix} \qquad (19.64a)$$

$$\mathbf{R}^T = \begin{bmatrix} \lambda_{11} & \lambda_{21} & \lambda_{31} \\ \lambda_{12} & \lambda_{22} & \lambda_{32} \\ \lambda_{13} & \lambda_{23} & \lambda_{33} \end{bmatrix} \qquad (19.64b)$$

where $\mathbf{R}$ is called the *rotation matrix*, and $\mathbf{R}^T$ is called the *transpose* of $\mathbf{R}$. It is well to recognize that the sets of elements in the rows of $\mathbf{R}$ represent the sets of direction cosines of x', y', and z' axes in $Oxyz$, respectively; while the sets of elements in the rows of $\mathbf{R}^T$ represent the sets of direction cosines of x, y, and z axes in $Ox'y'z'$, respectively. Of course, the direction cosines λ_{ij} $(i, j = 1, 2, 3)$ of $\boldsymbol{\lambda}_i$ $(i = 1, 2, 3)$ must be subject to the constraints that $\boldsymbol{\lambda}_i$ ($i = 1, 2, 3$) are unit vectors which are orthogonal to each other.

By using Eqs. (19.63) and the definitions for moments and products of inertia of a mass in Sec. 19.4, one can find $I_{x'x'}$, $I_{y'y'}$, $I_{z'z'}$, $I_{x'y'}$, $I_{y'z'}$, and $I_{z'x'}$ in terms of I_{xx}, I_{yy}, I_{zz}, I_{xy}, I_{yz}, I_{zx}, and the direction cosines λ_{ij} (i, $j = 1, 2, 3$) in the rotation matrix **R**. Nevertheless, the transformation of the moments and products of inertia of a body associated with $Oxyz$ into those associated with $Ox'y'z'$, as shown in Fig. 19.8, is best studied using *matrix algebra* or *tensor analysis*. A detailed coverage of such a study is beyond the scope of this chapter. Without dwelling on the details of matrix algebra or tensor analysis, we shall here simply point out that the arrays (or matrices) of inertia quantities

$$\mathbf{I} = \begin{bmatrix} I_{xx} & -I_{xy} & -I_{xz} \\ -I_{yx} & I_{yy} & -I_{yz} \\ -I_{zx} & -I_{zy} & I_{zz} \end{bmatrix} \tag{19.65a}$$

$$\mathbf{I'} = \begin{bmatrix} I_{x'x'} & -I_{x'y'} & -I_{x'z'} \\ -I_{y'x'} & I_{y'y'} & -I_{y'z'} \\ -I_{z'x'} & -I_{z'y'} & I_{z'z'} \end{bmatrix} \tag{19.65b}$$

are called the *inertia tensors* (or *inertia matrices*) at the origin O for a body. These two inertia tensors are second-order tensors and can be expressed in terms of each other by the laws of transformation for second-order tensors or the laws of rotation of axes for matrices. In terms of products of matrices, it can be shown that[†]

$$\boxed{\mathbf{I'} = \mathbf{R}\mathbf{I}\mathbf{R}^T} \qquad \boxed{\mathbf{I} = \mathbf{R}^T\mathbf{I'}\mathbf{R}} \tag{19.66}$$

where **R** and $\mathbf{R}^T$ are defined in Eqs. (19.64). It should be pointed out that the *negative* signs associated with the products of inertia in Eqs. (19.65) are needed when the products of inertia themselves are defined without a negative sign as those in Eqs. (19.57); otherwise, the laws relating **I'** to **I** in Eqs. (19.66) would be inapplicable.[††] Since tensors must obey the laws for their transformation under rotation of axes, we *cannot* call an arbitrary array of inertia quantities an ''inertia tensor.''

To explore the scalar equations implied in the first of Eqs. (19.66), we may first carry out the multiplications of the matrices in $\mathbf{R}\mathbf{I}\mathbf{R}^T$ and apply Eqs. (19.58) to simplify the result. Then we equate the corresponding elements in **I'** and the resulting product of $\mathbf{R}\mathbf{I}\mathbf{R}^T$; this will yield *six* independent scalar equations because the inertia tensor is symmetric. For example, in equating the element in the first row and first column of **I'** to the element in the first row and first column of the resulting product of $\mathbf{R}\mathbf{I}\mathbf{R}^T$, we find that

$$I_{x'x'} = \lambda_{11}^2 I_{xx} + \lambda_{12}^2 I_{yy} + \lambda_{13}^2 I_{zz}$$
$$- 2\lambda_{11}\lambda_{12}I_{xy} - 2\lambda_{12}\lambda_{13}I_{yz} - 2\lambda_{13}\lambda_{11}I_{zx} \tag{19.67}$$

As a check of Eq. (19.67) against Eq. (19.56), we may let the x' axis in

---

[†]See Gere, J. M. and W. Weaver, Jr., *Matrix Algebra for Engineers*, 2nd ed., Brooks/Cole, Monterey, CA, 1983. These laws of transformation may be *skipped* without prejudice to the understanding of the rest of the text.

[††]Cf. the footnote on p. 237.

Fig. 19.8 be the axis OA in Fig. 19.6. In this case, we have $\lambda_1 = \lambda$ and

$$\lambda_{11} = \lambda_x \qquad \lambda_{12} = \lambda_y \qquad \lambda_{13} = \lambda_z \qquad (19.68)$$

Substituting Eqs. (19.68) into Eq. (19.67) and comparing the result with Eq. (19.56), we see that the expressions for $I_{x'x'}$ and I_{OA} are identical (Q.E.D.). Thus, Eq. (19.56) is simply a subcase of Eqs. (19.66).

Note that the quantities in each array in Eqs. (19.65) are called the *components* of the inertia tensor at O. If the coordinate axes are oriented in such a way that the products of inertia of the body *all* vanish, then the coordinate axes are called the *principal axes of inertia* at O, and the moments of inertia about these principal axes are called the *principal moments of inertia* at O. When this is the case, the inertia tensor is "diagonalized" and may be written in the simplified form

$$\mathbf{I} = \begin{bmatrix} I_x & 0 & 0 \\ 0 & I_y & 0 \\ 0 & 0 & I_z \end{bmatrix} \qquad (19.69)$$

where $I_x = I_{xx}$, $I_y = I_{yy}$, and $I_z = I_{zz}$. Of these three principal moments of inertia at O, one will be the *maximum* and another a *minimum* of the body's moments of inertia about axes through the origin O.

Developmental Exercises

D19.19 Refer to Fig. 19.8 as well as Eqs. (19.61) and (19.64). (a) What do the sets of elements in the rows of the rotation matrix $\mathbf{R}$ represent? (b) What do the sets of elements in the rows of $\mathbf{R}^T$ represent? (c) Is the determinant of $\mathbf{R}$ given by $|\mathbf{R}| = \lambda_1 \cdot (\lambda_2 \times \lambda_3) = 1$? (d) Is the determinant of $\mathbf{R}^T$ given by $|\mathbf{R}^T| = \mathbf{i} \cdot (\mathbf{j} \times \mathbf{k}) = 1$?

D19.20 Define the principal axes and principal moments of inertia at a point for a body.

PROBLEMS

NOTE: Unless otherwise specified, the formulas given in Fig. 16.7 as well as the parallel-axis theorems may be applied in solving the following problems of mass moments and products of inertia.

19.17 and 19.18* Determine the products of inertia I_{xy}, I_{yz}, and I_{zx} of the body made of slender rods as shown.

Fig. P19.17

Fig. P19.18*

Fig. P19.19

19.19 A slender rod weighing 1.61 lb/ft is used to form a composite body as shown. Determine I_{xy}, I_{yz}, and I_{zx} of the body.

Fig. P19.20

Fig. P19.21

Fig. P19.22*

19.20 through 19.22* Determine I_{xy}, I_{yz}, and I_{zx} of the sheet metal part shown.

19.23 and 19.24* A piece of sheet metal is cut and bent into a machine component as shown. If its mass density is $\rho_A = 16$ kg/m², determine I_{xy}, I_{yz}, and I_{zx} of the component.

Fig. P19.25

Fig. P19.23

Fig. P19.24*

19.25 A rectangular parallelepiped of mass m is shown. Determine I_{xy}, I_{yz}, and I_{zx} of the parallelepiped.

19.26 through 19.28 The mass density of the steel machine element is $\rho = 7.85$ Mg/m³. Determine I_{xy}, I_{yz}, and I_{zx} of the machine element.

Fig. P19.26

Fig. P19.27*

Fig. P19.28

19.29 and 19.30* Determine by integration I_{xy} of the body, as shown, whose mass is m. (*Hint.* Choose dm to be a first-order element parallel to the yz plane.)

19.31 A tetrahedron of mass density $\rho = 15$ slugs/ft³ is bounded by the coordinate planes and the plane $3x + 6y + 4z = 12$, where the lengths are in feet. Determine I_{xy} of the tetrahedron.

Fig. P19.29

Fig. P19.30

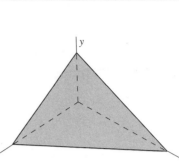

Fig. P19.31

19.32* Refer to Prob. 19.31. Determine I_{yz} and I_{zx} of the tetrahedron.

19.33 Refer to Prob. 19.17. Determine the moment of inertia of the body about the axis joining the points O and A.

19.34 Refer to Prob. 19.17. Determine the moment of inertia of the body about the axis joining the points A and B.

19.35 The slender bent rod weighs 3.22 lb/ft. Determine the moment of inertia of the bent rod about the axis joining the points O and C.

19.36 Refer to Prob. 19.35. Determine the moment of inertia of the bent rod about the axis joining the points A and C.

19.37 The solid right circular cone as shown has a mass m. Determine its moment of inertia about the axis OA which is the line of intersection of the surface of the cone and the xy plane.

Fig. P19.35

Fig. P19.37

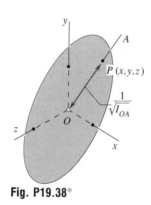

Fig. P19.38*

19.38* Suppose that the components of the inertia tensor at the origin O as well as I_{OA}, the moment of inertia about the axis OA, for a body are known. Furthermore, a point $P(x, y, z)$ is plotted on the axis OA at a distance $\overline{OP} = 1/\sqrt{I_{OA}}$ from O. Show that the locus of the point $P(x, y, z)$ is the surface of a *quadric* defined by

$$I_{xx}x^2 + I_{yy}y^2 + I_{zz}z^2 - 2I_{xy}xy - 2I_{yz}yz - 2I_{zx}zx = 1$$

which is known as the *ellipsoid of inertia*.

19.39* Refer to the sheet metal part in Fig. P19.20. Determine its moment of inertia about the axis joining the points P and Q.

19.40 Refer to the sheet metal part in Fig. P19.21. Determine its moment of inertia about the axis joining the points P and Q.

★19.6 Momentum of a Rigid Body

Suppose that a rigid body of mass m rotates with an angular velocity $\boldsymbol{\omega}$ and its mass center G moves with a velocity $\bar{\mathbf{v}}$ in the *fixed* reference frame $OXYZ$ as shown in Fig. 19.9, where $Gxyz$ is a *nonrotating* reference frame moving with G. Referring to this figure, we see that

$$\mathbf{r} = \bar{\mathbf{r}} + \mathbf{r}' \tag{19.70}$$

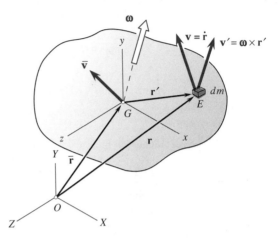

Fig. 19.9

Thus, the velocity $\mathbf{v}$ of the differential mass element dm at E in $OXYZ$ is given by

$$\mathbf{v} = \dot{\mathbf{r}} = \dot{\bar{\mathbf{r}}} + \dot{\mathbf{r}}' = \bar{\mathbf{v}} + [(\dot{\mathbf{r}}')_{Gxyz} + \mathbf{0} \times \mathbf{r}']$$
$$= \bar{\mathbf{v}} + [\boldsymbol{\omega} \times \mathbf{r}' + \mathbf{0}]$$
$$\mathbf{v} = \bar{\mathbf{v}} + \boldsymbol{\omega} \times \mathbf{r}' \tag{19.71}$$

Applying Eq. (19.71) and the *principle of moments*, we compute the *linear momentum* $\mathbf{L}$ of the rigid body in $OXYZ$ as follows:

$$\mathbf{L} = \int \mathbf{v} \, dm = \int (\bar{\mathbf{v}} + \boldsymbol{\omega} \times \mathbf{r}') \, dm$$
$$= \bar{\mathbf{v}} \int dm + \boldsymbol{\omega} \times \int \mathbf{r}' \, dm = \bar{\mathbf{v}}m + \boldsymbol{\omega} \times \mathbf{0}$$
$$\boxed{\mathbf{L} = m\bar{\mathbf{v}}} \tag{19.72}$$

Furthermore, we refer to Fig. 19.9 and apply Eq. (19.71) as well as the *principle of moments* to compute the angular momentum $\mathbf{H}_G$ of the rigid body about its mass center G in $OXYZ$ as follows:

$$\mathbf{H}_G = \int \mathbf{r}' \times \mathbf{v} \, dm = \int \mathbf{r}' \times (\bar{\mathbf{v}} + \boldsymbol{\omega} \times \mathbf{r}') \, dm$$
$$= \left(\int \mathbf{r}' \, dm \right) \times \bar{\mathbf{v}} + \int \mathbf{r}' \times (\boldsymbol{\omega} \times \mathbf{r}') \, dm$$
$$= \mathbf{0} \times \bar{\mathbf{v}} + \int \mathbf{r}' \times (\boldsymbol{\omega} \times \mathbf{r}') \, dm$$
$$\boxed{\mathbf{H}_G = \int \mathbf{r}' \times (\boldsymbol{\omega} \times \mathbf{r}') \, dm} \tag{19.73}$$

We see in Fig. 19.9 that the velocity of dm at E in $Gxyz$ is

$$\mathbf{v}' = \boldsymbol{\omega} \times \mathbf{r}' \tag{19.74}$$

Substituting Eq. (19.74) into Eq. (19.73), we find that

$$\mathbf{H}_G = \int \mathbf{r'} \times \mathbf{v'} \, dm = \mathbf{H}_G' \qquad (19.75)$$

which reveals that the angular momentum of a rigid body about its mass center G computed in the fixed reference frame $OXYZ$ and that computed in the nonrotating central reference frame $Gxyz$ are equal.

Noting that $Gxyz$ does not rotate, we may express $\mathbf{H}_G$, $\mathbf{r'}$, and $\boldsymbol{\omega}$ in rectangular components in $Gxyz$ as follows:

$$\mathbf{H}_G = H_x\mathbf{i} + H_y\mathbf{j} + H_z\mathbf{k} \qquad (19.76)$$

$$\mathbf{r'} = x\mathbf{i} + y\mathbf{j} + z\mathbf{k} \qquad (19.77)$$

$$\boldsymbol{\omega} = \omega_x\mathbf{i} + \omega_y\mathbf{j} + \omega_z\mathbf{k} \qquad (19.78)$$

Substituting Eqs. (19.76) through (19.78) into Eq. (19.73), carrying out the cross products, and combining terms, we get

$$H_x\mathbf{i} + H_y\mathbf{j} + H_z\mathbf{k}$$

$$= [\omega_x \int (y^2 + z^2) \, dm - \omega_y \int xy \, dm - \omega_z \int xz \, dm]\mathbf{i}$$

$$+ [-\omega_x \int yx \, dm + \omega_y \int (z^2 + x^2) \, dm - \omega_z \int yz \, dm]\mathbf{j}$$

$$+ [-\omega_x \int zx \, dm - \omega_y \int zy \, dm + \omega_z \int (x^2 + y^2) \, dm]\mathbf{k}$$

Recognizing that the integrals in this equation represent the central moments and products of inertia, we equate the coefficients of $\mathbf{i}$, $\mathbf{j}$, and $\mathbf{k}$ to obtain the following three equations:

$$\begin{aligned}
H_x &= \bar{I}_{xx}\omega_x - \bar{I}_{xy}\omega_y - \bar{I}_{xz}\omega_z \\
H_y &= -\bar{I}_{yx}\omega_x + \bar{I}_{yy}\omega_y - \bar{I}_{yz}\omega_z \\
H_z &= -\bar{I}_{zx}\omega_x - \bar{I}_{zy}\omega_y + \bar{I}_{zz}\omega_z
\end{aligned} \qquad (19.79)$$

These three equations expressing the components of $\mathbf{H}_G$ may be written in terms of arrays (or matrices) as follows:

$$\begin{bmatrix} H_x \\ H_y \\ H_z \end{bmatrix} = \begin{bmatrix} \bar{I}_{xx} & -\bar{I}_{xy} & -\bar{I}_{xz} \\ -\bar{I}_{yx} & \bar{I}_{yy} & -\bar{I}_{yz} \\ -\bar{I}_{zx} & -\bar{I}_{zy} & \bar{I}_{zz} \end{bmatrix} \begin{bmatrix} \omega_x \\ \omega_y \\ \omega_z \end{bmatrix} \qquad (19.80)$$

Using symbolic representation, we may write Eq. (19.80) as

$$\mathbf{H}_G = \bar{\mathbf{I}}\boldsymbol{\omega} \qquad (19.81)$$

where $\mathbf{H}_G$, $\bar{\mathbf{I}}$, and $\boldsymbol{\omega}$ are, respectively, the *angular momentum matrix*, the *inertia matrix*, and the *angular velocity matrix*, which are defined to represent the following arrays:

$$\mathbf{H}_G = \begin{bmatrix} H_x \\ H_y \\ H_z \end{bmatrix}$$

$$\bar{\mathbf{I}} = \begin{bmatrix} \bar{I}_{xx} & -\bar{I}_{xy} & -\bar{I}_{xz} \\ -\bar{I}_{yx} & \bar{I}_{yy} & -\bar{I}_{yz} \\ -\bar{I}_{zx} & -\bar{I}_{zy} & \bar{I}_{zz} \end{bmatrix} \qquad (19.82)$$

$$\boldsymbol{\omega} = \begin{bmatrix} \omega_x \\ \omega_y \\ \omega_z \end{bmatrix}$$

Fig. 19.10

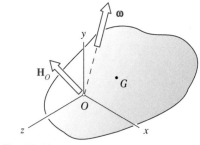

Fig. 19.11

By comparison, we clearly see that Eq. (19.81) resembles Eq. (18.13), and Eqs. (19.81) and (19.82) together are equivalent to Eqs. (19.79).

The foregoing analysis leads to the fact that the system of momenta of the particles of a rigid body in general motion is equivalent to a *momentum system* consisting of the linear momentum $\mathbf{L} = m\bar{\mathbf{v}}$ at the mass center G and the angular momentum $\mathbf{H}_G$ about G, as shown in Fig. 19.10, where the components of $\mathbf{H}_G$ are given by Eqs. (19.79). Once the momentum system at G has been determined, it can be replaced by an *equivalent momentum system* at another point.

For instance, if the momentum system at G in Fig. 19.10 were to be replaced by an equivalent momentum system at the fixed point O, the replacement would consist of the same linear momentum $\mathbf{L} = m\bar{\mathbf{v}}$ at O and a new angular momentum $\mathbf{H}_O$ given by

$$\mathbf{H}_O = \mathbf{H}_G + \bar{\mathbf{r}} \times m\bar{\mathbf{v}} \tag{19.83}$$

On the other hand, it can be shown that the angular momentum $\mathbf{H}_O$ of a rigid body rotating with $\boldsymbol{\omega} = \omega_x\mathbf{i} + \omega_y\mathbf{j} + \omega_z\mathbf{k}$ about the origin O of a fixed reference frame $Oxyz$, as shown in Fig. 19.11, is given by

$$\mathbf{H}_O = H_x\mathbf{i} + H_y\mathbf{j} + H_z\mathbf{k} \tag{19.84}$$

where

$$
\begin{aligned}
H_x &= I_{xx}\omega_x - I_{xy}\omega_y - I_{xz}\omega_z \\
H_y &= -I_{yx}\omega_x + I_{yy}\omega_y - I_{yz}\omega_z \\
H_z &= -I_{zx}\omega_x - I_{zy}\omega_y + I_{zz}\omega_z
\end{aligned}
\tag{19.85}
$$

and the moments of inertia I_{xx}, I_{yy}, I_{zz} and the products of inertia I_{xy}, I_{yz}, I_{zx} are computed with respect to the fixed reference frame $Oxyz$.

Knowing how to compute the momentum system for a rigid body at its mass center G as shown in Fig. 19.10, we may solve appropriate problems by the *principle of impulse and momentum* in Sec. 18.3, which states that

The *resultant system of momenta* of a rigid body at the time t_1	$+$	The *linear* and *angular impulses* exerted on the rigid body from t_1 to t_2	$=$	The *resultant system of momenta* of the rigid body at the time t_2	(18.15) (repeated)

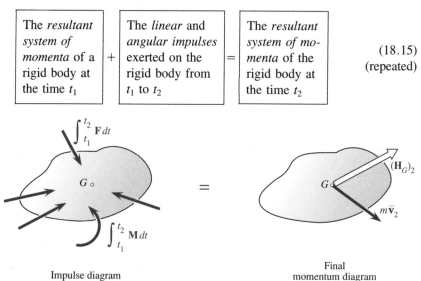

Initial
momentum diagram
(a)

Impulse diagram
(b)

Final
momentum diagram
(c)

Fig. 19.12

Note that Fig. 19.12 is a graphical representation of Eq. (18.15) for a rigid body in general motion.

EXAMPLE 19.8

A 96.6-lb rectangular plate is falling with a downward velocity of 7 ft/s and zero angular velocity when its corner C strikes the corner O of a post as shown. If the impact is perfectly plastic, determine immediately after the impact (a) the angular velocity $\boldsymbol{\omega}$ of the plate, (b) the velocity $\overline{\mathbf{v}}$ of its mass center G.

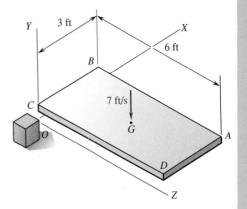

Solution. The mass of the plate is $m = 96.6/32.2$ slugs $= 3$ slugs. Letting the axes of the nonrotating central reference frame $Gxyz$ be parallel to the axes of the fixed reference frame $OXYZ$ and noting that the plate is symmetrical with respect to the xy and yz planes, we see that the x, y, z axes are the principal axes and we have, from Fig. 16.7,

$$\bar{I}_{xx} = \tfrac{1}{12} mc^2 = \tfrac{1}{12}(3)(6)^2 = 9 \text{ (slug·ft}^2)$$

$$\bar{I}_{yy} = \tfrac{1}{12} m(c^2 + a^2) = \tfrac{1}{12}(3)(6^2 + 3^2)$$

$$= 11.25 \text{ (slug·ft}^2)$$

$$\bar{I}_{zz} = \tfrac{1}{12} ma^2 = \tfrac{1}{12}(3)(3)^2 = 2.25 \text{ (slug·ft}^2)$$

$$\bar{I}_{xy} = \bar{I}_{yz} = \bar{I}_{zx} = 0$$

The components of the angular momentum about G are, from Eqs. (19.79),

$$H_x = 9\omega_x - 0 - 0 = 9\omega_x$$

$$H_y = -0 + 11.25\omega_y - 0 = 11.25\omega_y$$

$$H_z = -0 - 0 + 2.25\omega_z = 2.25\omega_z$$

Therefore, applying the *principle of impulse and momentum*, we have

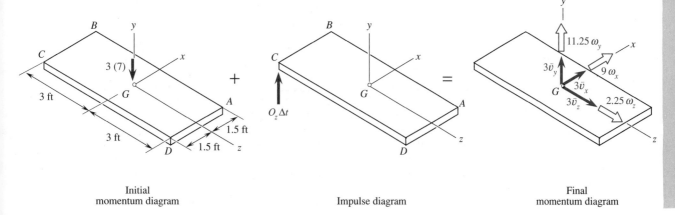

| Initial momentum diagram | Impulse diagram | Final momentum diagram |

Referring to the vector-diagram equation, we write

$$\Sigma V_x: \qquad 0 + 0 = 3\bar{v}_x \qquad \bar{v}_x = 0$$

$$\Sigma V_z: \qquad 0 + 0 = 3\bar{v}_z \qquad \bar{v}_z = 0$$

$$\Sigma \mathbf{M}_C: \qquad (1.5\mathbf{i} + 3\mathbf{k}) \times [-3(7)\mathbf{j}] + \mathbf{0}$$

$$= (1.5\mathbf{i} + 3\mathbf{k}) \times (0\mathbf{i} + 3\bar{v}_y\mathbf{j} + 0\mathbf{k})$$

$$+ (9\omega_x\mathbf{i} + 11.25\omega_y\mathbf{j} + 2.25\omega_z\mathbf{k})$$

Carrying out the cross products in this equation and then equating the coefficients of the unit vectors, we get

$$63 = -9\bar{v}_y + 9\omega_x \tag{1}$$

$$0 = 11.25\omega_y \qquad \omega_y = 0$$

$$-31.5 = 4.5\bar{v}_y + 2.25\omega_z \tag{2}$$

Because of perfectly plastic impact at C, we have $\mathbf{v}_C = \mathbf{0}$ and

$$\mathbf{v}_G = 0\mathbf{i} + \bar{v}_y\mathbf{j} + 0\mathbf{k}$$

$$= (\omega_x\mathbf{i} + 0\mathbf{j} + \omega_z\mathbf{k}) \times (1.5\mathbf{i} + 3\mathbf{k})$$

which yields a nontrivial equation

$$\bar{v}_y = -3\omega_x + 1.5\omega_z \tag{3}$$

Solving Eqs. (1), (2), and (3) simultaneously, we obtain

$$\bar{v}_y = -6 \qquad \omega_x = 1 \qquad \omega_z = -2$$

Thus, we have

$$\mathbf{\omega} = \mathbf{i} - 2\mathbf{k} \text{ rad/s} \qquad \bar{\mathbf{v}} = -6\mathbf{j} \text{ ft/s} \qquad \blacktriangleleft$$

Developmental Exercises

D19.21   Refer to Example 19.8. Determine the velocities of the corners A, B, and D of the plate immediately after the impact.

D19.22   Refer to Fig. 19.9. If the mass of the body is symmetrical with respect to the xy and yz planes, write the components of the angular momentum $\mathbf{H}_G$ of the body about G in $OXYZ$.

EXAMPLE 19.9

A thin disk of mass $m = 45$ kg and radius $r = 300$ mm is mounted on a horizontal shaft with which the disk makes an angle of $\phi = 60°$ as shown. If the shaft rotates with $\mathbf{\omega} = 10\mathbf{I}$ rad/s, determine (a) the angular momentum $\mathbf{H}_G$ of the disk about its mass center G, (b) the angle θ between $\mathbf{H}_G$ and $\mathbf{\omega}$.

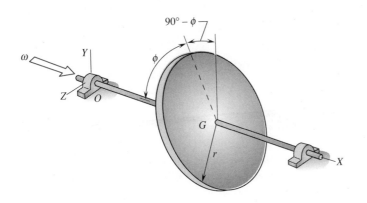

Solution. Let $Gxyz$ be a nonrotating reference frame oriented as shown. Then, we write

$$\omega_x = \omega \sin\phi \qquad \omega_y = -\omega \cos\phi \qquad \omega_z = 0$$

$$\bar{I}_{xx} = \tfrac{1}{2} mr^2 \qquad \bar{I}_{yy} = \bar{I}_{zz} = \tfrac{1}{4} mr^2$$

$$\bar{I}_{xy} = \bar{I}_{yz} = \bar{I}_{zx} = 0$$

Applying Eqs. (19.79) and (19.76), we get

$$H_x = \tfrac{1}{2} mr^2(\omega \sin\phi) - 0 - 0 = \tfrac{1}{2} mr^2 \omega \sin\phi$$

$$H_y = -0 + \tfrac{1}{4} mr^2(-\omega \cos\phi) - 0 = -\tfrac{1}{4} mr^2 \omega \cos\phi$$

$$H_z = -0 - 0 + 0 = 0$$

$$\mathbf{H}_G = H_x\mathbf{i} + H_y\mathbf{j} + H_z\mathbf{k} = \tfrac{1}{4} mr^2\omega(2 \sin\phi\, \mathbf{i} - \cos\phi\, \mathbf{j})$$

Expressing in terms of the unit vectors in $OXYZ$, we write

$$\mathbf{i} = \sin\phi\, \mathbf{I} + \cos\phi\, \mathbf{J} \qquad \mathbf{j} = -\cos\phi\, \mathbf{I} + \sin\phi\, \mathbf{J}$$

$$\mathbf{H}_G = \tfrac{1}{4} mr^2\omega[2 \sin\phi\, (\sin\phi\, \mathbf{I} + \cos\phi\, \mathbf{J})$$

$$- \cos\phi\, (-\cos\phi\, \mathbf{I} + \sin\phi\, \mathbf{J})]$$

$$= \tfrac{1}{4} mr^2\omega[(2 \sin^2\phi + \cos^2\phi)\mathbf{I} + \sin\phi \cos\phi\, \mathbf{J}]$$

Substituting $m = 4$ kg, $r = 0.3$ m, $\omega = 10$ rad/s, and $\phi = 60°$ into the expression for $\mathbf{H}_G$, we obtain

$$\mathbf{H}_G = 1.575\mathbf{I} + 0.390\mathbf{J} \text{ kg·m}^2\text{/s} \quad \blacktriangleleft$$

The angle θ between $\mathbf{H}_G$ and $\boldsymbol{\omega}$ is determined as follows:

$$\cos\theta = \frac{\boldsymbol{\omega} \cdot \mathbf{H}_G}{\omega H_G} = \frac{15.75}{16.225}$$

$$\theta = 13.90° \quad \blacktriangleleft$$

REMARK. The result shows that $\mathbf{H}_G$ and $\boldsymbol{\omega}$ for a rigid body in general motion are generally *not* in the same direction.

Developmental Exercise

D19.23 Refer to Example 19.9. Determine (a) the value of the angle ϕ for which the angle θ between $\mathbf{H}_G$ and $\boldsymbol{\omega}$ is maximum, (b) the corresponding maximum value of θ.

★19.7 Kinetic Energy of a Rigid Body

Let us consider again the case in which a rigid body of mass m rotates with an angular velocity $\boldsymbol{\omega}$ and its mass center G moves with a velocity $\bar{\mathbf{v}}$ in the *fixed* reference frame $OXYZ$ as shown in Fig. 19.9, where $Gxyz$ is a *nonrotating* reference frame moving with G. The differential mass dm, as shown in this figure, has a velocity $\mathbf{v}$ in $OXYZ$, which is given by Eq. (19.71). Applying Eq. (19.71) and the vector identity $\mathbf{A} \cdot \mathbf{P} \times \mathbf{Q} = \mathbf{A} \times \mathbf{P} \cdot \mathbf{Q}$, we write[†]

$$v^2 = \mathbf{v} \cdot \mathbf{v} = (\bar{\mathbf{v}} + \boldsymbol{\omega} \times \mathbf{r}') \cdot (\bar{\mathbf{v}} + \boldsymbol{\omega} \times \mathbf{r}')$$

$$= \bar{v}^2 + 2\bar{\mathbf{v}} \cdot (\boldsymbol{\omega} \times \mathbf{r}') + (\boldsymbol{\omega} \times \mathbf{r}') \cdot (\boldsymbol{\omega} \times \mathbf{r}')$$

$$v^2 = \bar{v}^2 + 2\bar{\mathbf{v}} \cdot (\boldsymbol{\omega} \times \mathbf{r}') + \boldsymbol{\omega} \cdot \mathbf{r}' \times (\boldsymbol{\omega} \times \mathbf{r}') \qquad (19.86)$$

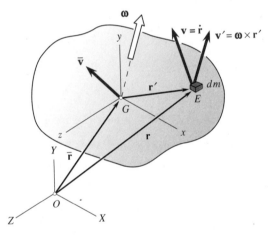

Fig. 19.9 (repeated)

Applying Eqs. (19.86) and (19.73) as well as the *principle of moments*, we compute the kinetic energy T of the rigid body in $OXYZ$ as follows:

$$T = \tfrac{1}{2} \int v^2 \, dm$$

$$= \tfrac{1}{2} \bar{v}^2 \int dm + \bar{\mathbf{v}} \cdot \boldsymbol{\omega} \times \int \mathbf{r}' \, dm$$

$$+ \tfrac{1}{2} \boldsymbol{\omega} \cdot \int \mathbf{r}' \times (\boldsymbol{\omega} \times \mathbf{r}') \, dm$$

$$= \tfrac{1}{2} \bar{v}^2 m + \bar{\mathbf{v}} \cdot \boldsymbol{\omega} \times \mathbf{0} + \tfrac{1}{2} \boldsymbol{\omega} \cdot \mathbf{H}_G$$

$$T = \tfrac{1}{2} m \bar{v}^2 + \tfrac{1}{2} \boldsymbol{\omega} \cdot \mathbf{H}_G \qquad (19.87)$$

[†] Here we let $\mathbf{A} = \boldsymbol{\omega}$, $\mathbf{P} = \mathbf{r}'$, and $\mathbf{Q} = \boldsymbol{\omega} \times \mathbf{r}'$. Cf. Eq. (4.32).

where $\mathbf{H}_G$ is the angular momentum of the rigid body about G. Since $\frac{1}{2} m\bar{v}^2 = \frac{1}{2} \bar{\mathbf{v}} \cdot (m\bar{\mathbf{v}})$, the kinetic energy in Eq. (19.87) may be written in terms of dot products as

$$T = \tfrac{1}{2} \bar{\mathbf{v}} \cdot \mathbf{L} + \tfrac{1}{2} \boldsymbol{\omega} \cdot \mathbf{H}_G \qquad (19.88)$$

where $\mathbf{L}$ is the linear momentum defined in Eq. (19.72). Substituting Eqs. (19.76), (19.78), and (19.79) into Eq. (19.87), we obtain the explicit form of the kinetic energy T of a rigid body in $OXYZ$ as

$$\begin{aligned} T = \tfrac{1}{2} m\bar{v}^2 + \tfrac{1}{2} (\bar{I}_{xx}\omega_x^2 &+ \bar{I}_{yy}\omega_y^2 + \bar{I}_{zz}\omega_z^2 \\ &- 2\bar{I}_{xy}\omega_x\omega_y - 2\bar{I}_{yz}\omega_y\omega_z - 2\bar{I}_{zx}\omega_z\omega_x) \end{aligned} \qquad (19.89)$$

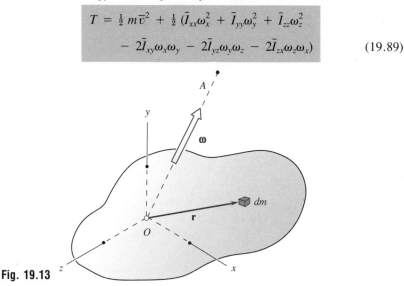

Fig. 19.13

In the particular case of a rigid body rotating with $\boldsymbol{\omega} = \omega_x\mathbf{i} + \omega_y\mathbf{j} + \omega_z\mathbf{k}$ about the origin O of a fixed reference frame $Oxyz$, as shown in Fig. 19.13, it can similarly be shown that the kinetic energy T of the rigid body is given by

$$\begin{aligned} T = \tfrac{1}{2} (I_{xx}\omega_x^2 &+ I_{yy}\omega_y^2 + I_{zz}\omega_z^2 - 2I_{xy}\omega_x\omega_y \\ &- 2I_{yz}\omega_y\omega_z - 2I_{zx}\omega_z\omega_x) \end{aligned} \qquad (19.90)$$

where the moments of inertia I_{xx}, I_{yy}, I_{zz} and the products of inertia I_{xy}, I_{yz}, I_{zx} are computed with respect to the fixed reference frame $Oxyz$. Furthermore, suppose that OA is the axis of rotation or the *velocity axis* (i.e., the instantaneous axis of zero velocity) of the rigid body, as indicated in Fig. 19.13, at the instant considered. Then, $\boldsymbol{\omega}$ is directed along the axis OA, and the velocities of all particles of the rigid body are perpendicular to the radii of their circular arcs of paths centered on the axis OA. In this case, the kinetic energy T of the rigid body can readily be shown to be given by

$$T = \tfrac{1}{2} I_{OA}\omega^2 \qquad (19.91)$$

where I_{OA} is the moment of inertia of the body about its *velocity axis OA*.

Knowing how to compute the kinetic energy of a rigid body in general motion, we may solve appropriate problems by the *principle of work and energy*, which is represented by the equation

$$T_1 + U_{1\to2} = T_2 \qquad (19.92)$$

where T_1 and T_2 are the kinetic energies of the rigid body in the positions 1 and 2, respectively, and $U_{1\rightarrow2}$ is the work of the force system acting on the rigid body during its motion from the position 1 to the position 2. Note that the kinetic energy of a body has different values in different inertial reference frames moving with different constant velocities. It is important to use the *same* inertial reference frame in computing the kinetic energy and the work of a force or moment in a given kinetic problem.

EXAMPLE 19.10

A solid circular cone has a base of radius $r = 200$ mm, a height of $h = 480$ mm, and a mass of $m = 160$ kg. The cone is welded to a horizontal shaft AB which can rotate freely in its bearings as shown. If the cone is at rest in the upright position when it is given a slight push, determine its angular speed ω after it has rotated through $180°$.

Solution. The slant height of the cone is $L = (r^2 + h^2)^{1/2}$. We may draw the side view of the cone as shown, where G is its mass center. Referring

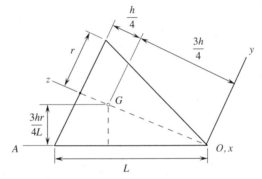

to this drawing and the formulas in Fig. 16.7, we write

$$I_{xx} = I_{yy} = \tfrac{3}{20}\,m(r^2 + 4h^2) \qquad I_{zz} = \tfrac{3}{10}\,mr^2$$

$$I_{xy} = I_{yz} = I_{zx} = 0$$

$$\lambda_{OA} = -\frac{r}{L}\,\mathbf{j} + \frac{h}{L}\,\mathbf{k}$$

$$\lambda_x = 0 \qquad \lambda_y = -\frac{r}{L} \qquad \lambda_z = \frac{h}{L}$$

Applying Eq. (19.56) to compute the moment of inertia, we write

$$I_{OA} = 0 + \frac{3}{20}\,m(r^2 + 4h^2)\left(-\frac{r}{L}\right)^2$$

$$+ \frac{3}{10}\,mr^2\left(\frac{h}{L}\right)^2 - 0 - 0 - 0$$

$$I_{OA} = \frac{3mr^2}{20L^2}\,(r^2 + 6h^2)$$

Since the problem directly involves displacement and velocity, it may be solved by using the *principle of work and energy*. As the cone rotates from rest through $180°$ to its lowest position, its mass center G will have descended a distance of $2(3hr)/(4L) = 3hr/(2L)$ and only its weight force $\mathbf{W} = mg \downarrow$ will do work during the rotation. Thus, we have $U_{1\to 2} = 3mghr/(2L)$. We note that the shaft is also the *velocity axis* of the cone. Therefore, we apply Eq. (19.91) to write

$$T_1 = 0 \qquad T_2 = \tfrac{1}{2} I_{OA}\omega^2 = \frac{3mr^2}{40L^2}(r^2 + 6h^2)\omega^2$$

Applying the *principle of work and energy*, $T_1 + U_{1\to 2} = T_2$, we write

$$0 + \frac{3mghr}{2L} = \frac{3mr^2}{40L^2}(r^2 + 6h^2)\omega^2$$

$$\omega^2 = \frac{20ghL}{r(r^2 + 6h^2)} = \frac{20gh(r^2 + h^2)^{1/2}}{r(r^2 + 6h^2)}$$

Substituting $g = 9.81$ m/s^2, $h = 0.48$ m, and $r = 0.2$ m into the expression for ω^2, we obtain

$$\omega^2 = 172.14 \qquad \omega = 13.12 \text{ rad/s} \blacktriangleleft$$

Developmental Exercises

D19.24   Refer to Example 19.10. Determine the kinetic energy of the cone when it reaches its lowest position.

D19.25   Refer to the solid circular cone in Example 19.10. Determine the angular speed ω of the cone after it has rotated through $90°$.

D19.26   Write the kinetic energy of a rigid body in general motion in terms of dot products.

PROBLEMS

19.41   The bent rod $OABC$ has a total mass m and revolves about the Z axis with an angular velocity $\boldsymbol{\omega}$. For the position shown, determine (a) the angular momentum of the rod about the origin O, (b) the kinetic energy of the rod.

19.42   A 2.4-m slender rod AB of mass 4 kg is welded at its midpoint G to a 2.4-m horizontal shaft OC. In the position shown, the shaft rotates with an angular velocity $\boldsymbol{\omega} = 120\mathbf{I}$ rpm and the rod lies in the ZX plane. Determine (a) the angular momentum of the rod about G, (b) the kinetic energy of the rod.

Fig. P19.41

Fig. P19.42

19.43* A thin rectangular plate weighing 10 lb is mounted on a horizontal shaft with which the plate makes an angle of 50° as shown. If the shaft rotates with $\boldsymbol{\omega} = 6\mathbf{I}$ rad/s, determine (a) the angular momentum $\mathbf{H}_G$ of the plate about G, (b) the angle between $\mathbf{H}_G$ and $\boldsymbol{\omega}$, (c) the kinetic energy of the plate.

Fig. P19.43*

Fig. P19.44

19.44 Four slender rods, each weighing 3 lb, are welded to a horizontal shaft as shown. If the shaft rotates at $\boldsymbol{\omega} = 90\mathbf{K}$ rpm, determine (a) the angular momentum $\mathbf{H}_O$ of the rods about the origin O, (b) the angle between $\mathbf{H}_O$ and $\boldsymbol{\omega}$, (c) the kinetic energy of the rods.

19.45* Two triangular plates, each of mass 5 kg, are welded to a horizontal shaft as shown. If the shaft rotates at $\boldsymbol{\omega} = 4\mathbf{K}$ rad/s in the position shown, determine (a) the angular momentum of the plates about the origin O, (b) the kinetic energy of the plates.

19.46 The 5-kg disk D rotates about the pin at A at the constant rate $\boldsymbol{\omega}_1 = -4\mathbf{I}$ rad/s relative to the frame F, while the frame F rotates about the fixed shaft at the constant rate $\boldsymbol{\omega}_2 = 8\mathbf{K}$ rad/s. For the instant shown, determine (a) the angular momentum of the disk about A, (b) the kinetic energy of the disk.

Fig. P19.45*

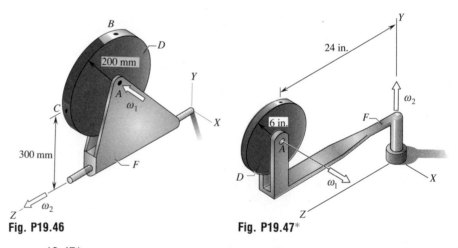

Fig. P19.46

Fig. P19.47*

19.47* The 12-lb disk D rotates about the pin at A at the constant rate $\boldsymbol{\omega}_1 = 6\mathbf{I}$ rad/s relative to the arm F, while the arm F rotates at the constant rate $\boldsymbol{\omega}_2 = 4\mathbf{J}$ rad/s relative to the ground. For the instant shown, determine (a) the angular momentum of the disk about A, (b) the kinetic energy of the disk.

19.48 A 6-kg cone C rolls without slipping inside a fixed conical cavity F as shown, where $\overline{OB} = 200$ mm. The cone C moves with a constant rate of spin $\boldsymbol{\omega}_1 = 2$ rad/s about its axis BO. For the instant shown, determine (a) the angular momentum of the cone about the origin O, (b) the kinetic energy of the cone.

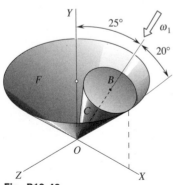

Fig. P19.48

19.49 At the instant shown, the 20-lb disk D rotates about its axis relative to the motor M at the rate $\omega_{D/M} = 2$ rad/s, the motor M rotates about the mounting pin relative to the bracket B at the rate $\omega_{M/B} = 1.5$ rad/s, while the bracket B rotates about the vertical shaft relative to the ground at the rate $\omega_B = 4$ rad/s. For this instant, determine (a) the angular momentum of the disk about its mass center, (b) the kinetic energy of the disk.

Fig. P19.49

Fig. P19.50

19.50 The 8-kg triangular plate shown is in equilibrium when a 4-g bullet is fired into it with a velocity $\mathbf{v}_0 = -1800\mathbf{I}$ m/s and becomes embedded at B. Determine the angular velocity of the plate (a) immediately after the plate is hit by the bullet, (b) at the instant the plate has rotated 180°.

19.51 A 20-lb plate is suspended by a wire fastened at its mass center G. The plate is originally at rest. If an impulse of $\mathbf{I} = -12\mathbf{J}$ lb·s is momentarily applied at the corner A, determine the angular velocity of the plate immediately after the impact.

19.52* A 5-Mg space vehicle is translating at the velocity $\bar{\mathbf{v}} = 800\mathbf{J}$ m/s when it is struck at the point $A(1, 3, -2)$ m on its surface by a meteoroid of 0.8 kg which is traveling with a velocity $\mathbf{v}_0 = -200\mathbf{I} - 600\mathbf{J} + 300\mathbf{K}$ m/s. If the radii of gyration of the space vehicle are $k_X = k_Z = 1.2$ m and $k_Y = 0.6$ m, determine the angular velocity of the space vehicle after the meteoroid becomes embedded in it.

Fig. P19.51

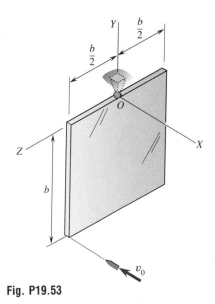

19.53 A heavy square plate of mass m is suspended from a ball-and-socket joint at O as shown. If a bullet of mass m_0 strikes the corner A of the plate with a velocity $\mathbf{v}_0 = -v_0\mathbf{I}$ and becomes embedded in the plate, determine the angular velocity of the plate immediately after the impact.

Fig. P19.53

Fig. P19.52*

★19.8   Equations of Motion for a Rigid Body

We know that a rigid body may be viewed as a special system of particles which keep fixed distances from each other at any time during the motion. In other words, the equations of motion expressed in terms of the time derivatives of the momenta of a system of particles in Sec. 14.4 are also applicable to a rigid body in general motion. Thus, for a rigid body moving in space, the motion must satisfy the following fundamental equations of motion

$$\Sigma \mathbf{F} = \dot{\mathbf{L}} \tag{19.93}$$

$$\Sigma \mathbf{M}_G = \dot{\mathbf{H}}_G \tag{19.94}$$

$$\Sigma \mathbf{M}_O = \dot{\mathbf{H}}_O \tag{19.95}$$

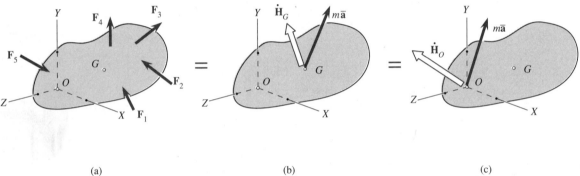

(a)   (b)   (c)

Fig. 19.14

I. TRANSLATIONAL MOTION.   Since $\dot{\mathbf{L}} = m\bar{\mathbf{a}}$ for a rigid body of mass m whose mass center G is moving with an acceleration $\bar{\mathbf{a}}$, we see that the force system in Fig. 19.14(a) is equivalent to the effective force-moment system consisting of $m\bar{\mathbf{a}}$ at G and $\dot{\mathbf{H}}_G$ about G as shown in Fig. 19.14(b), or the effective force-moment system consisting of $m\bar{\mathbf{a}}$ at O and $\dot{\mathbf{H}}_O$ about O as shown in Fig. 19.14(c). In any case, the *translational motion* of the rigid body is governed by Eq. (19.93) or the vector equation

$$\Sigma \mathbf{F} = m\bar{\mathbf{a}} \tag{19.96}$$

which entails the following three scalar equations:

$$\Sigma F_x = m\bar{a}_x \qquad \Sigma F_y = m\bar{a}_y \qquad \Sigma F_z = m\bar{a}_z \tag{19.97}$$

where $\Sigma \mathbf{F} = \Sigma F_x \mathbf{i} + \Sigma F_y \mathbf{j} + \Sigma F_z \mathbf{k}$ is the resultant force, and $\bar{\mathbf{a}} = \bar{a}_x \mathbf{i} + \bar{a}_y \mathbf{j} + \bar{a}_z \mathbf{k}$ is the acceleration of the mass center G.

II. ROTATIONAL MOTION ABOUT THE MASS CENTER G.   We recognize that $\dot{\mathbf{H}}_G$, the time derivative of the angular momentum about G, is simply equal to $\bar{I}\alpha$ in plane motion, where $\bar{I}$ is the moment of inertia about the central axis perpendicular to the plane of motion and α is the angular acceleration of the rigid body. In general motion, $\dot{\mathbf{H}}_G$ is much more involved. We may employ a *moving reference frame Gxyz*, which is not embedded in the rigid body but whose origin moves with the mass center G of the rigid body as shown in Fig. 19.15. Applying Eq. (19.8), we write

$$\dot{\mathbf{H}}_G = (\dot{\mathbf{H}}_G)_{Gxyz} + \boldsymbol{\Omega} \times \mathbf{H}_G \qquad (19.98)$$

where $\boldsymbol{\Omega}$ is the angular velocity of $Gxyz$ in $OXYZ$. Substituting Eq. (19.98) into Eq. (19.94), we see that the *rotational motion* of a rigid body must satisfy the vector equation

$$\Sigma \mathbf{M}_G = (\dot{\mathbf{H}}_G)_{Gxyz} + \boldsymbol{\Omega} \times \mathbf{H}_G \qquad (19.99)$$

Fig. 19.15

If we let $Gxyz$ in Fig. 19.15 be *embedded* in the rigid body whose angular velocity is $\boldsymbol{\omega}$ in the fixed reference frame $OXYZ$, then the moments and products of inertia in $Gxyz$ become constant and $\boldsymbol{\Omega}$ and $\boldsymbol{\omega}$ are equal. We have

$$\boldsymbol{\Omega} = \boldsymbol{\omega} = \omega_x \mathbf{i} + \omega_y \mathbf{j} + \omega_z \mathbf{k} \qquad (19.100)$$

$$\mathbf{H}_G = H_x \mathbf{i} + H_y \mathbf{j} + H_z \mathbf{k} \qquad (19.76)$$
$$\text{(repeated)}$$

$$(\dot{\mathbf{H}}_G)_{Gxyz} = \dot{H}_x \mathbf{i} + \dot{H}_y \mathbf{j} + \dot{H}_z \mathbf{k} \qquad (19.101)$$

where, by virtue of Eq. (19.79),

$$\dot{H}_x = \bar{I}_{xx}\dot{\omega}_x - \bar{I}_{xy}\dot{\omega}_y - \bar{I}_{xz}\dot{\omega}_z \qquad (19.102a)$$

$$\dot{H}_y = -\bar{I}_{yx}\dot{\omega}_x + \bar{I}_{yy}\dot{\omega}_y - \bar{I}_{yz}\dot{\omega}_z \qquad (19.102b)$$

$$\dot{H}_z = -\bar{I}_{zx}\dot{\omega}_x - \bar{I}_{zy}\dot{\omega}_y + \bar{I}_{zz}\dot{\omega}_z \qquad (19.102c)$$

Expressing the resultant moment $\Sigma \mathbf{M}_G$ about G in $Gxyz$, we write

$$\Sigma \mathbf{M}_G = \Sigma M_x \mathbf{i} + \Sigma M_y \mathbf{j} + \Sigma M_z \mathbf{k} \qquad (19.103)$$

Substituting Eqs. (19.100), (19.76), and (19.101) through (19.103) into Eq. (19.99), carrying out the cross product, and equating coefficients of the unit vectors $\mathbf{i}$, $\mathbf{j}$, and $\mathbf{k}$, we obtain the following three scalar equations:

$$\Sigma M_x = \bar{I}_{xx}\dot{\omega}_x - (\bar{I}_{yy} - \bar{I}_{zz})\omega_y\omega_z - \bar{I}_{xy}(\dot{\omega}_y - \omega_z\omega_x)$$
$$- \bar{I}_{yz}(\omega_y^2 - \omega_z^2) - \bar{I}_{zx}(\dot{\omega}_z + \omega_x\omega_y) \qquad (19.104a)$$

$$\Sigma M_y = \bar{I}_{yy}\dot{\omega}_y - (\bar{I}_{zz} - \bar{I}_{xx})\omega_z\omega_x - \bar{I}_{yz}(\dot{\omega}_z - \omega_x\omega_y)$$
$$- \bar{I}_{zx}(\omega_z^2 - \omega_x^2) - \bar{I}_{xy}(\dot{\omega}_x + \omega_y\omega_z) \qquad (19.104b)$$

$$\Sigma M_z = \bar{I}_{zz}\dot{\omega}_z - (\bar{I}_{xx} - \bar{I}_{yy})\omega_x\omega_y - \bar{I}_{zx}(\dot{\omega}_x - \omega_y\omega_z)$$
$$- \bar{I}_{xy}(\omega_x^2 - \omega_y^2) - \bar{I}_{yz}(\dot{\omega}_y + \omega_z\omega_x) \qquad (19.104c)$$

These are the *general rotational equations of motion* in terms of a body-axis coordinate system at the mass center G.

In the case the x, y, and z body axes in Fig. 19.15 are chosen to coincide with the *principal axes of inertia* at G of the rigid body, Eqs. (19.104) reduce to

$$\Sigma M_x = \bar{I}_x \dot{\omega}_x - (\bar{I}_y - \bar{I}_z)\omega_y\omega_z$$

$$\Sigma M_y = \bar{I}_y \dot{\omega}_y - (\bar{I}_z - \bar{I}_x)\omega_z\omega_x \qquad (19.105)$$

$$\Sigma M_z = \bar{I}_z \dot{\omega}_z - (\bar{I}_x - \bar{I}_y)\omega_x\omega_y$$

where $\bar{I}_x = \bar{I}_{xx}$, $\bar{I}_y = \bar{I}_{yy}$, and $\bar{I}_z = \bar{I}_{zz}$ are the principal moments of inertia at G. Equations (19.105) are called *Euler's equations of motion* for $\boldsymbol{\Omega} = \boldsymbol{\omega}$ for a rigid body about its mass center, after the Swiss mathematician Leonhard Euler (1707–1783).[†] In the present case, we do have $\boldsymbol{\Omega} = \boldsymbol{\omega}$ and

$$\dot{\boldsymbol{\omega}} = (\dot{\boldsymbol{\omega}})_{OXYZ} = (\dot{\boldsymbol{\omega}})_{Gxyz} + \boldsymbol{\Omega} \times \boldsymbol{\omega}$$

$$= (\dot{\boldsymbol{\omega}})_{Gxyz} + \boldsymbol{\omega} \times \boldsymbol{\omega} = (\dot{\boldsymbol{\omega}})_{Gxyz}$$

$$= \dot{\omega}_x \mathbf{i} + \dot{\omega}_y \mathbf{j} + \dot{\omega}_z \mathbf{k}$$

This special relation means that, when $\boldsymbol{\Omega} = \boldsymbol{\omega}$, the quantities $\dot{\omega}_x$, $\dot{\omega}_y$, and $\dot{\omega}_z$, which are the rectangular components of $(\dot{\boldsymbol{\omega}})_{Gxyz}$, can be obtained by first finding the angular acceleration $\dot{\boldsymbol{\omega}}$ of the rigid body in the fixed reference frame $OXYZ$, via Eq. (19.50), and then resolving it into components in the x, y, and z directions of the body-axis coordinate system $Gxyz$. Note that $(\dot{\boldsymbol{\omega}})_{Gxyz}$ is the time derivative, taken in the moving reference frame $Gxyz$, of the angular velocity $\boldsymbol{\omega}$ of the body in the fixed reference frame $OXYZ$. Do *not* assume that $\dot{\omega}_x = 0$ when $\omega_x = 0$ in $Gxyz$, etc. Moreover, note that Euler's equations of motion apply *only* for moments summed about the principal axes of inertia.

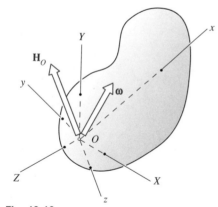

III. ROTATIONAL MOTION ABOUT A FIXED POINT O. When a rigid body is constrained to rotate about a fixed point, we may first let the fixed point be the origin of the fixed reference frame $OXYZ$ and then let the angular velocity of the rigid body in $OXYZ$ be $\boldsymbol{\omega}$, as shown in Fig. 19.16. Furthermore, we may employ a rotating reference frame $Oxyz$ whose angular velocity in $OXYZ$ is $\boldsymbol{\Omega}$. In general, $\boldsymbol{\Omega}$ is not equal to $\boldsymbol{\omega}$. Using Eq. (19.95) and applying Eq. (19.8), we readily see that the *rotational motion* of a rigid body must also satisfy the vector equation

$$\Sigma \mathbf{M}_O = (\dot{\mathbf{H}}_O)_{Oxyz} + \boldsymbol{\Omega} \times \mathbf{H}_O \qquad (19.106)$$

If we let $Oxyz$ in Fig. 19.16 be *embedded* in the rigid body, then the moments and products of inertia in $Oxyz$ become constant and $\boldsymbol{\Omega}$ and $\boldsymbol{\omega}$ are equal. In this case, Eq. (19.106) can similarly be shown to entail the following three scalar equations:

Fig. 19.16

[†]Leonhard Euler left his birthplace Basel, Switzerland, for St. Petersburg, Russia, in 1727. From then on, his life and scientific work were closely associated with the St. Petersburg Academy and with Russia.

$$\Sigma M_x = I_{xx}\dot{\omega}_x - (I_{yy} - I_{zz})\omega_y\omega_z - I_{xy}(\dot{\omega}_y - \omega_z\omega_x)$$
$$- I_{yz}(\omega_y^2 - \omega_z^2) - I_{zx}(\dot{\omega}_z + \omega_x\omega_y) \qquad (19.107a)$$

$$\Sigma M_y = I_{yy}\dot{\omega}_y - (I_{zz} - I_{xx})\omega_z\omega_x - I_{yz}(\dot{\omega}_z - \omega_x\omega_y)$$
$$- I_{zx}(\omega_z^2 - \omega_x^2) - I_{xy}(\dot{\omega}_x + \omega_y\omega_z) \qquad (19.107b)$$

$$\Sigma M_z = I_{zz}\dot{\omega}_z - (I_{xx} - I_{yy})\omega_x\omega_y - I_{zx}(\dot{\omega}_x - \omega_y\omega_z)$$
$$- I_{xy}(\omega_x^2 - \omega_y^2) - I_{yz}(\dot{\omega}_y + \omega_z\omega_x) \qquad (19.107c)$$

These are the *general rotational equations of motion* in terms of a body-axis coordinate system at the fixed point O as shown in Fig. 19.16. If the x, y, and z body axes are chosen to coincide with the *principal axes of inertia* at O for the rigid body, Eqs. (19.107) reduce to

$$\Sigma M_x = I_x\dot{\omega}_x - (I_y - I_z)\omega_y\omega_z$$
$$\Sigma M_y = I_y\dot{\omega}_y - (I_z - I_x)\omega_z\omega_x \qquad (19.108)$$
$$\Sigma M_z = I_z\dot{\omega}_z - (I_x - I_y)\omega_x\omega_y$$

where $I_x = I_{xx}$, $I_y = I_{yy}$, $I_z = I_{zz}$ are the principal moments of inertia at the fixed origin O. Equations (19.108) are called *Euler's equations of motion* for $\boldsymbol{\Omega} = \boldsymbol{\omega}$ for a rigid body about a fixed point.

IV. EULER'S EQUATIONS OF MOTION FOR $\boldsymbol{\Omega} \neq \boldsymbol{\omega}$.

At times, a body of revolution (such as a spinning top or gyroscope) may be made to rotate about a fixed point O on its axis of symmetry, as shown in Fig. 19.17, where the z axis of the moving reference frame $Oxyz$ is chosen to be *always* directed along the axis of symmetry of the body. If $Oxyz$ is *not embedded* in the body, the angular velocity $\boldsymbol{\Omega}$ of $Oxyz$ in the fixed reference frame $OXYZ$ may be different from the angular velocity $\boldsymbol{\omega}$ of the body in $OXYZ$; i.e., $\boldsymbol{\Omega} \neq \boldsymbol{\omega}$. Because of symmetry, the x, y, and z axes are principal axes of inertia at O and the principal moments of inertia I_x, I_y, and I_z all remain *constant* during the motion of the body. In this case, we have

$$\boldsymbol{\omega} = \omega_x\mathbf{i} + \omega_y\mathbf{j} + \omega_z\mathbf{k}$$
$$\boldsymbol{\Omega} = \Omega_x\mathbf{i} + \Omega_y\mathbf{j} + \Omega_z\mathbf{k}$$
$$\mathbf{H}_O = I_x\omega_x\mathbf{i} + I_y\omega_y\mathbf{j} + I_z\omega_z\mathbf{k}$$

Substituting the above equations into Eq. (19.106), carrying out the indicated differentiation and the cross product, and equating the coefficients of the unit vectors, we obtain

$$\Sigma M_x = I_x\dot{\omega}_x - I_y\Omega_z\omega_y + I_z\Omega_y\omega_z$$
$$\Sigma M_y = I_y\dot{\omega}_y - I_z\Omega_x\omega_z + I_x\Omega_z\omega_x \qquad (19.109)$$
$$\Sigma M_z = I_z\dot{\omega}_z - I_x\Omega_y\omega_x + I_y\Omega_x\omega_y$$

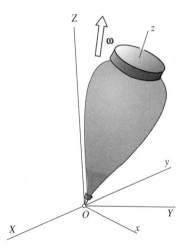

Fig. 19.17

where $\dot{\omega}_x$, $\dot{\omega}_y$, and $\dot{\omega}_z$ are the rectangular components of $(\dot{\omega})_{Oxyz}$. Note that $(\dot{\omega})_{Oxyz}$ is the time derivative, taken in the moving reference frame $Oxyz$, of the angular velocity ω of the body in the fixed reference frame $OXYZ$. *In this particular case*, we have $\Omega \neq \omega$ and

$$\dot{\omega} = (\dot{\omega})_{Oxyz} + \Omega \times \omega \neq (\dot{\omega})_{Oxyz}$$

Equations (19.109) are called *Euler's equations of motion* for $\Omega \neq \omega$ for a body of revolution.

Note that the usual steps to be taken prior to applying *Euler's equations of motion* in Eqs. (19.105), (19.108), or (19.109) are as follows:

a. Embed the x, y, and z axes in the rigid body (or a reference frame) so that they are, and remain to be, the principal axes of inertia at a fixed point O or the mass center G of the rigid body during the motion.
b. Compute ω in $OXYZ$ then find ω_x, ω_y, and ω_z. [Cf. Eq. (19.40).]
c. If $\Omega = \omega$, compute $\dot{\omega}$ in $OXYZ$ then find $\dot{\omega}_x$, $\dot{\omega}_y$, and $\dot{\omega}_z$. [Cf. Eq. (19.50).]
d. If $\Omega \neq \omega$, compute $(\dot{\omega})_{Oxyz}$ or $(\dot{\omega})_{Gxyz}$, *not* $\dot{\omega}$, then find $\dot{\omega}_x$, $\dot{\omega}_y$, and $\dot{\omega}_z$.
e. Compute the principal moments of inertia at O or G, as appropriate.
f. Draw the free-body diagram of the rigid body.

If the principal axes of inertia at a fixed point O or the mass center G are not readily identifiable for the rigid body, its rotational motion may be studied by applying, instead, the general equations such as Eqs. (19.99) and (19.106), or (19.104) and (19.107).

EXAMPLE 19.11

A slender rod OA of length $L = 0.5$ m and mass $m = 2$ kg can rotate freely about the pin in a clevis which remains at the position O and is rotated with the shaft about the vertical axis with a constant angular velocity ω as shown. If the rod OA and the vertical axis form a constant angle $\theta = 30°$, determine the magnitude of ω.

Solution. At the instant considered, the rod OA may be assumed to lie in the XY plane of the fixed reference frame $OXYZ$ as shown, where G is the mass center of the rod. We first let the reference frame $Oxyz$ be *embedded in the rod*, where the x, y, and z axes are the principal axes of inertia. The angular velocity of the rod can readily be written as

$$\omega = \omega \sin\theta \, \mathbf{i} + \omega \cos\theta \, \mathbf{j}$$

which is constant in $OXYZ$. Clearly, the angular acceleration of the rod in $OXYZ$ is $\dot{\omega} = \mathbf{0}$. Thus, we have

$$\omega_x = \omega \sin\theta \qquad \omega_y = \omega \cos\theta$$

$$\omega_z = \dot{\omega}_x = \dot{\omega}_y = \dot{\omega}_z = 0$$

For the rod and the axes shown, we have

$$I_x = I_z = \tfrac{1}{3} mL^2$$

$$I_y = I_{xy} = I_{yz} = I_{zx} = 0$$

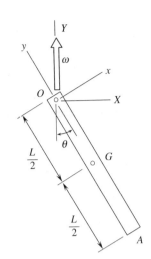

Referring to the free-body diagram drawn and applying *Euler's equations of motion* in Eqs. (19.108), we write

$$\Sigma M_z = I_z\dot{\omega}_z - (I_x - I_y)\omega_x\omega_y:$$

$$-(\tfrac{1}{2}L\sin\theta)(mg) = 0 - (\tfrac{1}{3}mL^2 - 0)(\omega\sin\theta)(\omega\cos\theta)$$

which simplifies to

$$\tfrac{1}{6}mL\sin\theta\,(3g - 2L\omega^2\cos\theta) = 0$$

Since $\sin\theta = \sin 30° \neq 0$, we obtain

$$\omega^2 = \frac{3g}{2L\cos\theta} = \frac{3(9.81)}{2(0.5)\cos 30°}$$

$$\omega = 5.83 \text{ rad/s} \quad \blacktriangleleft$$

Developmental Exercises

D19.27   Refer to Example 19.11. Determine (a) the acceleration $\bar{\mathbf{a}}$ of the mass center G, (b) the reaction components $\mathbf{O}_X$ and $\mathbf{O}_Y$.

D19.28   Refer to Example 19.11. Determine the maximum value of ω for which the rod will remain vertical (i.e., $\theta = 0$).

D19.29   Refer to Example 19.11. Determine the kinetic energy of the rod OA in $OXYZ$.

EXAMPLE 19.12

A thin disk having a radius $r = 1$ ft and weighing 16.1 lb rotates with a constant angular speed $\omega_1 = 10$ rad/s relative to the bent axle OAG, which has a negligible weight and is itself rotating with a constant angular velocity $\boldsymbol{\omega}_2 = 5\mathbf{J}$ rad/s about the Y axis of the fixed reference frame $OXYZ$ as shown. Determine the force-moment system, $\mathbf{O}$ and $\mathbf{M}_O$, representing the dynamic reaction from the support at O.

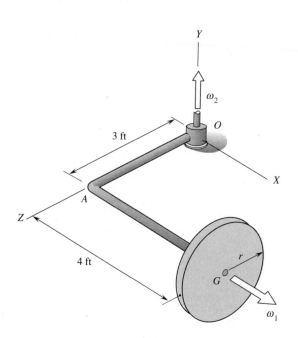

Solution. We first let the reference frame $Gxyz$ be *embedded in the disk* as shown (below left), where the x, y, and z axes are the principal axes at the mass center G of the disk. By the addition theorem for angular velocities [cf. Eq. (19.38)], the angular velocity of the disk in $OXYZ$ is

$$\boldsymbol{\omega} = \boldsymbol{\omega}_1 + \boldsymbol{\omega}_2 = \omega_1\mathbf{i} + \omega_2\mathbf{j}$$

By the addition theorem for angular accelerations [cf. Eq. (19.48)], we find that the angular acceleration of the disk in $OXYZ$ is

$$\dot{\boldsymbol{\omega}} = \boldsymbol{\alpha}_{\text{disk/axle}} + \boldsymbol{\alpha}_{\text{axle}} + \boldsymbol{\omega}_{\text{axle}} \times \boldsymbol{\omega}_{\text{disk/axle}}$$

$$= \mathbf{0} + \mathbf{0} + \boldsymbol{\omega}_2 \times \boldsymbol{\omega}_1 = \omega_2\mathbf{j} \times \omega_1\mathbf{i} = -\omega_1\omega_2\mathbf{k}$$

Thus, we have

$$\omega_x = \omega_1 \qquad \omega_y = \omega_2$$

$$\omega_z = \dot{\omega}_x = \dot{\omega}_y = 0 \qquad \dot{\omega}_z = -\omega_1\omega_2$$

For the disk and the axles shown, we write

$$\bar{I}_x = \tfrac{1}{2} mr^2 \qquad \bar{I}_y = \bar{I}_z = \tfrac{1}{4} mr^2$$

$$\bar{I}_{xy} = \bar{I}_{yz} = \bar{I}_{zx} = 0$$

Since the problem is mainly concerned with the dynamic reaction, the free-body diagram (bottom right p. 884) of the disk with its connecting bent axle may be drawn as shown, where the weight force **W** of the disk is a static force and is therefore neglected. We note that $\boldsymbol{\omega}_2$ is constant and the mass center G of the disk moves along a circular path of radius $\overline{OG} = 5$ ft at a constant speed. Thus, the acceleration $\bar{\mathbf{a}}$ of G is

$$\bar{\mathbf{a}} = \overline{OG}\, \omega_2^2\, \boldsymbol{\lambda}_{GO} = \omega_2^2\, \overrightarrow{GO} = -\omega_2^2(4\mathbf{I} + 3\mathbf{K})$$

Referring to the free-body diagram and applying the governing equation for the *translational motion*, we write

$$\Sigma\mathbf{F} = m\bar{\mathbf{a}}: \qquad \mathbf{O} = m\bar{\mathbf{a}} = -m\omega_2^2(4\mathbf{I} + 3\mathbf{K})$$

$$\mathbf{O} = -\frac{16.1}{32.2}\,(5)^2(4\mathbf{I} + 3\mathbf{K})$$

$$\mathbf{O} = -50\mathbf{I} - 37.5\mathbf{K} \text{ lb} \quad \blacktriangleleft$$

Applying *Euler's equations of motion* in Eqs. (19.105), we write

$$\Sigma M_x = 0 - 0 = 0 \qquad \Sigma M_y = 0 - 0 = 0$$

$$\Sigma M_z = \bar{I}_z\dot{\omega}_z - (\bar{I}_x - \bar{I}_y)\omega_x\omega_y$$

$$\Sigma M_z = \tfrac{1}{4} mr^2\,(-\omega_1\omega_2) - (\tfrac{1}{2} mr^2 - \tfrac{1}{4} mr^2)\omega_1\omega_2$$

$$= -\tfrac{1}{2} mr^2\omega_1\omega_2$$

$$\Sigma\mathbf{M}_G = \Sigma M_x\mathbf{i} + \Sigma M_y\mathbf{j} + \Sigma M_z\mathbf{k} = -\tfrac{1}{2} mr^2\omega_1\omega_2\mathbf{k} \qquad (1)$$

Referring to the free-body diagram, we write

$$\Sigma\mathbf{M}_G = \overrightarrow{GO} \times \mathbf{O} + \mathbf{M}_O = \mathbf{0} + \mathbf{M}_O = \mathbf{M}_O \qquad (2)$$

Comparing Eq. (1) with Eq. (2), we see that

$$\mathbf{M}_O = -\tfrac{1}{2} mr^2\omega_1\omega_2\mathbf{k} = -\frac{1}{2}\left(\frac{16.1}{32.2}\right)(1)^2(10)(5)\mathbf{k}$$

$$= -12.5\mathbf{k}$$

Since $\mathbf{k} = \mathbf{K}$, we obtain

$$\mathbf{M}_O = -12.5\mathbf{K} \text{ lb·ft} \quad \blacktriangleleft$$

Developmental Exercises

D19.30   Refer to Example 19.12. Modify the procedure in the solution by letting *Gxyz* be *embedded in the bent axle* and applying Eqs. (19.109).

D19.31   Refer to Example 19.12. Determine the force-moment system at *O* representing the *combined* static and dynamic reaction.

D19.32   Refer to Example 19.12. Determine the kinetic energy of the disk in *OXYZ*.

D19.33   Describe the usual steps to be taken prior to applying Euler's equations of motion.

EXAMPLE 19.13

Two slender arms *BC* and *DEF* are welded to a horizontal shaft *OA* as shown, where $b = 0.3$ m and $L = 1.2$ m. The segments *BC*, *DE*, and *EF* have equal masses of $m = 2$ kg. If a torque $\mathbf{M}_O = 18\mathbf{K}$ N·m acts on the shaft and the shaft is rotating at $\boldsymbol{\omega} = 10\mathbf{K}$ rad/s in the position shown, determine the angular acceleration $\boldsymbol{\alpha}$ of the shaft and the dynamic reaction from the bearing at *A*.

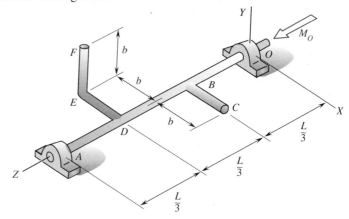

Solution.   Since the principal axes of inertia at either the fixed point *O* or the mass center of the system are not readily known, we simply let the

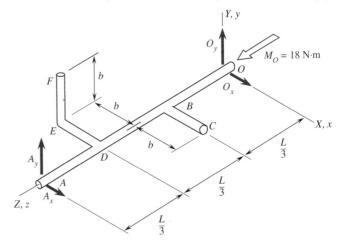

reference frame $Oxyz$ be *embedded in the shaft* as shown. We have $\mathbf{\Omega} = \mathbf{\omega}$ and

$$\mathbf{\omega} = 10\mathbf{K} = 10\mathbf{k} = \omega_z\mathbf{k} \qquad \omega_x = \omega_y = 0 \qquad \omega_z = 10$$

$$\dot{\mathbf{\omega}} = \alpha\mathbf{K} = \alpha\mathbf{k} = \dot{\omega}_z\mathbf{k} \qquad \dot{\omega}_x = \dot{\omega}_y = 0 \qquad \dot{\omega}_z = \alpha$$

Applying Eqs. (19.85) and (19.106), we write

$$H_x = -I_{xz}\omega_z \qquad H_y = -I_{yz}\omega_z \qquad H_z = I_{zz}\omega_z$$

$$\mathbf{H}_O = (-I_{xz}\mathbf{i} - I_{yz}\mathbf{j} + I_{zz}\mathbf{k})\omega_z$$

$$\Sigma\mathbf{M}_O = (\dot{\mathbf{H}}_O)_{Oxyz} + \mathbf{\Omega} \times \mathbf{H}_O$$

$$= (-I_{xz}\mathbf{i} - I_{yz}\mathbf{j} + I_{zz}\mathbf{k})\dot{\omega}_z$$

$$+ \omega_z\mathbf{k} \times (-I_{xz}\mathbf{i} - I_{yz}\mathbf{j} + I_{zz}\mathbf{k})\omega_z$$

$$\Sigma\mathbf{M}_O = (-I_{xz}\dot{\omega}_z + I_{yz}\omega_z^2)\mathbf{i}$$

$$+ (-I_{yz}\dot{\omega}_z - I_{xz}\omega_z^2)\mathbf{j} + I_{zz}\dot{\omega}_z\mathbf{k} \qquad (1)$$

Note that Eq. (1) would have been obtained if we directly applied Eqs. (19.107). Referring to the free-body diagram, where weight forces are neglected for the purpose of determining dynamic reactions, we write

$$\Sigma\mathbf{M}_O = M_O\mathbf{k} + L\mathbf{k} \times (A_x\mathbf{i} + A_y\mathbf{j})$$

$$\Sigma\mathbf{M}_O = -LA_y\mathbf{i} + LA_x\mathbf{j} + M_O\mathbf{k} \qquad (2)$$

Comparing Eq. (1) with Eq. (2), we see that

$$-LA_y = -I_{xz}\dot{\omega}_z + I_{yz}\omega_z^2 \qquad (3)$$

$$LA_x = -I_{yz}\dot{\omega}_z - I_{xz}\omega_z^2 \qquad (4)$$

$$M_O = I_{zz}\dot{\omega}_z \qquad (5)$$

Noting that $m_{BC} = m_{DE} = m_{EF} = m$ and applying the parallel-axis theorem, we write

$$I_{xz} = \left[0 + m\left(\frac{b}{2}\right)\left(\frac{L}{3}\right)\right] + \left[0 + m\left(-\frac{b}{2}\right)\left(\frac{2L}{3}\right)\right]$$

$$+ \left[0 + m(-b)\left(\frac{2L}{3}\right)\right]$$

$$I_{yz} = 0 + 0 + \left[0 + m\left(\frac{b}{2}\right)\left(\frac{2L}{3}\right)\right]$$

$$I_{zz} = \tfrac{1}{3}mb^2 + \tfrac{1}{3}mb^2 + [\tfrac{1}{12}mb^2 + m(b^2 + \tfrac{1}{4}b^2)]$$

$$I_{xz} = -\tfrac{5}{6}mbL \qquad I_{yz} = \tfrac{1}{3}mbL \qquad I_{zz} = 2mb^2 \qquad (6)$$

Substituting Eqs. (6) into Eqs. (3) through (5) and noting that $m = 2$ kg, $b = 0.3$ m, $L = 1.2$ m, $\omega_z = 10$ rad/s, and $M_O = 18$ N·m, we write

$$\alpha = \dot{\omega}_z = \frac{M_O}{2mb^2} = \frac{18}{2(2)(0.3)^2} = 50$$

$$A_x = -\frac{M_O}{6b} + \frac{5}{6}\,mb\omega_z^2 = -\frac{18}{6(0.3)} + \frac{5}{6}(2)(0.3)(10)^2 = 40$$

$$A_y = -\frac{5M_O}{12b} - \frac{1}{3}\,mb\omega_z^2 = -\frac{5(18)}{12(0.3)} - \frac{1}{3}(2)(0.3)(10)^2$$

$$= -45$$

$$\alpha = 50\mathbf{k} \text{ rad/s}^2 \qquad \mathbf{A} = 40\mathbf{i} - 45\mathbf{j} \text{ N} \quad \blacktriangleleft$$

Developmental Exercise

D19.34   Refer to Example 19.13. Determine the dynamic reaction from the support at O. (*Hint.* Consider $\Sigma \mathbf{M}_A$.)

★19.9   Gyroscopic Motion: Steady Precession

I. MOTION OF A GYROSCOPE.   The usual concept of a *gyroscope* is an axisymmetrical body (i.e., a body of revolution) that spins rapidly about its geometric axis of symmetry. When mounted in a Cardan's suspension, as shown in Fig. 19.18, a gyroscope can spin freely about its geometric axis and may assume any orientation; however, its mass center always remains at a fixed point in space. For convenience, we let the reference frame $Oxyz$ be *embedded in the inner gimbal*, with the z axis directed along the axis of spin of the rotor and the y axis directed along the axis joining the supporting

Fig. 19.18

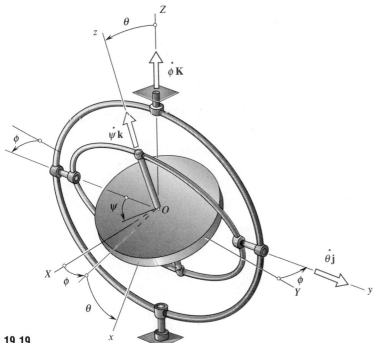

Fig. 19.19

bearings of the inner gimbal, so that the Z, z, and x axes are always co-planar. The position of the gyroscope in the fixed reference frame $OXYZ$ is completely characterized by the following three angles: (1) the angle ϕ specifying the rotation of the *outer gimbal* about the Z axis, (2) the angle θ specifying the rotation of the *inner gimbal* about the y axis which always lies in the XY plane of $OXYZ$, (3) the angle ψ specifying the rotation of the *rotor* about the z axis. The angles ϕ, θ, and ψ, which completely define the position of the gyroscope at any given time, are called the *Eulerian angles*. The Eulerian angles ϕ, θ, and ψ are also called the *precession angle*, the *nutation angle*, and the *spin angle* of the gyroscope, respectively.[†] Their time derivatives $\dot{\phi}$, $\dot{\theta}$, and $\dot{\psi}$, as indicated in Fig. 19.19, define, respectively, the *rate of precession*, the *rate of nutation*, and the *rate of spin* of the gyroscope; and the Z, y, and z axes are, respectively, called the *precession axis*, the *nutation axis*, and the *spin axis* of the gyroscope.

Clearly, the x, y, and z axes in Fig. 19.19 are the principal axes of inertia at the fixed point O of the rotor. Letting i.g. and o.g. denote the inner and outer gimbals, respectively, and applying Eqs. (19.38) and (19.40), we find that the angular velocity $\boldsymbol{\Omega}$ of $Oxyz$ (embedded in the inner gimbal) and the angular velocity $\boldsymbol{\omega}$ of the rotor in $OXYZ$ are

$$\boldsymbol{\Omega} = \boldsymbol{\Omega}_{\text{i.g./o.g.}} + \boldsymbol{\Omega}_{\text{o.g.}}$$

$$\boldsymbol{\Omega} = \dot{\theta}\mathbf{j} + \dot{\phi}\mathbf{K} \tag{19.110}$$

$$\boldsymbol{\omega} = \boldsymbol{\omega}_{\text{rotor/i.g.}} + \boldsymbol{\omega}_{\text{i.g./o.g.}} + \boldsymbol{\omega}_{\text{o.g.}}$$

$$\boldsymbol{\omega} = \dot{\psi}\mathbf{k} + \dot{\theta}\mathbf{j} + \dot{\phi}\mathbf{K} \tag{19.111}$$

[†]Unfortunately, the Eulerian angles (ϕ, θ, ψ) have *not* been represented by the same symbols in all books. It is important to choose a set to work with and be consistent.

Expressing the unit vector $\mathbf{K}$ in $Oxyz$, we have

$$\mathbf{K} = -\sin\theta\,\mathbf{i} + \cos\theta\,\mathbf{k} \qquad (19.112)$$

Substituting Eq. (19.112) into Eqs. (19.110) and (19.111), we write

$$\boldsymbol{\Omega} = -\dot{\phi}\,\sin\theta\,\mathbf{i} + \dot{\theta}\mathbf{j} + \dot{\phi}\,\cos\theta\,\mathbf{k} \qquad (19.113)$$

$$\boldsymbol{\omega} = -\dot{\phi}\,\sin\theta\,\mathbf{i} + \dot{\theta}\,\mathbf{j} + (\dot{\psi} + \dot{\phi}\,\cos\theta)\mathbf{k} \qquad (19.114)$$

$$(\dot{\boldsymbol{\omega}})_{Oxyz} = -(\ddot{\phi}\,\sin\theta + \dot{\phi}\dot{\theta}\,\cos\theta)\mathbf{i} + \ddot{\theta}\mathbf{j}$$
$$+ (\ddot{\psi} + \ddot{\phi}\,\cos\theta - \dot{\phi}\dot{\theta}\,\sin\theta)\mathbf{k} \qquad (19.115)$$

The principal moments of inertia of the rotor may be designated

$$I_x = I_y = I' \qquad I_z = I \qquad (19.116)$$

Extracting Ω_x, Ω_y, Ω_z, ω_x, ω_y, ω_z, $\dot{\omega}_x$, $\dot{\omega}_y$, and $\dot{\omega}_z$ from Eqs. (19.113), (19.114), and (19.115), using Eq. (19.116), and applying *Euler's equations of motion* for $\boldsymbol{\Omega} \neq \boldsymbol{\omega}$ in Eqs. (19.109), we write

$$\Sigma M_x = I'\,(-\ddot{\phi}\,\sin\theta - \dot{\phi}\dot{\theta}\,\cos\theta) - I'\,(\dot{\phi}\,\cos\theta)(\dot{\theta})$$
$$+ I\dot{\theta}(\dot{\psi} + \dot{\phi}\,\cos\theta)$$

$$\Sigma M_y = I'\ddot{\theta} - I(-\dot{\phi}\,\sin\theta)(\dot{\psi} + \dot{\phi}\,\cos\theta)$$
$$+ I'\,(\dot{\phi}\,\cos\theta)(-\dot{\phi}\,\sin\theta)$$

$$\Sigma M_z = I(\ddot{\psi} + \ddot{\phi}\,\cos\theta - \dot{\phi}\dot{\theta}\,\sin\theta) - I'\dot{\theta}(-\dot{\phi}\,\sin\theta)$$
$$+ I'(-\dot{\phi}\,\sin\theta)(\dot{\theta})$$

These three equations may be simplified and written as

$$\Sigma M_x = -I'\,(\ddot{\phi}\,\sin\theta + 2\dot{\phi}\dot{\theta}\,\cos\theta) + I\dot{\theta}(\dot{\psi} + \dot{\phi}\,\cos\theta)$$

$$\Sigma M_y = I'\,(\ddot{\theta} - \dot{\phi}^2\sin\theta\,\cos\theta) + I\dot{\phi}\,\sin\theta(\dot{\psi} + \dot{\phi}\,\cos\theta) \qquad (19.117)$$

$$\Sigma M_z = I\frac{d}{dt}\,(\dot{\psi} + \dot{\phi}\,\cos\theta)$$

Thus, the motion of a gyroscope subjected to an external force system is defined by Eqs. (19.117) when the mass of its gimbals is neglected.

Note that Eqs. (19.117) are useful in studying the motion of a *body of revolution* about its mass center or about a fixed point which lies on its axis of symmetry. For instance, a top spinning about its axis of symmetry with a high angular speed $\dot{\psi}$ and supported at the fixed point O may be depicted as shown in Fig. 19.20. In this figure, $OXYZ$ is the fixed reference frame, the z axis of the moving reference frame $Oxyz$ is chosen to be *always* directed along the axis of the top, the Z axis is the precession axis, the y axis is the nutation axis, the z axis is the spin axis, and the x, z, and Z axes are always coplanar. Thus, $Oxyz$ follows the motion of the top only in nutation and precession. Clearly, the position of the top is defined by the *Eulerian angles*: (a) the angle ϕ specifying its rotation about the precession axis, (b) the angle θ specifying its rotation about the nutation axis, (c) the angle ψ

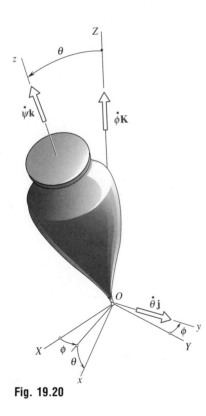

Fig. 19.20

specifying its rotation about the spin axis. A similar derivation will show that the motion of the top in Fig. 19.20 is also defined by Eqs. (19.117).

II. STEADY PRECESSION.

The *steady precession* of a gyroscope is a special case of gyroscopic motion in which the nutation angle θ, the rate of precession $\dot{\phi}$, and the rate of spin $\dot{\psi}$ all remain constant; i.e.,

$$\dot{\theta} = \ddot{\theta} = \ddot{\phi} = \ddot{\psi} = 0 \tag{19.118}$$

By Eqs. (19.110), (19.113), (19.114), and (19.118), we write

$$\mathbf{\Omega} = \dot{\phi}\mathbf{K} = -\dot{\phi}\sin\theta\,\mathbf{i} + \dot{\phi}\cos\theta\,\mathbf{k} \tag{19.119}$$

$$\mathbf{\omega} = -\dot{\phi}\sin\theta\,\mathbf{i} + \omega_z\mathbf{k} \tag{19.120}$$

where $\omega_z = \dot{\psi} + \dot{\phi}\cos\theta$. Substituting Eqs. (19.118) into Eqs. (19.117), we get

$$\Sigma M_x = \Sigma M_z = 0$$

$$\Sigma M_y = [I(\dot{\psi} + \dot{\phi}\cos\theta) - I'\dot{\phi}\cos\theta]\dot{\phi}\sin\theta$$

which are equivalent to the vector equation

$$\mathbf{\Sigma M}_O = [I(\dot{\psi} + \dot{\phi}\cos\theta) - I'\dot{\phi}\cos\theta]\dot{\phi}\sin\theta\,\mathbf{j} \tag{19.121}$$

or

$$\mathbf{\Sigma M}_O = (I\omega_z - I'\dot{\phi}\cos\theta)\dot{\phi}\sin\theta\,\mathbf{j} \tag{19.122}$$

Since the mass center of the gyroscope is fixed in space, we have, by Eq. (19.96), $\Sigma\mathbf{F} = \mathbf{0}$; therefore, the force system to be applied to the gyroscope to maintain a steady precession is equivalent to a *torque*, which is equal to the moment given by the right-hand member of Eq. (19.121) or (19.122). Note that the torque required to maintain a (forced) steady precession must act on the rotor about the nutation axis which is perpendicular to the plane containing the precession axis and the spin axis, as illustrated in Fig. 19.21.

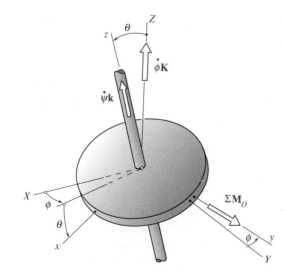

Fig. 19.21

In the particular case where the spin axis is perpendicular to the precession axis, we have $\theta = 90°$ and the required torque in Eq. (19.121) for steady precession reduces to

$$\Sigma\mathbf{M}_O = I\dot{\phi}\dot{\psi}\mathbf{j} \tag{19.123}$$

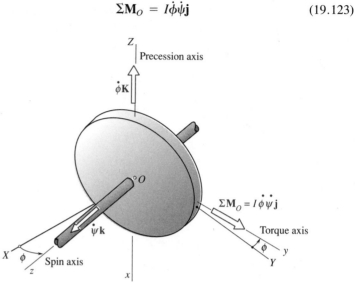

Fig. 19.22

In this case, the spin axis, the torque axis, and the precession axis form a right-handed trihedral as shown in Fig. 19.22. We can see from Eq. (19.123) that a large torque would be required to change the orientation of the axle of a rotor spinning at a high speed about its axis. This gyroscopic effect has many applications. For example, gyroscopes are used to stabilize ships and torpedos and to retain a fixed direction in inertial guidance systems.

III. TORQUE-FREE STEADY PRECESSION.   When the external force system exerts no moment about the mass center of a body, the motion of the body is referred to as a *torque-free motion*. We shall here consider the torque-free motion of an axisymmetrical body. Examples of this type of motion are furnished by artificial satellites and space vehicles, which may spin and precess during free flight.

A torque-free motion of an axisymmetrical body may be represented in Fig. 19.23. Since the resultant moment about the mass center G is zero, Eq. (19.94) yields $\dot{\mathbf{H}}_G = \mathbf{0}$. Thus, $\mathbf{H}_G$ has a constant magnitude and a fixed direction in space. We may first let this fixed direction be the direction of the Z axis, or the axis of precession, then let the z axis of the moving reference frame $Gxyz$ be always directed along the axis of symmetry of the body and the x axis be always in the plane defined by the z and Z axes, as shown in Fig. 19.23. We have

$$H_x = -H_G \sin\theta \qquad H_y = 0 \qquad H_z = H_G \cos\theta \tag{19.124}$$

where θ is the angle between the z and Z axes, and H_G is the constant magnitude of $\mathbf{H}_G$. Clearly, the x, y, and z axes are the principal axes of inertia of the axisymmetrical body. Designating the principal moments of inertia of the body as those in Eqs. (19.116), we write

Fig. 19.23

$$H_x = I'\omega_x \qquad H_y = I'\omega_y \qquad H_z = I\omega_z \qquad (19.125)$$

It follows from Eqs. (19.124) and (19.125) that

$$\omega_x = -\frac{H_G \sin\theta}{I'} \qquad \omega_y = 0 \qquad \omega_z = \frac{H_G \cos\theta}{I} \qquad (19.126)$$

We see from Eqs. (19.111) and (19.126) that

$$\dot{\theta} = \omega_y = 0 \qquad (19.127)$$

Thus, the angle θ between the z and Z axes is a constant and *the body is in steady precession about the Z axis* under no torque.

Substitution of $\dot{\theta} = 0$ into Eqs. (19.110) and (19.111) yields

$$\boldsymbol{\Omega} = \dot{\phi}\mathbf{K} \qquad \boldsymbol{\omega} = \dot{\psi}\mathbf{k} + \dot{\phi}\mathbf{K} \qquad (19.128)$$

The angle β in Fig. 19.23 is defined by

$$\tan\beta = -\frac{\omega_x}{\omega_z} \qquad (19.129)$$

Thus, dividing the first by the third in Eqs. (19.126), we obtain

$$\tan\beta = \left(\frac{I}{I'}\right)\tan\theta \qquad (19.130)$$

For torque-free motion, we set $\Sigma \mathbf{M}_O$ to zero in Eq. (19.121) to yield

$$\dot{\phi} = \frac{I\dot{\psi}}{(I' - I)\cos\theta} \qquad (19.131)$$

It is clear from Eqs. (19.129) through (19.131) that the directions of the spin and of the steady precession of an axisymmetrical body in torque-free motion depend on the given values of ω_x and ω_z as well as the relative magnitudes of I and I'. If the body is set to spin about its axis of symmetry, we have $\omega_x = 0$, $\omega_z \neq 0$, and $\beta = \theta = 0$; and the motion involves *no precession*. If the body is set to spin about a transverse axis, we have $\omega_x \neq 0$, $\omega_z = 0$, $\beta = \theta = 90°$; and the motion again involves *no precession*. However, there are two general cases which may be observed as follows:

1. If $I < I'$, then $\beta < \theta$, as shown in Fig. 19.24. This is the case of a body elongated in the direction of the spin axis (e.g., a space vehicle). The vector $\boldsymbol{\omega}$ is directed along the *velocity axis* (cf. Sec. 15.7) of the body and lies inside the angle ZGz, and the precession is said to be *direct*. Such a motion may be modeled by a *body cone* rolling around the outer surface of a fixed *space cone* without slipping, where the line of tangency of the two cones is the velocity axis.
2. If $I > I'$, then $\beta > \theta$, as shown in Fig. 19.25. This is the case of a body flattened in the direction of the spin axis (e.g., a satellite). The vector $\boldsymbol{\omega}$ is directed along the *velocity axis* of the body and lies outside the angle ZGz, and the precession is said to be *retrograde*, where the sense of the spin points in the negative direction of the z axis. The motion may be modeled by a *body cone* whose inside surface rolls on the outside surface of a fixed *space cone* without slipping, where the line of tangency of the two cones is the velocity axis.

Fig. 19.24

Fig. 19.25

EXAMPLE 19.14

A spinning top of mass $m = 0.4$ kg is precessing about the vertical axis at a constant rate and is supported at the fixed point O as shown, where $b = 60$ mm and $\theta = 30°$. The radii of gyration of the top about its axis of symmetry and about a transverse axis through O are $k = 25$ mm and $k' = 45$ mm, respectively. If the top spins at the rate $\dot{\psi} = 120$ rad/s, determine its rate of precession.

Solution. The moments of inertia of the top are

$$I = mk^2 = 0.4(0.025)^2$$

$$I = 250 \times 10^{-6} \text{ kg·m}^2$$

$$I' = mk'^2 = 0.4(0.045)^2$$

$$I' = 810 \times 10^{-6} \text{ kg·m}^2$$

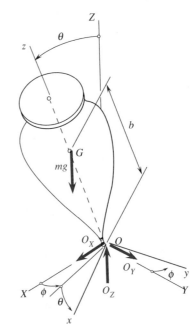

Since the motion is a *steady precession*, we use the reference frames as shown and apply Eq. (19.121) to write

$$\Sigma \mathbf{M}_O = [I(\dot{\psi} + \dot{\phi} \cos\theta) - I'\dot{\phi} \cos\theta]\dot{\phi} \sin\theta \, \mathbf{j} \qquad (1)$$

Referring to the free-body diagram shown, we write

$$\Sigma \mathbf{M}_O = (b \sin\theta)mg\mathbf{j} \qquad (2)$$

Comparing Eq. (1) with Eq. (2), we see that

$$[I(\dot{\psi} + \dot{\phi} \cos\theta) - I'\dot{\phi} \cos\theta]\dot{\phi} = mgb$$

or

$$(I' - I) \cos\theta \, \dot{\phi}^2 - I\dot{\psi}\dot{\phi} + mgb = 0 \qquad (3)$$

Substituting $I' = 810 \times 10^{-6}$ kg·m², $I = 250 \times 10^{-6}$ kg·m², $\theta = 30°$, $\dot{\psi} = 120$ rad/s, $m = 0.4$ kg, $b = 0.06$ m, and $g = 9.81$ m/s² into Eq. (3), we get

$$0.00048497\dot{\phi}^2 - 0.03\dot{\phi} + 0.23544 = 0$$

whose two roots are 9.22 and 52.6. Thus, we obtain two possible rates of precession:

$$\dot{\phi} = 9.22 \text{ rad/s} \qquad \dot{\phi} = 52.6 \text{ rad/s} \blacktriangleleft$$

REMARK. The lower rate of precession is more likely to be observed in reality because of the lower kinetic energy required of the top.

Developmental Exercises

D19.35 Refer to Example 19.14. Determine the two possible rates of precession of the top if the nutation angle is $\theta = 60°$.

D19.36 Refer to the gyroscope in Fig. 19.19. Suppose that the reference frame $Oxyz$ is *embedded in the rotor* and has an orientation as shown at the instant considered. (a) Applying Eq. (19.50), determine the angular acceleration $\dot{\omega}$ of the rotor in $OXYZ$. (b) Derive Eqs. (19.117) by applying Euler's equations of motion for $\mathbf{\Omega} = \boldsymbol{\omega}$ in Eq. (19.108).

D19.37 Refer to the gyroscope in Fig. 19.19. Suppose that the reference frame $Oxyz$ is embedded in the inner gimbal. (a) Determine $\boldsymbol{\omega}_{\text{rotor}/Oxyz}$ which is the angular velocity of the rotor in $Oxyz$. (b) Determine $(\dot{\boldsymbol{\omega}}_{\text{rotor}/Oxyz})_{Oxyz}$ which is the angular acceleration of the rotor in $Oxyz$. (c) Does the quantity $(\dot{\boldsymbol{\omega}})_{Oxyz}$, as given by Eq. (19.115), represent the angular acceleration of the rotor in $Oxyz$?

D19.38 Describe (a) direct precession, (b) retrograde precession.

PROBLEMS

19.54 A uniform slender rod is bent into the shape of a square and is attached by a collar at C to a vertical shaft which rotates with a constant angular velocity $\boldsymbol{\omega}$ as shown. Determine (a) the constant angle θ that the plane of the square forms with the vertical, (b) the maximum value of ω for which the square will remain vertical (i.e., $\theta = 0$).

Fig. P19.54

19.55　The 20-kg slender rod *AB* is supported by a wire at *C* and pinned at *A* to a vertical axle *AD* which rotates with a constant angular velocity of 20 rad/s. Determine the tension in the wire and the magnitude of the reaction from the pin at *A*.

19.56* A thin rectangular plate weighing 10 lb is mounted on a horizontal shaft with which the plate makes an angle of 50° as shown. If the shaft rotates at the constant rate $\boldsymbol{\omega} = 6\mathbf{I}$ rad/s, determine the moment exerted at *G by* the plate *on* the shaft due to the wobble of the plate.

Fig. P19.55　　　　**Fig. P19.56***　　　　　　**Fig. P19.57**

19.57　A 2.4-m slender rod *AB* of mass 4 kg is welded at its midpoint *G* to a 2.4-m horizontal shaft *OC*. If a torque $\mathbf{M}_O = 20\mathbf{I}$ N·m acts on the shaft and the shaft is rotating at $\boldsymbol{\omega} = 120\mathbf{I}$ rpm in the position shown, determine the angular acceleration of the shaft and the dynamic reaction from the bearing at *C*.

19.58　Two triangular plates, each of mass 5 kg, are welded to a horizontal shaft as shown. If the shaft is rotating with $\boldsymbol{\omega} = 5\mathbf{K}$ rad/s and $\boldsymbol{\alpha} = 10\mathbf{K}$ rad/s² in the position shown, determine the torque acting on the shaft.

Fig. P19.60　　　　　**Fig. P19.58**　　　　　**Fig. P19.59**

19.59　The 5-kg disk *D* rotates about the pin at the constant rate $\boldsymbol{\omega}_1 = -4\mathbf{I}$ rad/s relative to the frame *F*, while the frame *F* rotates about the fixed shaft at the constant rate $\boldsymbol{\omega}_2 = 8\mathbf{K}$ rad/s. For the instant shown, determine the moment exerted by the disk on the pin at *A*.

19.60　The top consists of a thin disk of mass 3 kg and a slender rod of length 150 mm and negligible mass as shown, where $\theta = 30°$. If the top spins with $\dot{\psi} = 900$ rpm and precesses at a constant rate, determine the rate of precession.

19.61* Solve Prob. 19.60 if $\theta = 60°$.

19.62   A 10-lb sphere of radius 1 ft is attached to the end of a 3-ft rod of negligible mass which is supported by a ball-and-socket joint at the fixed position O. If the sphere precesses at the constant rate $\dot{\phi} = 15$ rpm, determine its rate of spin $\dot{\psi}$.

19.63*   Solve Prob. 19.62 if $\dot{\phi} = 30$ rpm.

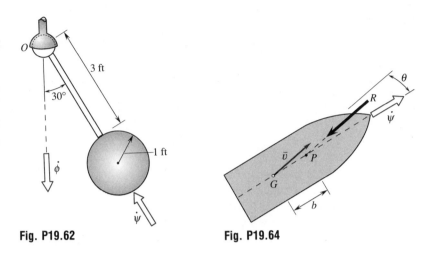

Fig. P19.62   **Fig. P19.64**

19.64   A 40-lb projectile is fired to move with a velocity $\bar{\mathbf{v}}$ of 2000 ft/s which makes an angle $\theta = 5°$ with its axis of symmetry as shown. The projectile has a rate of spin $\dot{\psi} = 500$ rad/s, and the air drag is equivalent to a force $\mathbf{R}$ of 30 lb that is parallel to $\bar{\mathbf{v}}$ and passes through the point P which is a distance $b = 5$ in. from the mass center G. The radii of gyration of the projectile about its axis of symmetry and about a transverse axis through G are $k = 1.5$ in. and $k' = 4.5$ in., respectively. Determine its rate of steady precession.

19.65   Since the earth is an oblate spheroid, the gravitational system acting on the earth is equivalent to a force $\mathbf{F}$ and a torque $\mathbf{M}$ as shown in Fig. (a). The effect of the torque $\mathbf{M}$ is to cause the earth to have a forced precession at the rate $\dot{\phi} = 1$ revolution per 25,800 years. The spin axis of the earth is now pointing to the proximity of Polaris, the North Star, as shown in Fig. (b). Thus, the spin axis of the earth will be pointing to near Vega some 12,900 years from now, as shown in Fig. (c). Determine the average magnitude of the torque $\mathbf{M}$. [*Hint.* Use Eqs. (12.29) and (12.59) and assume that $\bar{I} = 2mR^2/5$.]

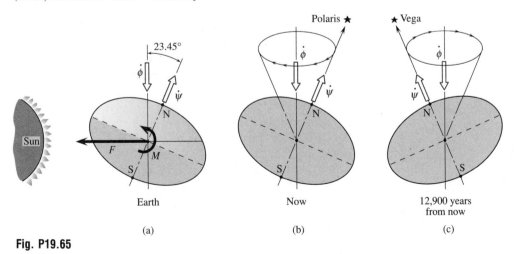

Fig. P19.65

★19.10   Concluding Remarks

The concept of reference frames in connection with the time derivatives of vectors is very important. In general motion, the velocity of a body in a reference frame is the time derivative of the position in that reference frame, of the body, taken (or measured) in that same reference frame. Similarly, the acceleration of a body in a reference frame is the time derivative of the velocity in that reference frame, of the body, taken in that same reference frame. For studying rigid bodies in motion, it is exceedingly important to know the reference frame in which the time derivative of any vector is taken. For instance, note that

$$\dot{\boldsymbol{\omega}} = (\dot{\boldsymbol{\omega}})_{OXYZ} \qquad \mathbf{v}_{B/Axyz} = (\dot{\mathbf{r}}_{B/A})_{Axyz} \neq (\dot{\mathbf{r}}_{B/A})_{OXYZ}$$

$$\mathbf{a}_{B/Axyz} = (\dot{\mathbf{v}}_{B/Axyz})_{Axyz} \neq (\dot{\mathbf{v}}_{B/Axyz})_{OXYZ}$$

where $Axyz$ and $OXYZ$ denote moving and fixed reference frames, respectively. In particular, note that the vector $(\dot{\mathbf{r}}_{B/A})_{OXYZ}$ represents simply the time derivative of $\mathbf{r}_{B/A}$ in $OXYZ$ and carries the units of a velocity, but it is *not the velocity* of B in $Axyz$ nor $OXYZ$; the vector $(\dot{\mathbf{v}}_{B/Axyz})_{OXYZ}$ represents simply the time derivative of $\mathbf{v}_{B/Axyz}$ in $OXYZ$ and carries the units of an acceleration, but it is *not the acceleration* of B in $Axyz$ nor $OXYZ$. However, $(\dot{\mathbf{r}}_{B/Axyz})_{Axyz}$ is the velocity of B in $Axyz$, and $(\dot{\mathbf{v}}_{B/Axyz})_{Axyz}$ is the acceleration of B in $Axyz$.

In Eq. (19.30), we have shown that the angular velocity of a moving reference frame in a fixed reference frame is definable in terms of the time rates of change of the cartesian unit vectors embedded in the moving reference frame as measured in the fixed reference frame. It represents a *general* definition for the angular velocity of a reference frame or a rigid body. The addition theorems for angular velocities and angular accelerations are particularly useful in studying the motion of a system which consists of two or more movable rigid parts. These two theorems become identical in form only in plane motion where angular velocities of all rigid bodies in the system are parallel to each other.

After studying Sec. 19.4 and 19.5, one should be able to apply the parallel-axes theorems to compute the moments and products of inertia of a body. Besides, it is important to remember the following property: *If the mass of a body is symmetrical about two coordinate planes, then all products of inertia of the body vanish and the coordinate axes are principal axes of inertia at the origin.* Concepts and skills related to mass moments and products of inertia are needed in studying the kinetics of rigid bodies in space.

Since the translational motion of a rigid body is similar to the motion of a particle, the main effort in studying the kinetics of rigid bodies in space lies in the study of their rotational motion. One can always apply the concise equations for rotational motion in Eqs. (19.99) and (19.106). If appropriate, Euler's equation of motion in one of the three special situations in Eqs. (19.105), (19.108), and (19.109) may be applied. For ease in getting started, one may find it useful to review the usual steps to be taken prior to applying Euler's equations of motion as outlined at the end of Sec. 19.8.

The motion of a gyroscope is an important type of the general motion of a rigid body. Unfortunately, the equations governing such a motion are nonlinear simultaneous differential equations whose solutions may require the use of numerical methods. In studying gyroscopic motion, we focus our attention mainly on problems related to steady precession, which is a special physical phenomenon exhibited by some dynamical systems.

REVIEW PROBLEMS

19.66   At the instant considered, the disk C rotates relative to the arm at the rate $\omega_1 = 5\mathbf{I}$ rad/s and $\alpha_1 = 2\mathbf{I}$ rad/s^2, while the arm rotates relative to the ground at the rate $\omega_2 = 4\mathbf{J}$ rad/s and $\alpha_2 = 3\mathbf{J}$ rad/s^2. Determine (a) the angular velocity and angular acceleration of the disk C, (b) the velocity and acceleration of the point P at the top of the disk C.

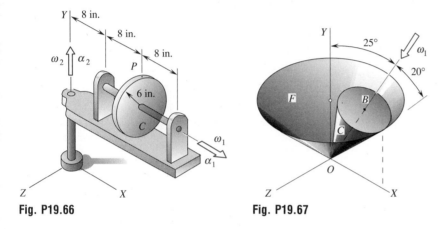

Fig. P19.66   Fig. P19.67

19.67   The right circular cone C rolls at a constant rate on the inside of the fixed right circular conic cavity F without slipping as shown. If the cone C makes one complete trip around the conic cavity F every 5 s, determine the angular velocity and angular acceleration of the cone C.

19.68*   Two slender rods, each of length L and mass m, are welded together to form a T bar. The T bar is attached to the pin of a clevis which rotates with the vertical shaft at a constant angular velocity ω as shown. Determine (a) the angle θ that the plane of the T bar forms with the vertical, (b) the maximum value of ω for which the T bar will remain vertical.

19.69   A thin disk of mass 45 kg and radius 300 mm is mounted on a horizontal shaft with which the disk makes an angle of 60° as shown. If the shaft rotates at the constant rate $\omega = 10\mathbf{I}$ rad/s, determine the moment exerted at G by the disk on the shaft due to the wobble of the disk.

Fig. P19.68*

Fig. P19.69

19.70 Four slender rods, each weighing 3 lb, are welded to a horizontal shaft as shown. If a torque $\mathbf{M}_O = 15\mathbf{K}$ lb·ft acts on the shaft and the shaft is rotating at $\boldsymbol{\omega} = 90\mathbf{K}$ rpm in the position shown, determine the angular acceleration of the shaft and the dynamic reactions from the bearings at O and A.

Fig. P19.70

Fig. P19.71*

Fig. P19.72*

19.71* The 12-lb disk D rotates about the pin at A at the constant rate $\omega_1 = 6\mathbf{I}$ rad/s relative to the arm F. At the instant shown, the arm F rotates at the rate $\omega_2 = 4\mathbf{J}$ rad/s and $\alpha_2 = -2\mathbf{J}$ rad/s^2 relative to the ground. Determine the moment exerted by the disk on the pin at A.

19.72* A slender rod weighing 2 lb/ft is bent to form a shaft as shown. If the shaft rotates at a constant rate of $\boldsymbol{\omega} = 10\mathbf{K}$ rad/s, determine the dynamic reactions from the supports at O and A.

19.73 A cylinder is connected by a wire OA to a support as shown, where $h > 2r$, the wire has a steady precession about the vertical at the rate $\dot{\phi}$, and the cylinder spins at the constant rate $\dot{\psi}$ about its axis of symmetry. Derive the expression for the angle θ which the cylinder makes with the vertical if $\gamma = \theta$.

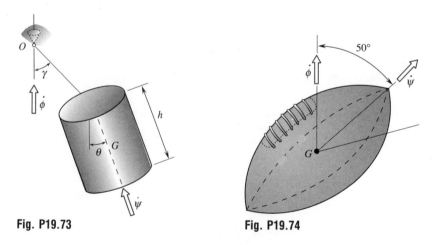

Fig. P19.73

Fig. P19.74

19.74 The football shown spins at the rate $\dot{\psi} = 6$ rad/s and has a steady precession about the vertical. If air resistance is negligible and the ratio of the axial to transverse moments of inertia is $I/I' = 1/3$, determine (a) the rate of precession $\dot{\phi}$, (b) the angular velocity $\boldsymbol{\omega}$ of the football.

Vibrations

Chapter 20

KEY CONCEPTS

- Free and forced vibrations, damping, undamped and damped vibrations, amplitude, and simple harmonic motion.
- Natural frequency, natural circular frequency, natural cyclic frequency, and period of vibration.
- Viscous damper, dashpot, damped natural frequency, and damped period of vibration.
- Overdamping, critical damping, underdamping, damping factor, and logarithmic decrement.
- Forcing frequency, steady-state amplitude, phase angle, magnification factor, resonance, and transmissibility.

The study of vibratory motion of a body about a certain position of equilibrium is an important and special area of dynamics. Examples include studies of the response of a structure to earthquake, the vibration of a rotating machine, the flutter of aircraft wings, and the vibration of a musical instrument. We can see that the analysis of vibrations is a very extensive and well-developed subject. In this chapter, we limit our study to mainly the fundamentals of vibrations of single-degree-of-freedom systems of particles and rigid bodies. A free vibration is an oscillatory motion without the action of a recurrent excitation. Free vibrations are studied in the first part of the chapter using the method of force and acceleration as well as the energy method. A forced vibration is an oscillatory motion under the action of a recurrent excitation; forced vibrations are studied in the latter part of the chapter.

FREE VIBRATIONS

*20.1   Undamped Free Vibrations

The *vibration* of a body is an oscillatory motion of the body about a certain position of equilibrium. The swinging of a pendulum and the flutter of the wings of an aircraft are common examples of vibration. The subject of vibration deals with the study of oscillatory motions of bodies and the forces associated with them.

Vibration can be dichotomized in several ways; e.g., free and forced vibrations, undamped and damped vibrations, linear and nonlinear vibrations, and deterministic and random vibrations. The *free vibration* of a body is an oscillatory motion, caused by an initial disturbance to the body, without the action of a recurrent excitation. If a recurrent excitation acts on a body, the resulting vibration is called a *forced vibration*. The phenomenon of gradual loss or dissipation of mechanical energy due to friction or rolling resistance during oscillation is known as *damping*. An *undamped vibration* is an oscillatory motion of an idealized system experiencing no damping.[†] If damping in the system is taken into account, the resulting vibration is called a *damped vibration*. A body is said to be in *linear vibration* if the differential equation governing its oscillation is linear; otherwise, it is in *nonlinear vibration*. If a vibratory system is subjected to an excitation which is known at any given time, the resulting vibration is called a *deterministic vibration*. If the excitation is random or nondeterministic, the resulting vibration is called *random vibration*.

Let us first consider the undamped free vibration of a spring-mass system as shown in Fig. 20.1, where a body with mass m is suspended by a linear spring having a modulus k and a free length L. The spring is taken to have negligible mass compared with the mass m. In part (a) of Fig. 20.1, the spring is unloaded. It is seen in part (b) that the lower end of the spring moves from P to O to undergo a static deflection δ_{st} when it is loaded to hold the body in equilibrium. Clearly, the spring force of magnitude $k\delta_{st}$ is balanced by the weight force of magnitude mg; i.e.,

$$mg = k\delta_{st} \tag{20.1}$$

The coordinate x defining the position of the displaced body as measured from its equilibrium position is called the *position coordinate* of the body.

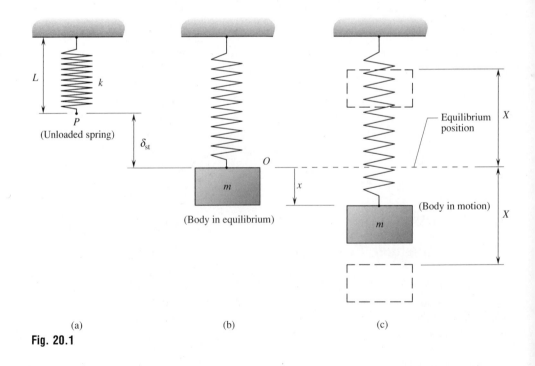

(a)   (b)   (c)

Fig. 20.1

[†]Unless otherwise indicated, damping is generally assumed to be negligible in the study.

If the body is displaced to the position $x = X$ and then released with no initial velocity, the body will move back and forth through the equilibrium position O. This generates a vibration of the body with a maximum displacement X from O as shown in part (c) of Fig. 20.1. The maximum displacement of a vibrating body from its equilibrium position is called the *amplitude*. Of course, the vibration can also be generated by imparting the body an initial velocity beside an initial displacement.

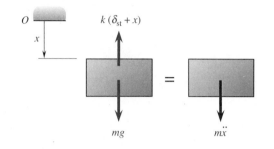

Fig. 20.2

The vibratory behavior of the body can be predicted by examining the dynamic condition of the body at a general time t in a general position x as shown in Fig. 20.2. Note that the unknown effective force is mathematically assumed to point in the positive direction of the position coordinate x. Referring to this figure (or applying Newton's second law), we write

$$+\downarrow \Sigma F: \qquad mg - k(\delta_{st} + x) = m\ddot{x}$$

Substituting Eq. (20.1) into this equation and rearranging, we get

$$m\ddot{x} + kx = 0 \qquad (20.2)$$

which is the differential equation of motion of the body. Setting

$$\frac{k}{m} = \omega_n^2 \qquad \text{or} \qquad \omega_n = \sqrt{k/m} \qquad (20.3)$$

we can write Eq. (20.2) in the form

$$\ddot{x} + \omega_n^2 x = 0 \qquad (20.4)$$

The physical significance of ω_n as defined will be clarified shortly.

We note that Eq. (20.4) is similar to Eq. (12.42); therefore, we can similarly assume that the general solution of Eq. (20.4) is

$$x = A \cos\omega_n t + B \sin\omega_n t \qquad (20.5)$$

where A and B are two arbitrary constants to be determined from the initial conditions of the body. Taking the time derivative of Eq. (20.5), we write

$$\dot{x} = -A\omega_n \sin\omega_n t + B\omega_n \cos\omega_n t \qquad (20.6)$$

Suppose that the body is situated at the initial position x_0 and has an initial velocity $\dot{x}_0$ at the time $t = 0$. Imposing these two initial conditions on Eqs. (20.5) and (20.6), we find that the values of the arbitrary constants are

$$A = x_0 \qquad (20.7)$$

$$B = \dot{x}_0/\omega_n \qquad (20.8)$$

Substituting these values of A and B into Eq. (20.5), we get

$$x = x_0 \cos \omega_n t + (\dot{x}_0/\omega_n) \sin \omega_n t \qquad (20.9)$$

which is the general solution expressed in terms of the initial conditions. Moreover, it can be verified that Eq. (20.9) is equivalent to the equation

$$x = X \sin(\omega_n t + \phi) \qquad (20.10)$$

where X is the *amplitude* and ϕ is the *phase angle*. Their values are

$$X = \sqrt{x_0^2 + (\dot{x}_0/\omega_n)^2} \qquad (20.11)$$

$$\phi = \tan^{-1}(x_0 \omega_n/\dot{x}_0) \qquad (20.12)$$

The parameters X, ϕ, A, B, x_0, $\dot{x}_0$, and ω_n may geometrically be related to each other by a right triangle as shown in Fig. 20.3. If a pen is attached to the lower end of the spring, which suspends the body, and a strip of paper can be moved to the left at a constant speed v near the body to continuously record the position of the body, a sinusoidal curve given by Eq. (20.10) will be generated on the paper as shown in Fig. 20.4.

Fig. 20.3

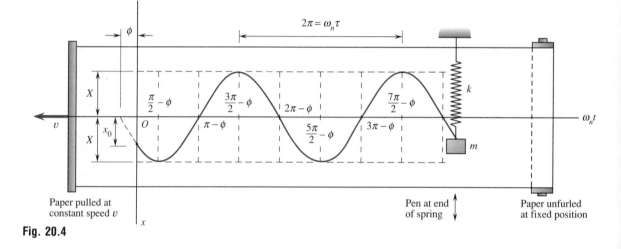

Fig. 20.4

For an additional geometrical representation, let a radius vector with a length equal to the amplitude X be drawn from the equilibrium position O of the body as shown in Fig. 20.5(a). If this radius vector is made to rotate clockwise about O at a constant angular velocity ω_n, the projection of this radius vector onto a vertical axis becomes identically equal to the position coordinate x of the body as given by Eq. (20.10). In Fig. 20.5(a), the magnitudes of the component vectors **A** and **B** are given by Eqs. (20.7) and (20.8). Clearly, the projection of the *tip* of the radius vector **X** onto a vertical axis corresponds to the position of the vibrating body. When this projection of the tip is plotted against the time t or the $\omega_n t$ axis, a sinusoidal curve is again generated as shown in Fig. 20.5(b).

The undamped free vibration described by Eq. (20.9) or (20.10) is often called *simple harmonic motion* and is characterized by the fact that the ac-

(a)   (b)

Fig. 20.5

celeration of the body is proportional to its displacement and of opposite direction; i.e., $\ddot{x} = -\omega_n^2 x$. It is readily seen from Figs. 20.4 and 20.5 that the *phase angle* ϕ represents the amount by which the sinusoidal curve is displaced from the origin on the $\omega_n t$ axis. The constant angular velocity ω_n with which the radius vector $\mathbf{X}$ rotates about the equilibrium position O is called the *natural circular frequency*. Usually, ω_n is expressed in radians per second (rad/s). The time required for the body to complete one cycle of vibration is denoted by τ and is called the *period* of vibration. We see from Figs. 20.4 and 20.5 that

$$2\pi = \omega_n \tau \tag{20.13}$$

$$\tau = \frac{2\pi}{\omega_n} \tag{20.14}$$

The number of cycles of vibration per unit of time is denoted by f_n and is called the *natural cyclic frequency*. We write

$$f_n = \frac{1}{\tau} = \frac{\omega_n}{2\pi} \tag{20.15}$$

Since ω_n is proportional to f_n, the words *circular* and *cyclic* in the names for ω_n and f_n are often *omitted* for simplicity. In other words, the term *natural frequency* in vibrations may refer to either ω_n or f_n as appropriate. The period τ is usually expressed in seconds (s). However, the units for the natural frequency f_n are given the special name *hertz* (Hz), after the German physicist Heinrich Rudolf Hertz (1857–1894). Note that 1 Hz = 1 cycle per second.[†]

[†]In 1965, the Institute of Electrical and Electronics Engineers (IEEE) adopted the unit hertz (Hz), which is now also used in vibration studies.

EXAMPLE 20.1

A pendulum used in impact tests is shown, where G is its mass center and $\bar{r} = 0.8$ m. If the period of small-amplitude vibration of the pendulum about the pivot O is $\tau = 2$ s, determine the radius of gyration $\bar{k}_G$ of the pendulum about G.

Solution. Since no damping is stated, the pendulum is taken to have undamped free vibration. Denoting the mass of the pendulum by m, we draw the free-body diagram and the effective-force diagram of the pendulum in a general angular position θ as shown, where $\dot{\theta}$ and $\ddot{\theta}$ are the angular velocity and angular acceleration of the pendulum, respectively.

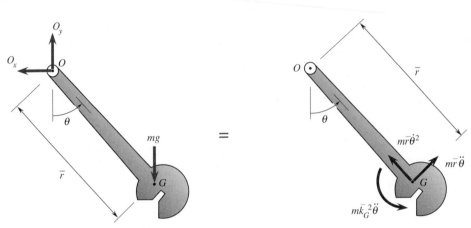

We write

$$+\circlearrowleft \; \Sigma M_O: \qquad -(\bar{r}\,\sin\theta)mg = \bar{r}(m\bar{r}\ddot{\theta}) + m\bar{k}_G^2\ddot{\theta}$$
$$(\bar{r}^2 + \bar{k}_G^2)\ddot{\theta} + \bar{r}g\,\sin\theta = 0$$

which is a nonlinear differential equation of motion. This means that the swinging of a pendulum is inherently a *nonlinear vibration*. However, $\sin\theta \approx \theta$ in small-amplitude vibration and the equation may be linearized. We write

$$(\bar{r}^2 + \bar{k}_G^2)\ddot{\theta} + \bar{r}g\theta = 0$$
$$\ddot{\theta} + \frac{\bar{r}g}{\bar{r}^2 + \bar{k}_G^2}\,\theta = 0 \qquad (1)$$

Comparing Eq. (1) with Eq. (20.4), we readily see that the natural frequency ω_n of the pendulum is given by

$$\omega_n^2 = \frac{\bar{r}g}{\bar{r}^2 + \bar{k}_G^2} \qquad (2)$$

By Eq. (20.14), we have

$$\omega_n^2 = \frac{4\pi^2}{\tau^2} \qquad (3)$$

For $\tau = 2$ s, $\bar{r} = 0.8$ m, and $g = 9.81$ m/s^2, we write, from Eqs. (2) and (3),

$$\frac{4\pi^2}{\tau^2} = \frac{\bar{r}g}{\bar{r}^2 + \bar{k}_G^2}$$

$$\frac{4\pi^2}{(2)^2} = \frac{0.8(9.81)}{(0.8)^2 + \bar{k}_G^2}$$

$$\bar{k}_G = 0.394 \text{ m} \qquad \bar{k}_G = 394 \text{ mm} \quad \blacktriangleleft$$

Developmental Exercises

D20.1 Refer to Example 20.1. (a) Determine the natural circular frequency ω_n and the natural cyclic frequency f_n of the pendulum. (b) Does the term *natural frequency* in vibrations refer to either ω_n or f_n as appropriate?

D20.2 A slender rod OA oscillates about the pivot at O as shown. If $L = 1$ m, determine the natural frequency ω_n and the period τ of the rod.

D20.3 Solve D20.2 if $L = 1$ ft.

Fig. D20.2

EXAMPLE 20.2

A circular cylinder of weight $W = 40$ lb and radius $r = 6$ in. is suspended from a slender rod OA as shown. The cylinder is rotated, thus twisting the rod, through 90° and then released. If the observed period of torsional vibration is $\tau = 1.5$ s, determine (a) the torsional spring modulus K of the rod, (b) the maximum angular velocity reached by the cylinder.

Solution. Let θ denote the angular displacement of the cylinder from its equilibrium position. We note that the weight force of the cylinder is balanced by the tensile force in the rod all the time. The unbalanced action on the cylinder during its torsional vibration is the restoring moment **M** exerted

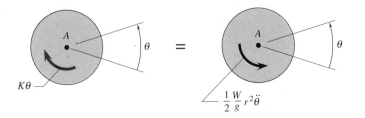

by the rod on the cylinder, where $M = K\theta$. Since the mass center of the cylinder does not move, the effective force on the cylinder is zero. However, the effective moment on the cylinder is $\bar{I}\alpha = \frac{1}{2}(W/g)r^2\ddot{\theta}$. Thus, the free-body diagram and the effective-force diagram of the cylinder may be drawn as shown. We write

$$+ \circlearrowleft \ \Sigma M_A: \qquad -K\theta = \frac{1}{2}\frac{W}{g}r^2\ddot{\theta}$$

$$\ddot{\theta} + \frac{2gK}{Wr^2}\theta = 0 \tag{1}$$

Comparing Eq. (1) with Eq. (20.4), we readily see that the natural frequency ω_n of the cylinder is given by

$$\omega_n^2 = \frac{2gK}{Wr^2} \tag{2}$$

By Eq. (20.14), we have

$$\omega_n^2 = \frac{4\pi^2}{\tau^2} \tag{3}$$

For $\tau = 1.5$ s, $r = 6$ in. $= 0.5$ ft, $W = 40$ lb, and $g = 32.2$ ft/s^2, we write, from Eqs. (2) and (3),

$$\frac{4\pi^2}{\tau^2} = \frac{2gK}{Wr^2}$$

$$\frac{4\pi^2}{(1.5)^2} = \frac{2(32.2)K}{40(0.5)^2}$$

$$K = 2.72 \qquad K = 2.72 \text{ lb} \cdot \text{ft/rad} \ \blacktriangleleft$$

According to Eqs. (20.10) through (20.12), the general solution of Eq. (1) is given by

$$\theta = \Theta \sin(\omega_n t + \phi) \tag{4}$$

$$\Theta = \sqrt{\theta_0^2 + (\dot{\theta}_0/\omega_n)^2} \tag{5}$$

$$\phi = \tan^{-1}(\theta_0\omega_n/\dot{\theta}_0) \tag{6}$$

where θ_0 and $\dot{\theta}_0$ are the initial angular position and initial angular velocity of the cylinder. For the cylinder as given, we have $\theta_0 = 90° = \pi/2$ rad, $\dot{\theta}_0 = 0$, and $\omega_n = 2\pi/\tau = 2\pi/1.5$ rad/s. Taking the time derivative of Eq. (4), we write

$$\dot{\theta} = \Theta\omega_n \cos(\omega_n t + \phi)$$

Thus, the maximum angular velocity reached by the cylinder is

$$\dot{\theta}_{max} = \Theta\omega_n = \sqrt{(\pi/2)^2 + 0}\ (2\pi/1.5) = 6.58$$

$$\dot{\theta}_{max} = 6.58 \text{ rad/s} \ \blacktriangleleft$$

Developmental Exercise

D20.4 Refer to Example 20.2. Suppose that the vibration is generated by imparting the cylinder an initial angular velocity of $\dot{\theta}_0 = 0.5$ rad/s beside the initial angular displacement of $\theta_0 = 90°$. Determine the maximum angular velocity reached by the cylinder.

EXAMPLE 20.3

Suppose that the cylinder of weight $W = 40$ lb and radius $r = 6$ in. in Example 20.2 is detached from the rod and placed on a cylindrical surface of radius $R = 2$ ft as shown. If the cylinder rolls without slipping on the surface, determine its period τ of small-amplitude vibration.

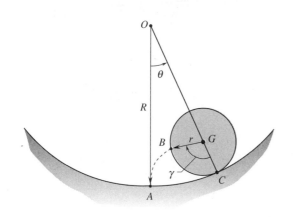

Solution. When the cylinder rolls without slipping, the point B (of the cylinder) which was coincident with the point A (of the surface) must have moved in such a way that the arcs of contact AC and CB are equal; i.e., $R\theta = r\gamma$. However, the angular displacement (or the change of orientation) of the cylinder in the inertial reference frame (or with respect to the ground) is β, where $\gamma = \theta + \beta$. We write

$$R\theta = r(\theta + \beta) \qquad \beta = \frac{1}{r}(R - r)\theta$$

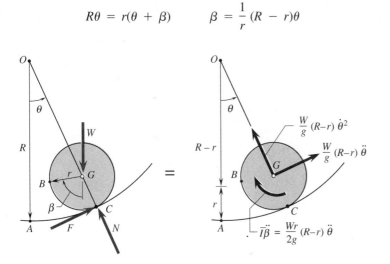

The angular acceleration of the cylinder in the inertial reference frame is $\ddot{\beta}$, not $\ddot{\gamma}$. The effective moment on the cylinder is given by

$$\bar{I}\ddot{\beta} = \frac{1}{2}\frac{W}{g}r^2 \cdot \frac{1}{r}(R - r)\ddot{\theta} = \frac{Wr}{2g}(R - r)\ddot{\theta}$$

Moreover, the mass center G of the cylinder is constrained to describe a circular path of radius $R - r$ about the center O. Its tangential and normal components of acceleration are

$$a_t = (R - r)\ddot{\theta} \qquad a_n = (R - r)\dot{\theta}^2$$

Therefore, the free-body diagram and the effective-force diagram of the cylinder may be drawn as shown. We write

$$+\,\circlearrowleft\; \Sigma M_C: \qquad -\,(r\sin\theta)W = r\left[\frac{W}{g}(R - r)\ddot{\theta}\right] + \frac{Wr}{2g}(R - r)\ddot{\theta}$$

$$\frac{3(R - r)}{2g}\ddot{\theta} + \sin\theta = 0$$

In small-amplitude vibration, we have $\sin\theta \approx \theta$. Thus, we write

$$\ddot{\theta} + \frac{2g}{3(R - r)}\theta = 0$$

which, as before, indicates that the natural frequency ω_n is given by

$$\omega_n^2 = \frac{2g}{3(R - r)}$$

For the given system, we have $R = 2$ ft, $r = 6$ in. $= 0.5$ ft, and $g = 32.2$ ft/s². The period τ of small-amplitude vibration of the cylinder is

$$\tau = \frac{2\pi}{\omega_n} = 2\pi\sqrt{\frac{3(R - r)}{2g}} = 2\pi\sqrt{\frac{3(2 - 0.5)}{2(32.2)}}$$

$$\tau = 1.661 \text{ s} \quad \blacktriangleleft$$

REMARK. An alternative solution for the problem in this example is presented in Example 20.4.

Developmental Exercise

D20.5   Work Example 20.3 if the cylinder is replaced by a sphere of the same radius and the same weight.

★20.2   Energy Method

We note in plane motion that the kinetic energy T of a rigid body of mass m is given by

$$T = \tfrac{1}{2}m\bar{v}^2 + \tfrac{1}{2}\bar{I}\omega^2 \tag{17.13}$$

(repeated)

or

$$T = \tfrac{1}{2}I_C\omega^2 \tag{17.14}$$

(repeated)

where $\bar{v}$ is the speed of the mass center, ω is the angular speed, $\bar{I}$ is the central mass moment of inertia, and I_C is the mass moment of inertia about the velocity center C of the rigid body. For a particle of mass m moving with a speed v, its kinetic energy T is simply given by

$$T = \tfrac{1}{2} mv^2 \qquad (13.22)$$
$$\text{(repeated)}$$

We know from Sec. 13.8 that the potential energy of a body in a conservative force field relative to an appropriate reference datum is defined as *the amount of work which the body will receive from the conservative force if the body travels from its present position to the reference datum.* Thus, the *gravitational potential energy* V_g of a body with a weight W near the surface of the earth at a position of altitude y above the reference datum is given by

$$V_g = Wy \qquad (13.32)$$
$$\text{(repeated)}$$

If the reference datum in an elastic force field is chosen to be the position of the body where the spring attached to the body is neutral, the *elastic potential energy* V_e of the body is given by

$$V_e = \tfrac{1}{2} kx^2 \qquad (13.36)$$
$$\text{(repeated)}$$

where k is the modulus and x is the change in length of the spring.

Note that the *total mechanical energy* E_{total} of a body is the sum of the kinetic energy T and the potential energy V of the body. In a conservative system, as defined in Sec. 13.8, the total mechanical energy of the body remains constant during its motion from one position to another; i.e.,

$$E_{\text{total}} = T + V = \text{constant} \qquad (20.16)$$

This property is referred to as the *principle of conservation of energy* in Secs. 13.12 and 17.5. The *energy method* is based on the principle of conservation of energy.

In Eq. (20.16), observe that the kinetic energy T is nonnegative, but, depending on the choice of reference datum, the potential energy V can be positive, zero, or negative during vibration; i.e., the minimum value of T is zero, but *the minimum value of V may be less than zero.* Thus, for a conservative vibratory system with $E_{\text{total}} = T + V$, we have

$$T_{\min} = 0 \qquad (20.17)$$

$$T_{\max} = E_{\text{total}} - V_{\min} \qquad (20.18)$$

$$V_{\max} = E_{\text{total}} \qquad (20.19)$$

$$T_{\max} \neq V_{\max} \quad \text{if} \quad V_{\min} \neq 0 \qquad (20.20)$$

$$T_{\max} = V_{\max} \quad \text{if} \quad V_{\min} = 0 \qquad (20.21)$$

However, the value of E_{total} may vary from one choice of reference datum to another.

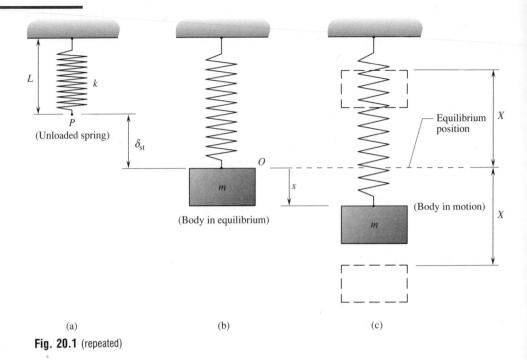

(a) (b) (c)

Fig. 20.1 (repeated)

I. REFERENCE DATUM P.

The spring-mass system in Fig. 20.1 is a conservative system. Let us choose the *unloaded position P* of the lower end of the spring as the *reference datum* for both the gravitational and the elastic potential energies of the body. When the body is displaced to the position x, its kinetic energy T and potential energy V are, by definition,

$$T = \tfrac{1}{2} m\dot{x}^2 \qquad (20.22)$$
$$V = V_g + V_e = -mg(\delta_{st} + x) + \tfrac{1}{2} k(\delta_{st} + x)^2$$

Since $mg = k\delta_{st}$ from Eq. (20.1), we may simplify V to the form

$$V = \tfrac{1}{2} k(x^2 - \delta_{st}^2) \qquad (20.23)$$

Clearly, it demonstrates that $V < 0$ whenever $x^2 < \delta_{st}^2$. By the energy method, the total mechanical energy E_{total} is

$$E_{total} = T + V = \tfrac{1}{2} m\dot{x}^2 + \tfrac{1}{2} k(x^2 - \delta_{st}^2)$$
$$= \text{constant}$$

Taking the time derivative to eliminate the constant, we write

$$\tfrac{1}{2} m(2\dot{x})\ddot{x} + \tfrac{1}{2} k(2x)\dot{x} = 0$$

$$\dot{x}(m\ddot{x} + kx) = 0 \qquad (20.24)$$

Since the velocity $\dot{x}$ of the body is not *always* zero during vibration, Eq. (20.24) yields the governing equation of motion

$$\boxed{m\ddot{x} + kx = 0} \qquad (20.25)$$

This result agrees with Eq. (20.2).

II. REFERENCE DATUM O. If we choose the *equilibrium position O* as the *reference datum* for both the gravitational and the elastic potential energies of the body in Fig 20.1, we write, by definition,

$$T = \tfrac{1}{2} m\dot{x}^2 \tag{20.26}$$

$$V = V_g + V_e = -mgx + \int_x^0 k(\delta_{st} + x)\,(-dx)$$

$$= -mgx + (k\delta_{st}x + \tfrac{1}{2} kx^2)$$

$$= (-mg + k\delta_{st})x + \tfrac{1}{2} kx^2$$

Substituting Eq. (20.1) into the above expression for V, we get

$$V = \tfrac{1}{2} kx^2 \tag{20.27}$$

This shows that the choice of the equilibrium position O as the reference datum does simplify the final expression of V. By the energy method, the total mechanical energy E_{total} is

$$E_{total} = T + V = \tfrac{1}{2} m\dot{x}^2 + \tfrac{1}{2} kx^2$$
$$= \text{constant}$$

Taking the time derivative to eliminate the constant, we write

$$\tfrac{1}{2} m(2\dot{x})\ddot{x} + \tfrac{1}{2} k(2x)\dot{x} = 0$$

$$\dot{x}(m\ddot{x} + kx) = 0 \tag{20.28}$$

Equation (20.28) agrees with Eq. (20.24) and will similarly yield

$$\boxed{m\ddot{x} + kx = 0} \tag{20.29}$$

We see that different choices of the reference datum for the potential energies in the energy method do not affect the final outcome for the equation of motion of a conservative system.

EXAMPLE 20.4

Using the *energy method*, solve for τ in Example 20.3.

Solution. Since the mass center G of the cylinder is constrained to describe a circular path of radius $R - r$ about the center O, its speed is

$$\bar{v} = (R - r)\dot{\theta}$$

As discussed in the solution in Example 20.3, the angular displacement of the cylinder is

$$\beta = \frac{1}{r}(R - r)\theta$$

Thus, the angular velocity of the cylinder is

$$\dot{\beta} = \frac{1}{r}(R - r)\dot{\theta}$$

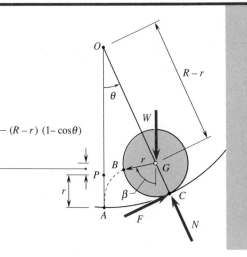

By Eq. (17.13), the kinetic energy T of this cylinder is

$$T = \frac{1}{2} \frac{W}{g} (R - r)^2 \dot{\theta}^2 + \frac{1}{2} \left(\frac{1}{2} \frac{W}{g} r^2 \right) \cdot \frac{1}{r^2} (R - r)^2 \dot{\theta}^2$$

$$= \frac{3}{4} \frac{W}{g} (R - r)^2 \dot{\theta}^2$$

Next, let the position of the mass center at its equilibrium position P, as shown, be chosen as the reference datum for the gravitational potential energy V_g. Noting that the reactions $\mathbf{N}$ and $\mathbf{F}$ from the support are nonworking forces, we find that the potential energy V of the conservative system is simply

$$V = V_g = W(R - r)(1 - \cos \theta)$$

For this conservative system, we set $E_{\text{total}} = T + V = \text{constant}$; i.e.,

$$\frac{3}{4} \frac{W}{g} (R - r)^2 \dot{\theta}^2 + W(R - r)(1 - \cos \theta) = \text{constant}$$

Taking the time derivative to eliminate the constant, we write

$$\frac{3}{4} \frac{W}{g} (R - r)^2 (2 \dot{\theta}) \ddot{\theta} + W(R - r)(\sin \theta) \dot{\theta} = 0$$

$$\frac{3W}{2g} (R - r)^2 \dot{\theta} \left[\ddot{\theta} + \frac{2g}{3(R - r)} \sin \theta \right] = 0$$

Since $\dot{\theta}$ is not always zero during vibration and $\sin \theta \approx \theta$ for small-amplitude vibration, we find that

$$\ddot{\theta} + \frac{2g}{3(R - r)} \theta = 0$$

$$\omega_n^2 = \frac{2g}{3(R - r)} \qquad \tau = \frac{2\pi}{\omega_n}$$

For the given system in Example 20.3, we have $R = 2$ ft, $r = 0.5$ ft, and $g = 32.2$ ft/s^2. Thus, the period is

$$\tau = 2\pi \sqrt{\frac{3(R - r)}{2g}} = 2\pi \sqrt{\frac{3(2 - 0.5)}{2(32.2)}}$$

$$\tau = 1.661 \text{ s} \quad \blacktriangleleft$$

Developmental Exercises

D20.6 Verify that the same result for the kinetic energy T in the solution in Example 20.4 can be obtained by applying Eq. (17.14).

D20.7 Using *energy method*, solve D20.2.

PROBLEMS

20.1 A particle in simple harmonic motion on the x axis has an amplitude of 2 in. and a period of 0.5 s. Determine its maximum velocity $\dot{x}_{\max}$ and maximum acceleration $\ddot{x}_{\max}$.

20.2 The forces in the springs are zero when the block of mass $m = 40$ kg is in equilibrium as shown. If the moduli of the two connecting springs are $k_1 = 640$ N/m and $k_2 = 800$ N/m, determine the natural frequency f_n of vibration of the block.

Fig. P20.2

Fig. P20.3* and P20.4

20.3* Determine the period of vibration of the 100-kg block B which is connected by a spring of modulus 2.5 kN/m as shown.

20.4 The 100-kg block B is connected by a spring of modulus 2.5 kN/m as shown. If the block is given an initial displacement of 20 mm $\searrow$ from its equilibrium position and an initial velocity of 75 mm/s $\searrow$, determine the amplitude and the phase angle of vibration of the block.

20.5 The spring supporting the slender rod AB, as shown, has a modulus of 600 lb/ft. If the period of vibration is 0.5 s, determine the weight of the rod.

Fig. P20.5

Fig. P20.6*

20.6* A 100-lb weight causes a cantilever beam to have a static deflection of 2 in. as shown, where the mass of the cantilever can be taken as negligible. Determine the natural frequency f_n of vibration of the weight.

20.7 Two springs of moduli $k_1 = 2$ kN/m and $k_2 = 3$ kN/m are used to connect a 40-kg block as shown. The block is given an initial displacement of 30 mm $\rightarrow$ from its equilibrium position and released. Determine the period and the maximum velocity of vibration if the springs are arranged (a) in parallel, (b) in series.

(a)

(b)

Fig. P20.7

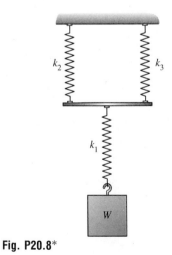

Fig. P20.8*

20.8* Three springs of moduli $k_1 = 20$ lb/in. and $k_2 = k_3 = 40$ lb/in. are arranged to suspend a cylinder of weight $W = 32$ lb as shown. If the amplitude of vibration of the cylinder is 6 in., determine the maximum velocity and maximum acceleration of the cylinder.

20.9 Two slender shafts in series with torsional moduli $K_1 = 30$ N·m/rad and $K_2 = 20$ N·m/rad are used to suspend an automobile wheel as shown. When the wheel is given an initial angular displacement and released, it makes 10 oscillations in 12 s. Determine the moment of inertia of the wheel about its central axis along the shafts.

$K_1 = 30$ N·m/rad

$K_2 = 20$ N·m/rad

θ

Fig. P20.9

Fig. P20.10

20.10 A sphere of radius r and mass m is welded to a slender rod of negligible mass as shown, where k is the modulus of the connecting spring. Derive an expression for the natural frequency ω_n of small-amplitude vibration.

20.11 A homogeneous slender rod is bent into the shape of a square as shown, where $b = 1$ ft. Determine the period of small-amplitude vibration.

20.12* Solve Prob. 20.11 if $b = 1$ m.

20.13 A homogeneous slender rod is bent into the shape of an angle as shown, where $L = 600$ mm. Determine the natural frequency f_n of small-amplitude vibration.

Fig. P20.11

L

$90°$

L

Fig. P20.13

Fig. P20.14

20.14 A homogeneous semicircular disk is suspended as shown. Determine the radius r for which the natural frequency of the disk is 1 Hz.

20.15* Solve Prob. 20.14 if the semicircular disk is replaced by a hemisphere of radius r.

20.16 A connecting rod weighing 10 lb is suspended as shown. If the period of small-amplitude oscillation is observed to be 1.5 s, determine the moment of inertia of the connecting rod about its mass center G.

20.17 A U-tube with uniform cross section contains a column of mercury as shown. Determine the length L of the mercury column for which the period of vibration is 2 s.

$\bar{r} = 10$ in.

G

Fig. P20.16

Fig. P20.17

Fig. P20.18

20.18   A spring of modulus k supports a disk A of mass m_A in equilibrium as shown. If another disk B of mass m_B is dropped from a height h and the impact is perfectly plastic, determine in terms of parameters of the system the subsequent motion of the disk since impact as defined by the position coordinate x which is measured from the original equilibrium position of the disk A.

20.19   An equilateral triangular plate ABC of mass m is suspended by three vertical wires as shown. If the plate is slightly rotated about its vertical central axis and released, determine the length L of the wires for which the period of vibration is 2 s.

Fig. P20.19

Fig. P20.20*

20.20*   A cylinder of mass m_1 and a block of mass m_2 are connected as shown, where k is the modulus of the connecting spring. Neglecting masses of the pulleys, derive an expression for the period of vibration of the system.

20.21   A cylinder of mass m and a double pulley are connected as shown, where k is the modulus of the connecting spring. If the moment of inertia of the double pulley about its axis of support is $\bar{I}$, derive an expression for the natural frequency f_n of vibration of the system.

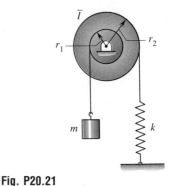

Fig. P20.21

*20.3   Damped Free Vibrations

In reality, every vibratory mechanical system experiences some degree of kinetic friction and suffers some dissipation of mechanical energy. The kinetic friction forces are the *damping forces*. These forces may be caused by *dry friction* between sliding surfaces, by *viscous friction* when a body moves

(a)
(b)

Fig. 20.6

on a lubricated surface or in a fluid, or by *internal friction* between molecules or elements of a deformable body.

Although the precise mechanism of damping is rather complex, the effects of damping on many vibratory systems can be simulated by the effect of a *viscous damper*, which is often referred to as a *dashpot*. A dashpot consists of a cylinder filled with a viscous fluid and a plunger with some tolerance or holes for the fluid to flow from one side to the other. A simple dashpot is illustrated in Fig. 20.6(a), which is usually schematically represented as shown in Fig. 20.6(b). The magnitude F_d of the damping force is proportional to the speed $\dot{x}$ of the plunger relative to the cylinder. We write

$$F_d = c\dot{x} \tag{20.30}$$

where the constant of proportionality c is known as the *coefficient of viscous damping* and has units of N·s/m or lb·s/ft. When a dashpot is connected to a body, the damping force exerted by the dashpot on the body is always directed opposite to the velocity of the body relative to the dashpot. In application, the mass of a dashpot is usually assumed to be negligible compared with the mass of the body to which it is connected.

To be specific, let us consider the damped free vibration of a body with mass m, which is suspended by a spring of modulus k and a simple dashpot having a coefficient of viscous damping c as shown in Fig. 20.7. In part (a) of Fig. 20.7, the spring and the dashpot are unloaded.[†] It is seen in part (b) that the lower ends of the spring and the dashpot move from P to O to undergo a static deflection δ_{st} when they are loaded to hold the body in equilibrium. Since the dashpot exerts no force on a body in equilibrium, the weight force of magnitude mg is balanced solely by the spring force of magnitude $k\delta_{st}$; i.e.,

$$mg = k\delta_{st} \tag{20.31}$$

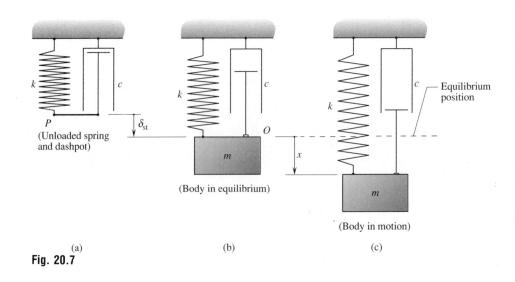

k c

P
(Unloaded spring
and dashpot)

δ_{st}

k c

O

m

(Body in equilibrium)

c — Equilibrium position

k

x

m

(Body in motion)

(a)
(b)
(c)

Fig. 20.7

[†]A dissipative restoring mechanism consisting of a linear spring in parallel with a simple dashpot, as shown in Fig. 20.7(a), is known as a *Voigt model* in viscoelasticity.

which is identical with Eq. (20.1). As before, we let the coordinate x defining the position of the displaced body be measured from its equilibrium position O, as shown in Fig. 20.7(b) and (c).

Suppose that the body is set in motion by certain initial conditions. The dynamic behavior of the body can be predicted by examining the condition of the body at a general time t in a general position x as shown in Fig. 20.8. Referring to this figure (or applying Newton's second law), we write

$$+\downarrow \Sigma F: \qquad mg - k(\delta_{st} + x) - c\dot{x} = m\ddot{x}$$

Substituting Eq. (20.31) into this equation and rearranging, we get

$$m\ddot{x} + c\dot{x} + kx = 0 \qquad (20.32)$$

Fig. 20.8

which is the differential equation of motion of the body. Using Eq. (20.3) and setting

$$\frac{c}{m} = 2\zeta\omega_n \qquad \text{or} \qquad \zeta = \frac{c}{2\sqrt{km}} \qquad (20.33)$$

we can write Eq. (20.32) in the form

$$\ddot{x} + 2\zeta\omega_n\dot{x} + \omega_n^2 x = 0 \qquad (20.34)$$

The parameter ζ, as defined in Eq. (20.33), is called the *damping factor* or *damping ratio*. It can be shown that ζ is nondimensional and is a measure of the severity of damping.

Substituting a solution of the form $x = Ae^{\lambda t}$ into Eq. (20.34) and recognizing that $Ae^{\lambda t} \neq 0$, we obtain the *characteristic equation*

$$\lambda^2 + 2\zeta\omega_n\lambda + \omega_n^2 = 0 \qquad (20.35)$$

whose roots are

$$\lambda_{1,2} = \left(-\zeta \pm \sqrt{\zeta^2 - 1}\right)\omega_n \qquad (20.36)$$

By superposition, the general solution for Eq. (20.34) is

$$x = A_1 e^{\lambda_1 t} + A_2 e^{\lambda_2 t} \qquad (20.37)$$

where A_1 and A_2 are arbitrary constants to be determined from the initial conditions $x(0) = x_0$ and $\dot{x}(0) = \dot{x}_0$. The characteristics of the damped motion are highly dependent on whether the radicand $\zeta^2 - 1$ in Eqs. (20.36) is positive, zero, or negative.

1. CASE OF $\zeta > 1$ (*overdamping*). Both roots λ_1 and λ_2 in Eqs. (20.36) are negative and have distinct values. Thus, Eq. (20.37) may be written as

$$x = A_1 e^{-(\zeta + \sqrt{\zeta^2 - 1})\omega_n t} + A_2 e^{-(\zeta - \sqrt{\zeta^2 - 1})\omega_n t} \qquad (20.38)$$

It can be shown by imposing the initial conditions on Eq. (20.38) that

$$A_1 = \frac{\dot{x}_0 + \left(\zeta + \sqrt{\zeta^2 - 1}\right)\omega_n x_0}{2\omega_n\sqrt{\zeta^2 - 1}} \qquad (20.39)$$

$$A_2 = -\frac{\dot{x}_0 + \left(\zeta - \sqrt{\zeta^2 - 1}\right)\omega_n x_0}{2\omega_n\sqrt{\zeta^2 - 1}} \qquad (20.40)$$

The motion as given by Eq. (20.38) diminishes exponentially as the time t increases. The motion is nonvibratory regardless of the initial conditions imposed on the system and is referred to as *aperiodic*.

2. CASE OF $\zeta = 1$ (*critical damping*).

Both roots λ_1 and λ_2 in Eqs. (20.36) become equal to $-\omega_n$. The general solution for the special case of equal roots is given by

$$x = (A_1 + A_2 t)e^{-\omega_n t} \tag{20.41}$$

Imposition of initial conditions on Eq. (20.41) yields

$$A_1 = x_0 \tag{20.42}$$

$$A_2 = \dot{x}_0 + \omega_n x_0 \tag{20.43}$$

Again, the motion as given by Eq. (20.41) is *aperiodic* and diminishes exponentially as the time t increases. Nevertheless, a critically damped system regains its equilibrium position faster than an otherwise overdamped system, as illustrated in Fig. 20.9. Note that, for certain initial conditions (e.g., $x_0 > 0$ and $\dot{x}_0 < 0$), the body in the damped system with $\zeta \geq 1$ may cross the equilibrium position at most once.

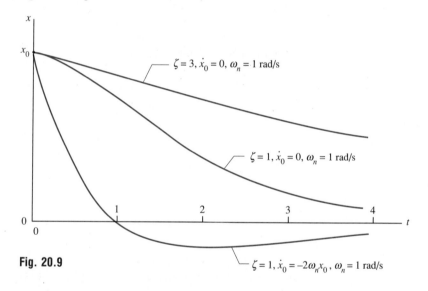

Fig. 20.9

3. CASE OF $\zeta < 1$ (*underdamping*).

The radicand $\zeta^2 - 1$ in Eqs. (20.36) is negative, and Eq. (20.37) may be written as

$$x = e^{-\zeta \omega_n t}(A_1 e^{i\omega_d t} + A_2 e^{-i\omega_d t}) \tag{20.44}$$

where $i = \sqrt{-1}$ and

$$\omega_d = \sqrt{1 - \zeta^2}\,\omega_n \tag{20.45}$$

Applying Euler's formula $e^{\pm i\theta} = \cos\theta \pm i\sin\theta$, we can rewrite Eq. (20.44) as

$$x = e^{-\zeta \omega_n t}(C_1 \cos\omega_d t + C_2 \sin\omega_d t) \tag{20.46}$$

where $C_1 = A_1 + A_2$ and $C_2 = i(A_1 - A_2)$. By imposing the initial conditions on Eq. (20.46), we get

$$C_1 = x_0 \qquad C_2 = (\dot{x}_0 + \zeta\omega_n x_0)/\omega_d \qquad (20.47)$$

It can be verified that Eq. (20.46) is equivalent to the equation

$$x = X\, e^{-\zeta\omega_n t} \sin(\omega_d t + \phi) \qquad (20.48)$$

where

$$X = \sqrt{x_0^2 + [(\dot{x}_0 + \zeta\omega_n x_0)/\omega_d]^2} \qquad (20.49)$$

$$\phi = \tan^{-1}[x_0\omega_d/(\dot{x}_0 + \zeta\omega_n x_0)] \qquad (20.50)$$

Fig. 20.10

We note that Eqs. (20.48) through (20.50) degenerate into Eqs. (20.10) through (20.12), respectively, if the damping factor $\zeta = 0$. Unlike preceding cases of $\zeta \geq 1$, the motion as given by Eq. (20.48) is *vibratory* with exponentially decaying amplitudes as shown in Fig. 20.10. The dashed curves defined by

$$x = \pm X\, e^{-\zeta\omega_n t}$$

are the envelopes of the motion curve. Although this motion does not exactly repeat itself, the parameter ω_d defined in Eq. (20.45) is called the *damped natural frequency* and the time

$$\tau_d = \frac{2\pi}{\omega_d} = \frac{2\pi}{\sqrt{1 - \zeta^2}\,\omega_n} \qquad (20.51)$$

is called the *damped period*.

Developmental Exercises

D20.8   What is the value or range of values of the *damping factor* ζ if a system is (a) critically damped, (b) overdamped, (c) underdamped?

D20.9   If the system in Fig. 20.7 is critically damped, show that the coefficient of viscous damping has the value

$$c = c_c = 2\sqrt{km}$$

★20.4 Logarithmic Decrement

We saw in Sec. 20.3 that a damped system may have free oscillations only when it is *underdamped* (i.e., $\zeta < 1$), and its amplitudes decay exponentially as shown in Fig. 20.10. The larger the damping, the greater will be the rate of decay of the amplitudes.

Suppose that x_1 is the first peak value of x occurring at $t = t_1$ and x_2 is the second peak value of x occurring at $t = t_2 = t_1 + \tau_d$ as indicated in Fig. 20.10. These peak values x_1 and x_2 are the amplitudes of the body in the first and second cycles of damped vibration. Referring to Fig. 20.10 and applying Eqs. (20.48) and (20.51), we write

$$x_1 = X e^{-\zeta \omega_n t_1} \sin(\omega_d t_1 + \phi) \tag{20.52}$$

$$
\begin{aligned}
x_2 &= X e^{-\zeta \omega_n (t_1 + \tau_d)} \sin[\omega_d(t_1 + \tau_d) + \phi] \\
&= e^{-\zeta \omega_n \tau_d} \{X e^{-\zeta \omega_n t_1} \sin[2\pi + (\omega_d t_1 + \phi)]\} \\
&= e^{-\zeta \omega_n \tau_d} \{X e^{-\zeta \omega_n t_1} \sin(\omega_d t_1 + \phi)\} \\
&= e^{-\zeta \omega_n \tau_d} \{x_1\}
\end{aligned}
$$

$$x_2 = x_1 e^{-\zeta \omega_n \tau_d} \tag{20.53}$$

By Eqs. (20.52) and (20.53), we write

$$\frac{x_1}{x_2} = e^{\zeta \omega_n \tau_d}$$

which can be shown to be also the ratio of the nth and $(n+1)$th amplitudes x_n and x_{n+1}. The *logarithmic decrement* δ is defined as the natural logarithm of the ratio of any two successive amplitudes of a body in damped vibration; i.e.,

$$\delta = \ln\left(\frac{x_n}{x_{n+1}}\right) \tag{20.54}$$

For $n = 1$, we have

$$\delta = \ln\left(\frac{x_1}{x_2}\right) = \ln(e^{\zeta \omega_n \tau_d}) = \zeta \omega_n \tau_d$$

Applying Eq. (20.51), we write

$$\delta = \frac{2\pi\zeta}{\sqrt{1 - \zeta^2}} \tag{20.55}$$

which can be solved to yield

$$\zeta = \frac{\delta}{\sqrt{4\pi^2 + \delta^2}} \tag{20.56}$$

This means that the damping factor ζ, which reflects the amount of damping in a system, can conveniently be determined by measuring the rate of decay of free oscillations of the damped system. If a system is slightly damped with $\zeta \ll 1$, then $\delta \approx 2\pi\zeta$ or $\zeta \approx \delta/(2\pi)$.

EXAMPLE 20.5

A 10-kg mass is connected by a spring of modulus 160 N/m and a dashpot as shown. It is observed that the ratio of any two successive amplitudes in free vibration is 1:0.6. Determine (a) the logarithmic decrement, (b) the damping factor, (c) the damped natural frequency of the system, (d) the coefficient of viscous damping.

Solution. We know from the given data that $m = 10$ kg, $k = 160$ N/m, and the motion is governed by Eq. (20.32). The logarithmic decrement is

$$\delta = \ln(x_n/x_{n+1}) = \ln(1/0.6) = 0.5108$$

$$\delta = 0.511 \quad \blacktriangleleft$$

By Eq. (20.56), we find that the damping factor is

$$\zeta = \frac{\delta}{\sqrt{4\pi^2 + \delta^2}} = \frac{0.5108}{\sqrt{4\pi^2 + (0.5108)^2}} = 0.08103$$

$$\zeta = 0.0810 \quad \blacktriangleleft$$

The undamped natural frequency is

$$\omega_n = \sqrt{\frac{k}{m}} = \sqrt{\frac{160}{10}} \qquad \omega_n = 4 \text{ rad/s}$$

Therefore, the damped natural frequency is

$$\omega_d = \sqrt{1 - \zeta^2}\, \omega_n = \sqrt{1 - (0.08103)^2}\, (4) = 3.987$$

$$\omega_d = 3.99 \text{ rad/s} \quad \blacktriangleleft$$

By Eq. (20.33), we find that the coefficient of viscous damping is

$$c = 2\zeta\sqrt{km} = 2(0.08103)\sqrt{160(10)}$$

$$c = 6.48 \text{ N·s/m} \quad \blacktriangleleft$$

Developmental Exercises

D20.10 Using Eqs. (20.48) and (20.51), show that the ratio of the amplitudes x_n and x_{n+1} in the nth and $(n+1)$th cycles is

$$\frac{x_n}{x_{n+1}} = e^{\zeta\omega_n\tau_d}$$

D20.11 According to Eq. (20.54), the ratio of any two successive amplitudes must have the same value of e^δ, where δ is the *logarithmic decrement*; i.e.,

$$\frac{x_1}{x_2} = \frac{x_2}{x_3} = \frac{x_3}{x_4} = \cdots = \frac{x_n}{x_{n+1}} = e^\delta$$

Using this fact, show that

$$\delta = \frac{1}{n} \ln\left(\frac{x_1}{x_{n+1}}\right)$$

PROBLEMS

20.22   A 10-kg block is connected by a spring and a dashpot as shown. The block vibrates with a damped natural frequency of 2 Hz. If the dashpot is removed, the block is observed to vibrate with a natural frequency of 4 Hz. Determine (a) the modulus k of the spring, (b) the coefficient of viscous damping c of the dashpot.

Fig. P20.22   **Fig. P20.23 and P20.26**

20.23   The 64.4-lb block in the spring-mass-dashpot system, as shown, is given an initial displacement of $x_0 = 3$ in. at the time $t = 0$ and then released. Determine the damping factor ζ and the position x of the block as a function of t if the dashpot has a coefficient of viscous damping $c = 60$ lb·s/ft.

20.24*   Solve Prob. 20.23 if $c = 40$ lb·s/ft.

20.25   Solve Prob. 20.23 if $c = 20$ lb·s/ft.

20.26   An impulse of 3 lb·s $\rightarrow$ is exerted at the time $t = 0$ on the 64.4-lb block in the spring-mass-dashpot system, which is at rest as shown. If the dashpot has a coefficient of viscous damping $c = 30$ lb·s/ft, determine the position x of the block as a function of t.

20.27*   A critically damped oscillator of natural frequency ω_n is set in motion with the initial conditions $x_0 > 0$ and $\dot{x}_0 > 0$ at the time $t = 0$. Determine the time t_m at which the displacement of the oscillator is maximum.

20.28   The oscillator shown in part (a) has a damped period of $\tau_d = 0.5$ s. If two successive amplitudes a full cycle apart are measured to be 17 mm and 12 mm as shown in part (b), determine the modulus k of the spring and the coefficient of viscous damping c of the dashpot.

(a)   (b)

Fig. P20.28

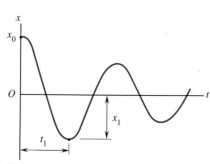

Fig. P20.29

20.29   A spring-mass-dashpot system has a modulus $k = 160$ N/m, a mass $m = 2.5$ kg, and a coefficient of viscous damping $c = 7$ N·s/m. The system is released from rest from an initial position x_0 as indicated. Determine (a) the time t_1 at which the overshoot distance x_1 is greatest, (b) the ratio x_1/x_0.

20.30   The slender rod AB of mass m is supported by a spring of modulus k and a dashpot with coefficient of viscous damping c. Determine the damping factor ζ and the damped period τ_d in terms of the parameters of the system.

20.31*   A spring-mass-dashpot system is arranged as shown. Neglecting the mass of the crank AOB, derive the equation of motion of the block in terms of the parameters of the system.

Fig. P20.30

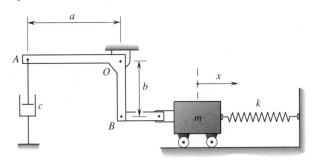

Fig. P20.31*

FORCED VIBRATIONS

*20.5   Harmonic Excitation

We saw in preceding sections that a mechanical system is set in free vibration by a certain initial disturbance. If a recurrent excitation acts on a body, the resulting vibration is a *forced vibration*.

The recurrent excitation may be externally applied or may be generated within the system by a rotating unbalance. Moreover, the recurrent excitation may also be applied to the support or foundation of the system. For fundamental study of forced vibrations, let us consider the motion of a spring-mass-dashpot system acted on by a harmonic force $F_0 \sin \omega t$ applied as shown in Fig. 20.11(a), where ω is the frequency of excitation and is

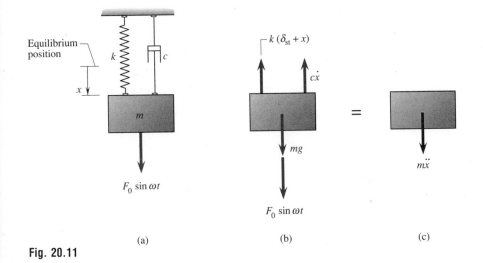

Fig. 20.11

called the *forcing frequency*. The free-body and effective-force diagrams of the body are shown in parts (b) and (c) of Fig. 20.11, where δ_{st} is the static deflection of the spring due to the weight mg. Referring to parts (b) and (c) of the figure, we write

$$+\downarrow \Sigma F: \qquad F_0 \sin \omega t + mg - k(\delta_{st} + x) - c\dot{x} = m\ddot{x}$$

Substituting Eq. (20.31) into this equation and rearranging, we get

$$m\ddot{x} + c\dot{x} + kx = F_0 \sin \omega t \tag{20.57}$$

which is the differential equation of motion of the body. Using Eqs. (20.3) and (20.33), we can write Eq. (20.57) in the form

$$\ddot{x} + 2\zeta\omega_n\dot{x} + \omega_n^2 x = \frac{F_0}{m} \sin \omega t \tag{20.58}$$

which is a nonhomogeneous second-order differential equation with constant coefficients.

The general solution of Eq. (20.58) is obtained by adding a particular solution x_p of Eq. (20.58) to the complementary solution x_c, which is equivalent to the general solution of the homogeneous equation in Eq. (20.34); i.e.,

$$x = x_c + x_p \tag{20.59}$$

The complementary solution x_c is given by one of Eqs. (20.38), (20.41), and (20.48), depending on the case of damping. If the system is underdamped (as in most vibratory systems), we write

$$x_c = X_1 e^{-\zeta\omega_n t} \sin(\omega_d t + \phi_1) \tag{20.60}$$

where X_1 and ϕ_1 are arbitrary constants, which are generally different from the values of X and ϕ given in Eqs. (20.49) and (20.50). In any case, the complementary solution x_c decays exponentially and will eventually be *damped out*. Thus, the complementary solution x_c for a damped system is significant only when a *transient solution* is being sought.

The particular solution x_p is *any* solution of the nonhomogeneous equation in Eq. (20.58). In the present case, x_p represents a *steady-state solution* which may be assumed to resemble the excitation in frequency but differ from it in phase; i.e.,

$$x_p = X \sin(\omega t - \varphi) \tag{20.61}$$

where the *steady-state amplitude* X and the *phase angle* φ are to be determined in terms of parameters in the system. Substituting x_p for x in Eq. (20.58) and rearranging, we obtain

$$(\omega_n^2 - \omega^2)X \sin(\omega t - \varphi) + 2\zeta\omega_n\omega X \cos(\omega t - \varphi)$$

$$= \frac{F_0}{m} \sin \omega t \tag{20.62}$$

Letting $\omega t - \varphi = \pi/2$ (or $\omega t = \pi/2 + \varphi$) in Eq. (20.62), we get

$$(\omega_n^2 - \omega^2)X = \frac{F_0}{m} \cos \varphi \tag{20.63}$$

Letting $\omega t - \varphi = 0$ (or $\omega t = \varphi$) in Eq. (20.62), we get

$$2\zeta\omega_n\omega X = \frac{F_0}{m}\sin\varphi \tag{20.64}$$

Squaring both members of Eqs. (20.63) and (20.64) and adding, we have

$$[(\omega_n^2 - \omega^2)^2 + (2\zeta\omega_n\omega)^2]X^2 = (F_0/m)^2 \tag{20.65}$$

Dividing Eqs. (20.64) and (20.63) member by member, we have

$$\frac{2\zeta\omega_n\omega}{\omega_n^2 - \omega^2} = \tan\varphi \tag{20.66}$$

Using Eq. (20.3) to substitute k/ω_n^2 for m, we can write Eqs. (20.65) and (20.66) in nondimensional forms as follows:

$$\frac{kX}{F_0} = \frac{1}{\sqrt{[1 - (\omega/\omega_n)^2]^2 + [2\zeta(\omega/\omega_n)]^2}} \tag{20.67}$$

$$\tan\varphi = \frac{2\zeta(\omega/\omega_n)}{1 - (\omega/\omega_n)^2} \tag{20.68}$$

Since F_0/k is the deflection of the spring by a static force F_0, the nondimensional ratio $kX/F_0 = X/(F_0/k)$ is known as the *amplitude ratio* or *magnification factor*. Equations (20.67) and (20.68) are plotted for several values of ζ in Figs. 20.12 and 20.13. Results pertaining to the undamped forced vibration can be obtained by simply setting $\zeta = 0$.

Fig. 20.12

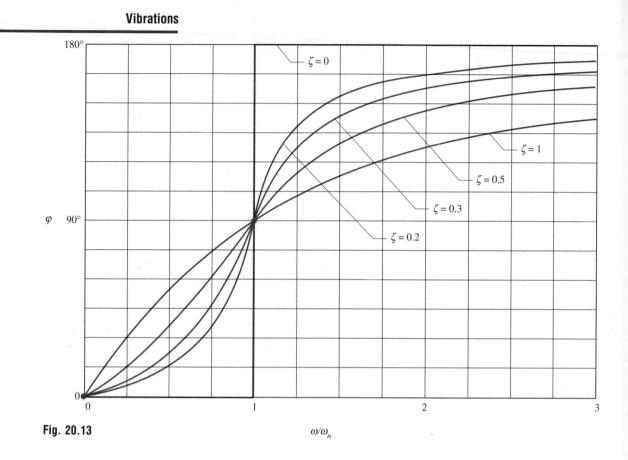

Fig. 20.13 ω/ω_n

Note in Fig. 20.12 that the *amplitude* of the steady-state response has a peak value when damping is small and the forcing frequency ω is near the value of the natural frequency ω_n. The phenomenon of exhibiting a peak amplitude is referred to as *resonance*. We see that the resonant peak is higher for smaller values of ζ; the resonant peak approaches infinity when ζ approaches zero. Thus, damping is beneficial in limiting the resonant peak. We also see that the steady-state amplitude is smaller when ω is much larger than ω_n.

The *phase angle* φ as shown in Fig. 20.13 ranges from 0 to 180°. It varies with the forcing frequency and the damping in the system. Without damping, the phase angle is zero if $\omega < \omega_n$ and is 180° if $\omega > \omega_n$. At $\omega = \omega_n$, the phase angle for all cases (with or without damping) is equal to 90°.

Fig. D20.13

Developmental Exercises

D20.12 Define the terms: (a) forcing frequency, (b) magnification factor, (c) resonance.

D20.13 Verify that the abscissa and ordinate of the resonant peak in Fig. 20.12 are, respectively, $\sqrt{1 - 2\zeta^2}$ and $1/(2\zeta\sqrt{1 - \zeta^2})$.

D20.14 Refer to D20.13. Determine the threshold value of the damping factor ζ below which the magnification factor may exceed 1 for a certain range of the frequency ratio ω/ω_n.

★20.6 Transmissibility

Since $mg = k\delta_{st}$ as given in Eq. (20.31), the free-body and effective-force diagrams in parts (b) and (c) of Fig. 20.11 may be simplified as shown in Fig. 20.14. Referring to this figure, we write

$$+\downarrow \Sigma F: \qquad F_0 \sin\omega t + (-kx) + (-c\dot{x}) = m\ddot{x} \qquad (20.69)$$

In steady-state vibration, x_c in Eq. (20.59) will be damped out; i.e., $x_c \to 0$. By Eqs. (20.59) and (20.61), we have

$$x = X \sin(\omega t - \varphi)$$

$$\dot{x} = \omega X \cos(\omega t - \varphi)$$

$$\ddot{x} = -\omega^2 X \sin(\omega t - \varphi)$$

Using these equations, we write

$$-kx = -kX \sin(\omega t - \varphi) = kX \sin[\pi + (\omega t - \varphi)]$$

$$-c\dot{x} = -c\omega X \cos(\omega t - \varphi) = c\omega X \sin\left[\frac{3\pi}{2} + (\omega t - \varphi)\right]$$

$$m\ddot{x} = -m\omega^2 X \sin(\omega t - \varphi) = m\omega^2 X \sin[\pi + (\omega t - \varphi)]$$

Substituting the above equations into Eq. (20.69), we obtain

$$F_0 \sin\omega t + kX \sin[\pi + (\omega t - \varphi)]$$

$$+ c\omega X \sin\left[\frac{3\pi}{2} + (\omega t - \varphi)\right]$$

$$= m\omega^2 X \sin[\pi + (\omega t - \varphi)] \qquad (20.70)$$

According to Eq. (20.70) and noting the phase differences among the applied force F_0, the spring force kX, the damping force $c\omega X$, and the effective force $m\omega^2 X$ on the body, we can depict the relationship among the vectors for these forces in steady-state forced vibration as shown in Fig. 20.15.

The force transmitted to the support of the body, as shown in Fig. 20.11(a), is clearly the vector sum of the spring force and the damping force. This transmitted force is denoted by the dashed vector $\mathbf{F}_T$ in Fig. 20.15. Referring to this figure and applying Pythagorean theorem and Eqs. (20.3), (20.33), and (20.67), we write

$$F_T^2 = (kX)^2 + (c\omega X)^2$$

$$= (kX)^2\left[1 + \left(\frac{c\omega}{m}\frac{m}{k}\right)^2\right]$$

$$= (kX)^2[1 + (2\zeta\omega_n\omega/\omega_n^2)^2]$$

$$= \frac{F_0^2}{[1 - (\omega/\omega_n)^2]^2 + [2\zeta(\omega/\omega_n)]^2}\{1 + [2\zeta(\omega/\omega_n)]^2\}$$

$$\boxed{\frac{F_T}{F_0} = \sqrt{\frac{1 + [2\zeta(\omega/\omega_n)]^2}{[1 - (\omega/\omega_n)^2]^2 + [2\zeta(\omega/\omega_n)]^2}}} \qquad (20.71)$$

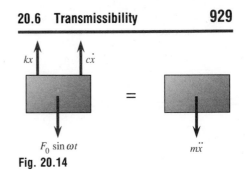

$F_0 \sin\omega t$

Fig. 20.14

Fig. 20.15

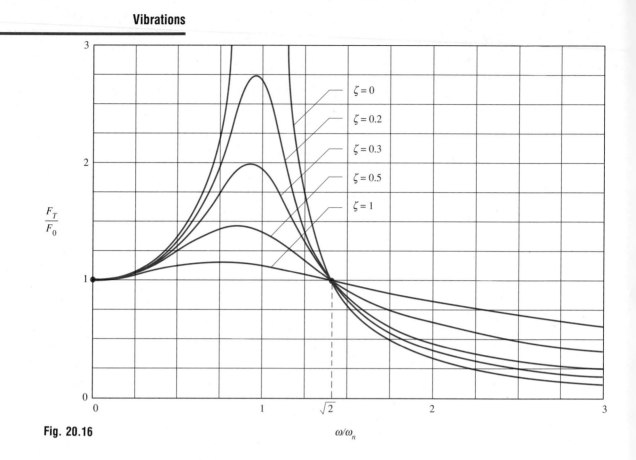

Fig. 20.16 ω/ω_n

The ratio of the maximum value F_T of the transmitted force to the maximum value F_0 of the disturbing force is known as the *transmissibility*.[†] Equation (20.71) is plotted for several values of ζ in Fig. 20.16. Results pertaining to the undamped forced vibration can be obtained by simply setting $\zeta = 0$. In particular, note in Fig. 20.16 that the transmissibility is less than 1 only when the ratio ω/ω_n is greater than $\sqrt{2}$. If $\omega/\omega_n < \sqrt{2}$, more damping results in less transmissibility; however, more damping results in more transmissibility if $\omega/\omega_n > \sqrt{2}$. As expected, the curves show that transmissibility is greatest at resonance.

Besides avoiding resonance, the transmission of vibratory forces generated by machines and other causes can be minimized by a proper isolator design. For instance, the ratio ω/ω_n can be increased to reduce transmissibility by using a softer spring or a larger mass in the system for a given forcing frequency ω.

[†]This is also known as the force transmissibility. A displacement transmissibility is similarly defined in vibration analysis. Cf. Prob. 20.36*.

EXAMPLE 20.6

A piece of equipment with a mass $m = 40$ kg is supported by springs with a total modulus $k = 100$ kN/m and a dashpot as shown. The equipment is acted on by harmonic excitations ranging in frequencies from 30 Hz to 60 Hz. If the isolation of force transmission is known to be 80%, determine the damping factor ζ and the corresponding coefficient of viscous damping c of the dashpot.

Solution. We know from the given data that $m = 40$ kg and $k = 100$ kN/m $= 10^5$ N/m. Thus, we have

$$\omega_n^2 = \frac{k}{m} = \frac{10^5}{40} \qquad \omega_n = 50 \text{ rad/s}$$

Since the smaller forcing frequency is more likely to incur more transmissibility, we choose to use the 30-Hz frequency to write

$$\omega = 2\pi f = 2\pi(30) \qquad \omega = 60\pi \text{ rad/s}$$

$$\frac{\omega}{\omega_n} = \frac{60\pi}{50} = \frac{6\pi}{5} > \sqrt{2}$$

For 80% isolation, the transmissibility is $1 - 0.8 = 0.2$. Applying Eq. (20.71), we write

$$0.2 = \sqrt{\frac{1 + [2\zeta(6\pi/5)]^2}{[1 - (6\pi/5)^2]^2 + [2\zeta(6\pi/5)]^2}}$$

which can be solved to yield

$$\zeta = 0.331 \blacktriangleleft$$

By Eq. (20.33), we write

$$c = 2\zeta\sqrt{km} = 2(0.331)\sqrt{10^5(40)} = 1324$$

$$c = 1.324 \text{ kN·s/m} \blacktriangleleft$$

Developmental Exercises

D20.15 Define *transmissibility* in forced vibration.

D20.16 Refer to Example 20.6. Determine (a) whether the isolation is increased or decreased if the dashpot is removed, (b) the isolation when the dashpot is removed.

PROBLEMS

20.32 For the system shown, determine the amplitude of the steady-state response if (a) $c = 10$ lb·s/ft, (b) $c = 0$.

20.33 For the system shown, determine the magnification factor if (a) $c = 5$ lb·s/ft, (b) $c = 0$.

Fig. P20.32 and P20.33

20.34   A harmonic force $F_0 \sin\omega t$ acts on a given spring-mass-dashpot system, where F_0 is constant. When the ratio ω/ω_n is increased from 1 to 2, the amplitude of the steady-state response is observed to decrease by 80%. Determine the damping factor ζ of the system.

20.35*   Verify that the abscissa and ordinate of the resonant peak in Fig. 20.12 are, respectively,

$$[(1 + 8\zeta^2)^{1/2} - 1]^{1/2}/(2\zeta) \quad \text{and}$$

$$2\sqrt{2}\zeta^2/[(1 + 8\zeta^2)^{1/2} - 1 - 4\zeta^2 + 8\zeta^4]^{1/2}$$

Fig. P20.35*   **Fig. P20.36***

20.36*   The base of a viscously damped spring-mass system is given a harmonic excitation of $y = Y \sin\omega t$, and the displacement x of the mass m is measured from an equilibrium position. Show that (a) the equation of motion of the mass is given by

$$m\ddot{x} + c\dot{x} + kx = Y\sqrt{k^2 + c^2\omega^2} \sin(\omega t + \beta)$$

where $\beta = \tan^{-1}(c\omega/k)$, (b) the ratio of the steady-state amplitude X of the mass to the amplitude Y of the excitation, called the *displacement transmissibility*, is identical with the expression for the transmissibility F_T/F_0 in Eq. (20.71); i.e.,

$$\frac{X}{Y} = \sqrt{\frac{1 + [2\zeta(\omega/\omega_n)]^2}{[1 - (\omega/\omega_n)^2]^2 + [2\zeta(\omega/\omega_n)]^2}}$$

20.37   A spring-dashpot-supported vehicle travels at $v = 60$ mi/h over a sinusoidal road surface as shown, where $W = 2000$ lb, $k = 1000$ lb/in., and it is critically damped. Derive the equation of motion of W and determine the steady-state amplitude X of W.

Fig. P20.37   **Fig. P20.38**

20.38   A slender rod of mass m is subjected to a harmonic excitation as shown. Assuming small-amplitude vibration, determine the steady-state angular amplitude and phase angle of the rod in terms of the parameters of the system.

*20.7 Concluding Remarks

Except in finding the *period* of orbit in central-force motion in the latter part of Chap. 12, we were mostly concerned with the motion and the force system associated with the motion of particles and rigid bodies *at a given instant*, rather than over a period of time, in preceding chapters. In this chapter, we turn our attention to the study of the vibratory motion of a system *over a period of time*, rather than just at a given instant.

The analysis of vibrations involves, among other things, the derivation of the equation of motion valid for anytime during the period of motion of the system. There are several methods which can be used to accomplish this task. For fundamental study, two methods are used; they are (a) the *method of force and acceleration* for both conservative and nonconservative systems, and (b) the *energy method* for conservative systems. In either method, an inertial reference frame must be used in writing the velocity and acceleration of the body. In the energy method, special care should be exercised in computing the potential energy with respect to a properly chosen reference datum. The working definition for the potential energy as given in the text should facilitate the computation. Once the equation of motion for a particular system is established and written in the "standard" form, the natural frequency and other characteristics of the steady-state vibration can usually be ascertained from the coefficients of that equation.

Excessive vibrations and damping in mechanical systems are generally undesirable because of the noise and energy dissipation which accompany them. With an understanding of the vibratory behavior of the system, engineers can eliminate or reduce them by appropriately designing the parameters of the system. The increasing use of high-speed machines and high-strength structures in modern engineering certainly makes the analysis of vibrations increasingly important.

REVIEW PROBLEMS

20.39   The 20-kg flywheel shown has a radius of gyration of 300 mm about its mass center G. Determine its period of small-amplitude vibration.

Fig. P20.39

400 mm

Fig. P20.40

20.40   Determine the frequency of small-amplitude vibration of the hemisphere shown, where $R = 300$ mm.

20.41 A section of semicircular shell of mass m and radius r rests at the bottom of a cylindrical surface of radius R as shown. If the shell rocks without slipping on the surface, determine its period of small-amplitude vibration.

20.42 Determine the weight W of a cylinder, as shown, whose period of vibration is 0.2 s.

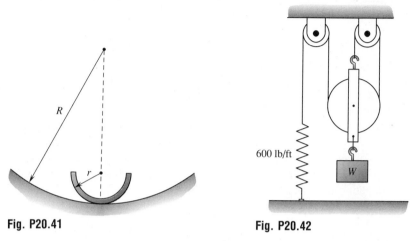

Fig. P20.41   **Fig. P20.42**

20.43 The right end of the connecting spring in the system shown is given a harmonic motion $x_2 = X_2 \sin \omega t$, and the position of the mass m is defined by the coordinate x_1. Derive the equation of motion and determine the steady-state amplitude and the phase angle of the mass in terms of the parameters of the system.

Fig. P20.43

20.44 A slender bar of mass m is subjected to a harmonic excitation as shown. Assuming small-amplitude vibration, determine the steady-state angular amplitude and phase angle of the bar in terms of the parameters of the system.

Fig. P20.44

Prior Basic Mathematics

Appendix A

Prior mathematical skills needed for studying Chaps. 1 through 5 involve an ability to perform calculation with a calculator and a proficiency in basic algebra, geometry, and trigonometry. However, some prior background in basic analytic geometry and elementary calculus will be needed starting with Chap. 6. The following problems are designed to reflect the typical prior skills required. Answers to these problems are provided at the end of the text. Some review materials on the aforementioned basic mathematics are presented in App. B.

Choose the correct or best item to complete the sentence or answer the question in each of the following:

1. The solution for x in the equation $3x + 3 = -2$ is
   (a) 3/5.   (b) $-3/5$.   (c) 5/3.
   (d) $-5/3$.   (e) -1.

2. The simultaneous equations $2x + 5y = 21$ and $-3x + y = 11$ have a solution given by
   (a) $(-2, 5)$.   (b) $(-5, -2)$.   (c) $(2, 5)$.
   (d) $(-5, 2)$.   (e) $(-2, -5)$.

3. The roots of the quadratic equation $x^2 + x - 12 = 0$ are
   (a) -1 and 12.   (b) 1 and -12.   (c) -2 and 6.
   (d) 3 and -4.   (e) 3 and 4.

4. The roots of the quadratic equation $x^2 + x - 9 = 0$ are
   (a) -1 and 9.   (b) 1 and -9.   (c) 2.54 and -3.54.
   (d) 5.08 and -7.08.   (e) -5.08 and 7.08.

5. Expansion of $(a + b)^3$ yields
   (a) $a^3 + b^3$.   (b) $a^3 - 2a^2b + 2ab^2 - b^3$.
   (c) $a^3 + 2a^2b + 2ab^2 + b^3$.   (d) $a^3 - 3a^2b + 3ab^2 - b^3$.
   (e) $a^3 + 3a^2b + 3ab^2 + b^3$.

6. The term $(x^2)^3$ is equivalent to
   (a) x^5.   (b) x^2.   (c) x^6.
   (d) x^3.   (e) $x^{2.5}$.

7. The term $(a/b)^{-3/2}$ is equivalent to
   (a) $(b/a)^{3/2}$.   (b) $-a^3/b^2$.   (c) $a^{3/2}/b^{3/2}$.
   (d) $\sqrt{(a/b)^3}$.   (e) $-\sqrt{(a/b)^3}$.

8. The term $\frac{1}{2} \ln (a^{1/3})$ is equivalent to
   (a) $\ln (a^{5/6})$.   (b) $\ln (a^{6/5})$.   (c) $\frac{1}{3} \ln (a^{1/2})$.
   (d) $\ln (a^{2/3})$.   (e) $\ln (a^{3/2})$.

9. The term $\frac{1}{3}$ ln (a/b) is equivalent to
   (a) ln $(a^{1/3}b)$.   (b) ln $(a^{1/3})$ + ln $(b^{1/3})$.   (c) $\frac{1}{3}$ (ln a + ln b).
   (d) $\frac{1}{3}$ (ln a)(ln b).   (e) $\frac{1}{3}$ (ln a − ln b).

10. If $3.45^x = 8$, then the value of x is
    (a) 1.679.   (b) 1.583.   (c) 1.402.   (d) 1.338.
    (e) 1.238.

11. The value of the determinant $\begin{vmatrix} 5 & -2 \\ 8 & -2 \end{vmatrix}$ is

    (a) 26.   (b) −26.   (c) 9.   (d) 6.   (e) −6.

12. The value of the determinant $\begin{vmatrix} 5 & -7 & 9 \\ 2 & 0 & 0 \\ 6 & -4 & 1 \end{vmatrix}$ is

    (a) −14.   (b) −54.   (c) 54.   (d) 58.   (e) −58.

13. The value of [6.44 + $(4.75)^{1/3}(e^{0.4\pi})$ ln 3.65 − 3 sin 35° cot 215°] is
    (a) 7.30.   (b) 8.90.   (c) 11.63   (d) 12.88.   (e) 16.54.

Fig. P14

14. The shortest distance from the point A to the line which is tangent to the circle at the point B, as shown, is
    (a) 9.5 m.   (b) 9 m.   (c) 8.5 m.   (d) 8 m.   (e) 7.5 m.

Fig. P15

15. The length of the line segment $\overline{AB}$, as shown, is
    (a) 5.29 ft.   (b) 5.66 ft.   (c) 6.29 ft.   (d) 6.66 ft.
    (e) 6.92 ft.

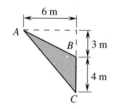

Fig. P16

16. The area of the triangle ABC shown is
    (a) 9 m².   (b) 12 m².   (c) 18 m².   (d) 21 m².   (e) 24 m².

17. The area of the trapezoid shown is
    (a) 28 in².   (b) 42 in².   (c) 48 in².   (c) 63 in².   (e) 84 in².

Fig. P17   **Fig. P18**   **Fig. P19**

18. If θ = 3/5 rad, the length of the arc AB, as shown, is
    (a) 0.6 m.   (b) 1.667 m.   (c) 9 m.   (d) 12 m.   (e) 8 m.

19. The circumference of the circle shown is
    (a) 3π m².   (b) 6π m².   (c) 9π m².   (d) 3π m.   (e) 6π m.

Fig. P20

20. The area of the quarter circular sector shown is
    (a) 2π m².   (b) 4π m².   (c) 8π m².   (d) 2π m.   (e) 4π m.

21. In the figure shown, the angle α is
    (a) 44°.   (b) 45°.   (c) 46°.   (d) 47°.   (e) 60°.

Fig. P21

22. The length of the side $\overline{BC}$ of the right triangle shown is
    (a) 8 ft.   (b) 6 ft.   (c) 7.5 ft.   (d) 4.12 ft.   (e) 8.5 ft.

23. In the triangle shown, $\tan\theta$ is equal to
    (a) 3/5.   (b) 5/3.   (c) 3/4.   (d) 4/3.   (e) 4/5.

Fig. P22

Fig. P23   **Fig. P24**

24. The length of the side $\overline{AB}$ of the triangle shown is
    (a) 13 m.   (b) 11.27 m.   (c) 9.22 m.   (d) 8.95 m.
    (e) 6.48 m.

25. The length of the side $\overline{AB}$ of the right triangle shown is
    (a) 12.6 m.   (b) 11.8 m.   (c) 10.5 m.   (d) 9.24 m.
    (e) 8.51 m.

26. The angle θ of the right triangle shown is
    (a) 35.0°.   (b) 36.9°.   (c) 44.4°.   (d) 45.6°.   (e) 53.1°.

Fig. P25

Fig. P26   **Fig. P27**

27. The angle θ of the triangle shown is
    (a) 0.785 rad.   (b) 0.813 rad.   (c) 1.05 rad.
    (d) 1.57 rad.   (e) 3.14 rad.

28. The length of the side $\overline{AC}$ of the triangle shown is
    (a) 5.00 in.   (b) 5.08 in.   (c) 5.32 in.   (d) 5.53 in.
    (e) 5.77 in.

Fig. P28

29. The term $\cos^2\theta$ is equivalent to
    (a) $(1 - \sin 2\theta)/2$.   (b) $(1 - \cos 2\theta)/2$.   (c) $(1 + \sin 2\theta)/2$.
    (d) $(1 + \cos 2\theta)/2$.   (e) $-\sin 2\theta$.

30. The expression $1 - \sin^2\theta$ is equivalent to
    (a) $\tan^2\theta$.   (b) $\sec^2\theta$.   (c) $1 + \cot^2\theta$.   (d) $\cos^2\theta - 1$.
    (e) $\cos^2\theta$.

Fig. P31

31. The equation of the line AB shown is
    (a) $3x - 4y = 0$.   (b) $4x - 3y = 0$.
    (c) $3x - 4y + 12 = 0$.   (d) $3x - 4y - 12 = 0$.
    (e) $4x - 3y + 12 = 0$.

32. The length of the line segment $\overline{AB}$ shown is
    (a) 22 m.   (b) 24 m.   (c) 26 m.   (d) 28 m.   (e) 38 m.

Fig. P32

Fig. P33

Fig. P34

33. The equation of the parabola shown is
    (a) $y^2 = 16x$.   (b) $x^2 = 2y$.   (c) $x^2 = y + 8$.
    (d) $y^2 = x + 60$.   (e) $4y = 8x$.

34. The equation of the ellipse shown is
    (a) $\dfrac{x}{4} + \dfrac{y}{3} = 1$.   (b) $\dfrac{x}{4} - \dfrac{y}{3} = 1$.   (c) $\dfrac{x^2}{4} + \dfrac{y^2}{3} = 1$.
    (d) $\dfrac{x^2}{16} - \dfrac{y^2}{9} = 1$.   (e) $\dfrac{x^2}{16} + \dfrac{y^2}{9} = 1$.

35. Which of the following represents an ellipse in polar coordinates?
    (a) $\dfrac{1}{r} = 3\cos\theta$.   (b) $\dfrac{1}{r} = 3\theta$.   (c) $\dfrac{1}{r} = 3 + 3\cos\theta$.
    (d) $\dfrac{1}{r} = 3 + 4\cos\theta$.   (e) $\dfrac{1}{r} = 4 + 3\cos\theta$.

36. The differential area element dA in polar coordinates, as shown, is
    (a) $u\,du\,d\theta$.   (b) $\theta\,du$.   (c) $du\,d\theta$.   (d) $u\,d\theta$.   (e) $dx\,dy$.

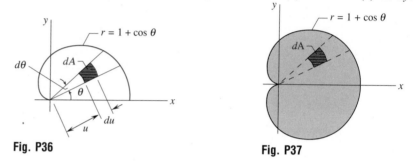

Fig. P36   **Fig. P37**

37. The area enclosed by the cardioid shown is given by
    (a) $\displaystyle\int_0^{2\pi} (1 + \cos\theta)\,d\theta$.   (b) $\displaystyle 2\int_{\theta=0}^{\pi}\int_{u=0}^{1+\cos\theta} du\,d\theta$.
    (c) $\displaystyle\int_{\theta=0}^{2\pi}\int_{u=0}^{1+\cos\theta} du\,d\theta$.   (d) $\displaystyle 2\int_{\theta=0}^{\pi}\int_{u=0}^{1+\cos\theta} u\,du\,d\theta$.
    (e) $\displaystyle\int_{\theta=0}^{2\pi}\int_{u=0}^{1} (1 + \cos\theta)\,du\,d\theta$.

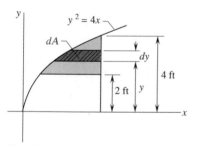

Fig. P38

38. The differential area element dA, as shown, is
    (a) $(y^2/4)\,dy$.   (b) $2x^{1/2}\,dx$.   (c) $[(y^2 - 16)/4]\,dy$.
    (d) $[(16 - y^2)/4]\,dy$.   (e) $(4 - y)\,dy$.

39. If $\dfrac{dx}{dt} > 0$, which of the following statements is *always* true?
    (a) The magnitude of x is increasing.
    (b) The algebraic value of x is increasing.
    (c) The magnitude of x is decreasing.
    (d) The algebraic value of x is decreasing.
    (e) The value of x is positive.

40. The slope of the normal to the curve $2x^2 + 9y = 0$ at the point $(3, -2)$ is
    (a) 3/4.   (b) 4/3.   (c) $-4/3$.   (d) $-3/4$.   (e) 3/5.

41. The radius of curvature of the curve $2x^2 + 9y = 0$ at the point $(3, -2)$ is
    (a) 0.096.   (b) 1.25.   (c) 3.75.   (d) 10.42.   (e) 15.52.

42. The differential area element dA, as shown, is
    (a) $2x^{1/2}\, dx$.   (b) $\frac{1}{4} y^2\, dy$.   (c) $(2x^{1/2} - 2)\, dx$.
    (d) $(4 - 2x^{1/2})\, dx$.   (e) $(2x^{1/2} - 4)\, dx$.

43. The area of the shaded region shown is given by
    (a) $\int_1^3 2y^{1/2}\, dy$.   (b) $\int_1^4 2y^{1/2}\, dy$.   (c) $\int_0^4 \frac{x^2}{4}\, dx$.
    (d) $\int_2^4 \frac{x^2}{4}\, dx$.   (e) $\int_1^3 (4 - y)dy$.

Fig. P42

Fig. P43

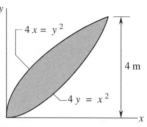

Fig. P44

44. The area of the shaded region shown is given by
    (a) $\int_0^4 \left(\frac{1}{4} x^2 - 2x^{1/2} \right) dx$.   (b) $\int_0^4 \frac{1}{4} x^2\, dx$.
    (c) $\int_0^4 \left(2y^{1/2} - \frac{1}{4} y^2 \right) dy$.   (d) $\int_0^4 2y^{1/2}\, dy$.   (e) $\int_0^4 x\, dy$.

45. The volume of the right circular cone shown is given by
    (a) $\int_0^h (2\pi b/h)x\, dx$.   (b) $\int_0^h 2\pi x\, dx$.   (c) $\int_0^h \pi x^2\, dx$.
    (d) $\int_0^h (\pi h^2/b^2)x^2\, dx$.   (e) $\int_0^h (\pi b^2/h^2)x^2\, dx$.

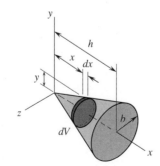

Fig. P45

46. The value of $0.234 \int_1^9 (2x - x^{-1} - x^{-1/2})\, dx$ is
    (a) 76.0.   (b) 73.8.   (c) 17.78.   (d) 17.27.   (e) -2.20.

47. Which of the following is *false*?
    (a) $\sin^2 x + \cos^2 x = 1$.   (b) $\sinh^2 x + \cosh^2 x = 1$.
    (c) $\sec^2 x - \tan^2 x = 1$.   (d) $\text{sech}^2 x + \tanh^2 x = 1$.
    (e) $\csc^2 x - \cot^2 x = 1$.

48. Which of the following represents the shape of the curve defined by $y = \cosh x$:

       (a)   (b)   (c)   (d)   (e)

Fig. P48

49. If $e^x = y + (1 + y^2)^{1/2}$, then y is equal to
    (a) $e^x + e^{-x}$.   (b) $e^x - e^{-x}$.   (c) $\tanh x$.
    (d) $\cosh x$.   (e) $\sinh x$.

50. The value of $\displaystyle\int_0^1 \sinh x \, dx$ is
    (a) -1.543.   (b) -0.1752.   (c) 0.1752.   (d) 0.543.
    (e) 1.175.

Review of Basic Mathematics

Appendix B

B1   ALGEBRA

B1.1   Fundamental Principles

The following principles allow us to solve a great many equations:

1. *Addition principle*. If the equation $a = b$ is true, then $a + c = b + c$ is true, for any number or expression c.
2. *Multiplication principle*. If the equation $a = b$ is true, then $ac = bc$ is true for any number or expression c.
3. *Principle of zero products*. For any numbers or expressions a and b, if $ab = 0$, then $a = 0$ or $b = 0$, and if $a = 0$ or $b = 0$, then $ab = 0$.
4. *Principle of powers*. If the equation $a = b$ is true, then $a^n = b^n$ is true for any number n.

B1.2   Roots of a Quadratic Equation

If $ax^2 + bx + c = 0$ and $a \neq 0$, then

$$x = \frac{-b \pm \sqrt{b^2 - 4ac}}{2a}$$

B1.3   Solution of *n* Linear Simultaneous Equations

A set of n linear simultaneous equations in n unknowns $x_1, x_2, \ldots, x_n$ may be written in the form

$$a_{11}x_1 + a_{12}x_2 + \cdots + a_{1n}x_n = b_1 \qquad (1)$$
$$a_{21}x_1 + a_{22}x_2 + \cdots + a_{2n}x_n = b_2 \qquad (2)$$
$$\cdots \quad \cdots \quad \cdots \quad \cdots \quad \cdots \qquad \cdots$$
$$a_{n1}x_1 + a_{n2}x_2 + \cdots + a_{nn}x_n = b_n \qquad (n)$$

If these n equations are independent, nonhomogeneous, and consistent, then they have a unique nontrivial solution. Such a solution may be obtained by the following steps:

1. From one of the n given equations, express the unknown x_1 in terms of the remaining $n - 1$ unknowns $x_2, x_3, \ldots, x_n$.
2. Substitute the expression of x_1 obtained in step (1) into the remaining $n - 1$ equations to obtain a set of $n - 1$ linear simultaneous equations in $n - 1$ unknowns $x_2, x_3, \ldots, x_n$.
3. From one of the $n - 1$ equations obtained in step (2), express the unknown x_2 in terms of the remaining $n - 2$ unknowns $x_3, x_4, \ldots, x_n$.
4. Substitute the expression of x_2 obtained in step (3) into the remaining $n - 2$ equations to obtain a set of $n - 2$ linear simultaneous equations in $n - 2$ unknowns $x_3, x_4, \ldots, x_n$.
5. Continue the reduction process similar to the above four steps until the size of the set is reduced to one linear equation in one unknown.
6. Solve for the single unknown in the last equation and perform the back substitutions of determined values into the expressions of other unknowns to obtain the values of all other unknowns.

B1.4   Second- and Third-Order Determinants

Second- and third-order determinants can be expanded by special schemes as follows:

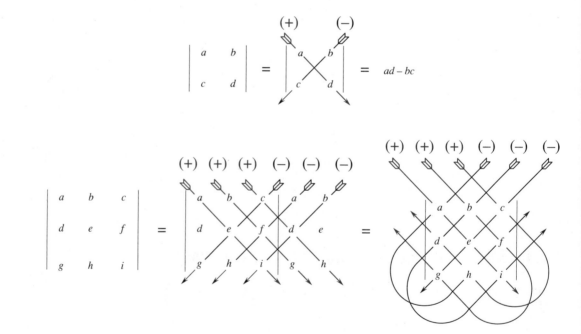

$$= aei + bfg + cdh - ceg - bdi - afh$$

Note that the sign to be used with each individual product is indicated at the tail of each individual arrow. Furthermore, the above schematic methods of expansion *cannot* be used with determinants of order higher than three. However, any determinant may be expanded by using the cofactors of its row elements or column elements. Using the cofactors of the elements of the first row of a third-order determinant, we write

$$
\begin{vmatrix} a & b & c \\ d & e & f \\ g & h & i \end{vmatrix} = a\begin{vmatrix} e & f \\ h & i \end{vmatrix} - b\begin{vmatrix} d & f \\ g & i \end{vmatrix} + c\begin{vmatrix} d & e \\ g & h \end{vmatrix}
$$

$$
= a(ei - fh) - b(di - fg) + c(dh - eg)
$$

If *any* two rows or columns of a determinant are interchanged, the value of the determinant will change its sign. For example, we write

$$
\begin{vmatrix} a & b & c \\ d & e & f \\ g & h & i \end{vmatrix} = -\begin{vmatrix} d & e & f \\ a & b & c \\ g & h & i \end{vmatrix} = \begin{vmatrix} d & e & f \\ g & h & i \\ a & b & c \end{vmatrix} = -\begin{vmatrix} g & h & i \\ d & e & f \\ a & b & c \end{vmatrix}
$$

Thus, by interchanging the rows twice, we write

$$
\begin{vmatrix} a & b & c \\ d & e & f \\ g & h & i \end{vmatrix} = \begin{vmatrix} d & e & f \\ g & h & i \\ a & b & c \end{vmatrix} = \begin{vmatrix} g & h & i \\ a & b & c \\ d & e & f \end{vmatrix}
$$

This property is useful in studying the volume of a parallelepiped using the scalar triple product in Sec. 4.15.

B1.5   Binomials

(a) $a^2 - b^2 = (a + b)(a - b)$
(b) $a^3 - b^3 = (a - b)(a^2 + ab + b^2)$
(c) $(a + b)^2 = a^2 + 2ab + b^2$
(d) $(a - b)^2 = a^2 - 2ab + b^2$
(e) $(a + b)^3 = a^3 + 3a^2b + 3ab^2 + b^3$
(f) $(a - b)^3 = a^3 - 3a^2b + 3ab^2 - b^3$

(g) $(a + b)^n = a^n + na^{n-1}b + \dfrac{n(n - 1)}{2!} a^{n-2}b^2$

$$
+ \frac{n(n - 1)(n - 2)}{3!} a^{n-3}b^3 + \cdots + b^n
$$

(binomial theorem)

B1.6   Exponents

(a) $A^a A^b = A^{a+b}$
(b) $(A^a)^b = A^{ab}$
(c) $A^{-a} = \dfrac{1}{A^a}$
(d) $A^{a/b} = (A^a)^{1/b} = \sqrt[b]{A^a}$

B1.7   Logarithms

The logarithm of a number N to a base b is the exponent x indicating the power to which b must be raised to produce the number N. Thus, we write $\log_b N = x$ if $b^x = N$. The base of the common logarithm is 10, while that of the natural logarithm is e, where

$$e = 1 + \frac{1}{1!} + \frac{1}{2!} + \frac{1}{3!} + \frac{1}{4!} + \cdots = 2.718281828 \cdots$$

The natural logarithm is usually abbreviated as ln. We have

(a) $\ln (AB) = \ln A + \ln B$
(b) $\ln (A/B) = \ln A - \ln B$
(c) $\ln (A^a) = a \ln A$
(d) $A^a = e^{a \ln A}$

B2   GEOMETRY

B2.1   Useful Postulates and Theorems

Fig. [a]

Fig. [b]

Fig. [c]

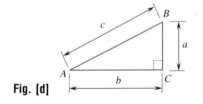

Fig. [d]

1. *Vertical angles*: Vertical angles formed by two intersecting lines are congruent. In Fig. [a], we have $\theta_1 = \theta_3$ and $\theta_2 = \theta_4$.
2. *Corresponding angles*: When a transversal intersects two parallel lines, corresponding angles are congruent, and conversely. In Fig. [b], we have $\theta_1 = \theta_3$ and $\theta_2 = \theta_4$.
3. *Angle sum for triangles*: The sum of the measures of the angles of a triangle is 180°. In Fig. [c], we have $\theta_1 + \theta_2 + \theta_3 = 180°$.
4. *Pythagorean theorem*: In a right triangle, the square of the length of its hypotenuse is equal to the sum of the squares of the lengths of its legs.[†] In Fig. [d], we have $c^2 = a^2 + b^2$.
5. *Congruent angles*: If the two lines including an angle are respectively perpendicular to the two lines including another angle, then these two angles must be either congruent angles or supplementary angles. In Fig. [e], we have $\theta_1 = \theta_2$ and $\theta_1 + \theta_3 = 180°$.

Fig. [e]   Center of curve   **Fig. [f]**

6. *Center of curve*: A line drawn perpendicular to the tangent to a curve and passing through the point of tangency will also pass through the

---

[†]The general formula $[\frac{1}{2}(m^2 + 1)]^2 = m^2 + [\frac{1}{2}(m^2 - 1)]^2$, where m is any odd integer, is attributed to the Greek philosopher and mathematician Pythagoras (ca. 580–500 B.C.).

center of the curve at that point. In Fig. [f], the point of tangency is P and $\overline{CP} \perp \overline{AB}$.

7. *Shortest distance*: The shortest distance from a point to a line is equal to the length of the perpendicular segment drawn from the point to the line. In Fig. [g], d_s is the shortest distance from the point P to the line AB.

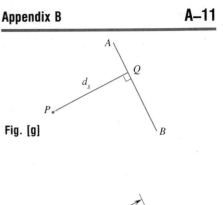

Fig. [g]

B2.2   Radian Measure

One radian is defined as the angle which subtends a circular arc whose length is equal to the length of the radius of the circular arc. As shown in Fig. [h], we have

$$\theta = \frac{s}{R} \qquad s = R\theta$$

where θ is measured in radians.

Fig. [h]

B2.3   Mensuration Formulas for Some Common Shapes

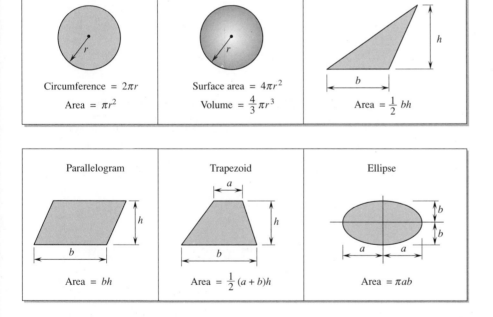

Circle	Sphere	Triangle
Circumference $= 2\pi r$	Surface area $= 4\pi r^2$	
Area $= \pi r^2$	Volume $= \frac{4}{3}\pi r^3$	Area $= \frac{1}{2}bh$

Parallelogram	Trapezoid	Ellipse
Area $= bh$	Area $= \frac{1}{2}(a+b)h$	Area $= \pi ab$

B3   TRIGONOMETRY

B3.1   Definitions with Mnemonics

Let a right triangle be oriented as shown in Fig. [a], where the angle θ is on the left and the right angle is on the right. Then, the values of $\sin\theta$,

Fig. [a]

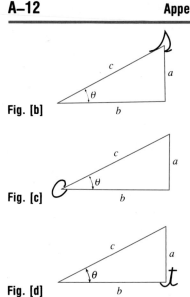

Fig. [b]

Fig. [c]

Fig. [d]

Fig. [e]

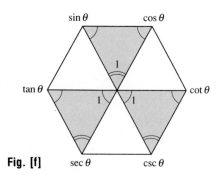

Fig. [f]

$\cos\theta$, and $\tan\theta$ are, respectively, equal to those obtained by *dividing the first side touched into the second side touched* when the letters $\mathcal{s}$, $\mathcal{c}$, and $\mathcal{t}$ are written in *script* on the triangle as shown in Figs. [b], [c], and [d]. We have

$$\sin\theta = \frac{a}{c} = \frac{\text{Opposite side}}{\text{Hypotenuse}} = \theta - \frac{\theta^3}{3!} + \frac{\theta^5}{5!} - \frac{\theta^7}{7!} + \cdots \ (\theta \text{ in rad})$$

$$\cos\theta = \frac{b}{c} = \frac{\text{Adjacent side}}{\text{Hypotenuse}} = 1 - \frac{\theta^2}{2!} + \frac{\theta^4}{4!} - \frac{\theta^6}{6!} + \cdots \ (\theta \text{ in rad})$$

$$\tan\theta = \frac{a}{b} = \frac{\text{Opposite side}}{\text{Adjacent side}} = \theta + \frac{\theta^3}{3} + \frac{2\theta^5}{15} + \frac{17\theta^7}{315} + \cdots \ (\theta \text{ in rad})$$

B3.2   Some Useful Properties with Mnemonics

The quadrants in which the trigonometric functions have positive values may be indicated as shown in Fig. [e]. Furthermore, several trigonometric identities may readily be written from the hexagon shown in Fig. [f]. The mnemonic rules associated with Fig. [f] are as follows:[†]

1. The product of the functions at the ends of each diagonal is equal to 1. We have

   $$\sin\theta \csc\theta = 1 \qquad \cos\theta \sec\theta = 1 \qquad \tan\theta \cot\theta = 1$$

2. Any function is equal to the product of its two neighboring functions. We have

   $$\sin\theta = \cos\theta \tan\theta \qquad \csc\theta = \sec\theta \cot\theta$$

   $$\cos\theta = \sin\theta \cot\theta \qquad \sec\theta = \tan\theta \csc\theta$$

   $$\tan\theta = \sin\theta \sec\theta \qquad \cot\theta = \cos\theta \csc\theta$$

3. In each of the three shaded triangles, the square of the quantity at the lower angle is equal to the sum of the squares of the quantities at the upper two angles. We have

   $$\sin^2\theta + \cos^2\theta = 1$$

   $$\tan^2\theta + 1 = \sec^2\theta \qquad 1 + \cot^2\theta = \csc^2\theta$$

B3.3   Other Useful Trigonometric Identities

(a) $\sin(\alpha \pm \beta) = \sin\alpha \cos\beta \pm \cos\alpha \sin\beta$

(b) $\cos(\alpha \pm \beta) = \cos\alpha \cos\beta \mp \sin\alpha \sin\beta$

(c) $\tan(\alpha \pm \beta) = \dfrac{\tan\alpha \pm \tan\beta}{1 \mp \tan\alpha \tan\beta}$

(d) $\sin^2\theta = \frac{1}{2}(1 - \cos 2\theta)$

[†]Note that the order in which the trigonometric functions appear on the hexagon is that $\sin\theta$ and $\cos\theta$ are in the first row, $\tan\theta$ and $\cot\theta$ are in the second row, and $\sec\theta$ and $\csc\theta$ are in the third row as shown.

(e) $\cos^2\theta = \frac{1}{2}(1 + \cos 2\theta)$

(f) $\sin\alpha \pm \sin\beta = 2 \sin \frac{1}{2}(\alpha \pm \beta) \cos \frac{1}{2}(\alpha \mp \beta)$

(g) $\cos\alpha + \cos\beta = 2 \cos \frac{1}{2}(\alpha + \beta) \cos \frac{1}{2}(\alpha - \beta)$

(h) $\cos\alpha - \cos\beta = -2 \sin \frac{1}{2}(\alpha + \beta) \sin \frac{1}{2}(\alpha - \beta)$

(i) $\sin\alpha \cos\beta = \frac{1}{2}[\sin(\alpha + \beta) + \sin(\alpha - \beta)]$

(j) $\sin\alpha \sin\beta = \frac{1}{2}[\cos(\alpha - \beta) - \cos(\alpha + \beta)]$

(k) $\cos\alpha \cos\beta = \frac{1}{2}[\cos(\alpha + \beta) + \cos(\alpha - \beta)]$

B3.4 Law of Sines

For a triangle with angles and sides as shown in Fig. [g], we may write the law of sines as follows:

$$\frac{a}{\sin\alpha} = \frac{b}{\sin\beta} = \frac{c}{\sin\gamma}$$

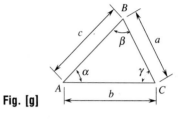

Fig. [g]

B3.5 Law of Cosines

For a triangle with angles and sides as shown in Fig. [g], we may write the law of cosines as follows:

$$a^2 = b^2 + c^2 - 2bc \cos\alpha \qquad b^2 = c^2 + a^2 - 2ca \cos\beta$$

$$c^2 = a^2 + b^2 - 2ab \cos\gamma$$

B3.6 Inverse Functions

(a) $\alpha = \sin^{-1} x$ if $x = \sin\alpha$

(b) $\theta = \csc^{-1} u$ if $u = \csc\theta$

(c) $\beta = \cos^{-1} y$ if $y = \cos\beta$

(d) $\phi = \sec^{-1} v$ if $v = \sec\phi$

(e) $\gamma = \tan^{-1} z$ if $z = \tan\gamma$

(f) $\psi = \cot^{-1} w$ if $w = \cot\psi$

B4 ANALYTIC GEOMETRY

B4.1 Equation of a Straight Line

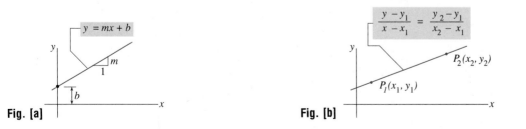

Fig. [a]   **Fig. [b]**

B4.2 Equations of Conic Sections

A *conic section* is any curve which is the locus of a point that moves in such a way that the ratio of its distance from a fixed point to its distance from a fixed line is constant. This ratio is called the *eccentricity*, the fixed point the *focus*, and the fixed line the *directrix* of the curve. These are illustrated as shown, where $\varepsilon = eccentricity$.

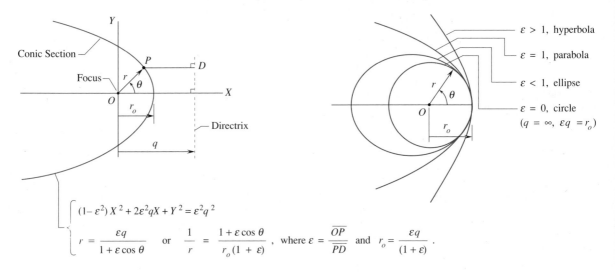

$$(1-\varepsilon^2)\,X^2 + 2\varepsilon^2 qX + Y^2 = \varepsilon^2 q^2$$

$$r = \frac{\varepsilon q}{1+\varepsilon\cos\theta} \quad \text{or} \quad \frac{1}{r} = \frac{1+\varepsilon\cos\theta}{r_o(1+\varepsilon)} \;, \quad \text{where } \varepsilon = \frac{\overline{OP}}{\overline{PD}} \text{ and } r_o = \frac{\varepsilon q}{(1+\varepsilon)}\;.$$

B4.3 Equation of a Circle

Note that $\varepsilon = 0$, $q = \infty$, and $\varepsilon q = R$.

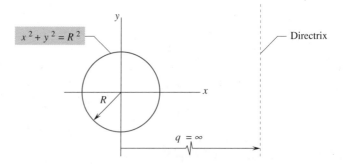

$$x^2 + y^2 = R^2$$

B4.4 Equation of an Ellipse[†]

Note that $a^2 - b^2 = c^2$, $\varepsilon = \dfrac{c}{a}$, $b = a(1-\varepsilon^2)^{1/2}$, and $q = \pm\dfrac{b^2}{c}$.

[†]If a conic section has two foci, the geometric center of the conic section does not lie between a focus and its associated directrix.

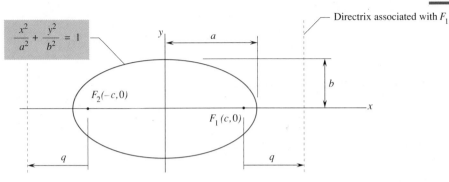

B4.5   Equation of a Parabola

Note that $\varepsilon = 1$ and $q = -2c$.

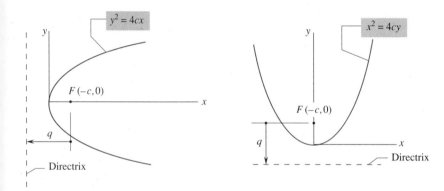

B4.6   Equation of a Hyperbola

Note that $a^2 + b^2 = c^2$, $\varepsilon = \dfrac{c}{a}$, $b = a(\varepsilon^2 - 1)^{1/2}$, and $q = \pm\dfrac{b^2}{c}$.

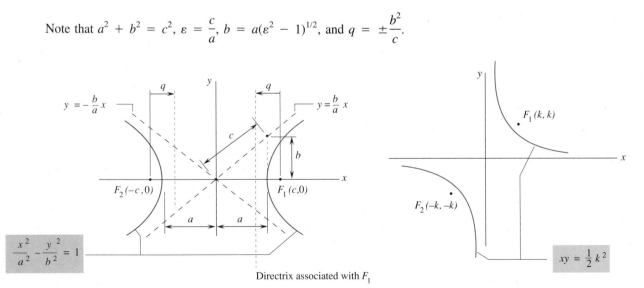

B5   CALCULUS

B5.1   Derivatives

(a) $\dfrac{dC}{dx} = 0$ (C is a constant.)

(b) $\dfrac{dx}{dx} = 1$

(c) $\dfrac{dy}{dt} = \dfrac{dy}{dx}\dfrac{dx}{dt}$

(d) $\dfrac{dx^n}{dx} = n\,x^{n-1}$

(e) $\dfrac{de^x}{dx} = e^x$

(f) $\dfrac{da^x}{dx} = (\ln a)\,a^x$

(g) $\dfrac{d}{dx}(\ln x) = \dfrac{1}{x}$

(h) $\dfrac{d}{dx}(\sin x) = \cos x$

(i) $\dfrac{d}{dx}(\cos x) = -\sin x$

(j) $\dfrac{d}{dx}(\tan x) = \sec^2 x$

(k) $\dfrac{d}{dx}(\cot x) = -\csc^2 x$

(l) $\dfrac{d}{dx}(\sec x) = \sec x \tan x$

(m) $\dfrac{d}{dx}(\csc x) = -\csc x \cot x$

(n) $\dfrac{d}{dx}(uv) = \dfrac{du}{dx}v + u\dfrac{dv}{dx}$

(o) $\dfrac{d}{dx}\left(\dfrac{u}{v}\right) = \dfrac{v\dfrac{du}{dx} - u\dfrac{dv}{dx}}{v^2}$

B5.2   Integrals

(a) $\int u\,dv = uv - \int v\,du$ (integration by parts)

(b) $\int x^n\,dx = \dfrac{1}{n+1}x^{n+1}$ ($n \neq -1$)

(c) $\int \dfrac{dx}{x} = \ln x$

(d) $\int \sin x\,dx = -\cos x$

(e) $\int \cos x\,dx = \sin x$

(f) $\int \tan x \, dx = -\ln(\cos x)$

(g) $\int \cot x \, dx = \ln(\sin x)$

(h) $\int \sec x \, dx = \ln(\sec x \tan x)$

(i) $\int \csc x \, dx = -\ln(\csc x + \cot x)$

(j) $\int \sin^n x \, dx = -\frac{1}{n} \sin^{n-1} x \cos x + \frac{1}{n}(n-1) \int \sin^{n-2} x \, dx$

(k) $\int \cos^n x \, dx = \frac{1}{n} \cos^{n-1} x \sin x + \frac{1}{n}(n-1) \int \cos^{n-2} x \, dx$

(l) $\int \sec^n x \, dx = \frac{1}{n-1} \sin x \sec^{n-1} x + \frac{n-2}{n-1} \int \sec^{n-2} x \, dx$

(m) $\int \csc^n x \, dx = -\frac{1}{n-1} \cos x \csc^{n-1} x + \frac{n-2}{n-1} \int \csc^{n-2} x \, dx$

(n) $\int e^x \, dx = e^x$

(o) $\int (a^2 - x^2)^{3/2} \, dx = \frac{1}{8}\left[2x(a^2 - x^2)^{3/2} \right.$
$$\left. + 3a^2 x(a^2 - x^2)^{1/2} + 3a^4 \sin^{-1}\frac{x}{a} \right]$$

(p) $\int (a^2 - x^2)^{-1/2} \, dx = \sin^{-1}\frac{x}{a}$

(q) $\int (x^2 \pm a^2)^{-1/2} \, dx = \ln[x + (x^2 \pm a^2)^{1/2}]$

(r) $\int x^{-1}(x^2 - a^2)^{-1/2} \, dx = \frac{1}{a} \cos^{-1}\frac{a}{x}$

(s) $\int x^2(a^2 - x^2)^{-1/2} \, dx = \frac{1}{2}\left[-x(a^2 - x^2)^{1/2} + a^2 \sin^{-1}\frac{x}{a} \right]$

(t) $\int (x^2 \pm a^2)^{1/2} \, dx = \frac{1}{2}\{x(x^2 \pm a^2)^{1/2} \pm a^2 \ln[x + (x^2 \pm a^2)^{1/2}]\}$

(u) $\int x^2(x^2 \pm a^2)^{1/2} \, dx = \frac{1}{8}\{2x(x^2 \pm a^2)^{3/2} \mp a^2 x(x^2 \pm a^2)^{1/2}$
$$- a^4\ln[x + (x^2 \pm a^2)^{1/2}]\}$$

(v) $\int x^3(a^2 \pm x^2)^{1/2} \, dx = \frac{1}{15}(\pm 3x^2 - 2a^2)(a^2 \pm x^2)^{3/2}$

(w) $\int (a^2 - x^2)^{1/2} \, dx = \frac{1}{2}\left[x(a^2 - x^2)^{1/2} + a^2 \sin^{-1}\frac{x}{a} \right]$

(x) $\int x(a^2 - x^2)^{1/2} \, dx = -\frac{1}{3}(a^2 - x^2)^{3/2}$

(y) $\int x^2(a^2 - x^2)^{1/2} \, dx = \frac{1}{8}\left\{ -2x(a^2 - x^2)^{3/2} \right.$
$$\left. + a^2 \left[x(a^2 - x^2)^{1/2} + a^2 \sin^{-1}\frac{x}{a} \right] \right\}$$

(z) $\int x^3(a^2 - x^2)^{1/2} \, dx = \frac{1}{15}(3x^2 + 2a^2)(a^2 - x^2)^{3/2}$

B5.3 Differential of a Function

For the purpose of this text, the first-order differential (or variation) δy of a function $y = y(x)$ may be obtained from the relationship

$$\delta y = \frac{dy}{dx} \delta x$$

For example, $\delta(x^2) = 2x \, \delta x$, $\delta(\sin x) = (\cos x) \, \delta x$, etc.

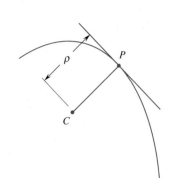

B5.4   Curvature κ and Radius of Curvature ρ

(a) In rectangular coordinates:

$$\kappa = \frac{\left|\dfrac{d^2y}{dx^2}\right|}{\left[1 + \left(\dfrac{dy}{dx}\right)^2\right]^{3/2}} \qquad \rho = \frac{1}{\kappa}$$

(b) In polar coordinates:

$$\kappa = \frac{\left|r^2 + 2\left(\dfrac{dr}{d\theta}\right)^2 - r\dfrac{d^2r}{d\theta^2}\right|}{\left[r^2 + \left(\dfrac{dr}{d\theta}\right)^2\right]^{3/2}} \qquad \rho = \frac{1}{\kappa}$$

B5.5   Expansion of $f(x)$ into Series

(a) Taylor's series:

$$f(x) = f(a) + (x - a)f'(a) + \frac{(x - a)^2}{2!}f''(a) + \cdots$$

$$= \sum_{n=0}^{\infty} \frac{(x - a)^n}{n!}f^{(n)}(a)$$

(b) Maclaurin's series:

$$f(x) = f(0) + xf'(0) + \frac{x^2}{2!}f''(0) + \cdots = \sum_{n=0}^{\infty}\frac{x^n}{n!}f^{(n)}(0)$$

B5.6   Hyperbolic Sine and Cosine Functions

(a) $\sinh x = \frac{1}{2}(e^x - e^{-x}) = x + \dfrac{x^3}{3!} + \dfrac{x^5}{5!} + \dfrac{x^7}{7!} + \cdots$

(b) $\cosh x = \frac{1}{2}(e^x + e^{-x}) = 1 + \dfrac{x^2}{2!} + \dfrac{x^4}{4!} + \dfrac{x^6}{6!} + \cdots$

(c) $\sinh^{-1} x = \ln\left(x + \sqrt{x^2 + 1}\right)$

(d) $\cosh^{-1} x = \ln\left(x + \sqrt{x^2 - 1}\right)$

(e) $\dfrac{d}{dx}(\sinh x) = \cosh x$

(f) $\dfrac{d}{dx}(\cosh x) = \sinh x$

(g) $\int \sinh x\, dx = \cosh x$

(h) $\int \cosh x\, dx = \sinh x$

B5.7 Some Hyperbolic Identities with Mnemonics

The mnemonics associated with the hexagon for hyperbolic functions shown consists of the following basic rules:

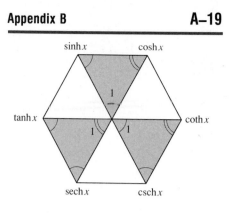

1. The product of the functions at the ends of each diagonal is equal to 1. We have

$$\sinh x \ \mathrm{csch}\, x = 1 \qquad \cosh x \ \mathrm{sech}\, x = 1 \qquad \tanh x \ \coth x = 1$$

2. Any function is equal to the product of its two neighboring functions.

$$\sinh x = \cosh x \ \tanh x \qquad \mathrm{csch}\, x = \mathrm{sech}\, x \ \coth x$$

$$\cosh x = \sinh x \ \coth x \qquad \mathrm{sech}\, x = \tanh x \ \mathrm{csch}\, x$$

$$\tanh x = \sinh x \ \mathrm{sech}\, x \qquad \coth x = \cosh x \ \mathrm{csch}\, x$$

3. In each of the three shaded triangles, the square of the quantity at the upper right side angle is equal to the sum of the squares of the quantities at the other two angles. We have

$$\cosh^2 x = 1 + \sinh^2 x \qquad 1 = \tanh^2 x + \mathrm{sech}^2 x$$
$$\coth^2 x = 1 + \mathrm{csch}^2 x$$

B5.8 Newton-Raphson Method

To extract a root of the equation $y(x) = 0$ by the Newton-Raphson method, we first assume a trial root $x = x_1$. Then the next improved trial root $x = x_2$ is given by $x_2 = x_1 - y(x_1)/y'(x_1)$ where $y(x_1) = y|_{x=x_1}$, and $y'(x_1) = dy/dx|_{x=x_1}$. The procedure is repeated until the correction term $y(x_i)/y'(x_i)$ for the $(i + 1)$th trial is a negligible fraction of the trial root $x = x_i$.

B5.9 Leibnitz's Rule

The taking of a derivative of a definite integral must be consistent with the rule that

$$\frac{d}{dx} \int_\alpha^\beta f(x,t)\, dt = \int_\alpha^\beta \frac{\partial}{\partial x} f(x,t)\, dt + f(x,\beta) \frac{d\beta}{dx} - f(x,\alpha)\frac{d\alpha}{dx}$$

where $\alpha = \alpha(x)$ and $\beta = \beta(x)$.

Appendix C

Infinitesimal Angular Displacements

Finite angular displacements (or finite rotations) are known to disobey the commutative law for the addition of vectors.[†] Thus, *finite* angular displacements do *not* qualify as vectors, despite the fact that they have magnitudes and directions given by the right-hand rule. However, we shall here show that *infinitesimal angular displacements are vectors*.

Let **r** be the position of the particle P of a rigid body, which initially occupies a region indicated by the dashed curve as shown in Fig. C.1. If the body undergoes successively an infinitesimal rotation $d\theta_1$ about the axis OA_1 and then an infinitesimal rotation $d\theta_2$ about the axis OA_2, the particle at P will go first to the point Q and then to the point P' as indicated, where $\overline{PQ} \approx \overline{C_1P}\, d\theta_1$ and $\overline{QP'} \approx \overline{C_2Q}\, d\theta_2$. If such successive infinitesimal rotations took place in reverse order, the particle at P would have gone first to the point R and then to the point P'' as indicated.

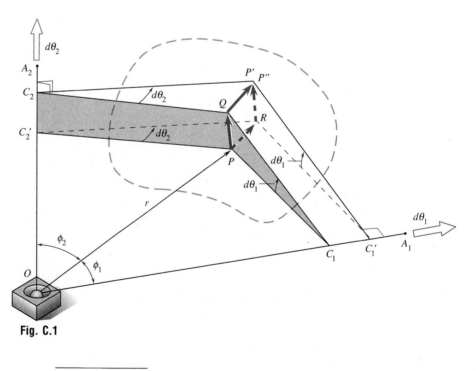

Fig. C.1

[†]Cf. D1.3.

A–20

Since the rotations $d\theta_1$ and $d\theta_2$ are infinitesimal, we infer that, within the framework of *first-order accuracy*, P'' must coincide with P', and

$$\overline{PQ} = \overline{RP'} = (r \sin\phi_1)d\theta_1$$

$$\overline{PR} = \overline{QP'} = (r \sin\phi_2)d\theta_2$$

Recalling that a rotation has a magnitude and a direction consistent with the right-hand rule, we may use the "mechanics" of cross products to write the differential displacements

$$\overrightarrow{PQ} = \overrightarrow{RP'} = \overrightarrow{d\theta}_1 \times \mathbf{r} \tag{C.1}$$

$$\overrightarrow{PR} = \overrightarrow{QP'} = \overrightarrow{d\theta}_2 \times \mathbf{r} \tag{C.2}$$

As the quadrilateral $PQP'R$ is a parallelogram, we have

$$\overrightarrow{PQ} + \overrightarrow{QP'} = \overrightarrow{PR} + \overrightarrow{RP'} \tag{C.3}$$

Substituting Eqs. (C.1) and (C.2) into Eq. (C.3), we write

$$\overrightarrow{d\theta}_1 \times \mathbf{r} + \overrightarrow{d\theta}_2 \times \mathbf{r} = \overrightarrow{d\theta}_2 \times \mathbf{r} + \overrightarrow{d\theta}_1 \times \mathbf{r}$$

$$(\overrightarrow{d\theta}_1 + \overrightarrow{d\theta}_2) \times \mathbf{r} = (\overrightarrow{d\theta}_2 + \overrightarrow{d\theta}_1) \times \mathbf{r} \tag{C.4}$$

Since Eq. (C.4) holds true for the position vector $\mathbf{r}$ of any point P, we must have

$$\overrightarrow{d\theta}_1 + \overrightarrow{d\theta}_2 = \overrightarrow{d\theta}_2 + \overrightarrow{d\theta}_1 \tag{C.5}$$

Thus, *infinitesimal angular displacements* (or infinitesimal rotations, such as virtual angular displacements) *are vectors* because they obey the commutative law for the addition of vectors as shown in Eq. (C.5) and are consistent with the parallelogram law. Since the time variable t is a scalar, we readily see that *angular velocities* $\boldsymbol{\omega}_1 = \overrightarrow{d\theta}_1/dt$, $\boldsymbol{\omega}_2 = \overrightarrow{d\theta}_2/dt$ *and angular accelerations* $\boldsymbol{\alpha}_1 = d\boldsymbol{\omega}_1/dt$, $\boldsymbol{\alpha}_2 = d\boldsymbol{\omega}_2/dt$ *are also vectors.*

Appendix D Simultaneous Equations Solver

```
10 REM   SOLUTION OF SIMULTANEOUS EQUATIONS    (File name: SOSE)
12 REM
14 REM   A program to solve N linear simultaneous equations
16 REM   in N unknowns, [A][X] = [B], via matrix inversion
18 REM
20 REM     This is a BASIC program which runs on IBM PC and
22 REM   compatibles.  With 60K or more bytes free for BASIC,
24 REM   it can accommodate the solution and checking of at
26 REM   least 45 equations in 45 unknowns if running in double
28 REM   precision, and 65 equations in 65 unknowns if running
30 REM   in single precision.  The program automatically prints
32 REM   the DETERMINANT of the coefficient matrix [A] and does
34 REM   the back substitution of the solution for [X] into the
36 REM   original equations to see how well the product [A][X]
38 REM   CHECKs the constant vector [B].
40 REM
42 REM           D I R E C T I O N S :
44 REM   (a) If it is desired to divert the output to a
46 REM       printer, use, instead, the program SOSEP,
48 REM       which is also stored on the diskette.
50 REM   (b) Enter data as illustrated in Lines 1000-4000,
52 REM       then run the program.
54 REM   --------------------------------------------------
56 REM   |           C A U T I O N !                      |
58 REM   |    This program is intended to find unique solu-|
60 REM   | tions.  Sometimes, the rounding-off process with-|
62 REM   | in the computer may prevent the computer from   |
64 REM   | obtaining an exact value of zero for the determi-|
66 REM   | nant of a singular matrix.  Thus, in the output |
68 REM   | of this program, if the product [A][X] in the   |
70 REM   | checking by back substitution differs greatly   |
72 REM   | from the original [B], then the solution for [X]|
74 REM   | should be discarded.  In this case, the matrix  |
76 REM   | [A] is possibly singular or the set of simulta- |
78 REM   | neous equations is ill-conditioned.             |
80 REM   --------------------------------------------------
82 REM   Programmed by: I. C. Jong, University of Arkansas
84 REM
300 DEFDBL A-H,O-Z: REM Omit this line for single precision.
310 CLS: PRINT "There are N equations in [A][X] = [B]."
320 B$ = "Rowwise, the matrix [A] =": G1$ = "Subroutine "
330 C$ = "The constant vector [B] =": G2$="execution time ="
340 PRINT "N =";: READ N: D$ = "The determinant of [A] = "
350 E$ = "In checking, [A][X] gives [B] =": J$ = ") = "
360 F$ = "There is no unique solution for [A][X] = [B]."
370 IF N=0 OR N <> ABS(INT(N)) THEN PRINT " Illegal N!": END
380 PRINT N: ND=N+1: IF N>1 THEN ND=N: P$ = "*IN PROGRESS..."
390 DIM  A(ND,ND), AP(N,N), U(ND,ND), X(N), B(N): PRINT B$
400 FOR I=1 TO N: FOR J=1 TO N: READ A(I,J): AP(I,J)=A(I,J)
410 PRINT A(I,J);: NEXT J: PRINT: NEXT I: PRINT C$
420 FOR I = 1 TO N: READ B(I): PRINT B(I);: NEXT I: PRINT
430 PRINT: PRINT "*PRESS SPACE BAR TO COMPUTE..."
440 SP$ = INKEY$: ON -(SP$ <> CHR$(32)) GOTO 440: PRINT P$
450 PRINT: PRINT D$;: GOSUB 900: C = T: GOSUB 700: GOSUB 900
460 E = T: PRINT D: IF D = 0 THEN PRINT F$: GOTO 540
470 PRINT "The solution yields [X] =": FOR I=1 TO N: X(I)=0
480 FOR J = 1 TO N: X(I) = X(I) + U(I,J)*B(J): NEXT J
490 PRINT SPC(8);"X (";  I; J$; X(I): NEXT I
500 ON -(N<10) GOTO 520: PRINT "*PRESS SPACE BAR TO CHECK.";
510 SP$ = INKEY$: ON -(SP$ <> CHR$(32)) GOTO 510: PRINT
520 PRINT E$: FOR I = 1 TO N: B(I) = 0: FOR J = 1 TO N
530 B(I)=B(I)+AP(I,J)*X(J):NEXT J:PRINT SPC(11);B(I):NEXT I
540 I$ = " second.": K$ = " seconds.": ET$ = " Less than one"
550 IF E >= C+1 THEN ET$ = STR$(E-C): IF E >= C+2 THEN I$=K$
```

```
560 PRINT G1$; G2$; ET$; I$: END
570 REM
700 REM  ..........  SUBROUTINE  DANDI  ..........  (10-27-86)
702 REM 1. To let this subroutine use maximum pivots in find-
704 REM    ing the Determinant and Inverse of [A], do the
706 REM    following: (a) omit Lines 740 and 750, and (b)
708 REM    delete the word REM in each of Lines 745 and 746.
710 REM 2. Upon return: D = Determinant, [U] = Inverse of [A]
720 FOR I=1 TO N:FOR J=1 TO N:U(I,J)=0:IF I=J THEN U(I,J)=1
730 NEXT J: NEXT I: D = 1: N1 = N - 1: FOR I = 1 TO N1
740 ON -(A(I,I) <> 0) GOTO 790: K = I: FOR J = I TO N1
745 REM    BG = ABS(A(I,I)): K = I: FOR J = I TO N1: JP=J+1
746 REM    IF BG < ABS(A(JP,I)) THEN K=JP: BG = ABS(A(K,I))
750 JP = J + 1: IF A(JP,I) <> 0 THEN K = JP: GOTO 770
760 NEXT J: IF A(K,I) = 0 THEN D = 0: RETURN
770 ON -(K=I) GOTO 790: FOR J=1 TO N:P=A(I,J):A(I,J)=A(K,J)
780 A(K,J)=P:P=U(I,J):U(I,J)=U(K,J):U(K,J)=P:NEXT J: D = -D
790 AI = A(I,I): D = D*AI: FOR J=1 TO N: A(I,J) = A(I,J)/AI
800 U(I,J)=U(I,J)/AI: NEXT J: NI=N-I: FOR J=1 TO NI: IJ=I+J
810 AJ = A(IJ,I): FOR K=1 TO N: A(IJ,K) = A(IJ,K)-AJ*A(I,K)
820 U(IJ,K) = U(IJ,K) - AJ*U(I,K): NEXT K: NEXT J: NEXT I
830 D = D*A(N,N): IF D = 0 THEN RETURN
840 FOR I=1 TO N: U(N,I) = U(N,I)/A(N,N): NEXT I: A(N,N)=1
850 FOR I=1 TO N1: NI=N-I: FOR J=1 TO NI: IJ=I+J:AK=A(I,IJ)
860 FOR K = 1 TO N: A(I,K) = A(I,K) - AK*A(IJ,K)
870 U(I,K)=U(I,K)-AK*U(IJ,K): NEXT K: NEXT J: NEXT I:RETURN
900 REM  ..............  SUBROUTINE  TIME  ..............
910 T$=TIME$:H$=MID$(T$,1,2):M$=MID$(T$,4,2):S$=MID$(T$,7,2)
920 T = VAL(H$)*3600 + VAL(M$)*60 + VAL(S$): RETURN
930 REM
1000 DATA 6:     REM Enter N, the number of equations, here.
2000 REM Enter, rowwise, the coefficient matrix [A] below:
2010 DATA .6, .96, -1, 0, 0, 0
2020 DATA .8, -.28, 0, 0, 0, 0
2030 DATA -.6, 0, 0, .384615, 0, 0
2040 DATA -.8, 0, 0, -.923077, 1, 0
2050 DATA 0, -.96, 0, -.384615, 0, .882353
2060 DATA 0, .28, 0, .923077, 0, .470588
3000 REM Enter the constant vector [B] below:
4000 DATA 0, 0, 0, 0, 0, 98.1
```

The preceding program, named SOSE, is to be run under BASIC and is stored on the computer diskette labeled TUTORIAL, as described in App. F. Following the DIRECTIONS contained in the first REMark segment of the program, we show two sample runs as follows:[†]

☐ SOLUTION OF SIMULTANEOUS EQUATIONS IN EXAMPLE 3.8

```
There are N equations in [A][X] = [B].
N = 6
Rowwise, the matrix [A] =
 .6   .96  -1   0   0   0
 .8  -.28   0   0   0   0
-.6   0    0    .384615  0   0
-.8   0    0   -.923077  1   0
 0   -.96  0   -.384615  0   .882353
 0    .28  0    .923077  0   .470588
The constant vector [B] =
 0   0   0   0   0   98.1

The determinant of [A] =  .382262138896008
The solution yields [X] =
        X ( 1 ) =   24.38565007683738
        X ( 2 ) =   69.6732859338211
        X ( 3 ) =   81.51774454257069
        X ( 4 ) =   38.04165216151848
        X ( 5 ) =   54.6238942137679
        X ( 6 ) =   92.38677098912871
In checking, [A][X] gives [B] =
        1.77635683940025D-15
       -4.440892098500626D-16
       -1.332267629550188D-15
        0
       -1.77635683940025D-15
       98.1
Subroutine execution time = 3 seconds.
```

[†]Notice the first item in the DIRECTIONS in the program.

☐ CASE OF SINGULAR COEFFICIENT MATRIX [A]

```
There are N equations in [A][X] = [B].
N = 3
Rowwise, the matrix [A] =
   1   2   3
   4   5   6
   7   8   9
The constant vector [B] =
   1   2   3

The determinant of [A] =  0
There is no unique solution for [A][X] = [B].
Subroutine execution time = Less than one second.
```

Digital Root Finder[†]   # Appendix E

```
10 REM    DIGITAL ROOT FINDER AND PLOTTER     (File name: DIGITAL)
12 REM
14 REM      A program to find the real roots of Y(X) = 0 to any
16 REM      desired digits of accuracy by plotting Y versus X
18 REM
20 REM        This is a BASIC program which runs on IBM PC and
22 REM    compatibles.  It plots in 80 columns from X = XI, with
24 REM    increment DX, to X = XF.
26 REM
28 REM                   D I R E C T I O N S:
30 REM    ----------------------------------------------------------
32 REM    1. If it is desired to divert the output to a printer,
34 REM       use, instead, the program DIGITALP, which is also
36 REM       stored on the diskette.
38 REM    2. To find a root of a given nonlinear equation defined
40 REM       by Y(X) = 0, do the following: (a) define Y = Y(X) in
42 REM       BASIC in Lines 1000-1999; (b) run the program to ex-
44 REM       plore the locations of the roots by making trial runs
46 REM       until the graph crosses the X axis; (c) read the two
48 REM       neighboring digital values of X which bound the root
50 REM       at the crossing of the graph over the X axis; (d) use
52 REM       a smaller increment (e.g., one-tenth of the preceding
54 REM       increment) to plot between those two values of X; (e)
56 REM       repeat steps (c) and (d) until the root is determined
58 REM       to as many digits of accuracy as desired or permitted.
60 REM    ----------------------------------------------------------
62 REM       Programmed by: I. C. Jong, University of Arkansas
64 REM
100 DEFDBL A-H, O-Z: REM Omit this line for single precision.
110 INPUT "XI = "; XI: INPUT "DX = "; DX: INPUT "XF = "; XF
120 N% = INT(ABS((XF - XI)/DX) + 1.5): DIM Y(N%): PRINT
130 PRINT "*IN PROGRESS": R=0: X = XI: FOR I=1 TO N%:  GOSUB 1000
140 Y(I) = Y: X = XI + I*DX: AY = ABS(Y(I)): IF AY>R THEN R = AY
150 NEXT I: SF=R/38: X = XI: FOR I = 1 TO 7: Y$ = Y$+":----!----"
160 NEXT I: Y$ = " !--" + Y$ + ":--!Y": I = 1: PRINT: PRINT
170 PRINT "XI =";XI; "    DX =";DX; "    XF =";XF; "    R =";R
180 PRINT: PRINT "-R"; SPC(37); "0"; SPC(37); "R": PRINT Y$
190 P% = INT(Y(I)/SF + 39 + .5): X% = P% - 40: LT = LEN(STR$(X))
200 Y% = INT(Y(I)/SF-LT-1.5): Z% = 38 - P%: IF P% >= 39 THEN 220
210 PRINT SPC(P%); "*"; SPC(Z%); "!"; X: GOTO 260
220 IF P% = 39 THEN PRINT SPC(P%); "*"; X: GOTO 260
230 IF 44>=P% AND P%>39 THEN PRINT SPC(39)"!"SPC(X%)"*":GOTO 260
240 IF Y% < 0 THEN PRINT SPC(39); "!"; X: GOTO 260
250 PRINT SPC(39); "!"; X; SPC(Y%); "*"
260 X = XI + I*DX: I = I + 1: IF X < XF + DX/3 THEN 190
270 PRINT SPC(39); "X": PRINT: LIST 1000-1999: END
1000 Y = 20 - X*((EXP(50/X)+EXP(-50/X))/2 - 1) ◄────────── Y = Y(X)
2000 RETURN
```

[†]This program is provided for the convenience of those who may find it helpful in finding the root or roots of any nonlinear equation $y(x) = 0$, as well as plotting curves, with a computer. In the plot, the y axis is proportionally calibrated with the computed value of R; however, the x axis is digitally calibrated with any increment chosen by the user.

SOLUTION OF A TRANSCENDENTAL EQUATION (CF. EXAMPLE 8.24.)

XI = 30   DX = 3   XF = 60   R = 32.25048303604126

```
-R                                      0                                        R
!--:----!----:----!----:----!----:----!----:----!----:----!----:----!----:--!Y
   *                                    ! 30
        *                               ! 33
           *                            ! 36
              *                         ! 39
                 *                      ! 42
                    *                   ! 45
                       *                ! 48
                         *              ! 51
                            *           ! 54
                               *        ! 57
                                  * !   ! 60
                                    X
```

1000 Y = 20 - X*((EXP(50/X)+EXP(-50/X))/2 - 1)

(a) A trial run which indicates that the root is located at X > 60

XI = 60   DX = 1   XF = 70   R = 2.067220211029053

```
-R                                      0                                        R
!--:----!----:----!----:----!----:----!----:----!----:----!----:----!----:--!Y
   *                                    ! 60
        *                               ! 61
            *                           ! 62
               *                        ! 63
                  *                     ! 64
                    *                   ! 65
                                        !  *
                                        ! 67    *
                                        ! 68       *
                                        ! 69          *
                                        ! 70             *.
                                    X
```

1000 Y = 20 - X*((EXP(50/X)+EXP(-50/X))/2 - 1)

(b) Next trial run

XI = 65   DX = .1   XF = 66   R = .1979368925094605

```
-R                                      0                                        R
!--:----!----:----!----:----!----:----!----:----!----:----!----:--!Y
   *                                    ! 65
        *                               ! 65.1
            *                           ! 65.2
               *                        ! 65.3
                  *                     ! 65.4
                     *                  ! 65.5
                                        !*
                                        ! 65.7  *
                                        ! 65.8       *
                                        ! 65.9          *
                                        ! 66               *
                                    X
```

1000 Y = 20 - X*((EXP(50/X)+EXP(-50/X))/2 - 1)

(c) Next run[†]

[†]Additional runs and outputs show that Y = 0 when X ≈ 65.586. Since $\cosh(50/x) = [\exp(50/x) + \exp(-50/x)]/2$, we may take $c = 65.59$ as the root of the equation $20 = c(\cosh(50/c) - 1)$ encountered in Example 8.24.

Supplementary Software (Dynamics)

Appendix F

As alluded to in the Preface, diskettes containing supplementary software for this text may be obtained from bookstores or the publisher. The software is designed for an IBM PC® or compatible computer with at least 384K of random access memory, a CGA or compatible graphics display card, and DOS 2.0 or higher.

The BASIC programs PROJECT, SPACFT, SOSE, and DIGITAL have been documented in Secs. 11.12, 12.11, and Apps. D and E of the text. These four BASIC programs plus a 40-column version of DIGITAL, named DIGIT40, and a compiled program named ISE40.EXE (which is an interactive version of SOSE for $N \le 40$), as well as the tutorial modules for all chapters, are stored in the Mech * E TUTORIAL diskette. Of course, the BASIC programs are to be run under a BASIC interpreter (e.g., BASICA.COM or GWBASIC.EXE). They all output to the screen of the monitor. If it is desired to output to a printer which is properly connected, the user may use, instead, the corresponding programs ending with a P (e.g., PROJECTP, SPACFTP, SOSEP, and DIGITALP) which are also stored on that diskette. To run the tutorial modules, the user may do the following: (a) set the computer in DOS, (b) insert the TUTORIAL diskette in drive A, (c) key in TUTOR at the prompt from the DOS (e.g., A:>TUTOR), and (d) make a selection from the menu as directed on the screen.

The four programs TRAJECT, ORBIT, LINKAGE, and VIBRATE are compiled programs designed to complement selected sections in the text. These compiled programs are stored on the Mech * E UTILITY diskette. They are described in this appendix and may be run by doing the following: (a) set the computer in DOS, (b) insert the UTILITY diskette in drive A, (c) key in DYNASOFT at the prompt from the DOS (e.g., A:>DYNASOFT), and (d) make a selection from the menu as directed on the screen.

Naturally, the principal sources of guidance in learning a subject are the *instructor* and the *textbook*. Students are advised to first study the text as assigned by the instructor. Then, as needed, they may utilize the software to strengthen their understanding of the concepts as covered in the following programs:

TUTOR: A Multiple-Choice Tutorial

This program contains ten modules covering Chaps. 11 through 20. It is designed to help students strengthen as well as assess their understanding of basic concepts in engineering mechanics by providing an interactive tutorial consisting of a series of multiple-choice questions related to one chapter at a time. As a user, you are to read each question carefully, go over the prospective answers thoughtfully, and key in the letter (e.g., A, B, C, D, or E) corresponding to the correct or best answer and then press the RETURN key. After your response, a window will appear to provide a comment or a hint. This process will continue until the correct answer is entered.

You receive a full credit for a question if you answer it correctly on the first try, half a credit if you answer it correctly on the second try, and no credit thereafter. Your score is continually updated on the screen. At the end of the quiz, a summary of credits earned for the individual questions, along with the total score, is shown. If you did not make 100% on the quiz, you may elect to retry those questions you missed on the first try. Note that the order of the prospective answers has been randomized and may not be the same when the question is posed again.

In responding to any question, you may enter END to terminate the quiz or SAVE to save the record of the quiz up to the current question for future restart. Entering HELP will recall the foregoing information.

TRAJECT: A Program to Demonstrate Trajectories of a Projectile

This program complements the material covered in Sec. 11.12 of the text. It is designed to demonstrate the free flights of a spherical projectile through a vacuum (as an idealized case) and through a fluid (e.g., air or water). The user can specify the launch parameters (e.g., the initial position and the initial velocity) and the medium in which the projectile will travel. The initial velocity vector of the projectile may be entered in either cartesian or polar form. If the spherical projectile is to travel through a fluid, the diameter of the sphere as well as other data should be entered at the prompt from the computer. The program plots the trajectory starting from an initial position and ending in a position which is established by specifying the x coordinate, y coordinate, or time. Then, a menu is presented for selections to plot multiple trajectories, change the launch parameters or medium, or exit the program. To access the help screen, press the F1 key.

ORBIT: A Program to Simulate Space Flight

This program complements the material covered in Secs. 12.8 through 12.12. It is designed to provide the user with a computer simulation of launching, controlling, and landing a spacecraft flying in the earth-moon system. The spacecraft with a given amount of fuel flies from the earth. The user controls the forward or reverse thrust engines by using the up, down, left, and right cursor keys. The spacecraft can be placed in orbit around the earth, flown to the moon, orbited around the moon or landed on it, and returned to the earth. A ''time accelerator'' is on board with which the user can make time pass up to 2000 times faster. For example, a 90-minute orbit around the earth would take only 2.7 seconds, and a 2-week lunar excursion can be done in about 10 minutes. The time accelerator is controlled by using the Pg Up and Pg Dn keys; it can be toggled on and off by using the END key.

The program constantly displays a control panel showing various flight parameters such as the settings of the thrusters, the elapsed time since launch, its speed and acceleration, the apogee and perigee in an orbit, the period of orbit, the current distance from the earth and moon surfaces, and the remaining fuel supply. Pressing the F1 key at any time brings up a help screen, and pressing the ESCAPE key exits the program.

LINKAGE: A Program to Demonstrate Motion of a Fourbar Linkage

This program complements the material covered in Secs. 15.5 through 15.10. It is designed to allow a user to create, modify, and analyze the plane motion of fourbar linkages. A fourbar linkage consists of a crank, a coupler, a follower, and the ground link. The user is asked to specify the initial coordinates of the four connecting pins in the system. The path, velocity, and acceleration of a chosen point on the coupler can be continuously monitored. After a pause to compute the kinematic quantities involved in the plane motion of the linkage, the program graphically displays the linkage and animates its motion as the crank rotates. If the mechanism locks up, the rotation of the crank is confined to the operable range. The user has several options while the mechanism is being displayed. For example, pressing the + or − keys speeds or slows the rotation of the crank, pressing the S key stops the motion and displays a kinematic analysis of the mechanism at the selected position, pressing the R key resumes the motion, and pressing the T key toggles a trace of the path of the chosen point on the coupler. The user may press the F1 key for help in entering any data or command, and press the ESCAPE key to exit the program.

VIBRATE: A Program to Plot Vibrations of a SDOF System

This program complements the material covered in Secs. 20.1 through 20.6. It is designed to allow the user to design and analyze the free and forced vibration of a single-degree-of-freedom (SDOF) system and to plot response curves. The user is asked to specify the mass, the modulus of the spring, the viscous damping coefficient, the initial displacement and velocity of the mass, and the parameters A, B, C, and D in the forcing function $F = A * \sin(B * t) + C * \cos(D * t)$. The program then calculates the damping factor and other characteristics associated with the vibration.

The program animates the motion of the mass and plots the response curves. By selecting from the menu, the user can compare the responses of previously defined systems by showing them on the same plot, redefine the current system, or exit the program. A help screen is available at any time by pressing the F1 key.

Answers to Selected Developmental Exercises and Problems

Chapter 11

Developmental Exercises

D11.5 (a) -48 ft/s, (b) 4 s, (c) 30 ft, (d) 124 ft. **D11.6** $\Delta x = 20$ m, $x_T = 50$ m, $v_{\text{avg}} = 4$ m/s, $|v|_{\text{avg}} = 10$ m/s. **D11.7** (a) $a_0 = 6$ ft/s^2, $a_6 = -30$ ft/s^2; (b) -12 ft/s^2. **D11.9** $v = v_0 + (A/\omega) \sin \omega t$, $x = x_0 + v_0 t - (A/\omega^2) \cos \omega t$. **D11.10** $v = -(A/\omega) \cos \omega t$, $x = -(A/\omega^2) \sin \omega t$. **D11.12** $x = (v_0/\omega) \sin \omega t$. **D11.13** $v = v_0/(1 + bv_0 t)$, $x = (1/b) \ln (1 + bv_0 t)$. **D11.14** (a) -40 ft/s, (b) 2.48 s. **D11.15** (a) $x_A < x_B$, (b) $v_C > v_D$, (c) $a_D < a_C$. **D11.16** 2. **D11.17** 2 m/s $\uparrow$. **D11.18** $\Delta x = 7.5$ ft, $x_T = 107.5$ ft. **D11.19** $\tau = 3.46$ s, $t_2 = 11.46$ s. **D11.20** (a) 62.5 m, (b) 77.5 m. **D11.25** a, d, e. **D11.26** (a) $2\mathbf{i} + 3\mathbf{j} - 3\mathbf{k}$, (b) -6, (c) $6\mathbf{i} - 9\mathbf{j} - 5\mathbf{k}$. **D11.27** $2t^3\mathbf{i}$. **D11.28** $v = 12.81$ m/s, $\theta_v = 128.7°$, $a = 10.77$ m/s^2, $\theta_a = 248.2°$. **D11.29** 3.1 ft/s^2 $\downarrow$. **D11.31** (a) 55.7 s, (b) 15.08 s. **D11.32** (a) 29.7 s, (b) 14.30 s. **D11.33** (a) $\theta_1 = 55.0°$, $\theta_2 = 49.0°$; (b) the projectile cannot reach the target. **D11.36** b. **D11.37** 30.6 m. **D11.38** 4.5 m/s^2. **D11.39** (a) No, (b) no. **D11.40** $\mathbf{v} = 4.44$ ft/s $\nearrow$ 45°, $\mathbf{a} = 22.1$ ft/s^2 $\searrow$ 26.6°. **D11.41** Yes. **D11.42** (a) $4\pi\mathbf{i} - 80\pi^2\mathbf{j} + (200\pi^3 - \pi)\mathbf{k}$ m^2/s^3, (b) $0.00201\mathbf{i} - 0.1264\mathbf{j} + 0.992\mathbf{k}$, (c) 7.26°, (d) 7.26°.

Problems

11.1 $t_1 = 2$ s, $x_1 = 10$ m, $a_1 = -6$ m/s^2, $t_2 = 4$ s, $x_2 = 6$ m, $a_2 = 6$ m/s^2. **11.3** $t = 3$ s, $x = 8$ m, $v = -3$ m/s. **11.4** $t = 4$ s, $x = 57$ mm, $v = -3$ mm/s. **11.6** (a) 3 s or 6 s, (b) 297 in. **11.7** $x = t^4 - 32t + 48$ m. **11.8** (a) 6 s, (b) 608 in. **11.11** (a) ± 4 s^{-1}, (b) $\pi/8$ s. **11.12** (a) ± 6 m/s, (b) 3.46 m, (c) 2 m, (d) 8 m/s. **11.14** (a) 12.77 km, (b) 1.593 Mm, (c) 25.5 Mm. **11.16** (a) 11.18 km/s, (b) 6.95 mi/s. **11.17** (a) 1.195×10^{-3} ft^{-1}, (b) 31.7 s. **11.19** (a) 1910 ft, (b) 21.7 s. **11.20** (a) 99.9 ft, (b) 81.6 ft/s. **11.21** 107.3 m. **11.24** (a) 11.25 s, (b) 309 m. **11.26** (a) -2.5 m/s^2, (b) 10 s. **11.27** 1.874 mi. **11.28** $v = [(b - c)g - bk]e^{-bt}/[b(b - c)] + ke^{-ct}/(b - c) - g/b$, $x = [(b - c)g - bk](1 - e^{-bt})/[b^2(b - c)] + k(1 - e^{-ct})/[c(b - c)] - gt/b$. **11.29** $\mathbf{v}_B = 1$ ft/s $\downarrow$, $\mathbf{a}_B = 1$ ft/s^2 $\uparrow$. **11.30** $\mathbf{v}_B = 0.75$ ft/s $\uparrow$, $\mathbf{a}_B = 0.25$ ft/s^2 $\downarrow$. **11.32** (a) 1.6 m/s $\leftarrow$; (b) $\mathbf{v}_C = 0.8$ m/s $\leftarrow$, $\mathbf{v}_D = 2.4$ m/s $\leftarrow$; (c) 0.4 m/s $\rightarrow$. **11.34** 100 mm/s $\rightarrow$. **11.35** (a) 8 m, (b) 8 m/s $\uparrow$. **11.36** $\mathbf{v}_D = 0.4$ m/s $\uparrow$, $\mathbf{a}_D = 9$ mm/s^2 $\uparrow$.

11.38 $t = 4.5$ s.

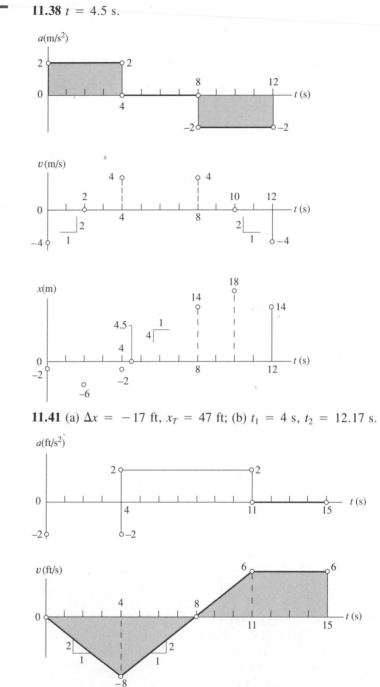

11.41 (a) $\Delta x = -17$ ft, $x_T = 47$ ft; (b) $t_1 = 4$ s, $t_2 = 12.17$ s.

11.44 (a) 3.08 ft/s^2, (b) 378 ft.   **11.45** (a) 0.5 m/s^2, (b) 20 m/s.
11.46 (a) To avoid collision, L_{min} = 227 m > L = 210 m; (b) 9.54 m/s
→; (c) $(x_T)_A$ = 121.9 m, $(x_T)_B$ = 88.1 m.   **11.47** (a) a_A = − 1.667
m/s^2, a_B = − 1.25 m/s^2 ; (b) $(x_T)_A$ = 120 m, $(x_T)_B$ = 90 m.   **11.48**
14.25 s.   **11.49** 6 s.   **11.51 v** = 50 ft/s ∡ 53.1°, **a** = 49.5 ft/s^2 ∡
76.0°.   **11.52 v** = − 12**i** + 3**j** m/s, **a** = − 12**i** m/s^2.   **11.55** v =
41 m/s, a = 40 m/s^2.   **11.56** 48 mm/s →.   **11.57** 30.6 mm/s^2 →.
11.59 − 1.358**i** − 1.019**j** mm/s^2.   **11.60** 6.26 m/s < v_0 < 9.40 m/s.
11.61 1.193 m < L < 3.193 m.   **11.62** 7.38 ft/s < v_0 < 16.82 ft/s.
11.63 (a) $v_0^2(\sin 2\theta)/g$, (b) 45°, (c) v_0^2/g, (d) $v_0^2/(4g)$.   **11.65** (a) 45°,
(b) 4h.   **11.66** (a) 45°, (b) 14.68 m.   **11.67** (a) 41.3°, (b) 22.7 m.
11.69 23.9°.   **11.70** 40.9°.   **11.72** 6.63 ft.   **11.73** (a) θ_1 =
55.0°, θ_2 = 31.4°; (b) t_1 = 69.7 s, t_2 = 46.9 s.   **11.74** 1.55 h.
11.76 (a) 15.17 min, (b) 29.6°.   **11.77** (a) 82.5 km/h ⦨ 49.1°,
(b) 45.8 m ⦨ 49.1°, (c) 90.1 m.   **11.79** 71.6° ⤵ .   **11.80** 79.7 km/h.
11.82 2.81 m/s^2.   **11.83** v_B = 198.1 m/s, θ = 60°.   **11.84** (a)
4.77 mi/s, (b) 1.542 h.   **11.86** 250 m.   **11.87** 44.8 ft/s.   **11.88**
$\mathbf{a}_t$ = 1.732 mm/s^2 ⦫ 30°, $\mathbf{a}_n$ = 1 mm/s^2 ⦨ 60°.   **11.89 v** = 400 m/s
↑ , **a** = 18.48 m/s^2 ↑ .   **11.90 v** = $L\dot{\theta}/\cos^2 \theta$ ↑ .   **11.92 v** = $\pi/3$
m/s ↓ , **a** = $\pi^2/12$ m/s^2 →.   **11.93 v** = 1.888 m/s ⦫ 33.7°, **a** =
1.371 m/s^2 ⦨ 36.9°.   **11.95** v = 1.047 ft/s, a = 2.21 ft/s^2.   **11.96**
v = 4.56 ft/s, a = 6.61 ft/s^2.   **11.98** $\dot{r}$ = 0.528 m/s, $\ddot{r}$ = − 0.1114
m/s^2, $\dot{\theta}$ = − 0.0616 rad/s, $\ddot{\theta}$ = − 0.1398 rad/s^2.   **11.99** $\dot{r}$ = 30 mm/s,
$\ddot{r}$ = 2.76 mm/s^2, $\dot{\theta}$ = − 0.0530 rad/s, $\ddot{\theta}$ = 0.00325 rad/s^2.   **11.101**
4.55°.   **11.103** v = 2.96 m/s, a = 0.956 m/s^2.   **11.104** v = 2.47
ft/s, a = 2.84 ft/s^2.   **11.105** (a) ±3 s^{-1}, (b) $\pi/6$ s.   **11.106** 0.5 m/s
↑ .   **11.107** Δx = 6 ft, x_T = 66 ft, t_1 = 5.16 s, t_2 = 8.75 s.   **11.108**
θ_1 = 43.3°, θ_2 = 58.0°.   **11.109** 21.0 Mm.   **11.110 v** = 231 m/s
↑ , **a** = 52.9 m/s^2 ↑ .

Chapter 12

Developmental Exercises

D12.3 (a) m, kg, s, N; (b) ft, lb, s, slug.   **D12.4** (a) 98.1 N, (b) 98.1
× 10^3 N, (c) 98.1 × 10^{-6} N, (d) 98.1 × 10^{-3} N, (e) 322 lb, (f) 10
lb.   **D12.6** 0.1147.   **D12.7** 8.61 ft/s^2 ↓ .   **D12.8** 17.21 kg.
D12.9 (a) 24**i** m/s^2, (b) 48**i** m/s, (c) 64 m.   **D12.10** (a) − 1280**k** m/s^2,
(b) − 1280**k** m/s, (c) − 1024 m.   **D12.11** 68.0 km/h.   **D12.12** 5.11
m/s.   **D12.13** (a) 0, (b) 0.   **D12.14** (a) (40/41)**e**$_\theta$ + (9/41)**e**$_z$, (b)
12.68°, (c) 42.0 m.   **D12.15** No.   **D12.16** (a) − 15**i** − 314**j** −
157.6**k** lb·ft, (b) no.   **D12.18** 5.97 × 10^{24} kg.   **D12.19** (a) Yes, (b)
no.   **D12.22** (a) 10.75 × 10^3 mi^2/s, (b) 4.96 × 10^3 mi, (c) 800 mi, (d)
4.90 × 10^3 mi, (e) 76.3 × 10^6 mi^2, (f) 1.970 h.   **D12.23** Yes.
D12.24 (a) Hyperbola, (b) parabola, (c) ellipse, (d) circle.   **D12.28** (a)
4.34 × 10^3 mi, (b) 4.33 × 10^3 mi, (c) 59.0 × 10^6 mi^2, (d) 10.18 × 10^3
mi^2/s, (e) 1.611 h.   **D12.29** (a) 1.981 h, (b) 7.82 h.   **D12.31** r_P =
28.6 × 10^6 mi, r_Q = 43.4 × 10^6 mi.

Problems

12.2 (a) 6.44 ft/s^2 ↓, (b) 2.5 lb.   **12.3** 0.2 N/m.   **12.4** (a) 32.2 ft/s^2 ↑, (b) 10.73 ft/s^2 ↑, (c) 1.4 ft/s^2 ↑.   **12.6** (a) 23.6 ft/s^2 ↙, (b) 14.19 ft/s^2 ↓, (c) 11.19 lb.   **12.9** (a) 7.36 m/s^2 →, (b) 490 N.   **12.10** 42.7 m.   **12.12** a_A = 3.21 m/s^2, a_B = 1.507 m/s^2.   **12.13** a_A = 1.512 m/s^2, a_B = 0.   **12.14** (a) 39.1 lb, (b) 35.5 lb.   **12.15** (a) 2.92 m/s^2 ↘, (b) 5.66 N.   **12.17** (a) 7.51 ft/s^2, (b) 66.7 lb ←, (c) 100 lb →.   **12.18** (a) $\mathbf{a}_A$ = 2.03 m/s^2 →, $\mathbf{a}_B$ = 2.71 m/s^2 →; (b) 30.5 N.   **12.21** (a) $\mathbf{a}_A$ = 3.41 ft/s^2 →, $\mathbf{a}_B$ = 6.81 ft/s^2 →; (b) 4.97 lb.   **12.22** (a) 47.4 lb; (b) T_1 = 83 lb, T_2 = 41.5 lb.   **12.24** (a) 150.7 lb; (b) T_1 = 88 lb, T_2 = 44 lb.   **12.25** (a) 170.8 kg; (b) T_1 = 889 N, T_2 = 445 N.   **12.26** (a) 9.33 kg; (b) T_1 = 97.1 N, T_2 = 194.2 N.   **12.28** a_A = 0.892 m/s^2 ↑, $\mathbf{a}_B$ = 0.892 m/s^2 ↑, $\mathbf{a}_C$ = 2.68 m/s^2 ↓, T_1 = 428 N, T_2 = 214 N.   **12.29** $\mathbf{a}_A$ = 6.99 ft/s^2 ↑, $\mathbf{a}_B$ = 2.81 ft/s^2 ↓, $\mathbf{a}_C$ = 12.61 ft/s^2 ↓, $\mathbf{a}_D$ = 26.6 ft/s^2 ↑, $\mathbf{a}_E$ = 26.3 ft/s^2 ↓, $\mathbf{a}_F$ = 17.51 ft/s^2 ↓, T_1 = 182.5 lb, T_2 = 91.3 lb, T_3 = 45.6 lb.   **12.32** x = 424 m, y = −912 m, z = 1578 m.   **12.34** 440**i** + 1824**j** − 1946**k** ft.   **12.36** 132.5 N.   **12.37** (a) 11.55 lb, (b) 7.47 ft/s.   **12.39** T_{AB} = 35.0 N, T_{BC} = 84.2 N.   **12.40** k_A = 0.5, k_B = 0.704, k_C = 0.296, k_D = 0.2.   **12.42** (a) 49.0 N; (b) T_{AB} = 62.8 N, **a** = 5.89 m/s^2 ⦨ 36.9°.   **12.43** (a) 4.57 mi, (b) 2.09° s^{-1}.   **12.44** 195.3 km/h.   **12.46** 1.115 rev/s.   **12.47** 0.657.   **12.49** 0.408.   **12.50** ρ = 1569 mi, a_t = 28.8 ft/s^2.   **12.52** $\mathbf{F}_f$ = 19.29 N ↙, $\mathbf{F}_s$ = 36.1 N ↙.   **12.54** (a) 1.5 m, (b) 2 N.   **12.56** (a) 3.25 lb, (b) 2.5 lb.   **12.58** (a) −54.4**i** − 55.4**j** − 7.5**k** N, (b) 68.7 N C.   **12.59** (a) −49.0**i** − 199.9**j** − 3**k** lb, (b) 195.8 lb C.   **12.61** (a) −1.2**i** − 0.8**j** + 0.9**k** m, (b) 1.6**i** + 0.7**j** − 2.9**k** m/s, (c) −**i** − 10**j** + 2.5**k** m/s^2.   **12.63** $\mathbf{a}_{1/G}$ = **i** + 12.5**k** m/s^2, $\mathbf{a}_{2/G}$ = −4**i** − 6.5**k** m/s^2, $\mathbf{a}_{3/G}$ = 8.5**i** − 2.5**k** m/s^2.   **12.65** $\mathbf{a}_{1/G}$ = 15.9**i** + 2**j** − 26**k** ft/s^2, $\mathbf{a}_{2/G}$ = 2.4**i** − 10**j** + 4**k** ft/s^2, $\mathbf{a}_{3/G}$ = −14.1**i** + 10**j** + 14**k** ft/s^2.   **12.66** (a) 0.1380, (b) 6.95 km/s.   **12.69** (a) 3.38 mi/s, (b) 11.77 × 10^3 mi^2/s, (c) 110.0 × 10^6 mi^2, (d) 2.60 h.   **12.70** (a) 563 mi, (b) 4.60 mi/s.   **12.71** 22.2 × 10^3 mi or 35.8 Mm.   **12.73** 10.59 × 10^3 mi.   **12.76** (a) 2.50 Gm; (b) r_P = 147.0 Gm, r_Q = 152.0 Gm; (c) 146.3 Gm.   **12.77** (a) 2.22 × 10^3 Mm^2/s, (b) 333 × 10^3.   **12.78** (a) 6.37 km/s, (b) 3.95 km/s, (c) 6.32 km/s.   **12.80** (a) 1.706 km/s, (b) 1.579 km/s, (c) 31.3 m/s.   **12.81** (a) 2.05 h, (b) 0.967 h.   **12.82** 1.487 km/s.   **12.84** 72.8°.   **12.86** (a) $\frac{1}{2}R\sqrt{(1 + \varepsilon)gr_P}$, (b) $R[(g/r_P)(1 + \varepsilon^2 + 2\varepsilon \cos\theta)/(1 + \varepsilon)]^{1/2}$, (c) $\tan^{-1}[(1 + \varepsilon\cos\theta)/(\varepsilon\sin\theta)]$.   **12.87** $[(2r_P^3/g)^{1/2}/(3R)][3 + \tan^2(\theta/2)]\tan(\theta/2)$.   **12.89** 2.17 h.   **12.90** (a) 19.26 kg, (b) 339 N.   **12.91** (a) 64 kN, (b) 32 kN T, (c) 14 kN T.   **12.92** 97.5 km/h.   **12.93** (a) 4.34 Mm, (b) 5.50 km/s, (c) 0.1849, (d) 29.5 × 10^3 km^2/s, (e) 2.38 h.   **12.94** (a) 1.412 mi/s, (b) 0.998 mi/s, (c) 2.06 h.   **12.96** 30.3 km/s.

Chapter 13

Developmental Exercises

D13.2 (a) −200 J, (b) 200 J, (c) −100 J.   **D13.3** 1000 ft·lb.   **D13.4** (a) dx, (b) $6x\,dx$, (c) 3 J.   **D13.5** (a) $2(1 − x)\,dx$, (b) $12\,x(1 −$

x) dx, (c) 2 J; no, **F** is not a conservative force.   **D13.7** No.   **D13.8** (a) Zero, (b) negative, (c) positive.   **D13.9** (a) 769 W, (b) 0.961. **D13.10** (a) Yes, (b) yes, (c) no, (d) no.   **D13.11** -53.1 MJ. **D13.12** $U_W = 117.7$ J, $U_s = 225$ J.   **D13.15** $-mv^2/2$.   **D13.16** a, c, d.   **D13.18** 6.56 Mm.   **D13.19** 56.4°.   **D13.20** 1.835 m $\downarrow$. **D13.21** (a) 16 ft·lb, (b) yes.   **D13.22** 25.2 ft·lb   **D13.26** At infinity.   **D13.30** Suddenly.   **D13.33** No.   **D13.35** 8.02 ft/s. **D13.38** 1.635 m/s² $\leftarrow$.

Problems

13.2 (a) $T = 16.18$ J, $v = 2.54$ m/s, (b) $T = 98.1$ J, $v = 6.26$ m/s. **13.4** (a) 10.80 m, (b) 4.32 m/s.   **13.5** 42.1 ft/s.   **13.6** (a) 13.09 ft/s, (b) 28.0 lb.   **13.8** 2.68 m/s.   **13.10** 6.33 mi/s.   **13.11** 4.54 mi/s. **13.14** 6.02 Mm.   **13.15** 4.35×10^3 mi.   **13.16** 6.66 Mm. **13.17** 8.88 m/s.   **13.19** 1.449 m/s.   **13.20** (a) 1 ft, (b) 4.01 ft/s. **13.21** (a) 3.61 m/s, (b) 0.986 m.   **13.24** (a) 13.19 in., (b) 4.60 ft/s. **13.25** 4.79 ft/s.   **13.27** 2.74 ft/s.   **13.28** 5.49 m/s.   **13.29** 4.15 m/s. **13.31** 2.47 m/s.   **13.33** 0.1462.   **13.34** 2.94 m/s² $\measuredangle$ 28.1°. **13.35** 39.5 mi/h.   **13.36** 38.6 mi/h.   **13.38** 0.114 m/s² $\uparrow$. **13.51** (a) $\sqrt{2gR}$, (b) no.   **13.53** (a) $gmR^2(r_Q - r_P)/[2r_P(r_P + r_Q)]$, (b) $gmR^2(r_Q - r_P)/[2r_Q(r_P + r_Q)]$, (c) $gmR^2(r_Q - r_P)/(2r_Pr_Q)$.   **13.55** (a) 9.77 m/s, (b) 17.70 ft/s.   **13.57** 1.274 m.   **13.58** (a) 0.709 m, (b) 2.81 m/s.   **13.59** 1.2 m.   **13.60** 1.414 m/s.   **13.62** 6.50 in. **13.63** 6.61 in.   **13.64** 2.02 m/s.   **13.65** (a) 89.2 N, (b) no. **13.67** 18.39 ft/s.   **13.78** 0.25.   **13.79** (a) 1.5 ft, (b) 4.91 ft/s. **13.81** 9.00 m/s.   **13.84** 37.4 hp.

Chapter 14

Developmental Exercises

D14.2 $9\mathbf{i} - 6\mathbf{j} - 27\mathbf{k}$ N·s.   **D14.3** (a) $12.42\mathbf{i} - 15.53\mathbf{j} + 18.63\mathbf{k}$ lb·s, (b) $45\mathbf{j} - 24\mathbf{k}$ N·s.   **D14.5** 2.73 s.   **D14.8** $-155.2\mathbf{i}$ kg·m²/s. **D14.9** 40 kN.   **D14.10** 2.68 m/s.   **D14.11** $F_1 = 235$ N, $F_2 = 117.7$ N, $\mathbf{v}_A = 4.89$ m/s $\uparrow$, $\mathbf{v}_B = 7.89$ m/s $\downarrow$, $\mathbf{v}_C = 3.77$ m/s $\downarrow$. **D14.12** $t_1 = 0.5$ s, $x_1 = 378$ m.   **D14.14** 98.3 lb $\measuredangle$ 33.9°. **D14.15** 668×10^9 lb·ft·s.   **D14.16** $v_{max} = 5.53$ mi/s, $v_{min} = 2.45$ mi/s. **D14.17** 8.66 kg·m²/s.   **D14.18** $v_{max} = 12.25$ m/s, $v_{min} = 7.07$m/s. **D14.21** 34.9°.   **D14.22** 0.5.   **D14.24** 18.44%.   **D14.25** 0.577. **D14.27** 3.6 m/s $\uparrow$.   **D14.31** 160 N $\measuredangle$ 60°.   **D14.32** 6.37×10^3 hp.   **D14.33** $\mathbf{B}_x = 12.44$ lb $\rightarrow$, $\mathbf{B}_y = 68.7$ lb $\uparrow$.   **D14.34** (a) $\frac{1}{2}\rho_L g L^2 + \rho_L L v_0^2$, (b) $\frac{1}{2}\rho_L L v_0^2$, (c) $\frac{1}{2}\rho_L g L^2$, (d) $\frac{1}{2}\rho_L L v_0^2$, (e) $\rho_L g(L - y)$. **D14.36** 30 kN.

Problems

14.1 615 lb.   **14.3** 29.6 km/h.   **14.4** 7.95 s.   **14.6** (a) 4 s, (b) 7 s, (c) 11.27 ft/s, (d) 9.65 s.   **14.7** (a) $-12\mathbf{i} + 6\mathbf{j} - 4\mathbf{k}$ N, (b) 7 m/s², (c) 18 m/s.   **14.9** (a) $v_y = 2$, $v_z = -8$, (b) $84\mathbf{j}$ kg·m²/s.   **14.11** $-231\mathbf{i} + 93\mathbf{j} + 48\mathbf{k}$ kg·m²/s.   **14.12** $\theta_A = 52.0°$, $\theta_B = 47.7°$.

14.13 0.4 m/s.   **14.15** (a) 4.75 ft/s, (b) 1.475 lb, (c) 151.1 lb.
14.16 (a) 0.8 m/s, (b) 15 kN, (c) 8.57%.   **14.17** v_B = 3.55 m/s, v_C =
1.383m/s.   **14.19** 1498 ft/s.   **14.20** 43.0 m/s.   **14.23** 14.71 kips
∇ 84.0°.   **14.24** 31.7 N $\measuredangle$ 10.89°.   **14.25** 3.75 m/s.   **14.28**
17.04 ft/s.   **14.29** 29.4 lb.   **14.31** 89.3 lb.   **14.32** 107.2 kg.
14.33 18.62 kg.   **14.35** T_1 = 215 N, T_2 = 431 N, v_A = 5.25 m/s ↓,
v_B = 2.38 m/s ↓, v_C = 5.01 m/s ↑.   **14.36** T_1 = 19.19 lb, T_2 =
38.4 lb, T_3 = 19.19 lb, v_A = 9.77 ft/s ↑, v_B = 18.01 ft/s ↑, v_C =
2.59 ft/s ↓, v_D = 6.24 ft/s ↑, v_E = 33.5 ft/s ↓, v_F = 23.2 ft/s ↓.
14.37 v_A = 112.3 ft/s, v_B = 122.0 ft/s, v_C = 146.3 ft/s.   **14.39** (a)
30**i** m/s; (b) 111**i** + 45**j** m; (c) 1.5 s; (d) $(v_A)_y$ = −34, $(v_B)_y$ = 16, $(v_C)_y$
= 15.   **14.42** (a) 76.5°, (b) 5.07 km/s.   **14.43** (a) 449 m, (b) 82.8°
< ϕ_S < 97.2°.   **14.44** (a) 80.7° or 99.3°, (b) 0.1822, (c) 2.01 h.
14.46 4.88 km/s.   **14.47** (a) 14 m/s, (b) 0.742 m.   **14.49** 12.15
m/s.   **14.50** (a) 13.65 ft/s, (b) 8.89 in.   **14.52** (a) 0.949 m, (b)
1.896 m/s.   **14.53** 6.39 m/s.   **14.55** (a) v'_A = 6.2 ft/s ←, v'_B = 5.8
ft/s →, (b) 35.3%.   **14.56** (a) 0.904 m/s ←, (b) 1.021 m →.   **14.58**
(a) 0.633, (b) 2.55 ft/s →.   **14.59** v'_A = 6.5 m/s $\measuredangle$ 60°, v'_B = 3.9
m/s ←.   **14.60** v'_A = 8.76 m/s ∇ 51.2°, v'_B = 2.20 m/s →.   **14.61**
v'_A = 3.99 m/s $\measuredangle$ 11.23°, v'_B = 0.978 m/s →.   **14.63** 1.091 m.
14.65 53.8 kN.   **14.76** 475 lb.   **14.78** 52.8 lb.   **14.79** 1.8 kN.
14.80 1 kN.   **14.83** **B** = −160**i** − 162**j** N, **C** = 214**i** N.   **14.84** **B**
= −300**i** − 1106**j** N, **C** = 1.344**j** kN.   **14.85** 2.03 in.   **14.87** (a)
900 kW, (b) 657 kW, (c) 73.0%.   **14.88** 15.80 kN.   **14.90** 771
km/h.   **14.91** (i) 55.8 kN, (ii) 828 km/h.   **14.92** $3\rho_L g(L - y)$.
14.93 (a) $\sqrt{2gy/3}$, (b) $gt^2/6$, (c) $\rho_L g L^2/6$.   **14.94** $3\rho_L gy$.   **14.96** 3.58
tons/s.   **14.97** 11.65 kg/s.   **14.99** (a) 1.141 km/s, (b) 6.54 km/s.
14.100 (a) 10.74 m/s →, (b) 10.96 s.   **14.101** 702 m/s.   **14.102** (a)
0.270 s, (b) 0.289 m →, (c) 0.504 m →.   **14.103** (a) 0.324 s, (b)
0.295 m →, (c) 0.605 m →.   **14.105** 3.01 s.   **14.106** r_A = 99**i** +
18**j** ft, v_A = 23**i** + 6**j** ft/s.   **14.107** 1.25 rad/s.   **14.108** a_{max} =
9.48 × 10³ mi, a_{min} = 520 mi.   **14.109** 0.371 kg.   **14.110** (a) 2.5
s, (b) −0.2 ft/s².   **14.111** 0.841.

Chapter 15

Developmental Exercises

D15.2 (a) Rotation, (b) translation, (c) general plane motion, (d)
translation.   **D15.3** No.   **D15.4** (a) 375 rev, (b) 20 s.   **D15.5** $\Delta\theta_A$
= 75 rad $\curvearrowleft$, $\Delta\theta_B$ = 25 rad $\curvearrowright$.   **D15.6** 0.942 m/s.   **D15.7** ω = 4
rad/s $\curvearrowright$, α = 30 rad/s² $\curvearrowright$, a_P = 102 in./s² ∇ 61.9°.   **D15.11** The
equation linking the motions of the key points of the system.   **D15.12**
(a) Zero, (b) $\omega_{AB} \times \overrightarrow{AB}$.   **D15.13** 90 rpm $\curvearrowleft$.   **D15.16** Yes.
D15.17 Generally, a_C ≠ **0**.   **D15.18** 0.433 m/s $\measuredangle$ 33.7°.   **D15.19**
α = 0.32 rad/s² $\curvearrowright$, a_A = 1.55 m/s² ↓.   **D15.20** 302 ft/s² $\measuredangle$ 83.2°.
D15.21 156.1 ft/s² $\measuredangle$ 64.7°.   **D15.22** 1.505 m/s² $\measuredangle$ 23.5°.   **D15.23**
0.918 m/s² ∇ 78.7°.   **D15.26** 32 ft/s² →.   **D15.27** No.   **D15.28**
(a) Z at C; (b) a_B = 0.48 m/s² ←, a_D = 0.339 m/s² $\measuredangle$ 45°.   **D15.29**
ω = 0.3 rad/s $\curvearrowright$, α = 0.32 rad/s² $\curvearrowright$, v_A = 1.2 m/s ↓, a_A = 1.55
m/s² ↓.   **D15.30** (a) 20 cos θ − 15 sin θ; (b) ω = 0.24 rad/s $\curvearrowright$, α

$= 0.0168$ rad/s² $\mathbf{\mathsf{5}}$.   **D15.31** $\alpha_{BD} = \alpha_{DE} = 0$.   **D15.32** (a) A fixed reference frame, (b) a moving reference frame, (c) the angular velocity of $Axyz$ in $OXYZ$, (d) the time derivative of $\mathbf{i}$ taken in $OXYZ$, (e) the time derivative of $\mathbf{i}$ taken in $Axyz$, (f) the same as $(\dot{\mathbf{i}})_{OXYZ}$.   **D15.33** a, c. **D15.34** $\Omega \times \mathbf{Q}$.   **D15.37** $\boldsymbol{\omega}_{OD} = 0.56$ rad/s $\mathbf{\mathsf{5}}$, $\mathbf{v}_{B/OD} = 0.96$ m/s $\measuredangle$ $36.9°$.   **D15.38** (a) Yes, (b) no.   **D15.42** $\boldsymbol{\alpha}_{OD} = 5.99$ rad/s² $\mathbf{\mathsf{2}}$, $\mathbf{a}_{B/OD} = 0.717$ m/s² $\measuredangle$ $36.9°$.   **D15.43** $\mathbf{v}_{A/B} = 5\mathbf{j}$ m/s, $\mathbf{a}_{A/B} = -4\mathbf{i} - 20\mathbf{j}$ m/s².   **D15.44** $\mathbf{v}_{B/Axyz} = -5\mathbf{j}$ m/s, $\mathbf{a}_{B/Axyz} = 4\mathbf{i} + 20\mathbf{j}$ m/s².

Problems

15.1 (a) 4 s, (b) 45 s.   **15.2** $\omega_{max} = 6.58$ rad/s, $\alpha_{max} = 41.3$ rad/s². **15.4** $\alpha_c = \pi$ rad/s² $\mathbf{\mathsf{5}}$, $\Delta t = 20$ s.   **15.5** $\alpha = 15.59$ rad/s² $\mathbf{\mathsf{2}}$, $a_Q = 3.6$ m/s².   **15.7** 0.478 m.   **15.8** $\boldsymbol{\omega}_B = 15\mathbf{i}$ rad/s, $\boldsymbol{\alpha}_B = 7.5\mathbf{i}$ rad/s². **15.10** Estimate by applying Eq. (15.8); thus, (a) 3.0 rad/s², (b) 2.5 rad/s².   **15.12** 7.68 ft/s².   **15.14** $h(\omega_B)_0^2 (r_A^2 + r_B^2)/(2\pi r_A^3)$.   **15.15** $\mathbf{v}_P = -24\mathbf{i} + 66\mathbf{j} - 10\mathbf{k}$ m/s, $\mathbf{a}_P = -792\mathbf{i} - 338\mathbf{j} - 330\mathbf{k}$ m/s². **15.16** (a) 5.46 m; (b) $v = 70.9$ m/s, $a_t = 0$, $a_n = 922$ m/s², $a = 922$ m/s².   **15.17** $\omega = 2$ rad/s $\mathbf{\mathsf{5}}$, $\mathbf{v}_A = 2.6$ m/s $\measuredangle$ $67.4°$.   **15.19** $\mathbf{v}_A = 0.6$ m/s $\rightarrow$, $\mathbf{v}_B = 0.849$ m/s $\measuredangle$ $45°$.   **15.20** $\omega = 1.5$ rad/s $\mathbf{\mathsf{5}}$, $(v_Q)_y = 3.9$.   **15.22** $\omega_{AB} = 21.9$ rad/s $\mathbf{\mathsf{5}}$, $\mathbf{v}_P = 42.6$ ft/s $\uparrow$.   **15.23** $\omega_P = 3150$ rpm $\mathbf{\mathsf{5}}$, $\omega_C = 1260$ rpm $\mathbf{\mathsf{5}}$.   **15.25** $\omega_{AB} = 0.196$ rad/s $\mathbf{\mathsf{5}}$, $\omega_O = 1.054$ rad/s $\mathbf{\mathsf{2}}$.   **15.26** (a) $\omega_{AB} = 1.075$ rad/s $\mathbf{\mathsf{2}}$, $\mathbf{v}_B = 0.215$ m/s $\downarrow$; (b) $\omega_{AB} = 0.866$ rad/s $\mathbf{\mathsf{2}}$, $\mathbf{v}_B = 0.310$ m/s $\downarrow$.   **15.28** (a) Yes, (b) 4.2 m/s $\rightarrow$.   **15.29** (a) Yes, (b) 2.1 m/s $\measuredangle$ $36.9°$.   **15.31** $\omega_{BD} = 3.75$ rad/s $\mathbf{\mathsf{2}}$, $\omega_{DE} = 7.5$ rad/s $\mathbf{\mathsf{5}}$.   **15.32** $\mathbf{v}_B = 210$ mm/s $\leftarrow$, $\omega_{CD} = 1.125$ rad/s $\mathbf{\mathsf{2}}$.   **15.34** 28 in./s $\rightarrow$.   **15.36** 2.56 m/s $\measuredangle$ $69.4°$. **15.51** (a) 3.10 in. above B, (b) 5.53 mi/h $\leftarrow$.   **15.53** (a) 150 mm, (b) 510 mm/s $\measuredangle$ $61.9°$, (c) 300 mm/s $\measuredangle$ $36.9°$.   **15.54** (a) 70 mm, (b) 2 rad/s $\mathbf{\mathsf{5}}$, (c) 278 mm/s $\measuredangle$ $59.7°$.   **15.55** $\omega = 0.6$ rad/s $\mathbf{\mathsf{2}}$, $\mathbf{v}_B = 50.9$ mm/s $\measuredangle$ $45°$.   **15.57** $\omega_{AB} = 0.2$ rad/s $\mathbf{\mathsf{5}}$, $\mathbf{v}_B = 0.52$ m/s $\measuredangle$ $59.5°$. **15.58** (a) 0.12 rad/s $\mathbf{\mathsf{5}}$, (b) 0.9 ft/s $\measuredangle$ $53.1°$, (c) 1.02 ft/s $\measuredangle$ $81.2°$. **15.60** $\omega_{AB} = 0.8$ rad/s $\mathbf{\mathsf{2}}$, $\mathbf{v}_B = 0.532$ m/s $\measuredangle$ $74.3°$.   **15.61** $\mathbf{v}_B = 56$ in./s $\uparrow$, $v_M = 50.3$ in./s $\measuredangle$ $72.6°$.   **15.63** $\omega_A = \omega_B = 20$ rad/s $\mathbf{\mathsf{5}}$. **15.65** $\omega_{DE} = 1.6$ rad/s $\mathbf{\mathsf{5}}$, $\mathbf{v}_F = 488$ mm/s $\measuredangle$ $10.39°$.   **15.66** (a) 1.8 rad/s, (b) 366 mm/s.   **15.67** $\alpha = 3$ rad/s² $\mathbf{\mathsf{5}}$, $\mathbf{a}_A = 7.8$ m/s² $\measuredangle$ $67.4°$.   **15.68** $\mathbf{a}_B = 4.45$ m/s² $\measuredangle$ $57.4°$, $\mathbf{a}_C = 3.75$ m/s² $\uparrow$, $\mathbf{a}_E = 2.82$ m/s² $\measuredangle$ $25.2°$.   **15.70** (a) $\omega = 2$ rad/s $\mathbf{\mathsf{5}}$, $\alpha = 5$ rad/s² $\mathbf{\mathsf{5}}$; (b) 9.06 m/s² $\measuredangle$ $83.7°$.   **15.71** $x = 4$ m, $y = 1$ m.   **15.73** $\alpha_P = 35$ rad/s² $\mathbf{\mathsf{5}}$, $\alpha_C = 14$ rad/s² $\mathbf{\mathsf{5}}$, $\mathbf{a}_T = 27.2$ m/s² $\measuredangle$ $84.1°$.   **15.74** $\alpha_{AB} = 0.336$ rad/s² $\mathbf{\mathsf{5}}$, $\alpha_O = 1.939$ rad/s² $\mathbf{\mathsf{2}}$.   **15.75** $\alpha_{AB} = 9$ rad/s² $\mathbf{\mathsf{5}}$, $\mathbf{a}_B = 660$ mm/s² $\uparrow$.   **15.77** $\alpha_{BD} = 81$ rad/s² $\mathbf{\mathsf{5}}$, $\mathbf{a}_D = 63$ m/s² $\measuredangle$ $36.9°$.   **15.78** $\alpha_{BD} = 4.2$ rad/s² $\mathbf{\mathsf{5}}$, $\mathbf{a}_D = 70.6$ in./s² $\measuredangle$ $87.4°$. **15.80** $\alpha_{BD} = 3$ rad/s² $\mathbf{\mathsf{2}}$, $\mathbf{a}_D = 63$ in./s² $\rightarrow$.   **15.82** (a) $\mathbf{a}_B = 372$ mm/s² $\leftarrow$; (b) $\alpha_{AD} = 1.282$ rad/s² $\mathbf{\mathsf{2}}$, $\alpha_{BD} = 11.23$ rad/s² $\mathbf{\mathsf{2}}$, $\alpha_{CD} = 8.37$ rad/s² $\mathbf{\mathsf{2}}$.   **15.83** (a) $\mathbf{a}_B = 0.42$ m/s² $\leftarrow$, $\mathbf{a}_E = 1.449$ m/s² $\measuredangle$ $14.38°$; (b) $\alpha_{AD} = 7.2$ rad/s² $\mathbf{\mathsf{5}}$, $\alpha_{BD} = 4$ rad/s² $\mathbf{\mathsf{2}}$, $\alpha_{CD} = 2.25$ rad/s² $\mathbf{\mathsf{2}}$.   **15.85** (a) $\alpha_{AB} = 3$ rad/s² $\mathbf{\mathsf{2}}$, $\alpha_{BC} = 4$ rad/s² $\mathbf{\mathsf{2}}$; (b) 59.1 in./s² $\measuredangle$ $66.0°$.   **15.86** (a) 22 rad/s² $\mathbf{\mathsf{2}}$, (b) 48 rad/s² $\mathbf{\mathsf{5}}$, (c) 56.3 m/s² $\measuredangle$ $7.66°$.   **15.87** (a) 0.5 rad/s² $\mathbf{\mathsf{5}}$; (b) $\mathbf{a}_B = 1.3$ m/s² $\measuredangle$ $59.5°$, $\mathbf{a}_D = 1.2$ m/s² $\measuredangle$ $36.9°$.   **15.88** (a) 0.2 rad/s² $\mathbf{\mathsf{5}}$; (b) $\mathbf{a}_B = 1.7$ ft/s² $\measuredangle$ $81.2°$, $\mathbf{a}_D = 1.5$ ft/s² $\measuredangle$ $53.1°$.   **15.89** (a) 0.25 rad/s² $\mathbf{\mathsf{2}}$; (b) $\mathbf{a}_B = 134.2$ mm/s²

$\angle 63.4°$, $\mathbf{a}_D = 75$ mm/s^2 ↑ .   **15.90** 9**i** + 4**j** in.   **15.91** (a) 75 mm, (b) 0.8 rad/s^2 ⤵ , (c) 156 mm/s^2 $\angle 67.4°$.   **15.97** $\alpha_{AB} = 1.8$ rad/s^2 ⤵ , $\mathbf{a}_E = 207$ mm/s^2 $\angle 29.2°$.   **15.98** $\mathbf{a}_D = 721\alpha_{OB}$ ⦨ 19.44°, $\mathbf{a}_E = 300\alpha_{OB}$ mm/s^2 ⦨ 53.1°.   **15.99** $\mathbf{a}_G = 1.3\alpha$ m/s^2 ⦨ 22.6°, $\mathbf{a}_H = 0.7\alpha$ m/s^2 ←.   **15.101** $bv_A/(x^2 + b^2)$ ⤵ .   **15.102** $v_A/(b^2 - x^2)^{1/2}$ ⤸ .   **15.103** $bv_A^2(2x^2 - b^2)/[x^2(x^2 - b^2)^{3/2}]$ ⤸ .   **15.105** $v_A^2 x/(b^2 - x^2)^{3/2}$ ⤸ .   **15.107** $\omega_{BD} = 0.580$ rad/s ⤸ and $\mathbf{v}_D = 6.32$ ft/s ↓   (or $\omega_{BD} = 1.380$ rad/s ⤵ and $\mathbf{v}_D = 1.519$ ft/s ↑).   **15.108** $\omega_{AD} = 3$ rad/s ⤸ , $\mathbf{v}_{B/AD} = 8$ ft/s ←.   **15.109** $\omega_{AD} = 0.4$ rad/s ⤸ , $\mathbf{v}_{B/AD} = 1.35$ m/s →.   **15.111** $\alpha_{AD} = 41.5$ rad/s^2 ⤵ , $\mathbf{a}_{B/AD} = 38$ ft/s^2 ←.   **15.113** $\alpha_{AD} = 3.94$ rad/s^2 ⤸ , $\mathbf{a}_{B/AD} = 1.286$ m/s^2 $\angle 67.4°$.   **15.115** $\alpha_{AE} = 0.0745$ rad/s^2 ⤸ , $\mathbf{a}_{B/AE} = 16.22$ mm/s^2 $\angle 16.26°$.   **15.116** $\omega_D = 0.25$ rad/s ⤸ , $\alpha_D = 0.01$ rad/s^2 ⤵ .   **15.117** $\omega = 0.75$ rad/s ⤵ , $\alpha = 0.9$ rad/s^2 ⤸ .   **15.118** (a) 0.714 rad/s^2 ⤵ , (b) 12.18 in./s^2 ⦨ 62.4°.   **15.120** (a) 2.04 rad/s ⤸ , (b) 28.6 in./s $\angle 47.6°$.   **15.122** (a) 1.25 rad/s^2 ⤵ , (b) 3 ft/s.   **15.124** $\alpha_{AB} = 0.0135$ rad/s^2 ⤵ , $\alpha_{BD} = 0.024$ rad/s^2 ⤸ .   **15.125** $\mathbf{v}_B = 600$ mm/s ⦨ 53.1°, $\mathbf{a}_B = 1200$ mm/s^2 $\angle 53.1°$.   **15.126** $\mathbf{a}_E = 0.6$ ft/s^2 ↓ , $\alpha_{AB} = 1.31$ rad/s^2 ⤸ .   **15.128** (a) $\alpha_{BD} = 2$ rad/s^2 ⤸ , $\alpha_{DE} = 3.5$ rad/s^2 ⤸ ; (b) 2.82 m/s^2 $\angle 16.50°$.   **15.129** $\mathbf{a}_P = (2u\omega - 2b\omega^2 + u^2/b)\mathbf{i}$, $\mathbf{a}_Q = -4b\omega^2\mathbf{i} + (b\omega^2 - 2u\omega)\mathbf{k}$, $\mathbf{a}_R = -(5b\omega^2 + 2u\omega)\mathbf{i}$, $\mathbf{a}_S = -4b\omega^2\mathbf{i} + (2u\omega - b\omega^2)\mathbf{k}$.   **15.131** 3.47 rad/s^2 ⤵ .   **15.133** $\alpha_{CD} = 17$ rad/s^2 ⤸ , $\mathbf{a}_{A/CD} = 2$ m/s^2 ←.

Chapter 16

Developmental Exercises

D16.3 $\frac{1}{12}mc^2$.   **D16.4** $\frac{1}{2}mr^2$.   **D16.5** No.   **D16.6** 1088 slug·ft^2.   **D16.8** $0.548r$.   **D16.9** 8 kg·m^2.   **D16.10** No.   **D16.13** $\frac{1}{20}m(3r^2 + 2h^2)$.   **D16.14** $I_x = 1.89$ kg·m^2, $k_x = 410$ mm.   **D16.15** $I_z = 8.2$ kg·m^2, $k_z = 854$ mm.   **D16.16** $I_x = 0.310$ slug·ft^2, $k_x = 7.89$ in.   **D16.17** $I_z = 0.487$ slug·ft^2, $k_z = 9.89$ in.   **D16.18** $I_x = 9.31 \times 10^{-3}$ kg·m^2, $k_x = 56.6$ mm.   **D16.19** $I_y = 4.09 \times 10^{-3}$ kg·m^2, $k_y = 37.5$ mm.   **D16.20** $m\bar{\mathbf{a}}$ and $\bar{I}\alpha$ at G.   **D16.22** 2.45 in.   **D16.24** 818 mm.   **D16.25** (a) No, (b) yes.   **D16.26** $-40.1\mathbf{i} + 205\mathbf{j}$ N.   **D16.27** 270 N.   **D16.29** 116.1 N ↑ .   **D16.30** (a) $\mathbf{F} = 24$ lb →, $\mathbf{N} = 55.6$ lb ↑ ; (b) 0.432.   **D16.31** $O_x = 4.93$ lb →, $O_y = 10.72$ lb ↑ .   **D16.32** N_A, N_B, D_y, E_y.   **D16.33** 0.909 m/s^2 ←.

Problems

16.3 $I_x = \frac{1}{3}mb^2$, $I_y = \frac{1}{3}ma^2$, $I_z = \frac{1}{3}m(a^2 + b^2)$.   **16.10** $I_x = 21.5 \times 10^{-3}$ slug·ft^2, $k_x = 2.31$ in.   **16.11** $I_x = 326$ kg·mm^2, $k_x = 18.97$

mm.   **16.13** $\frac{1}{12}m(4b^2 + c^2)$.   **16.15** $I_y = \frac{1}{12}m(4a^2 + c^2)$, $k_y = (\sqrt{3}/6)(4a^2 + c^2)^{1/2}$.   **16.16** $I_y = \frac{1}{6}m(a^2 + c^2)$, $k_y = (\sqrt{6}/6)(a^2 + c^2)^{1/2}$. **16.17** $I_x = 19.44$ kg·m^2, $k_x = 367$ mm.   **16.19** $I_x = 0.944$ kg·m^2, $I_y = 0.512$ kg·m^2.   **16.21** $I_x = 60.9 \times 10^{-3}$ kg·m^2, $I_y = 143.2 \times 10^{-3}$ kg·m^2.   **16.23** $I_z = 2.4$ slug·ft^2, $k_z = 1.414$ ft.   **16.24** $I_z = 103.6 \times 10^{-3}$ kg·m^2, $k_z = 158.7$ mm.   **16.25** $I_x = 55 \times 10^{-3}$ slug·ft^2, $I_y = 185 \times 10^{-3}$ slug·ft^2.   **16.27** $I_x = 1.259$ slug·ft^2, $I_y = 2.32$ slug·ft^2.   **16.29** $I_z = 748$ kg·mm^2, $k_z = 69.3$ mm.   **16.30** $I_z = 1.179$ slug·ft^2, $k_z = 1.145$ ft.   **16.31** $I_y = 110.8$ kg·mm^2, $k_y = 35.4$ mm.   **16.33** $I_y = 419$ kg·mm^2, $k_y = 47.1$ mm.   **16.35** $I_x = 0.933$ slug·ft^2, $k_x = 11.04$ in.   **16.36** $I_x = 681$ kg·mm^2, $k_x = 60.1$ mm. **16.37** $I_y = 1.570 \times 10^{-3}$ kg·m^2, $k_y = 25.2$ mm.   **16.39** $I_y = 7.46 \times 10^{-3}$ kg·m^2, $k_y = 34.5$ mm.   **16.41** $I_x = 39.5$ slug·ft^2, $k_x = 10.36$ in.   **16.42** $I_x = 20.2 \times 10^{-3}$ kg·m^2, $k_x = 56.7$ mm.   **16.43** $I_z = 21.3 \times 10^{-3}$ slug·ft^2, $k_z = 2.46$ in.   **16.44** $I_x = 4.25$ kg·m^2, $k_x = 204$ mm.   **16.45** $I_y = 4.66 \times 10^{-3}$ kg·m^2, $I_z = 2.63 \times 10^{-3}$ kg·m^2.   **16.49** 2.86 m/s^2.   **16.50** 24.2 ft/s^2.   **16.53** 3.68 m/s^2. **16.54** 13.42 ft/s^2.   **16.56** $\mathbf{a}_B = 25.8$ ft/s^2 ⟋ 53.1°, $F_{BE} = 22.5$ lb T, $F_{CD} = 1.5$ lb T.   **16.57** (a) $(40 - 19.62 \sin\theta)$ rad/s^2 ↻, (b) 1.043 kN T.   **16.58** (a) 7.74 s, (b) 7.49 s, (c) 5.40 s.   **16.60** 1.150 rad/s. **16.61** $\mathbf{A} = 6.41$ kN ↑, $\mathbf{B} = 7.32$ kN ↑.   **16.63** (a) -37.2%, (b) -52.9%.   **16.64** (a) 4.8 rad/s^2 ↺, (b) 4.32 rad/s^2 ↺.   **16.65** (a) $\boldsymbol{\alpha} = 3g/(4b)$ ↻, $\mathbf{A} = \frac{3}{8}W\mathbf{i} + \frac{5}{8}W\mathbf{j}$; (b) $\boldsymbol{\alpha} = 3g/(5b)$ ↻, $\mathbf{A} = \frac{3}{10}W\mathbf{i} + \frac{7}{10}W\mathbf{j}$; (c) $\boldsymbol{\alpha} = 2g/(3r)$ ↻, $\mathbf{A} = \frac{1}{3}W\mathbf{j}$.   **16.66** (a) 5.25 lb, (b) 7.97 lb.   **16.68** $-466\mathbf{i} + 12.81\mathbf{j}$ lb.   **16.70** 0.931 N·m ↻.   **16.72** $-45.6\mathbf{i} + 70.8\mathbf{j}$ N.   **16.73** 21.8 kN ↑.   **16.74** 4.04 rad/s^2 ↺.   **16.75** (a) 89.2 rad/s^2 ↺, (b) 1.1 s.   **16.78** 208 N.   **16.79** 14.17 rev.   **16.81** 2.41 rad/s^2 ↻.   **16.83** (a) 0.8 m, (b) 10 m/s^2 ↗.   **16.84** $T_A = 1.949$ kN, $T_B = 2.91$ kN.   **16.87** (a) 1.6 rad/s^2 ↺, (b) $\mathbf{a}_A = 4.06$ m/s^2 ↑, $\mathbf{a}_B = 2.34$ m/s^2 ↓.   **16.88** $1.9W$ ⟍ 76.8°.   **16.90** 2.55 rad/s^2 ↺. **16.91** $0.787W$ ↑.   **16.93** 0.663.   **16.95** 1181 N ↑.   **16.97** 2.39 s.   **16.98** 0.903 s.   **16.99** $\frac{2}{5}r$.   **16.100** 3.11 s.   **16.102** (a) 4.02 ft/s^2 ←, (b) 0.236.   **16.105** (a) No slipping, (b) 1.28 m/s^2 →. **16.106** 2.10 rad/s ↻.   **16.108** $\boldsymbol{\alpha} = 15.32$ rad/s^2 ↻, $\mathbf{C} = -15.64\mathbf{i} + 88.9\mathbf{j}$ N.   **16.110** (i) (a) $\frac{1}{3}a$ →, (b) $1.5b$; (ii) (a) $\frac{2}{7}a$ →, (b) $1.4b$. **16.111** 162.6 N.   **16.113** $\boldsymbol{\alpha} = 8.06$ rad/s^2 ↺, $\mathbf{A} = 33.6$ N ↑, $\mathbf{B} = 0$.   **16.115** 23.3 ft/s^2 ↓.   **16.117** 19.95 N.   **16.118** 2.28 m/s^2 ⟍ 16.26°.   **16.120** (a) 41.5 rad/s^2 ↺, (b) 5.40 ft/s^2 ←.   **16.122** 6.85 rad/s^2 ↻, (b) 6.33 m/s^2 ⟋ 30°.   **16.123** $I_x = 26.6$ slug·ft^2, $k_x = 3.79$ ft.   **16.125** $I_x = 1.130 \times 10^{-3}$ kg·m^2, $k_x = 48.9$ mm.   **16.126** $I_y = 2.18 \times 10^{-3}$ kg·m^2, $k_y = 67.9$ mm.   **16.128** $I_y = 46.1 \times 10^{-3}$ kg·m^2, $k_y = 86.1$ mm.   **16.129** $\boldsymbol{\alpha} = 1.391$ rad/s^2 ↺, $T_{AD} = 3.23$ lb, $T_{BF} = 12.65$ lb.   **16.130** 1.951 m/s^2 ↓.   **16.133** 2.96 m/s^2 ⟍ 16.26°.   **16.134** $\boldsymbol{\alpha}_{OA} = 3.86$ rad/s^2 ↺, $\boldsymbol{\alpha}_{AB} = 3.86$ rad/s^2 ↻, $\mathbf{B} = 2.44\mathbf{i}$ lb, $\mathbf{O} = -6.33\mathbf{i} + 7.31\mathbf{j}$ lb.   **16.135** $0 < P \leq 62.8$ N. **16.136** 32 N.   **16.137** 4.12 ft/s^2 ←.   **16.138** 8.99 lb.

Chapter 17

Developmental Exercises

D17.2 (a) No, (b) no.   **D17.3** (a) Yes, (b) no.   **D17.5** (a) 15 J, (b) −15 J, (c) −5.24 J.   **D17.9** (a) 0.8 m, (b) 2.96 rad/s^2 $\circlearrowright$.   **D17.10** 16.38 in.   **D17.12** 2.68 m/s $\downarrow$.   **D17.13** 1.143 m $\rightarrow$.   **D17.14** 80.4 N.   **D17.15** 4.87 rad/s $\circlearrowright$.   **D17.19** (a) −37.7 J, (b) 0.   **D17.20** 175.1 lb·ft.   **D17.22** 0.851 m/s^2 $\leftarrow$, no change in $\mathbf{a}_C$.

Problems

17.1 10.96 rad/s $\circlearrowleft$.   **17.2** 4.59 rad/s $\circlearrowright$.   **17.4** 4.66 rev.   **17.5** Mass of gear should be 30 kg, not 50 kg; thus, $v_O = 1.225$ m/s.   **17.8** 63.9 km/h.   **17.9** 10.90 rad/s $\circlearrowleft$.   **17.11** 0.309.   **17.13** 20.3 N·m.   **17.14** 15.53 kN/m.   **17.16** 0.669 m/s.   **17.18** 20.4 rev.   **17.19** 17.67 rev.   **17.20** 591 ft.   **17.21** 16.77 rev.   **17.23** 337. N.   **17.24** 5.77 rev.   **17.26** 8.58 ft/s.   **17.27** (a) $\omega_{AB} = \omega_{BD} = 1.530 \sqrt{g/L}$ $\circlearrowleft$; (b) $\omega_{AB} = 0$, $\omega_{BD} = 3.46 \sqrt{g/L}$ $\circlearrowleft$.   **17.46** $\bar{\mathbf{v}}_{\text{cylinder}} = 3.62$ m/s $\rightarrow$, $\bar{\mathbf{v}}_{\text{sphere}} = 3.74$ m/s $\rightarrow$.   **17.48** (a) $9.27\sqrt{1 - \cos\theta}$ ft/s, (b) 75.5°.   **17.50** (a) 0.993 m/s $\downarrow$, (b) 3.45 m/s $\downarrow$.   **17.52** 4.90 rad/s $\circlearrowright$.   **17.53** 4.82 rad/s $\circlearrowright$.   **17.56** (a) 1.934 m/s $\rightarrow$, (b) 1.755 m/s $\rightarrow$.   **17.57** (a) 16.13 ft/s $\downarrow$, (b) 1.911 in.   **17.58** 1.501 m/s $\downarrow$.   **17.59** (a) 4.26 m/s $\downarrow$, (b) 6.61 m/s $\downarrow$.   **17.60** 53.1°.   **17.62** (a) $1.142 \sqrt{g/r}$, (b) $1.553mg$ $\uparrow$.   **17.64** (a) 5.81 kW, (b) 8.56 kW.   **17.66** Maximum belt tension should be 160 N, not 60 N; thus, $(\mu_s)_{\text{min}} = 0.307$.   **17.67** 33.2 N/m.   **17.68** $M_{MA} = 35.4$ N·m, $M_{BC} = 17.68$ N·m, $M_{DE} = 13.26$ N·m.   **17.69** $2\pi rma$.   **17.72** 4.77 rad/s^2 $\circlearrowright$.   **17.74** 8.11 rad/s^2 $\circlearrowleft$.   **17.75** 1.982 ft/s^2 $\leftarrow$.   **17.78** 4.44 ft/s^2 $\leftarrow$.   **17.79** 0.219 m/s^2 $\leftarrow$.   **17.81** (a) 4.29 rad/s $\circlearrowright$, (b) 6.42 rad/s $\circlearrowright$.   **17.83** 4.63 rad/s.   **17.84** 1.561 m/s $\downarrow$.   **17.86** 60°.   **17.87** 55.2°.   **17.90** 1.809 m/s $\uparrow$.   **17.92** *Mass*, not weight, of each rear wheel is 60 kg; thus, $\mathbf{v} = 3.41$ m/s $\leftarrow$.   **17.95** 0.602 rad/s^2 $\circlearrowleft$.

Chapter 18

Developmental Exercises

D18.1 (a) 600 N·s $\downarrow$, (b) 120 N·m·s $\circlearrowright$, (c) 100 N·m·s $\circlearrowleft$.   **D18.2** (a) 600 N·s $\downarrow$, (b) 120 N·m·s $\circlearrowright$, (c) 100 N·m·s $\circlearrowleft$.   **D18.4** (a) **0**, (b) 2π N·m·s $\circlearrowright$.   **D18.5** (a) 2.48 lb·s $\rightarrow$, (b) 0.932 lb·ft·s $\circlearrowleft$.   **D18.7** 1.568 s.   **D18.8** (a) No, (b) yes.   **D18.9** (a) 4.19 s, (b) 6.52 rad/s.   **D18.11** 1.577 m/s $\uparrow$.   **D18.12** 1.6 m/s $\leftarrow$.   **D18.15** 991**j** rpm.   **D18.16** 1.866**j** rad/s.   **D18.18** 45.9 ft·lb.   **D18.19** 63.5 J.   **D18.20** 5.14 rad/s.   **D18.22** $\omega'_{AB} = 3.47$ rad/s $\circlearrowright$, $\omega'_{CD} = 2.97$ rad/s $\circlearrowright$.   **D18.23** $\omega'_{AB} = 3.17$ rad/s $\circlearrowright$, $\omega'_{CD} = 2.38$ rad/s $\circlearrowright$.

Problems

18.1 19.13 rad/s $\circlearrowleft$.   **18.2** 5.03 rad/s $\circlearrowright$.   **18.4** 7.55 rad/s $\circlearrowright$.   **18.5** 102.6 rad/s $\circlearrowleft$.   **18.8** 1.175 s.   **18.9** Mass of gear should be 30 kg, not 50 kg; thus, $v_O = 13.93$ m/s.   **18.10** Mass of gear should be 30 kg, not 50 kg; thus, $v_O = 1.126$ m/s.   **18.11** 29.4 lb.   **18.14** $\mathbf{v}_2 = 7.36$ ft/s $\uparrow$, $\omega_2 = 14.35$ rad/s $\circlearrowright$.   **18.15** The cord should be

wrapped around a drum of radius 100 mm; thus, $v_G = 3.26$ m/s.
18.16 134.6 N.　　**18.17** 2.24 s.　　**18.20** 0.513 s.　　**18.22** 56.1**j**
rpm.　　**18.23** 52.1**j** rpm.　　**18.24** The rims of the disks A and B should
be in contact, not overlap; thus, $\omega_{CD} = 22.1$ rpm.　　**18.26** 0.821 m/s
←.　　**18.28** 0.771 m/s ∡ 16.26°.　　**18.30** 1501 rad/s.　　**18.31** 6.55
m/s ↑.　　**18.32** (a) 0.309 m/s ←, (b) **0**.　　**18.33** Rod AC should have
a dimension of 30 in.; thus, $\omega_2 = 5.43$**j** rad/s and $v_r = 10.12$ ft/s.
18.35 (a) 3.36 rad/s ↺, (b) 1.866 kN →.　　**18.36** (a) 2.91 rad/s ↺, (b)
1.616**i** − 7.50**j** kN.　　**18.39** (a) 8.85 rad/s ↻, (b) 16.50**i** + 33.3**j** kips.
18.40 (a) 8.85 rad/s ↻, (b) 16.50**i** − 33.3**j** kips.　　**18.41** 119.5 rad/s.
18.42 **ω** = 7.5 rad/s ↺, $\mathbf{v}_G$ = 7.5 m/s ↓.　　**18.43** (a) 10.92 rad/s ↺;
(b) $\mathbf{F}_B$ = 8.69 kips ⦜ 20.6°, $\mathbf{F}_C$ = 8.08 kips ↑.　　**18.44** 23.3 m/s →.
18.46 $0.594\sqrt{gh}$, h = height of crate.　　**18.48** $1.105\sqrt{gh}$, h = height
of crate.　　**18.49** 1.844 ft/s.　　**18.51** 2.　　**18.53** (a) 4.61 rad/s ↻;
(b) **C** = 73.7 kN ↑, **A** = 18.43 kN ↓.　　**18.56** (a) 2.30 rad/s ↻; (b)
C = 55.3 kN ↑, **A** = 13.83 kN ↓.　　**18.58** (a) 25 in., (b) 98.5 lb ↑
≥ **A** ≥ 80 lb ↑.　　**18.59** (a) 3.68 rad/s ↻, (b) 4.05 m/s ↑.　　**18.62**
1.065 ft.　　**18.64** 1.517 m/s ⦨ 28.1°.　　**18.65** (a) 4.26 rad/s ↺,
(b) **0**.　　**18.67** 0.207 rad/s.　　**18.69** ω'_{AB} = 0.802 rad/s ↺, ω'_{CD} =
3.21 rad/s ↺.　　**18.70** ω'_{AB} = 0.401 rad/s ↺, ω'_{CD} = 7.22 rad/s ↻.
18.72 2.38 ft/s ←.　　**18.75** 5.33 ft/s ←.　　**18.76** 0.263 m/s ←.
18.84 4.94 m/s ↓.　　**18.87** 2.57 s.　　**18.88** 0.924 s.　　**18.90** (a)
14.00 m/s ∡25°, (b) 12.94%.　　**18.92** 2.45 ft.　　**18.94** $\sqrt{75g/(26L)}$
↺.　　**18.96** $\sqrt{312g/(25L)}$ ↺.　　**18.97** 33.2° ↺.　　**18.99** **A** = 44.6
lb ↑, **B** = 22.3 lb ↑.　　**18.100** 9.21 rad/s ↻.　　**18.103** 11.24 m/s
⦨16.26°.　　**18.104** 3.52 m/s ⦨16.26°.

Chapter 19

Developmental Exercises

D19.1 $\dot{\mathbf{Q}}$ is taken in $OXYZ$.　　**D19.2** $\dot{\mathbf{Q}} = \boldsymbol{\Omega} \times \mathbf{Q}$, $\ddot{\mathbf{Q}} = \dot{\boldsymbol{\Omega}} \times \mathbf{Q} +$
$\boldsymbol{\Omega} \times (\boldsymbol{\Omega} \times \mathbf{Q})$.　　**D19.3** (a) **0**, (b) −19.33**J** rad/s.　　**D19.4** (a)
−18.56**J** m/s, (b) 3.2**I** + 4**J** − 8**K** rad/s.　　**D19.5** (a) **0**, (b) 8.29**J**
rad/s.　　**D19.7** $\boldsymbol{\Omega} = (\mathbf{j} \cdot \mathbf{k})\mathbf{i} + (\mathbf{k} \cdot \mathbf{i})\mathbf{j} + (\mathbf{i} \cdot \mathbf{j})\mathbf{k}$.　　**D19.9** −36**J** +
16**K** ft/s^2.　　**D19.10** −36**K** ft/s^2.　　**D19.11** $\mathbf{v}_E = (r + R)\omega_2\mathbf{I} − r\omega_1\mathbf{J}$,
$\mathbf{a}_E = (r + R)\alpha_2\mathbf{I} − r\alpha_1\mathbf{J} − [r\omega_1^2 + (r + R)\omega_2^2]\mathbf{K}$.　　**D19.13** $\boldsymbol{\omega}_{N/A} =$
I + 0.75**J** + 1.299**K** rad/s, $\boldsymbol{\alpha}_{N/A} = −0.299\mathbf{J} + 2.48\mathbf{K}$ rad/s^2.
D19.14 $\boldsymbol{\omega}_E$ = **I** + 3**J** rad/s, $\boldsymbol{\alpha}_E$ = 2.5**J** − 3**K** rad/s^2.　　**D19.15** (a)
23.0 slug·ft^2, (b) 18.08 slug·ft^2.　　**D19.16** (a) I_{yz} and I_{zx}, (b) I_{xy} and I_{zx},
(c) I_{xy} and I_{yz}.　　**D19.19** (a) The sets of direction cosines of x', y', and
z' in $Oxyz$; (b) the sets of direction cosines of x, y, z in $Ox'y'z'$; (c) yes;
(d) yes.　　**D19.21** $\mathbf{v}_A = −12\mathbf{j}$ ft/s, $\mathbf{v}_B = −6\mathbf{j}$ ft/s, $\mathbf{v}_D = −6\mathbf{j}$ ft/s.
D19.22 $H_x = \bar{I}_{xx}\omega_x$, $H_y = \bar{I}_{yy}\omega_y$, $H_z = \bar{I}_{zz}\omega_z$.　　**D19.23** (a) 35.3°, (b)
19.47°.　　**D19.24** 435 J.　　**D19.25** 9.28 rad/s.　　**D19.26** $T = \frac{1}{2}\bar{\mathbf{v}} \cdot$
$\mathbf{L} + \frac{1}{2}\boldsymbol{\omega} \cdot \mathbf{H}_G$.　　**D19.27** (a) −4.25**I** m/s^2; (b) $\mathbf{O}_X = −8.50\mathbf{I}$ N, $\mathbf{O}_Y =$
19.62**J** N.　　**D19.28** 5.42 rad/s.　　**D19.29** 0.708 J.　　**D19.31** **O** =
−50**I** + 16.1**J** − 37.5**K** lb, $\mathbf{M}_O$ = −48.3**I** + 51.9**K** lb·ft.　　**D19.32**
170.3 ft·lb.　　**D19.34** 5**i** − 15**j** N.　　**D19.35** 8.53 rad/s or 98.6 rad/s.
D19.37 (a) $\dot{\psi}\mathbf{k}$, (b) $\ddot{\psi}\mathbf{k}$, (c) no.

Problems

19.1 (a) $\mathbf{v}_P = -5.20\mathbf{I} + 3\mathbf{J} - 1.875\mathbf{K}$ m/s, $\mathbf{a}_P = -1.669\mathbf{I} - 2.08\mathbf{J} + 2.60\mathbf{K}$ m/s²; (b) $\boldsymbol{\omega} = 0.25\mathbf{J} + 0.4\mathbf{K}$ rad/s, $\boldsymbol{\alpha} = 0.1\mathbf{I}$ rad/s². **19.2** (a) $\mathbf{v}_P = -5.20\mathbf{I} + 3\mathbf{J} - 1.875\mathbf{K}$ m/s, $\mathbf{a}_P = -8.16\mathbf{I} + 1.672\mathbf{J} + 0.348\mathbf{K}$ m/s²; (b) $\boldsymbol{\omega} = 0.25\mathbf{J} + 0.4\mathbf{K}$ rad/s, $\boldsymbol{\alpha} = 0.1\mathbf{I} + 0.3\mathbf{J} + 0.5\mathbf{K}$ rad/s². **19.4** $\mathbf{v}_B = 1.92\mathbf{I} + 1.100\mathbf{J} - 0.56\mathbf{K}$ m/s, $\mathbf{a}_B = 2.88\mathbf{I} + 2.97\mathbf{J} + 2.61\mathbf{K}$ m/s². **19.5** $\mathbf{v}_C = 3.46\mathbf{I} - 2\mathbf{J} - 13.86\mathbf{K}$ ft/s, $\mathbf{a}_C = -23.4\mathbf{I} - 2.5\mathbf{J} - 34.6\mathbf{K}$ ft/s². **19.6** $\mathbf{v}_B = -2.4\mathbf{I}$ ft/s, $\boldsymbol{\omega}_{AB} = 0.258\mathbf{I} - 0.258\mathbf{J} - 0.1477\mathbf{K}$ rad/s. **19.9** $\mathbf{v}_B = -4\mathbf{I} - 0.8\mathbf{K}$ m/s, $\mathbf{v}_C = -2.4\mathbf{I} + 0.8\mathbf{J}$ m/s, $\mathbf{a}_B = -35.2\mathbf{J}$ m/s², $\mathbf{a}_C = -12.8\mathbf{I} - 19.2\mathbf{J} - 3.2\mathbf{K}$ m/s². **19.10** $\boldsymbol{\alpha}_D = -2\mathbf{J} - 24\mathbf{K}$ rad/s², $\mathbf{a}_B = 20\mathbf{I} - 18\mathbf{J} - 32\mathbf{K}$ ft/s², $\mathbf{a}_C = -3\mathbf{I} - 6\mathbf{K}$ ft/s². **19.11** $\boldsymbol{\alpha}_{AB} = 20\mathbf{I} + 3\mathbf{J} + 2\mathbf{K}$ rad/s², $\mathbf{v}_B = 2.4\mathbf{I} + 4.16\mathbf{J} - 5.20\mathbf{K}$ m/s, $\mathbf{a}_B = -41.4\mathbf{I} + 11.68\mathbf{J} - 27.1\mathbf{K}$ m/s². **19.12** (a) $0.967\mathbf{J}$ rad/s; (b) $\boldsymbol{\omega}_C = 0.845(-\mathbf{I} - \mathbf{J})$ rad/s, $\boldsymbol{\alpha}_C = 0.818\mathbf{K}$ rad/s². **19.13** $\boldsymbol{\omega} = 0.274\mathbf{I} + 2\mathbf{J} + 0.475\mathbf{K}$ rad/s, $\mathbf{v}_A = -0.316\mathbf{J} + 1.333\mathbf{K}$ ft/s. **19.15** $\boldsymbol{\omega}_D = 1.5\mathbf{I} + 5.53\mathbf{J} + 1.286\mathbf{K}$ rad/s, $\boldsymbol{\alpha}_D = 7.64\mathbf{I} + 5.37\mathbf{J} - 1.774\mathbf{K}$ rad/s². **19.17** $I_{xy} = I_{zx} = 0$, $I_{yz} = 1.8$ slug·ft². **19.19** $I_{xy} = I_{zx} = 0$, $I_{yz} = 1.6$ slug·ft². **19.20** $I_{xy} = -0.1125$ slug·ft², $I_{yz} = 0$, $I_{zx} = 0.0562$ slug·ft². **19.21** $I_{xy} = -116.6 \times 10^{-6}$ kg·m², $I_{yz} = I_{zx} = 0$. **19.23** $I_{xy} = -138.2 \times 10^{-6}$ kg·m², $I_{yz} = I_{zx} = 0$. **19.25** $I_{xy} = -\frac{1}{4} mab$, $I_{yz} = -\frac{1}{4} mbc$, $I_{zx} = \frac{1}{4} mca$. **19.26** $I_{xy} = I_{zx} = 0$, $I_{yz} = -6.03 \times 10^{-3}$ kg·m². **19.28** $I_{xy} = I_{zx} = 0$, $I_{yz} = 4.74 \times 10^{-3}$ kg·m². **19.29** $-mRh/(5\pi)$. **19.31** 24 slug·ft². **19.33** 58.9 slug·ft². **19.34** 15.39 slug·ft². **19.35** 10.15 slug·ft². **19.36** 24.2 slug·ft². **19.37** $(3/20)mR^2(R^2 + 6h^2)/(R^2 + h^2)$. **19.40** 392×10^{-6} kg·m². **19.41** (a) $mb^2\omega(-0.8\mathbf{I} - 0.4\mathbf{J} + 3\mathbf{K})$, (b) $1.5mb^2\omega^2$. **19.42** (a) $18.10\mathbf{I} + 10.45\mathbf{K}$ kg·m²/s, (b) 113.7 J. **19.44** (a) $1.317\mathbf{I} + 1.317\mathbf{J} + 2.63\mathbf{K}$ slug·ft²/s, (b) 35.3°, (c) 12.41 ft·lb. **19.46** (a) $-0.4\mathbf{I} + 0.4\mathbf{K}$ kg·m²/s, (b) 16.8 J. **19.48** (a) $-0.0597\mathbf{I} + 0.01599\mathbf{J}$ kg·m²/s, (b) 18.45 mJ. **19.49** (a) $0.0582\mathbf{I} + 0.365\mathbf{J} + 0.1763\mathbf{K}$ slug·ft²/s, (b) 46.4 ft·lb. **19.50** (a) $6\mathbf{K}$ rad/s, (b) $11.10\mathbf{K}$ rad/s. **19.51** $19.32\mathbf{I} - 38.6\mathbf{K}$ rad/s. **19.53** $-[3m_0v_;/(mb)](2\mathbf{J} + \mathbf{K})$. **19.54** (a) $\cos^{-1} 6g/(5b\omega^2)$ if $5b\omega^2 > 6g$, (b) $\sqrt{6g/(5b)}$. **19.55** $T = 3.54$ kN, $A = 800$ N. **19.57** $\boldsymbol{\alpha} = 13.89\mathbf{I}$ rad/s², $\mathbf{C} = 4.81\mathbf{J} + 54.7\mathbf{K}$ N. **19.58** $2.16\mathbf{K}$ N·m. **19.59** $3.2\mathbf{J}$ N·m. **19.60** 3.60 rad/s or 23.6 rad/s. **19.62** 150.6 rad/s. **19.64** 1.316 rad/s or 61.4 rad/s. **19.65** 21.7×10^{21} N·m. **19.66** (a) $\boldsymbol{\omega}_C = 5\mathbf{I} + 4\mathbf{J}$ rad/s, $\boldsymbol{\alpha}_C = 2\mathbf{I} + 3\mathbf{J} - 20\mathbf{K}$ rad/s²; (b) $\mathbf{v}_P = -34\mathbf{K}$ in./s, $\mathbf{a}_P = -16\mathbf{I} - 150\mathbf{J} - 36\mathbf{K}$ in./s². **19.67** $\boldsymbol{\omega}_C = 1.098(-\mathbf{I} - \mathbf{J})$ rad/s, $\boldsymbol{\alpha}_C = 1.380\mathbf{K}$ rad/s². **19.69** $-43.8\mathbf{K}$ N·m. **19.70** $\boldsymbol{\alpha} = 53.7\mathbf{K}$ rad/s², $\mathbf{O} = -3.32\mathbf{I} - 0.819\mathbf{J}$ N, $\mathbf{A} = 3.32\mathbf{I} + 0.819\mathbf{J}$ N. **19.73** $\theta = \cos^{-1} 6r^2\dot{\psi}/[(h^2 - 3r^2)\dot{\phi}]$. **19.74** (a) 4.67 rad/s, (b) $-3.58\mathbf{i} + 9\mathbf{k}$ rad/s.

Chapter 20

Developmental Exercises

D20.1 (a) $\omega_n = \pi$ rad/s, $f_n = 0.5$ Hz; (b) yes. **D20.2** $\omega_n = 3.84$ rad/s, $\tau = 1.638$ s. **D20.3** $\omega_n = 6.95$ rad/s, $\tau = 0.904$ s. **D20.4**

6.60 rad/s.   **D20.5** 1.605 s.   **D20.7** $\omega_n = 3.84$ rad/s, $\tau =$ 1.638 s.   **D20.8** (a) $\zeta = 1$, (b) $\zeta > 1$, (c) $\zeta < 1$.   **D20.14** $\sqrt{2}/2$. **D20.16** (a) Increased, (b) 92.4%.

Problems

20.1 $\dot{x}_{max} = 2.09$ ft/s, $\ddot{x}_{max} = 26.3$ ft/s^2.   **20.2** 0.955 Hz.   **20.4** $X = 25$ mm. $\phi = 53.1°$.   **20.5** 367 lb.   **20.7** (a) $\tau = 0.562$ s, $\dot{x}_{max} = 335$ mm/s; (b) $\tau = 1.147$ s, $\dot{x}_{max} = 164.3$ mm/s.   **20.9** 0.438 kg·m^2.   **20.10** $\{(5/m)[kb^2 + (L + r)mg]/[5(L + r)^2 + 2r^2]\}^{1/2}$. **20.11** 1.422 s.   **20.13** 0.627 Hz.   **20.14** 180.0 mm.   **20.16** 0.259 slug·ft^2.   **20.17** 1.988 m.   **20.18** $x = (m_B g/k)(1 - \cos \omega_n t) + [m_B\sqrt{2gh}/\sqrt{k(m_A + m_B)}] \sin \omega_n t$, where $\omega_n = \sqrt{k/(m_A + m_B)}$.   **20.19** 3.98 m.   **20.21** $[kr_2^2/(\bar{I} + mr_1^2)]^{1/2}/(2\pi)$.   **20.22** (a) 6.32 kN/m, (b) 435 N·s/m.   **20.23** $\zeta = 1.5$, $x = 3.512\,e^{-26.18t} - 0.512\,e^{-3.82t}$ in. **20.25** $\zeta = 0.5$, $x = 2\sqrt{3}\,e^{-5t} \sin (5\sqrt{3}t + \pi/3)$ in.   **20.26** $x = (3\sqrt{7}/35)\,e^{-7.5t} \sin 2.5\sqrt{7}t$ ft.   **20.28** $k = 792$ N/m, $c = 6.97$ N·s/m.   **20.29** (a) 0.399 s, (b) -0.572.   **20.30** $\zeta = \sqrt{3}c/(8\sqrt{km})$, $\tau_d = 16\pi m/\sqrt{3(64km - 3c^2)}$.   **20.32** (a) 0.564 in., (b) 0.683 in. **20.33** (a) 0.259, (b) 0.273.   **20.34** 0.327.   **20.37** $(W/g)\ddot{x} + c\dot{x} + kx = ky + c\dot{y}$, $X = 1.731$ in.   **20.38** $\Theta = (12F_0/L)/\sqrt{(3k - m\omega^2)^2 + (3c\omega)^2}$, $\varphi = \tan^{-1}[3c\omega/(3k - 4m\omega^2)]$.   **20.39** 1.586 s.   **20.40** 0.691 Hz.   **20.41** $2\pi[2(\pi - 2)r(R - r)]^{1/2}/\{[\pi r + 2(R - r)]g\}^{1/2}$.   **20.42** 176.2 lb.   **20.43** $X_1 = \frac{1}{2} kX_2/\sqrt{(k - m\omega^2)^2 + (c\omega)^2}$, $\varphi = \tan^{-1}c\omega/(k - m\omega^2)$.   **20.44** $\Theta = (36F_0/L)/[(3k - 7m\omega^2)^2 + (3c\omega)^2]^{1/2}$, $\varphi = \tan^{-1}3c\omega/(3k - 7m\omega^2)$.

Appendix A

1. d	**2.** a	**3.** d	**4.** c	**5.** e	**6.** c	**7.** a	**8.** c
9. e.	**10.** a	**11.** d	**12.** e	**13.** c	**14.** b	**15.** a	
16. b	**17.** b	**18.** c	**19.** e	**20.** b	**21.** a	**22.** a	
23. d	**24.** b	**25.** e	**26.** c	**27.** b	**28.** c	**29.** d	
30. e	**31.** d	**32.** c	**33.** b	**34.** e	**35.** e	**36.** a	
37. d	**38.** d	**39.** b	**40.** a	**41.** d	**42.** c	**43.** b	
44. c	**45.** e	**46.** d	**47.** b	**48.** a	**49.** e	**50.** d	

INDEX